Springer Undergraduate Mathematics Series

For further volumes:
www.springer.com/series/3423

Christopher Norman

Finitely Generated Abelian Groups and Similarity of Matrices over a Field

 Springer

Christopher Norman
formerly Senior Lecturer in Mathematics
Royal Holloway, University of London
London, UK

ISSN 1615-2085   Springer Undergraduate Mathematics Series
ISBN 978-1-4471-2729-1   e-ISBN 978-1-4471-2730-7
DOI 10.1007/978-1-4471-2730-7
Springer London Dordrecht Heidelberg New York

British Library Cataloguing in Publication Data
A catalogue record for this book is available from the British Library

Library of Congress Control Number: 2012930645

Mathematics Subject Classification: 15-01, 15A21, 20-01, 20K30

Printed on acid-free paper

Springer is part of Springer Science+Business Media (www.springer.com)

To Lucy, Tim and Susie

Preface

Who is this book for? The target reader will already have experienced *a first course in linear algebra* covering matrix manipulation, determinants, linear mappings, eigenvectors and diagonalisation of matrices. Ideally the reader will have met bases of finite-dimensional vector spaces, the axioms for groups, rings and fields as well as some set theory including equivalence relations. Some familiarity with elementary number theory is also assumed, such as the Euclidean algorithm for the greatest common divisor of two integers, the Chinese remainder theorem and the fundamental theorem of arithmetic. In the proof of Lemma 6.35 it is assumed that the reader knows how to resolve a permutation into cycles. With these provisos the subject matter is virtually self-contained. Indeed many of the standard facts of linear algebra, such as the multiplicative property of determinants and the dimension theorem (any two bases of the same finite-dimensional vector space have the same number of vectors), are proved in a more general context. Nevertheless from a didactic point of view it is highly desirable, if not essential, for the reader to be already familiar with these facts.

What does the book do? The book is in two analogous parts and is designed to be *a second course in linear algebra* suitable for second/third year mathematics undergraduates, or postgraduates. The first part deals with the theory of finitely generated (f.g.) abelian groups: the emerging homology theory of topological spaces was built on such groups during the 1870s and more recently the classification of elliptic curves has made use of them. The starting point of the abstract theory couldn't be more concrete if it tried! Row and column operations are applied to an arbitrary matrix having integer entries with the aim of obtaining a diagonal matrix with non-negative entries such that the $(1, 1)$-entry is a divisor of the $(2, 2)$-entry, the $(2, 2)$-entry is a divisor of the $(3, 3)$-entry, and so on; a diagonal matrix of this type is said to be in *Smith normal form* (*Snf*) after the 19th century mathematician HJS Smith. Using an extension of the Euclidean algorithm it is shown in Chapter 1 that the Snf can be obtained without resort to prime factorisation. In fact the existence of the Snf is the cornerstone of the decomposition theory.

Free abelian groups of finite rank have $\mathbb{Z}$-bases and behave in many ways like finite-dimensional vector spaces. Each f.g. abelian group is best described as a quotient group of such a free abelian group by a subgroup which is necessarily also free. In Chapter 2 some time is spent on the concept of quotient groups which no student initially finds easy, but luckily in this context turns out to be little more than working modulo a given integer. The quotient groups arising in this way are specified by matrices over $\mathbb{Z}$ and the theory of the Snf is exactly what is needed to analyse their structure. Putting the pieces together in Chapter 3 each f.g. abelian group is seen to correspond to a sequence of non-negative integers (its *invariant factors*) in which each integer is a divisor of the next. The sequence of invariant factors of an f.g. abelian group encapsulates its properties: two f.g. abelian groups are isomorphic (abstractly identical) if and only if their sequences of invariant factors are equal. So broadly, apart from important side-issues such as specifying the automorphisms of a given group G, this is the end of the story as far as f.g. abelian groups are concerned! Nevertheless these side-issues are thoroughly discussed in the text and through numerous exercises; complete solutions to all exercises are on the associated website.

In the second part of the book the ring $\mathbb{Z}$ of integers is replaced by the ring $F[x]$ of polynomials over a field F. Such polynomials behave in the same way as integers and in particular the Euclidean algorithm can be used to find the gcd of each pair of them. In Chapter 4 the theory of the Smith normal form is shown to extend, almost effortlessly, to matrices over $F[x]$, the non-zero entries in the Snf here being *monic* (leading coefficient 1) polynomials. To what end? A question which occupies centre stage in linear algebra concerns $t \times t$ matrices A and B over a field F: is there a systematic method of finding, where it exists, an *invertible* $t \times t$ matrix X over F with $XA = BX$? Should X exist then A and $B = XAX^{-1}$ are called *similar*. The answer to the question posed above is a resounding YES! The systematic method amounts to reducing the matrices $xI - A$ and $xI - B$, which are $t \times t$ matrices over $F[x]$, to their Smith normal forms; if these forms are equal then A and B are similar and X can be found by referring back to the elementary operations used in the reduction processes; if these forms are different then A and B are not similar and X doesn't exist. The matrix $xI - A$ should be familiar to the reader as $\det(xI - A)$ is the characteristic polynomial of A. The non-constant diagonal entries in the Snf of $xI - A$ are called the *invariant factors* of A. It is proved in Chapter 6 that A and B are similar if and only if their sequences of invariant factors are equal. The theory culminates in the rational canonical form (rcf) of A which is the simplest matrix having the same invariant factors as A. It's significant that the rcf is obtained in a constructive way; in particular there is no reliance on factorisation into irreducible polynomials.

The analogy between the two parts is established using *R-modules* where R is a commutative ring. Abelian groups are renamed $\mathbb{Z}$-modules and structure-preserving mappings (*homomorphisms*) of abelian groups are $\mathbb{Z}$-linear mappings. The terminology helps the theory along: for instance the reader comfortable with 1-dimensional subspaces should have little difficulty with *cyclic* submodules. Each $t \times t$ matrix A over a field F gives rise to an associated $F[x]$-module $M(A)$. The relationship between A and $M(A)$ is explained in Chapter 5 where *companion* matrices are introduced. Just as each finite abelian group G is a direct sum of cyclic groups, so each matrix A, as above, is similar to a direct sum of companion matrices; the polynomial analogue of the order $|G|$ of G is the characteristic polynomial $\det(xI - A)$ of A.

The theory of the two parts can be conflated using the overarching concept of a finitely generated module over a principal ideal domain, which is the stance taken by several textbooks. An exception is *Rings, Modules and Linear Algebra* by B. Hartley and T.O. Hawkes, Chapman and Hall (1970) which opened my eyes to the beauty of the analogy explained above. I willingly acknowledge the debt I owe to this classic exposition. The two strands are sufficiently important to merit individual attention; nevertheless I have adopted proofs which generalise without material change, that of *the invariance theorem* 3.7 being a case in point.

Mathematically there is nothing new here: it is a rehash of 19th and early 20th century matrix theory from Smith to Frobenius, ending with the work of Shoda on automorphisms. However I have not seen elsewhere the step-by-step method of calculating the matrix Q described in Chapter 1 though it is easy enough once one has stumbled on the basic idea. The book is an expansion of material from a lecture course I gave in the University of London, off and on, over a 30 year period to undergraduates first at Westfield College and latterly at Royal Holloway. Lively students forced me to rethink both theory and presentation and I am grateful, in retrospect, to them. Dr. W.A. Sutherland read and commented on the text and Dr. E.J. Scourfield helped with the number theory in Chapter 3; I thank both. Any errors which remain are my own.

Finally I hope the book will attract mathematics students to what is undoubtedly an important and beautiful theory.

London, UK   Christopher Norman

Contents

A Bird's-Eye View of Finitely Generated Abelian Groups

To get the general idea of the theory let's start with the smallest non-trivial abelian group, that is, a cyclic group of order 2 which we denote by G. We will use *additive* notation for all abelian groups meaning that the group operation is denoted by $+$. So G consists of just two elements 0 and $g \neq 0$ satisfying

$$0 + 0 = 0, \qquad 0 + g = g, \qquad g + 0 = g, \qquad g + g = 0.$$

Here 0 is the *zero element* of G. From the first and last of these equations we obtain $-0 = 0$ and $-g = g$, and so G in common with all additive groups is *closed* under *addition* and *negation*. The reader should feel comfortable with the group G as it's small enough to present no threat!

It is reasonable to express $g + g = 0$ as $2g = 0$ and in the same way $3g = (g + g) + g = 0 + g = g$ and $4g = 3g + g = g + g = 0$. By the *associative law of addition*, which holds in every additive group, there is no need for brackets in any sum of group elements. So for every positive integer n the group element ng is obtained by adding together n elements equal to g, that is,

$$ng = g + g + \cdots + g \quad (n \text{ terms}).$$

So $1g = g$ and it is also reasonable to define $(-n)g$ to be the group element $-(ng)$ and to define $0g$ to be the zero element 0 of G. Therefore we have given meaning to mg for *all* integers m (positive, negative and zero) and

$$mg \in G \quad \text{for all } m \in \mathbb{Z}, \ g \in G$$

that is, mg is a certain element of G; we define $m0$ to be the zero element of G, and so G is *closed* under *integer multiplication*. This procedure can be carried out for any additive group and gives it the structure of a $\mathbb{Z}$-*module*. For our particular G we have $mg = 0$ for even m, and $mg = g$ for odd m. We use $|G|$ for the number of elements (the *order* of G) in the finite group G. So $|G| = 2$ in our case. Also G is *cyclic* with *generator* g and we write $G = \langle g \rangle$ since the elements of G are precisely the integer multiples of the single element g. The reader will doubtless be familiar with the field $\mathbb{Z}_2$ having just the elements $\overline{0}$ and $\overline{1}$ working modulo 2. Ignoring multiplication, we see that the *additive group* of $\mathbb{Z}_2$ is *isomorphic* (abstractly identical) to G as

$$\overline{0} + \overline{0} = \overline{0}, \qquad \overline{0} + \overline{1} = \overline{1}, \qquad \overline{1} + \overline{0} = \overline{1}, \qquad \overline{1} + \overline{1} = \overline{0}$$

showing that $\overline{0}$ and $\overline{1}$ in $\mathbb{Z}_2$ behave in the same way as 0 and g in G. In fact the additive group of $\mathbb{Z}_2$ is the standard example of a cyclic group of order 2 and any such group is said to be of *isomorphism type* C_2. What is behind the close connection between $\mathbb{Z}_2$ and G? It's worth finding out because a vista of the whole theory of finitely generated (f.g.) abelian groups can be glimpsed from this standpoint.

Let $\theta : \mathbb{Z} \to G$ be the mapping defined by $(m)\theta = mg$ for all integers m. Then θ is a *homomorphism* from the additive group of $\mathbb{Z}$ to G, meaning

$$(m_1 + m_2)\theta = (m_1)\theta + (m_2)\theta \quad \text{for all } m_1, m_2 \in \mathbb{Z}.$$

As $(0)\theta = 0$ and $(1)\theta = g$ we see that θ is *surjective* (onto), that is, *the image* $\operatorname{im} \theta = \{(m)\theta : m \in \mathbb{Z}\}$ *of* θ equals G. So $G = \operatorname{im} \theta$. Which elements of $\mathbb{Z}$ does θ map to the zero element of G? These are the elements of *the kernel of* θ. We write $K = \ker \theta$ for this important subgroup of $\mathbb{Z}$. In our case

$$K = \{m \in \mathbb{Z} : (m)\theta = 0\} = \{m \in \mathbb{Z} : mg = 0\} = \{m \in \mathbb{Z} : m \text{ is even}\}.$$

So $\ker \theta = \langle 2 \rangle$ is the subgroup of all even integers. The *coset* $K + 1 = \{m + 1 : m \in K\}$ is the set of all odd integers. There are just two cosets of K in $\mathbb{Z}$, namely K and $K + 1$ corresponding to the two elements 0 and g of G. These two cosets are the elements of the quotient group $\mathbb{Z}/K = \mathbb{Z}/\langle 2 \rangle = \{K, K + 1\}$ with addition

$$K + K = K, \qquad K + (K + 1) = K + 1,$$

$$(K + 1) + K = K + 1, \qquad (K + 1) + (K + 1) = K.$$

But hang on a moment: we've seen these equations twice before! The reader will know that $K = \overline{0}$ and $K + 1 = \overline{1}$ are the two elements of $\mathbb{Z}_2$ and so $\mathbb{Z}/\langle 2 \rangle$ is the additive group of $\mathbb{Z}_2$. Also θ leads to the *isomorphism* (bijective homomorphism)

$$\tilde{\theta} : \mathbb{Z}/K \cong \operatorname{im} \theta \quad \text{where } (K + m)\tilde{\theta} = (m)\theta \text{ for all } m \in \mathbb{Z}.$$

So $\tilde{\theta}$ is an additive bijection between the groups $\mathbb{Z}/\langle 2 \rangle$ and G, that is,

$$(x + y)\tilde{\theta} = (x)\tilde{\theta} + (y)\tilde{\theta} \quad \text{for all } x, y \in \mathbb{Z}/\langle 2 \rangle.$$

In our case $(\bar{0})\tilde{\theta} = 0$, $(\bar{1})\tilde{\theta} = g$ and we write $\tilde{\theta} : \mathbb{Z}/\langle 2 \rangle \cong G$. Luckily, as we'll see, the inverse of every isomorphism is again an isomorphism.

We've deliberately made a fuss over the way θ gives rise to $\tilde{\theta}$ since this is a particular case of the *first isomorphism theorem* 2.16. As another (not entirely facetious) application of this important theorem, consider the identity isomorphism $\iota : \mathbb{Z} \to \mathbb{Z}$ of the additive group $\mathbb{Z}$ and so $(m)\iota = m$ for all m in $\mathbb{Z}$. As $\text{im}\,\iota = \mathbb{Z}$ and $\ker \iota = \langle 0 \rangle$, by the first isomorphism theorem $\tilde{\iota} : \mathbb{Z}/\langle 0 \rangle \cong \mathbb{Z}$ where $\tilde{\iota}$ maps the singleton coset $\ker \iota + m = \langle 0 \rangle + m = \{m\}$ to the integer m, that is, $(\{m\})\tilde{\iota} = m$ for all m in $\mathbb{Z}$. Now $\mathbb{Z} = \langle 1 \rangle$ is an *infinite* cyclic group being generated by the integer 1; so it's convenient to say that such groups are of isomorphism type C_0 as they are isomorphic to $\mathbb{Z}/\langle 0 \rangle$.

Let G' denote the abelian group having the 8 pairs (x, y) as its elements where $x \in \mathbb{Z}_2$ and $y \in \mathbb{Z}_4$, the group operation being *componentwise* addition, that is,

$$(x, y) + (x', y') = (x + x', y + y') \quad \text{for all } x, x' \in \mathbb{Z}_2 \text{ and } y, y' \in \mathbb{Z}_4.$$

The elements of G' are

$$(\bar{0}, \bar{0}), \quad (\bar{1}, \bar{0}), \quad (\bar{0}, \bar{1}), \quad (\bar{1}, \bar{1}), \quad (\bar{0}, \bar{2}), \quad (\bar{1}, \bar{2}), \quad (\bar{0}, \bar{3}), \quad (\bar{1}, \bar{3}).$$

The zero element of G' is $(\bar{0}, \bar{0})$ and $-(\bar{1}, \bar{1}) = (-\bar{1}, -\bar{1}) = (\bar{1}, \bar{3})$ since $-\bar{1} = \bar{1}$ in $\mathbb{Z}_2$ and $-\bar{1} = \bar{3}$ in $\mathbb{Z}_4$. The reader should realise that the group laws hold and so G' is an abelian group. In fact G' is the *(external) direct sum* of the additive group $\mathbb{Z}/\langle 2 \rangle$ of the field $\mathbb{Z}_2$ and the additive group $\mathbb{Z}/\langle 4 \rangle$ of the ring $\mathbb{Z}_4$ and we write $G' = \mathbb{Z}/\langle 2 \rangle \oplus \mathbb{Z}/\langle 4 \rangle$. The element $g_1 = (\bar{1}, \bar{0})$ has *order* 2 (the smallest positive integer multiple of g_1 equal to the zero element is the order of g_1) and $g_2 = (\bar{0}, \bar{1})$ has order 4. As $4g = (\bar{0}, \bar{0})$ for all g in G', we see that G' is not cyclic since G' has no element of order $|G'| = 8$. However every element of G' is *uniquely* expressible as a sum of two elements, one from the cyclic subgroup $H_1 = \langle g_1 \rangle$ of order 2 and one from the cyclic subgroup $H_2 = \langle g_2 \rangle$ of order 4, that is, G' is the *(internal) direct sum* of its cyclic subgroups H_1 and H_2 and we write $G' = H_1 \oplus H_2$. As H_1 has isomorphism type C_2 and H_2 has isomorphism type C_4, we say that G' has isomorphism type $C_2 \oplus C_4$. It is time to define exactly what we are talking about.

An additive abelian group G is called *finitely generated* (abbreviated to *f.g.*) if it contains a finite number t of elements $g_1, g_2, \ldots, g_t$ such that every element of G is expressible as an integer linear combination $m_1 g_1 + m_2 g_2 + \cdots + m_t g_t$ $(m_i \in \mathbb{Z})$, in which case we write $G = \langle g_1, g_2, \ldots, g_t \rangle$ and say that $g_1, g_2, \ldots, g_t$ *generate* G.

A group is cyclic if it is of the form $\langle g_1 \rangle$, that is, its elements consist purely and simply of all integer multiples of just one element g_1.

The fundamental theorem 3.4 concerning f.g. abelian groups states

> Every f.g. abelian group G has cyclic subgroups $H_1, H_2, \ldots, H_t$ such that
> $$G = H_1 \oplus H_2 \oplus \cdots \oplus H_t$$

meaning that every element g of G is expressible as $g = h_1 + h_2 + \cdots + h_t$ where $h_i \in H_i$ for $1 \leq i \leq t$ and (most crucially) the h_i are *unique*. So every element g of G is expressible in one and only one way as a sum of elements h_i from the cyclic subgroups H_i, that is, G is the (internal) direct sum of its cyclic subgroups H_i.

Let C_{d_i} be the isomorphism type of H_i for $1 \leq i \leq t$, and so H_i is cyclic of finite order d_i for $d_i > 0$ and H_i is infinite cyclic for $d_i = 0$. This theorem goes on to say that the subgroups H_i can be chosen so that the sequence of non-negative integers

$$d_1, d_2, \ldots, d_t \text{ satisfies } d_1 \neq 1 \text{ and } d_i | d_{i+1} \ (d_i \text{ is a divisor of } d_{i+1}) \text{ for } 1 \leq i < t.$$

Finally Corollary 3.5 and Theorem 3.7 deliver the 'clincher': $d_1, d_2, \ldots, d_t$ are *uniquely* determined by G, that is, for each f.g. abelian group G there is one and only one sequence $d_1, d_2, \ldots, d_t$ as above. The integers d_i are called the *invariant factors of G* and $C_{d_1} \oplus C_{d_2} \oplus \cdots \oplus C_{d_t}$ is called the *isomorphism type of G*.

The preceding example $G' = \mathbb{Z}/\langle 2 \rangle \oplus \mathbb{Z}/\langle 4 \rangle$ has invariant factors 2, 4. A non-trivial cyclic group has a single invariant factor. Trivial groups have no invariant factors (take $t = 0$) and, having one element only, belong to the isomorphism class C_1. A word of caution: although the invariant factors d_i of G are unique (they depend on G only) the subgroups H_i are not in general unique. For instance $G' = \mathbb{Z}/\langle 2 \rangle \oplus \mathbb{Z}/\langle 4 \rangle$ has the 'obvious' decomposition $G' = \langle g_1 \rangle \oplus \langle g_2 \rangle$ where $g_1 = (\overline{1}, \overline{0})$, $g_2 = (\overline{0}, \overline{1})$, but also $G' = \langle g_1 + 2g_2 \rangle \oplus \langle g_1 + g_2 \rangle$ is a less obvious but equally valid decomposition. So G' can be decomposed in two different ways as in the fundamental theorem and each way shows that $C_2 \oplus C_4$ is the isomorphism type of G'.

Now consider the homomorphism $\theta : \mathbb{Z}^2 \to G'$ where $(m_1, m_2)\theta = m_1 g_1 + m_2 g_2$ for all integers m_1, m_2. Then $(1, 0)\theta = 1g_1 + 0g_2 = g_1 = (\overline{1}, \overline{0})$ and $(0, 1)\theta = 0g_1 + 1g_2 = g_2 = (\overline{0}, \overline{1})$. As $G' = \langle g_1, g_2 \rangle$ we see that θ is surjective, that is, $\operatorname{im} \theta = G'$ and so G' is a *homomorphic image* of $\mathbb{Z}^2$. In fact every abelian group which can be generated by two of its elements is a homomorphic image of $\mathbb{Z}^2$, and the additive group $\mathbb{Z}^2$ of all pairs (m_1, m_2) of integers is the 'Big Daddy' of such groups. Notice $\mathbb{Z}^2 = \langle e_1, e_2 \rangle = \langle e_1 \rangle \oplus \langle e_2 \rangle$ where $e_1 = (1, 0)$ and $e_2 = (0, 1)$, showing that $\mathbb{Z}^2$ is the internal direct sum of its infinite cyclic subgroups $\langle e_1 \rangle$ and $\langle e_2 \rangle$. So $\mathbb{Z}^2$ has isomorphism type $C_0 \oplus C_0$. Although $\mathbb{Z}^2$ is not a vector space (we need hardly remind the reader – but we will to be on the safe side – that $\mathbb{Z}$ is *not* a field as $1/2$ (the inverse of the integer 2) is not an integer), nevertheless $\mathbb{Z}^2$ has some of the properties of a 2-dimensional vector space especially where bases are concerned.

The ordered pair of elements ρ_1, ρ_2 of $\mathbb{Z}^2$ is called a *$\mathbb{Z}$-basis of $\mathbb{Z}^2$* if each element of $\mathbb{Z}^2$ can be expressed in the form $m_1 \rho_1 + m_2 \rho_2$ for *unique* integers m_1 and m_2. It turns out that every $\mathbb{Z}$-basis of $\mathbb{Z}^2$ consists of the rows of an *invertible* 2×2 matrix

$Q = \begin{pmatrix} \rho_1 \\ \rho_2 \end{pmatrix}$ over $\mathbb{Z}$, that is, Q has an inverse Q^{-1} which also has integer entries. The condition for Q to have this property is that the determinant of Q is an invertible integer, in other words $\det Q = \pm 1$, as ± 1 are the only integers having integer inverses. Of course e_1, e_2 is a $\mathbb{Z}$-basis (*the standard basis*) of $\mathbb{Z}^2$ as e_1, e_2 are the rows of the identity matrix I over $\mathbb{Z}$ and $\det I = 1$. But $\rho_1 = (3, 5)$, $\rho_2 = (2, 3)$, for example, is also a $\mathbb{Z}$-basis of $\mathbb{Z}^2$ since $\begin{vmatrix} 3 & 5 \\ 2 & 3 \end{vmatrix} = 3 \times 3 - 5 \times 2 = -1$. How can the unique integers m_1 and m_2 be found in the case of the element $(7, 8)$ of $\mathbb{Z}^2$? The answer is found using Q^{-1}:

$$(7, 8) = (7, 8)I = (7, 8)Q^{-1}Q = (7, 8) \begin{pmatrix} -3 & 5 \\ 2 & -3 \end{pmatrix} \begin{pmatrix} \rho_1 \\ \rho_2 \end{pmatrix} = (-5, 11) \begin{pmatrix} \rho_1 \\ \rho_2 \end{pmatrix}$$

$$= -5\rho_1 + 11\rho_2.$$

So $m_1 = -5$, $m_2 = 11$ in this case.

Returning to $\theta : \mathbb{Z}^2 \to G'$ we turn our attention to the subgroup $K = \ker \theta$ of $\mathbb{Z}^2$. As K consists of the elements (m_1, m_2) of $\mathbb{Z}^2$ with $(m_1, m_2)\theta = (\overline{0}, \overline{0}) \in G'$ and $(m_1, m_2)\theta = m_1 g_1 + m_2 g_2 = m_1(\overline{1}, \overline{0}) + m_2(\overline{0}, \overline{1}) = (\overline{m}_1, \overline{0}) + (\overline{0}, \overline{m}_2) = (\overline{m}_1, \overline{m}_2)$, we see that (m_1, m_2) belongs to $K \Leftrightarrow (\overline{m}_1, \overline{m}_2) = (\overline{0}, \overline{0})$ in G', that is, $\overline{m}_1 = \overline{0}$ in $\mathbb{Z}_2$ and $\overline{m}_2 = \overline{0}$ in $\mathbb{Z}_4$. So

$$K = \{(m_1, m_2) \in \mathbb{Z}^2 : m_1 \equiv 0 \; (\text{mod } 2), \; m_2 \equiv 0 \; (\text{mod } 4)\}.$$

In other words K consists of pairs (m_1, m_2) in $\mathbb{Z}^2$ with m_1 even and m_2 divisible by 4, that is, $K = \langle 2e_1, 4e_2 \rangle$. In fact Theorem 3.1 says:

Every subgroup K of $\mathbb{Z}^t$ has a $\mathbb{Z}$-basis consisting of at most t elements of K

This apparently modest theorem allows the theory of f.g. abelian groups to be expressed in terms of matrices over $\mathbb{Z}$. In our case $2e_1, 4e_2$ is a $\mathbb{Z}$-basis of K. We construct the 2×2 matrix

$$D = \begin{pmatrix} 2e_1 \\ 4e_2 \end{pmatrix} = \begin{pmatrix} 2 & 0 \\ 0 & 4 \end{pmatrix}$$

over $\mathbb{Z}$. The significance of D will become clear later. For the moment notice that the rows of D generate the kernel K of θ. On dividing m_1 by 2 and m_2 by 4 a typical element (m_1, m_2) of $\mathbb{Z}^2$ is

$$(m_1, m_2) = (2q_1 + r_1, 4q_2 + r_2) = (q_1 2e_1 + q_2 4e_2) + (r_1, r_2) \in K + (r_1, r_2)$$

where $q_1, q_2, r_1, r_2 \in \mathbb{Z}$ and $0 \le r_1 < 2, 0 \le r_2 < 4$. So (m_1, m_2) belongs to the coset $K + (r_1, r_2)$. There are 8 cosets of K in $\mathbb{Z}^2$ namely

$$K + (0, 0), \quad K + (1, 0), \quad K + (0, 1), \quad K + (1, 1),$$

$$K + (0, 2), \quad K + (1, 2), \quad K + (0, 3), \quad K + (1, 3)$$

and these are the elements of the quotient group $\mathbb{Z}^2/K$. Applying the first isomorphism theorem to $\theta : \mathbb{Z}^2 \to G'$ gives the isomorphism

$$\tilde{\theta} : \mathbb{Z}^2/K \cong \mathbb{Z}/\langle 2\rangle \oplus \mathbb{Z}/\langle 4\rangle \quad \text{where } (K + (r_1, r_2))\tilde{\theta} = (\bar{r}_1, \bar{r}_2),$$

that is, $\tilde{\theta}$ amounts to no more than a small change in notation. It's something of an anti-climax to realise that $\tilde{\theta}$ is such a bland isomorphism, but the reason is not hard to find: we started with a group G' which was already *decomposed* (expressed as a direct sum) as in the fundamental theorem and so there's nothing for $\tilde{\theta}$ to do!

Our final example is of a 'mixed-up' group G'' which, in common with all f.g. abelian groups, can nevertheless be decomposed as above. Let $G'' = \mathbb{Z}^2/K'$ where $K' = \langle (4, 6), (8, 10)\rangle$. We construct $A = \begin{pmatrix} 4 & 6 \\ 8 & 10 \end{pmatrix}$ and so K' is the subgroup of $\mathbb{Z}^2$ generated by the rows of A. Then $G'' = \langle K' + e_1, K' + e_2\rangle$ is an f.g. abelian group. There are invertible matrices

$$P = \begin{pmatrix} 1 & 0 \\ -1 & 1 \end{pmatrix} \quad \text{and} \quad Q = \begin{pmatrix} 2 & 3 \\ 1 & 1 \end{pmatrix}$$

over $\mathbb{Z}$ such that $PAQ^{-1} = D = \begin{pmatrix} 2 & 0 \\ 0 & 4 \end{pmatrix}$. We will explain how P and Q are found from A in Chapter 1; for the moment the reader should check that $PA = DQ$. The diagonal matrix D is called *the Smith normal form of A* and reveals the structure of G''. In fact we'll see in a moment that $d_1 = 2$, $d_2 = 4$ are the invariant factors of G''. Denote the rows of Q by ρ_1 and ρ_2. So $\mathbb{Z}^2$ has $\mathbb{Z}$-basis $\rho_1 = (2, 3)$, $\rho_2 = (1, 1)$ as Q is invertible over $\mathbb{Z}$. The rows of A generate K' and, as P is invertible over $\mathbb{Z}$, it follows that the rows of PA also generate K'. The equation $PA = DQ = \begin{pmatrix} 2\rho_1 \\ 4\rho_2 \end{pmatrix}$ shows that $2\rho_1, 4\rho_2$ generate K'. In fact $2\rho_1, 4\rho_2$ is a $\mathbb{Z}$-basis of K' and so

$$\mathbb{Z}^2 = \langle \rho_1, \rho_2\rangle, \qquad K' = \langle 2\rho_1, 4\rho_2\rangle.$$

We've seen this type of thing before, the only difference being that e_1, e_2 have been replaced by ρ_1, ρ_2. Consider the *natural homomorphism*

$$\eta : \mathbb{Z}^2 \to \mathbb{Z}^2/K' = G''$$

defined by $(m_1, m_2)\eta = K' + (m_1, m_2)$ for all $m_1, m_2 \in \mathbb{Z}$. Then $\operatorname{im} \eta = G''$ and $\ker \eta = K'$. As in the case of G' we obtain

$$G'' = H'_1 \oplus H'_2$$

where $H'_1 = \langle (\rho_1)\eta\rangle$ is cyclic of order 2 and $H'_2 = \langle (\rho_2)\eta\rangle$ is cyclic of order 4. The crucial fact is that K' has a $\mathbb{Z}$-basis consisting of integer multiples (the integers being the invariant factors of G'') of the elements of a $\mathbb{Z}$-basis of $\mathbb{Z}^2$. The standard $\mathbb{Z}$-basis e_1, e_2 is good enough to reveal the structure of the preceding example $G' = \mathbb{Z}/\langle 2\rangle \oplus \mathbb{Z}/\langle 4\rangle$. The $\mathbb{Z}$-basis ρ_1, ρ_2 does the analogous job for $G'' =$

$\mathbb{Z}^2/\langle(4,6),(8,10)\rangle$. As G' and G'' have the same sequence of invariant factors, namely 2, 4, the groups G' and G'' are isomorphic.

EXERCISES

1. Show that $\rho_1 = (3,4)$, $\rho_2 = (4,5)$ is a $\mathbb{Z}$-basis of the additive group $\mathbb{Z}^2$. Find integers m_1 and m_2 such that $(10,7) = m_1\rho_1 + m_2\rho_2$. Do $(3,5)$, $(5,6)$ form a $\mathbb{Z}$-basis of $\mathbb{Z}^2$?

2. List the six cosets of $K = \langle 2e_1, 3e_2\rangle$ in $\mathbb{Z}^2$. Show that $\mathbb{Z}^2/K$ is cyclic with generator $g_0 = K + e_1 + e_2$. What is the invariant factor of $\mathbb{Z}^2/K$?

3. Show that $G' = \mathbb{Z}/\langle 2\rangle \oplus \mathbb{Z}/\langle 4\rangle$ has three elements of order 2 and four elements of order 4. What is the isomorphism type of the subgroup $H = \langle(\overline{1},\overline{0}),(\overline{0},\overline{2})\rangle$? State the isomorphism types of the other seven subgroups of G'.

   Hint: Except for G' itself they are all cyclic.

   Specify the four pairs of subgroups H_1, H_2 with $G' = H_1 \oplus H_2$ where H_1 and H_2 have isomorphism types C_2 and C_4 respectively.

1

Matrices with Integer Entries: The Smith Normal Form

We plunge in at the deep end by discussing the equivalence of rectangular matrices with whole number entries. The elegant and concrete conclusion of this theory was first published in 1861 by Henry J.S. Smith and it is exactly what is needed to analyse the abstract concept of a *finitely generated abelian group*, which is carried out in Chapter 3.

1.1 Reduction by Elementary Operations

Let A denote an $s \times t$ matrix over the ring $\mathbb{Z}$ of integers, that is, all the entries in A are whole numbers. All matrices in this chapter have integer entries. We consider the effect of applying *elementary row operations (eros) over* $\mathbb{Z}$ and *elementary column operations (ecos) over* $\mathbb{Z}$ to the matrix A, that is, operations of the following types:

(i) interchange of two rows or two columns
(ii) changing the sign of a row or a column
(iii) addition of an integer multiple of a row/column to a different row/column.

We use $r_1 \leftrightarrow r_2$ to denote the interchange of rows 1 and 2. We use $-c_3$ to mean: change the sign of column 3. Also $r_3 + 5r_1$ means: to row 3 add five times row 1, and so on. Notice that all these operations are invertible and the inverse operations are again of the same type: operations (i) and (ii) are self-inverse, and the inverse of $r_3 + 5r_1$, for example, is $r_3 - 5r_1$.

C. Norman, *Finitely Generated Abelian Groups and Similarity of Matrices over a Field*,
Springer Undergraduate Mathematics Series,
DOI 10.1007/978-1-4471-2730-7_1, © Springer-Verlag London Limited 2012

On applying a single *ero* over $\mathbb{Z}$ to the identity matrix I we obtain the *elementary matrix over* $\mathbb{Z}$ corresponding to the *ero*. For instance

$$\begin{pmatrix} 0 & 1 \\ 1 & 0 \end{pmatrix}, \quad \begin{pmatrix} 1 & 0 \\ 0 & -1 \end{pmatrix}, \quad \begin{pmatrix} 1 & 5 \\ 0 & 1 \end{pmatrix}, \quad \begin{pmatrix} 1 & 0 \\ -3 & 1 \end{pmatrix}$$

are the elementary 2×2 matrices which result on applying $r_1 \leftrightarrow r_2$, $-r_2$, $r_1 + 5r_2$, $r_2 - 3r_1$ respectively to the 2×2 identity matrix I. Every elementary matrix can be obtained equally well by applying a single *eco* to I; the above four matrices arise from I by $c_1 \leftrightarrow c_2$, $-c_2$, $c_2 + 5c_1$, $c_1 - 3c_2$. Elementary matrices themselves are unbiased: they do not prefer rows to columns or vice versa. An *ero* and an *eco* are *paired* if they produce the same elementary matrix. So $r_1 + 5r_2$ and $c_2 + 5c_1$ are paired elementary operations. Every elementary matrix over $\mathbb{Z}$ is invertible and its inverse is again an elementary matrix over $\mathbb{Z}$; for instance, applying the inverse pair of *eros* $r_2 + 3r_1$ and $r_2 - 3r_1$ to the 2×2 identity matrix I produces the inverse pair

$$\begin{pmatrix} 1 & 0 \\ 3 & 1 \end{pmatrix}, \quad \begin{pmatrix} 1 & 0 \\ -3 & 1 \end{pmatrix}$$

of elementary matrices.

Let

$$P_1 = \begin{pmatrix} 1 & 0 \\ 3 & 1 \end{pmatrix} \quad \text{and} \quad A = \begin{pmatrix} a & b \\ c & d \end{pmatrix}.$$

Then

$$P_1 A = \begin{pmatrix} a & b \\ 3a + c & 3b + d \end{pmatrix}.$$

This tells us that *premultiplying A* (multiplying A on the left) by the elementary matrix P_1 has the effect of applying the corresponding *ero* $r_2 + 3r_1$ to A. We say that A and $P_1 A$ are *equivalent* and use the notation

$$A \underset{r_2 + 3r_1}{\equiv} P_1 A.$$

Let $Q_1 = \begin{pmatrix} 0 & 1 \\ 1 & 0 \end{pmatrix}$. Then $A Q_1 = \begin{pmatrix} b & a \\ d & c \end{pmatrix}$.

Therefore *postmultiplication* (multiplication on the right) by the elementary matrix Q_1 carries out the corresponding *eco* $c_1 \leftrightarrow c_2$ on A. As above, we call A and $A Q_1$ equivalent matrices and write

$$A \underset{c_1 \leftrightarrow c_2}{\equiv} A Q_1.$$

The general principle

> *pre/postmultiplication by an elementary matrix carries out the corresponding ero/eco*

will be established in Lemma 1.4.

Now we wish to apply not just one elementary operation but a sequence of *eros* and *ecos* to an $s \times t$ matrix A. These operations are to be carried out in a particular order and so let $P_1, P_2, \ldots$ be the elementary $s \times s$ matrices corresponding to the first, second, ...of the *eros* we wish to apply, and let $Q_1, Q_2, \ldots$ be the elementary $t \times t$ matrices corresponding to the first, second, ...of the *ecos* we wish to apply. For simplicity, let's suppose there are just two *eros* and three *ecos*. Then A can be changed to $P_2 P_1 A Q_1 Q_2 Q_3$ by following one of the ten routes through the diagram

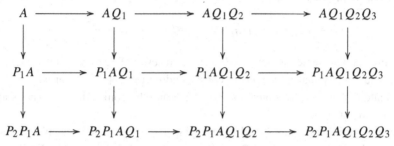

from top left to bottom right. The associative law of matrix multiplication ensures that we arrive at the same destination provided that the row operations are performed in the correct order amongst themselves, and similarly for the column operations.

We are now ready to start the main task of this chapter: to what extent can a matrix A over $\mathbb{Z}$ be simplified by applying elementary operations? You should know from your experience of eigenvectors and eigenvalues that *diagonal* matrices are often what one is aiming for. In this context, as we shall see, not only can each $s \times t$ matrix A be changed into a diagonal $s \times t$ matrix D say (all (i, j)-entries in D are zero for $i \neq j$), but also the diagonal entries $d_1, d_2, d_3, \ldots$ in D can be arranged to be non-negative and such that d_1 is a divisor of d_2, d_2 is a divisor of d_3, and so on. The matrix D is then *unique* and is known as the *Smith normal form of A*. The non-negative integers $d_1, d_2, d_3, \ldots$ are called the *invariant factors* of A.

Let us assume that we have reduced A to D by five elementary operations as above, that is, $P_2 P_1 A Q_1 Q_2 Q_3 = D$. Write $P = P_2 P_1$, and $Q = (Q_1 Q_2 Q_3)^{-1}$. We'll see later that Q is a particularly important matrix. For the moment notice that $Q = (Q_1 Q_2 Q_3)^{-1} = Q_3^{-1} Q_2^{-1} Q_1^{-1}$ expresses Q as a product of elementary matrices. So P and Q are invertible over $\mathbb{Z}$ and satisfy

$$PA = DQ.$$

The matrices A and D are said to be *equivalent over* $\mathbb{Z}$ and we write $A \equiv D$. The general definition of equivalence over $\mathbb{Z}$ is stated in Definition 1.5.

An *ero* and an *eco* are called *conjugate* if their corresponding elementary matrices are an inverse pair. Therefore $r_i \leftrightarrow r_j$ is conjugate to $c_i \leftrightarrow c_j$, and $-r_i$ is conjugate to $-c_i$. More importantly as we explain in a moment

$$r_i - mr_j \text{ is conjugate to } c_j + mc_i \quad \text{for } m \in \mathbb{Z}, \ i \neq j.$$

For instance, in the 2×2 case, $r_2 - 3r_1$ is conjugate to $c_1 + 3c_2$ since

$$\begin{pmatrix} 1 & 0 \\ -3 & 1 \end{pmatrix} \begin{pmatrix} 1 & 0 \\ 3 & 1 \end{pmatrix} = \begin{pmatrix} 1 & 0 \\ 0 & 1 \end{pmatrix}.$$

More generally let E_{ij} denote the $t \times t$ matrix with (i, j)-entry 1 and zeros elsewhere. For $i \neq j$ we see $E_{ij}^2 = 0$ as row i of E_{ij} fails to 'make contact' with column j of E_{ij} in the matrix product. Hence

$$(I - mE_{ij})(I + mE_{ij}) = I - mE_{ij} + mE_{ij} - m^2 E_{ij}^2 = I$$

showing that the elementary matrix $I - mE_{ij}$ corresponding to $r_i - mr_j$ is the inverse of the elementary matrix $I + mE_{ij}$ corresponding to $c_j + mc_i$, that is, $r_i - mr_j$ is conjugate to $c_j + mc_i$ as stated above. Notice that the conjugate of an *eco* is an *ero* and vice-versa.

The invertible matrices P and Q satisfying $PA = DQ$ can be calculated stage by stage as the reduction of A to D progresses. The equation $P = P_2 P_1 I$ tells us that the matrix P is the combined effect of applying the *eros* used in the reduction process to I. In the same way the equation $Q = Q_3^{-1} Q_2^{-1} Q_1^{-1} I$ tells us that the matrix Q is the cumulative effect of applying the *conjugates* of the *ecos* used in the reduction process to I.

Next, three reductions are worked through in detail.

Example 1.1

Let $A = \begin{pmatrix} 4 & 6 \\ 8 & 10 \end{pmatrix}$. We concentrate first on row 1 applying the following sequence of elementary operations to A:

$$A = \begin{pmatrix} 4 & 6 \\ 8 & 10 \end{pmatrix} \underset{c_2 - c_1}{\equiv} \begin{pmatrix} 4 & 2 \\ 8 & 2 \end{pmatrix} \underset{c_1 - 2c_2}{\equiv} \begin{pmatrix} 0 & 2 \\ 4 & 2 \end{pmatrix} \underset{c_1 \leftrightarrow c_2}{\equiv} \begin{pmatrix} 2 & 0 \\ 2 & 4 \end{pmatrix} \underset{r_2 - r_1}{\equiv} \begin{pmatrix} 2 & 0 \\ 0 & 4 \end{pmatrix} = D.$$

It is worth noticing that

the $(1, 1)$-entry in D is the greatest common divisor (gcd) of all the entries in A

that is, $2 = \gcd\{4, 6, 8, 10\}$ in this case. This fact will be used to establish the uniqueness of the Smith normal form in Section 1.3.

The matrix P resulting from the above sequence of operations is found by applying the above *ero* (there is only one) to I:

$$\begin{pmatrix} 1 & 0 \\ 0 & 1 \end{pmatrix} \underset{r_2 - r_1}{\equiv} \begin{pmatrix} 1 & 0 \\ -1 & 1 \end{pmatrix} = P.$$

The matrix Q resulting from the above reduction is found by applying, in order, the conjugates of the above *ecos* to I:

$$\begin{pmatrix} 1 & 0 \\ 0 & 1 \end{pmatrix}_{\underset{r_1+r_2}{\equiv}} \begin{pmatrix} 1 & 1 \\ 0 & 1 \end{pmatrix}_{\underset{r_2+2r_1}{\equiv}} \begin{pmatrix} 1 & 1 \\ 2 & 3 \end{pmatrix}_{\underset{r_1 \leftrightarrow r_2}{\equiv}} \begin{pmatrix} 2 & 3 \\ 1 & 1 \end{pmatrix} = Q.$$

You should now check that P and Q are invertible over $\mathbb{Z}$: in fact

$$P^{-1} = \begin{pmatrix} 1 & 0 \\ 1 & 1 \end{pmatrix} \quad \text{and} \quad Q^{-1} = \begin{pmatrix} -1 & 3 \\ 1 & -2 \end{pmatrix}.$$

The matrix equation

$$PA = \begin{pmatrix} 4 & 6 \\ 4 & 4 \end{pmatrix} = DQ$$

shows that P and Q do what is required of them. So $PAQ^{-1} = D = \mathrm{diag}(2, 4)$ is the Smith normal form of A, the invariant factors of A being 2 and 4.

The reader should realise that although every matrix A over $\mathbb{Z}$ has a unique Smith normal form D, the matrices P and Q (both invertible over $\mathbb{Z}$) satisfying $PA = DQ$ are by no means unique. You can check that the sequence $r_2 - 2r_1$, $r_1 + 3r_2$, $-r_2$, $c_1 \leftrightarrow c_2$, $r_1 \leftrightarrow r_2$ also reduces the above matrix A to D and leads to

$$P' = \begin{pmatrix} 2 & -1 \\ -5 & 3 \end{pmatrix} \quad \text{and} \quad Q' = \begin{pmatrix} 0 & 1 \\ 1 & 0 \end{pmatrix}$$

which are invertible over $\mathbb{Z}$ and satisfy $P'A = DQ'$.

Example 1.2

Consider the 3×3 matrix

$$A = \begin{pmatrix} 2 & 4 & 6 \\ 8 & 10 & 12 \\ 14 & 16 & 18 \end{pmatrix}.$$

In this case the gcd of the entries in A is already the $(1, 1)$-entry and so we begin by clearing (making zero) the other entries in row 1 and col 1. The reduction is then finished by applying the same method to the 2×2 submatrix obtained by deleting (in one's mind) row 1 and col 1.

$$A = \begin{pmatrix} 2 & 4 & 6 \\ 8 & 10 & 12 \\ 14 & 16 & 18 \end{pmatrix}_{\underset{c_3-3c_1}{\underset{c_2-2c_1}{\equiv}}} \begin{pmatrix} 2 & 0 & 0 \\ 8 & -6 & -12 \\ 14 & -12 & -24 \end{pmatrix}_{\underset{r_3-7r_1}{\underset{r_2-4r_1}{\equiv}}} \begin{pmatrix} 2 & 0 & 0 \\ 0 & -6 & -12 \\ 0 & -12 & -24 \end{pmatrix}$$

$$\underset{\substack{-r_2 \\ -r_3}}{\equiv} \begin{pmatrix} 2 & 0 & 0 \\ 0 & 6 & 12 \\ 0 & 12 & 24 \end{pmatrix} \underset{c_3-2c_2}{\equiv} \begin{pmatrix} 2 & 0 & 0 \\ 0 & 6 & 0 \\ 0 & 12 & 0 \end{pmatrix} \underset{r_3-2r_2}{\equiv} \begin{pmatrix} 2 & 0 & 0 \\ 0 & 6 & 0 \\ 0 & 0 & 0 \end{pmatrix} = D.$$

Note that two *ecos* have been applied apparently simultaneously at the first stage; this is unambiguous as these *ecos* commute: the order in which they are applied makes no difference. This is always the case when the gcd entry is used to clear the remaining entries in row 1. In the same way the *eros* used at stage 2 commute as do the *eros* used at stage 3. However, the reader is warned against carrying out too many elementary operations simultaneously, as generally the order in which they are applied is important.

The properties required of the Smith normal form $D = \mathrm{diag}(2, 6, 0)$ of A are more clearly in evidence: the invariant factors 2, 6, 0 of A are non-negative, 2 is a divisor of 6, and 6 is a divisor of 0. In fact *every* integer m is a divisor of 0 since $m \times 0 = 0$. As far as divisors are concerned, 0 is the *largest* integer!

We calculate P and Q as before. Applying the *eros* in the above sequence to I:

$$\begin{pmatrix} 1 & 0 & 0 \\ 0 & 1 & 0 \\ 0 & 0 & 1 \end{pmatrix} \underset{\substack{r_2-4r_1 \\ r_3-7r_1}}{\equiv} \begin{pmatrix} 1 & 0 & 0 \\ -4 & 1 & 0 \\ -7 & 0 & 1 \end{pmatrix} \underset{\substack{-r_2 \\ -r_3}}{\equiv} \begin{pmatrix} 1 & 0 & 0 \\ 4 & -1 & 0 \\ 7 & 0 & -1 \end{pmatrix}$$

$$\underset{r_3-2r_2}{\equiv} \begin{pmatrix} 1 & 0 & 0 \\ 4 & -1 & 0 \\ -1 & 2 & -1 \end{pmatrix} = P.$$

Applying the conjugates of the *ecos* in the above sequence to I:

$$\begin{pmatrix} 1 & 0 & 0 \\ 0 & 1 & 0 \\ 0 & 0 & 1 \end{pmatrix} \underset{\substack{r_1+2r_2 \\ r_1+3r_3}}{\equiv} \begin{pmatrix} 1 & 2 & 3 \\ 0 & 1 & 0 \\ 0 & 0 & 1 \end{pmatrix} \underset{r_2+2r_3}{\equiv} \begin{pmatrix} 1 & 2 & 3 \\ 0 & 1 & 2 \\ 0 & 0 & 1 \end{pmatrix} = Q.$$

The reader can now verify

$$PA = \begin{pmatrix} 2 & 4 & 6 \\ 0 & 6 & 12 \\ 0 & 0 & 0 \end{pmatrix} = DQ.$$

Incidentally, the above equation can be used to find all integer solutions x of the system of equations $xA = 0$, that is, all 1×3 matrices $x = (x_1, x_2, x_3)$ over $\mathbb{Z}$ satisfying

$$(x_1, x_2, x_3) \begin{pmatrix} 2 & 4 & 6 \\ 8 & 10 & 12 \\ 14 & 16 & 18 \end{pmatrix} = (0, 0, 0)$$

The above system can be transformed into a simpler system which can be solved on sight. To do this, put $y = xP^{-1}$. Then $x = yP$ and so $xA = 0$ becomes $yPA = 0$

which is the same as $yDQ = 0$. Postmultiplying by Q^{-1} produces $yD = 0$ which is the simpler system referred to. Putting $y = (y_1, y_2, y_3)$ we obtain

$$(y_1, y_2, y_3) \begin{pmatrix} 2 & 0 & 0 \\ 0 & 6 & 0 \\ 0 & 0 & 0 \end{pmatrix} = (0, 0, 0)$$

which means $2y_1 = 0$, $6y_2 = 0$, $0y_3 = 0$. The solution is $y_1 = 0$, $y_2 = 0$, y_3 arbitrary (any integer). So the general solution of $yD = 0$ is $y = (0, 0, y_3)$. Hence

$$x = yP = (0, 0, y_3)P = y_3(-1, 2, -1)$$

is the general solution in integers of the system $xA = 0$, where $y_3 \in \mathbb{Z}$. We have shown that the set of these solutions has $\mathbb{Z}$-*basis* $(-1, 2, -1)$ which is row 3 of P. This point will be taken up in Section 2.3. In general the matrix equation $PA = DQ$ shows that the last so many rows of P corresponding to the zero rows of D form a $\mathbb{Z}$-basis of the integer solutions x of $xA = 0$.

Example 1.3

The reduction method applies to $s \times t$ matrices A over $\mathbb{Z}$ which might not be square and the case $s = 1$, $t = 2$ is particularly significant.

Let $A = (204, 63)$. As A has only one row there is no need for *eros* and the reduction can be done using *ecos* alone. The following sequence of *ecos*:

$$A = (204, 63) \underset{c_1-3c_2}{\equiv} (15, 63) \underset{c_2-4c_1}{\equiv} (15, 3) \underset{c_1-5c_2}{\equiv} (0, 3) \underset{c_1 \leftrightarrow c_2}{\equiv} (3, 0) = D$$

shows that $D = (3, 0)$ is the Smith normal form of A and 3 (the only diagonal entry in D) is the only invariant factor of A. The reader will recognise this reduction as little more than the Euclidean algorithm, the sequence of divisions of one positive integer by another, showing $\gcd\{204, 63\} = 3$. As there are no row operations involved we take $P = I$, the 1×1 identity matrix. As before Q can be calculated by applying the conjugates of the above *ecos* to the 2×2 identity matrix I:

$$\begin{pmatrix} 1 & 0 \\ 0 & 1 \end{pmatrix} \underset{r_2+3r_1}{\equiv} \begin{pmatrix} 1 & 0 \\ 3 & 1 \end{pmatrix} \underset{r_1+4r_2}{\equiv} \begin{pmatrix} 13 & 4 \\ 3 & 1 \end{pmatrix} \underset{r_2+5r_1}{\equiv} \begin{pmatrix} 13 & 4 \\ 68 & 21 \end{pmatrix} \underset{r_1 \leftrightarrow r_2}{\equiv} \begin{pmatrix} 68 & 21 \\ 13 & 4 \end{pmatrix} = Q.$$

The matrix equation $PA = DQ$ becomes simply $A = DQ$ which, on comparing entries, gives the factorisations $204 = 3 \times 64$ and $63 = 3 \times 21$. Comparing leading entries in $AQ^{-1} = D$, that is,

$$(204, 63) \begin{pmatrix} -4 & 21 \\ 13 & -68 \end{pmatrix} = (3, 0),$$

gives

$$-4 \times 204 + 13 \times 63 = \gcd\{204, 63\}.$$

It is an important property of gcds that for each pair of integers l, m there are integers a, b satisfying

$$al + bm = \gcd\{l, m\}$$

(see Corollary 1.16). In general $a = q_{22} \det Q$ and $b = -q_{21} \det Q$ where $Q = (q_{ij})$ is an invertible 2×2 matrix over $\mathbb{Z}$ with $A = (l, m) = DQ$ and D in Smith normal form.

In the following parts of this chapter we show first that *every* $s \times t$ matrix A over $\mathbb{Z}$ can be reduced to a matrix D in Smith normal form using *eros* and *ecos*, and secondly that, no matter how the reduction is done, a given initial matrix A always leads to the same terminal matrix D.

EXERCISES 1.1

1. Write down the elementary 2×2 matrices corresponding to the following elementary operations: $r_1 + 2r_2,\ r_2 - 3r_1,\ c_2 - 2c_1,\ c_1 - 3c_2$.
   Are any two of the above operations (a) paired (b) conjugate?
   Which *eros* leave rows $2, 3, 4, \ldots$ unchanged?

2. Use *eros* and *ecos* to reduce the following matrices A to Smith normal form D. In each case determine matrices P and Q, which are products of elementary matrices over $\mathbb{Z}$, such that $PA = DQ$.

   (i) $\begin{pmatrix} 4 & 8 \\ 6 & 10 \end{pmatrix}$;   (ii) $\begin{pmatrix} 7 & 12 \\ 12 & 21 \end{pmatrix}$;   (iii) $\begin{pmatrix} 6 & 10 \\ 15 & 0 \end{pmatrix}$.

3. Answer Question 2 above in the case of the following matrices A:

   (i) $\begin{pmatrix} 1 & 2 & 3 \\ 3 & 1 & 2 \\ 2 & 3 & 1 \end{pmatrix}$;   (ii) $\begin{pmatrix} 1 & 2 & 3 \\ 2 & 3 & 4 \\ 3 & 4 & 5 \end{pmatrix}$;   (iii) $\begin{pmatrix} 3 & 0 & 0 \\ 0 & 2 & 0 \\ 0 & 0 & 4 \end{pmatrix}$.

   Hint: For (iii) begin with $r_1 + r_2, c_1 - c_2$.
   In each case find all integer solutions $x = (x_1, x_2, x_3)$ of the system of equations $xA = 0$.

4. For each of the following 1×2 matrices $A = (l, m)$ over $\mathbb{Z}$, use the Euclidean algorithm and *ecos* to find an invertible 2×2 matrix Q over $\mathbb{Z}$ satisfying $A = DQ$, where $D = (d, 0)$, $d = \gcd\{l, m\}$. Hence write down l/d and m/d as integers. Evaluate $\det Q$ and find integers a, b with $al + bm = d$ in each case:

   (i) $(72, 42)$;   (ii) $(34, 55)$;   (iii) $(7497, 5474)$.

5. (a) Change the 1×2 matrix $(0, d)$ into $(d, 0)$ using *ecos* of type (iii).

   (b) By applying the sequence: $c_1 + c_2$, $c_2 - c_1$, $c_1 + c_2$ once, twice, or three times as required, show that every 1×2 matrix A over $\mathbb{Z}$ can be changed into a matrix with non-negative entries using *ecos* of type (iii).

   (c) Using the Euclidean algorithm, show that every 1×2 matrix A over $\mathbb{Z}$ can be reduced to Smith normal form D using only *ecos* of type (iii). Deduce that $AQ = D$ where $\det Q = 1$.

   Hint: Type (iii) *ecos* leave determinants unchanged.

   (d) Let P be a 2×2 matrix $\mathbb{Z}$ with $\det P = 1$. Show that P can be reduced to Smith normal form I using only *ecos* of type (iii).

   Hint: Apply part (c) to row 1 of P.

   (e) Reduce $\left(\begin{smallmatrix} 3 & 4 \\ 5 & 7 \end{smallmatrix}\right)$ to I using only *ecos* of type (iii).

6. (a) Show that the matrix transpose P^T of an elementary matrix P is itself elementary. How are the corresponding *eros* related?

   (b) Let A be a square matrix over $\mathbb{Z}$. Suppose $PA = DQ$ where P and Q are products of elementary matrices and D is diagonal. By transposing $PA = DQ$ show that $Q^T PA$ is symmetric, and interpret this result in terms of *eros*.

   (c) Change $A = \left(\begin{smallmatrix} 2 & 7 \\ 8 & 5 \end{smallmatrix}\right)$ into a symmetric matrix using *eros* only.

1.2 Existence of the Smith Normal Form

The examples of the previous section suggest that every matrix A over $\mathbb{Z}$ is reducible to its Smith normal form D using *eros* and *ecos*. Although this is true, the reader should not be lulled into a false sense of security simply because it 'works out' in a few special cases. A mathematical proof is required to clinch the hunch you should now have, and the serious business of laying this proof out is now our concern. The reader should take heart as, in this instance, the transition from practice to theory is relatively smooth: most of the steps in the reduction of a general matrix have already been encountered in the numerical examples of the previous section.

In Chapters 4 and 5 we shall need to replace the ring $\mathbb{Z}$ of integers by the ring $F[x]$ of polynomials with coefficients in the field F. So it is important to be aware of the particular property of $\mathbb{Z}$ which allows the reduction process to work; in fact it's nothing more than the familiar *integer division property*:

> for each pair of integers m, n with $n > 0$ there are unique integers q, r with $m = nq + r$ and $0 \le r < n$.

In other words, the positive integer n divides q (the *quotient*) times into the arbitrary integer m with *remainder* r. A suitably modified division property involving degree of polynomials holds in $F[x]$, and this analogy between $\mathbb{Z}$ and $F[x]$ will allow us, in the second half of this book, to deal with matrices having polynomial entries in much the same way as matrices over $\mathbb{Z}$.

We begin the details with a closer look at the principle used repeatedly in Section 1.1, namely that pre/postmultiplication by an elementary matrix carries out the corresponding *ero/eco*. The reader is likely to have met this idea in the context of matrices with entries from a field F, where the operation of multiplying a row or column by any non-zero scalar (the invertible elements of F) is allowed; here, by contrast, rows and columns may be multiplied by ± 1 (the only invertible elements of $\mathbb{Z}$), that is, their signs may be changed, but that's all! In both contexts elementary operations are invertible and their inverses are also elementary.

Throughout we use e_i to denote row i of the identity matrix I. Using matrix transposition, column j of I is denoted by e_j^T. The number of entries in e_i and e_j^T should be clear from the context. For any matrix A

$$e_i A \text{ is row } i \text{ of } A, Ae_j^T \text{ is column } j \text{ of } A, e_i Ae_j^T \text{ is the } (i, j)\text{-entry in } A.$$

These useful facts are direct consequences of the matrix multiplication rule.

Lemma 1.4

Let A be an $s \times t$ matrix over $\mathbb{Z}$. Let P_1 be an elementary $s \times s$ matrix over $\mathbb{Z}$. Let Q_1 be an elementary $t \times t$ matrix over $\mathbb{Z}$. The matrix $P_1 A$ is the result of applying to A the *ero* corresponding to P_1. The matrix AQ_1 is the result of applying to A the *eco* corresponding to Q_1.

Proof

Let $r_j \leftrightarrow r_k$ be the *ero* corresponding to P_1 where $1 \le j, k \le s$, $j \ne k$. So P_1 is obtained from I by interchanging row j and row k. The equations

$$e_j P_1 = e_k, \qquad e_k P_1 = e_j, \qquad e_i P_1 = e_i \quad \text{for } i \ne j, k$$

describe all the rows of P_1 and hence P_1 itself. Postmultiplying by A gives

$$e_j P_1 A = e_k A, \qquad e_k P_1 A = e_j A, \qquad e_i P_1 A = e_i A, \quad i \ne j, k$$

which describe the s rows of $P_1 A$ in terms of the rows of A: row j of $P_1 A$ is row k of A, row k of $P_1 A$ is row j of A, row i of $P_1 A$ is row i of A for $i \ne j, k$. In other words, $P_1 A$ is obtained from A by applying $r_j \leftrightarrow r_k$.

We leave the reader to deal with *eros* of types (ii) and (iii) in the same way.

Now let $c_j + lc_k$ be the *eco* corresponding to Q_1 where $1 \leq j, k \leq t$, $j \neq k$, and l is any integer. So Q_1 is the result of applying $c_j + lc_k$ to I. The equations

$$Q_1 e_j^T = e_j^T + l e_k^T, \qquad Q_1 e_i^T = e_i^T \quad \text{for } i \neq j$$

describe all the columns of Q_1. Premultiplying by A gives

$$A Q_1 e_j^T = A e_j^T + l A e_k^T, \qquad A Q_1 e_i^T = A e_i^T \quad \text{for } i \neq j$$

which describe the t columns of AQ_1 in terms of the columns of A: col j of AQ_1 is col j of A plus l times col k of A, col i of AQ_1 is col i of A for $i \neq j$. So AQ_1 results on applying $c_j + lc_k$ to A.

As before we leave the reader to deal with *ecos* of types (i) and (ii). □

Suppose now that the $s \times t$ matrix A over $\mathbb{Z}$ is subjected to a finite number of *eros* and *ecos* carried out in succession. Let $P_1, P_2, \ldots, P_u$ be the elementary matrices corresponding to these *eros* and let $Q_1, Q_2, \ldots, Q_v$ be the elementary matrices corresponding to these *ecos*. Matrix multiplication is designed for this very situation! Applying the *eros* first we see that A changes to $P_1 A$, then $P_1 A$ changes to $P_2 P_1 A$, and so on until the matrix $P_u \cdots P_2 P_1 A$ is obtained; strictly speaking Lemma 1.4 has been used u times here. Let $P = P_u \cdots P_2 P_1$. Then PA is the result of applying these u *eros* to A. Secondly applying the *ecos*, we see that PA changes to PAQ_1, then PAQ_1 changes to $PAQ_1 Q_2$, until ultimately $PAQ_1 Q_2 \cdots Q_v$ is obtained, using Lemma 1.4 a further v times. Let $Q = Q_v^{-1} \cdots Q_2^{-1} Q_1^{-1}$. Then $Q^{-1} = Q_1 Q_2 \cdots Q_v$ and so

$$PAQ^{-1} = P_u \cdots P_2 P_1 A Q_1 Q_2 \cdots Q_v$$

is the result of applying the given sequence of u *eros* and v *ecos* to A. The reader will have to wait until Chapter 3 to fully appreciate the significance of the invertible matrix Q, although this was touched on in the overview before Chapter 1. In fact the rows of Q tell us how to decompose f.g. abelian groups, the matrix P being less important due to the formulation adopted.

Notice that the number of ways of interlacing the u *eros* in order with the v *ecos* in order is the binomial coefficient $\binom{u+v}{u}$. This is because each interlacing corresponds to a subset S of size u of $\{1, 2, \ldots, u + v\}$, the ith elementary operation applied to A being either an *ero* or an *eco* according as $i \in S$ or $i \notin S$. By the associative law of matrix multiplication, all these $\binom{u+v}{u}$ sequences lead to the same matrix, namely PAQ^{-1}. The diagram in Section 1.1 illustrates the case $u = 2$, $v = 3$.

As $P = PI = P_u \cdots P_2 P_1 I$, by Lemma 1.4 we see that P is obtained by applying the corresponding *eros*, in order, to the $s \times s$ identity matrix I. In almost the same way, $Q = QI = Q_v^{-1} \cdots Q_2^{-1} Q_1^{-1} I$, and so by Lemma 1.4 we see that Q is obtained by applying, in order, the conjugates of the corresponding *ecos* (these conjugates are *eros* in fact) to the $t \times t$ identity matrix I. In the following theory we show that A can be 'reduced' to its Smith normal form D by a sequence of *eros* and *ecos*. So both P and Q can be built up step by step as the reduction of A proceeds, as illustrated in the examples of Section 1.1.

Definition 1.5

The $s \times t$ matrices A and B over $\mathbb{Z}$ are called *equivalent* and we write $A \equiv B$ if there is an $s \times s$ matrix P and a $t \times t$ matrix Q, both invertible over $\mathbb{Z}$, such that

$$PAQ^{-1} = B.$$

We leave the reader to verify that the symbol $\equiv$ as defined in Definition 1.5 satisfies the three laws (reflexive, symmetric, transitive) of an equivalence relation (see Exercises 1.2, Question 1(d)). As elementary matrices and products of elementary matrices are invertible over $\mathbb{Z}$, the above discussion shows that A changes into an equivalent matrix B when *eros* and *ecos* are applied to A. In Theorem 1.18 the converse is shown to be true, that is, suppose $A \equiv B$. Then A can be changed into B by applying elementary row and column operations to A.

How can we decide whether the $s \times t$ matrices A and B over $\mathbb{Z}$ are equivalent or not? The reader should remember that *rank* is the all-important number when $\mathbb{Z}$ is replaced by a field F, that is, the $s \times t$ matrices A and B over F are equivalent if and only if rank A = rank B. What is more, an efficient method of determining rank A consists of applying elementary operations to A in order to find the *simplest* matrix over F which is equivalent to A. This simplest $s \times t$ matrix over F is $\text{diag}(1, 1, \ldots, 1, 0, 0, \ldots, 0)$ where the number of ones is rank A. We now introduce the corresponding concept for matrices over $\mathbb{Z}$.

Definition 1.6

Let D be an $s \times t$ matrix over $\mathbb{Z}$ such that
  (i) the (i, j)-entries in D are zero for $i \neq j$, that is, D is a diagonal matrix,
 (ii) each (i, i)-entry d_i in D is non-negative,
(iii) for each i with $1 \leq i < \min\{s, t\}$ there is an integer q_i with $d_{i+1} = q_i d_i$, that is, $d_i | d_{i+1}$ (d_i is a divisor of d_{i+1}).
Then D is said to be in *Smith normal form* and we write $D = \text{diag}(d_1, d_2, \ldots, d_{\min\{s,t\}})$.

Notice that d_1 is the gcd of the st entries in D. Also $d_1 d_2$ is the gcd of the 2-*minors* of D (the determinants of 2×2 submatrices of D). This theme is developed in Section 1.3 where it is proved Corollary 1.20 that each $s \times t$ matrix A over $\mathbb{Z}$ is equivalent to a *unique* matrix D in Smith normal form. It's therefore reasonable to refer to *the* Smith normal form D of A and adopt the notation $D = S(A)$. The integers d_i for $1 \leq i \leq \min\{s, t\}$ are called *the invariant factors of* A.

For $A = \begin{pmatrix} 4 & 6 \\ 8 & 10 \end{pmatrix}$ we see from Example 1.1 that $S(A) = \mathrm{diag}(2, 4)$ and the invariant factors of A are $2, 4$.

We shall prove $A \equiv B \Leftrightarrow S(A) = S(B)$, that is, the $s \times t$ matrices A and B over $\mathbb{Z}$ are equivalent if and only if their Smith normal forms are identical. There's a fair amount of work to be done to prove this basic fact – so let's get going!

First we describe a reduction process which changes every $s \times t$ matrix A over $\mathbb{Z}$ into a matrix D, as in Definition 1.6, using a finite number of *eros* and *ecos*. It turns out that the reduction of every such matrix, no matter what its size, boils down to the reduction of 1×2 matrices and diagonal 2×2 matrices over $\mathbb{Z}$.

Lemma 1.7

Let $A = (l, m)$ be a 1×2 matrix over $\mathbb{Z}$. There is a sequence of *ecos* over $\mathbb{Z}$ which reduces A to $D = (d, 0)$ where $d = \gcd\{l, m\}$.

Proof

Applying $-c_1, -c_2$ if necessary, we can assume $l \geq 0$, $m \geq 0$. Suppose first $l = d$ and so $l | m$. If $m = 0$ then $A = (d, 0) = D$, that is, A is already in Smith normal form and no *ecos* are needed. If $m > 0$, then $l > 0$ also and the *eco* $c_2 - (m/l)c_1$ reduces A to D.

Suppose now $l \neq d$. If $l = 0$ then $m = d > 0$ and the *eco* $c_1 \leftrightarrow c_2$ reduces A to D. If $l > 0$ then $m > 0$ also. In this case we let $r_1 = l$, $r_2 = m$ and carry out the *Euclidean algorithm* as follows to obtain $d = \gcd\{r_1, r_2\}$: let r_{i+2} be the remainder on dividing r_i by r_{i+1} for $i \geq 1$; these remainders form a decreasing sequence

$$r_2 > r_3 > \cdots > r_{i+2} > r_{i+3} > \cdots$$

of non-negative integers. So eventually a zero remainder r_{k+1} is obtained ($k \geq 2$) and the algorithm terminates. As $r_i = q_{i+1}r_{i+1} + r_{i+2}$ for some non-negative integer q_{i+1}, we deduce $\gcd\{r_i, r_{i+1}\} = \gcd\{r_{i+1}, r_{i+2}\}$ for $1 \leq i < k$, and hence $d = \gcd\{r_1, r_2\} = \gcd\{r_k, r_{k+1}\} = \gcd\{r_k, 0\} = r_k$. So d turns up as the last non-zero remainder r_k in the Euclidean algorithm. Applying $c_1 - q_{i+1}c_2$ to the 1×2 matrix (r_i, r_{i+1}) produces (r_{i+2}, r_{i+1}); applying $c_2 - q_{i+1}c_1$ to the 1×2 matrix (r_{i+1}, r_i) produces (r_{i+1}, r_{i+2}). So each of the $k - 1$ divisions in the algorithm gives rise to an *eco*, and applying these *ecos* to (r_1, r_2) eventually produces either (r_k, r_{k+1}) or (r_{k+1}, r_k) according as k is odd or even. As $c_1 \leftrightarrow c_2$ changes (r_{k+1}, r_k) into $D = (r_k, r_{k+1}) = (d, 0)$, we see A can be reduced to D using at most k *ecos*. $\square$

In fact the above reduction can be carried out using only *ecos* of type (iii) (see Exercises 1.1, Question 5). Notice that elementary matrices corresponding to *ecos* of type (iii) have determinant 1. The reader will be aware that the determinant function is multiplicative, that is,

$$\det Q_1 Q_2 = \det Q_1 \det Q_2$$

for all $t \times t$ matrices Q_1, Q_2 over $\mathbb{Z}$ and we review this important property in Theorem 1.18. It follows that every 1×2 matrix A over $\mathbb{Z}$ can be expressed $A = DQ$ where the 1×2 matrix D is in Smith normal form and the 2×2 matrix Q over $\mathbb{Z}$ satisfies $\det Q = 1$. We now give a direct proof of this fact.

Lemma 1.8

Every 1×2 matrix A over $\mathbb{Z}$ can be expressed $A = DQ$ where the 1×2 matrix D is in Smith normal form and Q is a 2×2 matrix over $\mathbb{Z}$ with $\det Q = 1$.

Proof

Let $A = (l, m)$ and take $D = (d, 0)$ where $d = \gcd\{l, m\}$. Then D is in Smith normal form as $d \geq 0$. If $d = 0$ then $l = m = 0$ and we take $Q = I$. So suppose $d > 0$. As d is a common divisor of l and m, there are integers l' and m' such that $l = dl'$, $m = dm'$. Also there are integers a, b with $al + bm = d$. Dividing this equation through by d gives $al' + bm' = 1$. Let $Q = \left(\begin{smallmatrix} l' & m' \\ -b & a \end{smallmatrix}\right)$. Then $\det Q = 1$ and $DQ = (d, 0)\left(\begin{smallmatrix} l' & m' \\ -b & a \end{smallmatrix}\right) = (dl', dm') = (l, m) = A$. $\square$

The reader will know that integers a and b as above can be calculated by reversing the steps in the Euclidean algorithm. We review their role from a more general perspective in Corollary 1.16.

The transposed version of Lemma 1.7 says: there is a sequence of *eros* which reduces

$$A = \binom{l}{m} \quad \text{to} \quad D = \binom{d}{0}$$

where $d = \gcd\{l, m\}$. We now use Lemma 1.7 and its transpose to establish the key step in the reduction process of a general matrix over $\mathbb{Z}$: all off-diagonal entries in the first row and first column can be cleared (replaced by zeros), using elementary operations. In other words, every $s \times t$ matrix A over $\mathbb{Z}$ can be changed into a matrix of the type

$$B = \left(\begin{array}{c|c} b_{11} & 0 \\ \hline 0 & B' \end{array}\right)$$

using *eros* and *ecos* where B' is an $(s-1) \times (t-1)$ matrix over $\mathbb{Z}$. The reduction process can then be completed by induction since B' is smaller than A.

Lemma 1.9

Using elementary operations every $s \times t$ matrix $A = (a_{ij})$ over $\mathbb{Z}$ can be changed into an $s \times t$ matrix $B = (b_{ij})$ over $\mathbb{Z}$ with $b_{11} \geq 0, b_{1j} = b_{i1} = 0$ for $2 \leq i \leq s, 2 \leq j \leq t$.

Proof

Suppose $A \neq B$ as otherwise there is nothing to do. We describe an algorithm based on Lemma 1.7 which changes A into B using elementary operations. Let $h_j = \gcd\{a_{11}, a_{12}, \ldots, a_{1j}\}$, that is, h_j is the gcd of the first j entries in row 1 of A. Then $h_1 = \pm a_{11}$ and $h_{j+1} = \gcd\{h_j, a_{1j+1}\}$ for $1 \leq j < t$. (There is a discussion of gcds of finite sets of integers at the beginning of Section 1.3.) By changing the sign of col 1 of A if necessary we obtain an $s \times t$ matrix A_1 with first row $(h_1, a_{12}, \ldots, a_{1t})$. Next apply Lemma 1.7 to the 1×2 submatrix (h_1, a_{12}) of A_1: there is a sequence of *ecos* changing (h_1, a_{12}) into $(h_2, 0)$; applying these *ecos* to A_1 produces an $s \times t$ matrix A_2 with first row $(h_2, 0, a_{13}, \ldots, a_{1t})$. Next we apply Lemma 1.7 to the 1×2 submatrix (h_2, a_{13}) of A_2: there is a sequence of *ecos* changing (h_2, a_{13}) into $(h_3, 0)$; applying these *ecos* to A_2 (they affect columns 1 and 3 only) produces an $s \times t$ matrix A_3 with first row $(h_3, 0, 0, a_{14}, \ldots, a_{1t})$. Suppose inductively that A_j with first row $(h_j, 0, \ldots, 0, a_{1j+1}, \ldots, a_{1t})$ has been obtained from A by *ecos* where $1 \leq j < t$. Applying Lemma 1.7 to the 1×2 submatrix (h_j, a_{1j+1}) of A_j produces a sequence of *ecos* changing (h_j, a_{1j+1}) into $(h_{j+1}, 0)$. We obtain A_{j+1} with first row $(h_{j+1}, 0, \ldots, 0, a_{1j+2}, \ldots, a_{1t})$ on applying these *ecos* to cols 1 and $j+1$ of A_j. By induction on j, after $t-1$ applications of Lemma 1.7, an $s \times t$ matrix A_t with first row $(h_t, 0, \ldots, 0)$ is obtained which has come from A by applying *ecos*. We have successfully completed the first step in the reduction of A to B. Let's write $B_1 = A_t$ and $b_1 = h_t$. So by applying a finite number of *ecos* to A we have obtained B_1 with first row $(b_1, 0, \ldots, 0)$ where b_1 is the gcd of the entries in row 1 of A. It's important to notice that none of the *ecos* used affect column 1 in the case $h_1 = h_t$. Should all $(i, 1)$-entries in B_1 be zero for $2 \leq i \leq s$, then the algorithm terminates with $B_1 = B$. Otherwise let b_2 denote the gcd of the entries in col 1 of B_1. Since b_1 is the $(1, 1)$-entry in B_1 we see

$$b_2 \text{ is a divisor of } b_1$$

and this fact is the key to the algorithmic proof. Notice $b_2 = 0$ implies $b_1 = 0$ in which case all entries in row 1 and col 1 of A are zero, that is, there's 'nothing to do' as we said at the beginning. So $b_2 > 0$ ($b_1 = 0$ is possible, that is, $e_1 A = 0$ but this doesn't seem to shorten the proof).

Next the spotlight is turned on the column 1 of B_1. Using the technique of the preceding paragraph but transposed, there is a sequence of *eros* changing B_1 into an $s \times t$ matrix, B_2 say, having $(1, 1)$-entry b_2 and zero $(i, 1)$-entries for $2 \leq i \leq s$. Should $b_1 = b_2$, that is, b_1 is the gcd of the entries in column 1 of B_1, then none of these *eros* change row 1 and the algorithm ends with $B_2 = B$. We therefore assume $b_1 \neq b_2$.

So far so good. However there is a snag. Clearing off-diagonal entries in column 1 may reintroduce non-zero off-diagonal entries in row 1 (if not then the algorithm ends as above with $B_2 = B$). When this happens, that is, when B_2 has non-zero off-diagonal entries in its first row, the whole process is started again with B_2 in place of the original matrix A. (The reader should now work through the numerical example following this proof.)

The reduction of A is completed by iterating the above procedure. Clearing off-diagonal entries in row 1 and column 1 alternately, we obtain matrices $B_2, B_3,$ $\ldots, B_k, \ldots$ their $(1, 1)$-entries forming a decreasing sequence

$$b_2, b_3, \ldots, b_k, \ldots$$

of positive integers such that $b_k | b_{k-1}$ for $k = 3, 4, \ldots$. For odd k each B_k is obtained by applying *ecos* to B_{k-1} and b_k is the gcd of the entries in row 1 of B_{k-1} and is the only non-zero entry in row 1 of B_k. For even k each B_k is obtained by applying *eros* to B_{k-1} and b_k is the gcd of the entries in column 1 of B_{k-1} and is the only non-zero entry in column 1 of B_k. Multiplying together the $k - 1$ inequalities $b_2/b_3 \geq 2$, $b_3/b_4 \geq 2, \ldots, b_{k-1}/b_k \geq 2, b_k \geq 1$ gives $b_2 \geq 2^{k-2}$ and so $k \leq \log_2 b_2 + 2$. Therefore $k \leq \lfloor \log_2 b_2 \rfloor + 2$ where $\lfloor \log_2 b_2 \rfloor$ denotes the integer part of the real number $\log_2 b_2$.

The process continues provided $B_k \neq B$, that is, the matrix B_k is not of the type we are looking for. On the other hand we have just seen that such k are bounded above by the integer $\lfloor \log_2 b_2 \rfloor + 2$. So the process must terminate with $B_l = B$ where $l \leq \lfloor \log_2 b_2 \rfloor + 3$. $\square$

To illustrate the above proof, consider

$$A = \begin{pmatrix} 130 & 260 \\ 110 & 221 \end{pmatrix}.$$

Using the method of Lemma 1.9 gives

$$A \equiv \begin{pmatrix} 130 & 0 \\ 110 & 1 \end{pmatrix} = B_1 \equiv \begin{pmatrix} 20 & -1 \\ 110 & 1 \end{pmatrix} \equiv \begin{pmatrix} 20 & -1 \\ 10 & 6 \end{pmatrix} \equiv \begin{pmatrix} 0 & -13 \\ 10 & 6 \end{pmatrix}$$

$$\equiv \begin{pmatrix} 10 & 6 \\ 0 & -13 \end{pmatrix} = B_2 \equiv \begin{pmatrix} 4 & 6 \\ 13 & -13 \end{pmatrix} \equiv \begin{pmatrix} 4 & 2 \\ 13 & -26 \end{pmatrix} \equiv \begin{pmatrix} 0 & 2 \\ 65 & -26 \end{pmatrix}$$

$$\equiv \begin{pmatrix} 2 & 0 \\ -26 & 65 \end{pmatrix} = B_3 \equiv \begin{pmatrix} 2 & 0 \\ 0 & 65 \end{pmatrix} = B_4.$$

We next show how the Smith normal form of diagonal matrices, such as B_4 above, can be obtained. The *Chinese remainder theorem* is discussed in Theorem 2.11. What follows is a matrix version of this theorem.

Lemma 1.10

Let A be a 2×2 diagonal matrix over $\mathbb{Z}$ with non-negative entries. Then A can be changed into Smith normal form D using at most five elementary operations of type (iii).

Proof

Write $A = \mathrm{diag}(l, m)$. In the case $l \mid m$ there is nothing to do as $A = D$. Otherwise let $d = \gcd\{l, m\}$. Then $d > 0$ and there are integers a, b with $al + bm = d$ (see Corollary 1.16). The following sequence of elementary operations of type (iii) changes A into D:

$$A = \begin{pmatrix} l & 0 \\ 0 & m \end{pmatrix} \underset{c_2 + ac_1}{\equiv} \begin{pmatrix} l & al \\ 0 & m \end{pmatrix} \underset{r_1 + br_2}{\equiv} \begin{pmatrix} l & d \\ 0 & m \end{pmatrix} \underset{c_1 - (l/d - 1)c_2}{\equiv} \begin{pmatrix} d & d \\ -lm/d + m & m \end{pmatrix}$$

$$\underset{c_2 - c_1}{\equiv} \begin{pmatrix} d & 0 \\ -lm/d + m & lm/d \end{pmatrix} \underset{r_2 + (m/d)(l/d - 1)r_1}{\equiv} \begin{pmatrix} d & 0 \\ 0 & lm/d \end{pmatrix} = D. \qquad \square$$

The integer lm/d is the *least common multiple* (lcm) of the positive integers l and m, that is, lm/d is the smallest positive integer which is divisible by both l and m. We write $lm/d = \mathrm{lcm}\{l, m\}$, and $\mathrm{lcm}\{0, m\} = \mathrm{lcm}\{0, 0\} = 0$.

For example of Lemma 1.10 take $l = 21$, $m = 35$. Then $2 \times 21 + (-1) \times 35 = 7 = d$ and so $a = 2$, $b = -1$. So $\gcd\{21, 35\} = 7$ and $\mathrm{lcm}\{21, 35\} = (21 \times 35)/7 = 105$. Also

$$A = \begin{pmatrix} 21 & 0 \\ 0 & 35 \end{pmatrix} \underset{c_2 + 2c_1}{\equiv} \begin{pmatrix} 21 & 42 \\ 0 & 35 \end{pmatrix} \underset{r_1 - r_2}{\equiv} \begin{pmatrix} 21 & 7 \\ 0 & 35 \end{pmatrix} \underset{c_1 - 2c_2}{\equiv} \begin{pmatrix} 7 & 7 \\ -70 & 35 \end{pmatrix}$$

$$\underset{c_2 - c_1}{\equiv} \begin{pmatrix} 7 & 0 \\ -70 & 105 \end{pmatrix} \underset{r_2 + 10r_1}{\equiv} \begin{pmatrix} 7 & 0 \\ 0 & 105 \end{pmatrix} = D.$$

We are now ready for the main theorem of Section 1.2.

Theorem 1.11 (The existence of the Smith normal form over $\mathbb{Z}$)

Every $s \times t$ matrix A over $\mathbb{Z}$ can be reduced to an $s \times t$ matrix D in Smith normal form using elementary operations over $\mathbb{Z}$.

Proof

We use induction on the positive integer $\min\{s, t\}$. If $\min\{s, t\} = 1$, then the matrix B in Lemma 1.9 is already in Smith normal form. Now suppose $\min\{s, t\} > 1$. By Lemma 1.9 there are elementary operations changing A into

$$B = \left(\begin{array}{c|c} b_1 & 0 \\ \hline 0 & B' \end{array} \right)$$

where $b_1 \geq 0$ and B' is an $(s - 1) \times (t - 1)$ matrix. By inductive hypothesis B' can be reduced to a matrix $D' = \mathrm{diag}(d'_2, d'_3, \ldots)$ in Smith normal form using elementary operations as $\min\{s - 1, t - 1\} = \min\{s, t\} - 1$. Hence A can be changed using elementary operations into a diagonal matrix $D_1 = \mathrm{diag}(b_1, d'_2, d'_3, \ldots)$ having non-negative entries such that $d'_i | d'_{i+1}$ for $1 < i < \min\{s, t\}$. By Lemma 1.10 there is a sequence of at most five elementary operations which changes the 2×2 matrix $\mathrm{diag}(b_1, d'_2)$ into $\mathrm{diag}(d_1, b_2)$ where $d_1 = \gcd\{b_1, d'_2\}$ and $b_2 = \mathrm{lcm}\{b_1, d'_2\}$. Applying this sequence of operations to D_1 produces $D_2 = \mathrm{diag}(d_1, b_2, d'_3, d'_4, \ldots)$. By inductive hypothesis there is a sequence of elementary operations changing the $(s - 1) \times (t - 1)$ matrix $D'_2 = \mathrm{diag}(b_2, d'_3, d'_4, \ldots)$ into $D' = \mathrm{diag}(d_2, d_3, d_4, \ldots)$ in Smith normal form. Hence D_2 can be changed into $D = (d_1, d_2, d_3, \ldots)$ using elementary operations. Is D in Smith normal form? As $d_1 = \gcd\{b_1, d'_2\}$ and we see $d_1 | d'_i$ for $2 \leq i \leq \min\{s, t\}$. As $b_2 = \mathrm{lcm}\{b_1, d'_2\}$ we see $d_1 | b_1$ and $b_1 | b_2$ which give $d_1 | b_2$. So d_1 is a divisor of all the entries in D'_2 and hence d_1 is a divisor of all the entries in D'. In particular $d_1 | d_2$ and so D *is* in Smith normal form. As A can be changed into D_1, D_1 into D_2 and D_2 into D using elementary operations, we see that A can be changed into D using elementary operations. The induction is now complete. $\square$

We have achieved our aim! So before going on it is perhaps worthwhile pausing to look back at the reduction process. Starting with any $s \times t$ matrix A over $\mathbb{Z}$ this process first uses Lemma 1.9 to systematically clear all off-diagonal entries, producing a diagonal matrix D'' say having non-negative entries. Secondly Lemma 1.10 is used starting with the *last* pair of diagonal entries in D'' (on unravelling the induction) and continuing 'up' the diagonal to produce ultimately the Smith normal form $D = S(A)$ of the original matrix A. Here is a numerical example.

Example 1.12

Let

$$A = \begin{pmatrix} 4 & 16 & 4 \\ 8 & 42 & 28 \\ 12 & 54 & 25 \end{pmatrix}.$$

First use Lemma 1.9 to clear all off-diagonal entries in row 1 and col 1. Then apply Lemma 1.9 to the 2×2 submatrix which remains on deleting row 1 and col 1 of resulting matrix to obtain D'':

$$\begin{pmatrix} 4 & 16 & 4 \\ 8 & 42 & 28 \\ 12 & 54 & 25 \end{pmatrix} \underset{\substack{c_2-4c_1 \\ c_3-c_1}}{\equiv} \begin{pmatrix} 4 & 0 & 0 \\ 8 & 10 & 20 \\ 12 & 6 & 13 \end{pmatrix} \underset{\substack{r_2-2r_1 \\ r_3-3r_1}}{\equiv} \begin{pmatrix} 4 & 0 & 0 \\ 0 & 10 & 20 \\ 0 & 6 & 13 \end{pmatrix}$$

$$\underset{c_3-2c_2}{\equiv} \begin{pmatrix} 4 & 0 & 0 \\ 0 & 10 & 0 \\ 0 & 6 & 1 \end{pmatrix} \underset{\substack{r_2-r_3 \\ r_3-r_2}}{\equiv} \begin{pmatrix} 4 & 0 & 0 \\ 0 & 4 & -1 \\ 0 & 2 & 2 \end{pmatrix}$$

$$\underset{\substack{r_2-2r_3 \\ r_2\leftrightarrow r_3}}{\equiv} \begin{pmatrix} 4 & 0 & 0 \\ 0 & 2 & 2 \\ 0 & 0 & -5 \end{pmatrix} \underset{\substack{c_3-c_2 \\ -r_3}}{\equiv} \begin{pmatrix} 4 & 0 & 0 \\ 0 & 2 & 0 \\ 0 & 0 & 5 \end{pmatrix} = D''.$$

Next apply Lemma 1.10 to the submatrix $\left(\begin{smallmatrix} 2 & 0 \\ 0 & 5 \end{smallmatrix}\right)$ of D'' obtaining $D_1 = \mathrm{diag}(4, 1, 10)$, $D_2 = \mathrm{diag}(1, 4, 10)$ and $S(A) = D = \mathrm{diag}(1, 2, 20)$ as in the proof of Theorem 1.11. The reader can check that this reduction of A to D uses 26 elementary operations; in fact the reduction can be done in less than half this number. Inevitably the general technique of Theorem 1.11 rarely yields the shortest reduction in a particular case (see Exercises 1.2, Question 6(b) for a particularly awkward matrix). The merit of the reduction method is not that the least number of *eros* and *ecos* is used, but rather that every matrix A over $\mathbb{Z}$, no matter what its entries are, can be reduced in this way to D as in Definition 1.6.

From Theorem 1.11 we see that d_1, the gcd of the entries in A, can always be created by applying elementary operations to A; in particular cases this may be easy to see – indeed d_1 may be present as an entry in A. In the latter case move d_1 into the $(1, 1)$-position, clear all other entries in row 1 and col 1 obtaining

$$\left(\begin{array}{c|c} d_1 & 0 \\ \hline 0 & A' \end{array} \right).$$

As d_1 is a divisor of all the entries in A', we see

$$S(A) = \left(\begin{array}{c|c} d_1 & 0 \\ \hline 0 & S(A') \end{array} \right)$$

showing that the reduction of A is completed by reducing the smaller matrix A'.

We close this section by making two deductions from Theorem 1.11 which will be useful later. The first deduction should come as no surprise to the reader.

Corollary 1.13

Let A be an $s \times t$ matrix over $\mathbb{Z}$. There are invertible matrices P and Q over $\mathbb{Z}$ such that $PAQ^{-1} = D$ where D is in Smith normal form.

Proof

By Theorem 1.11 there is a sequence of elementary operations reducing A to D in Smith normal form. Let $P = P_u \cdots P_2 P_1$ where P_i is the elementary matrix corresponding to the ith *ero* used in the reduction. Then P is invertible over $\mathbb{Z}$, being a product of invertible matrices over $\mathbb{Z}$. Similarly let $Q = Q_v^{-1} \cdots Q_2^{-1} Q_1^{-1}$ where Q_j is the elementary matrix corresponding to the jth *eco* used in the reduction. As each Q_j is invertible over $\mathbb{Z}$, so also is Q and $Q^{-1} = Q_1 Q_2 \cdots Q_v$. By Lemma 1.4 and the discussion following it, we deduce that $PAQ^{-1} = P_u \cdots P_2 P_1 A Q_1 Q_2 \cdots Q_v = D.$ $\square$

Can every invertible matrix P over $\mathbb{Z}$ be built up as a product of elementary matrices? We show next that the answer is: Yes! Further P can be reduced to the identity matrix I, which is its Smith normal form, using elementary operations over $\mathbb{Z}$ of just one kind, that is, using *eros* only or *ecos* only.

Corollary 1.14

Let P be an invertible $s \times s$ matrix over $\mathbb{Z}$. Then P is expressible as a product of elementary matrices over $\mathbb{Z}$. Also P can be reduced to the identity matrix I both using *eros* only and using *ecos* only.

Proof

By Corollary 1.13 there are invertible matrices P' and Q' over $\mathbb{Z}$ with $P'P(Q')^{-1} = D$, where $D = \operatorname{diag}(d_1, d_2, \ldots, d_s)$, $d_i \geq 0$. As D is a product of invertible matrices over $\mathbb{Z}$ we see that D itself is invertible over $\mathbb{Z}$. So each d_i is an invertible integer, that is, $d_i = \pm 1$. Hence $d_i = 1$ for $1 \leq i \leq s$ and $D = I$ as each d_i is non-negative. So $P'P(Q')^{-1} = I$ and hence

$$P = (P')^{-1}Q' = P_1^{-1}P_2^{-1} \cdots P_u^{-1} Q_v^{-1} \cdots Q_2^{-1} Q_1^{-1}$$

as in Corollary 1.13, which expresses the invertible matrix P as a product of elementary matrices over $\mathbb{Z}$.

The equation $P = (P')^{-1}Q'$ gives $P^{-1} = (Q')^{-1}P'$ and so $(Q')^{-1}P'P = I$, that is, $Q_1Q_2 \cdots Q_vP_u \cdots P_2P_1P = I$, which shows that P can be reduced to I using *eros* only: after using the *eros* in the reduction unchanged to obtain $P'P$, each *eco* used in the reduction is replaced by the *ero* paired to it. The reduction to I is completed by applying these *eros*, in the opposite order, to $P'P$.

In the same way, the equation $P(Q')^{-1}P' = I$, that is,

$$PQ_1Q_2 \cdots Q_vP_u \cdots P_2P_1 = I$$

shows that P can be reduced to I using *ecos* only.   □

As an illustration of Corollary 1.14 the matrix $P = \left(\begin{smallmatrix} 13 & 9 \\ 36 & 25 \end{smallmatrix}\right)$ is invertible over $\mathbb{Z}$ since $\det P = 1$ (the role of determinants is discussed in the next chapter). Using the method of Lemma 1.9, the sequence $c_1 - c_2, c_2 - 2c_1, c_1 - 4c_2, c_1 \leftrightarrow c_2, r_2 - 3r_1$, $-r_2$ reduces P to I, and so $P_2P_1PQ_1Q_2Q_3Q_4 = I$ in terms of the corresponding elementary matrices. Hence $P = P_1^{-1}P_2^{-1}Q_4^{-1}Q_3^{-1}Q_2^{-1}Q_1^{-1}$, that is,

$$\begin{pmatrix} 13 & 9 \\ 36 & 25 \end{pmatrix} = \begin{pmatrix} 1 & 0 \\ 3 & 1 \end{pmatrix}\begin{pmatrix} 1 & 0 \\ 0 & -1 \end{pmatrix}\begin{pmatrix} 0 & 1 \\ 1 & 0 \end{pmatrix}\begin{pmatrix} 1 & 0 \\ 4 & 1 \end{pmatrix}\begin{pmatrix} 1 & 2 \\ 0 & 1 \end{pmatrix}\begin{pmatrix} 1 & 0 \\ 1 & 1 \end{pmatrix}$$

showing explicitly that P is a product of elementary matrices. As $P^{-1} = Q_1Q_2Q_3Q_4P_2P_1$, the equation $P^{-1}P = I$ shows by Lemma 1.4 that the sequence $r_2 - 3r_1, -r_2, r_1 \leftrightarrow r_2, r_2 - 4r_1, r_1 - 2r_2, r_2 - r_1$, of *eros* reduces P to I. Similarly the equation $PP^{-1} = I$ produces the sequence $c_1 - c_2, c_2 - 2c_1, c_1 - 4c_2, c_1 \leftrightarrow c_2$, $-c_2, c_1 - 3c_2$ of *ecos* which also reduces P to I.

Let P be an invertible $s \times s$ matrix over $\mathbb{Z}$ and let Q be an invertible $t \times t$ matrix over $\mathbb{Z}$. Applying Corollary 1.14 to P and Q^{-1} we see that these matrices are expressible as products of elementary matrices. By Lemma 1.4 the matrix $B = PAQ^{-1}$ can be obtained from the arbitrary $s \times t$ matrix A over $\mathbb{Z}$ by a sequence of elementary operations, in other words using Definition 1.5 we see

$$A \equiv B \text{ if and only if } A \text{ is obtainable from } B \text{ by elementary operations.}$$

In Section 1.3 we continue our study of PAQ^{-1} concentrating on those properties which remain *unchanged* when elementary operations are carried out.

One final word. The reader will be aware of the *fundamental theorem of arithmetic*: every positive integer can be uniquely expressed in the form

$$p_1^{n_1}p_2^{n_2} \cdots p_k^{n_k}$$

where each p_i is *prime* ($p_i > 1$ and the only integer divisors of p_i are ± 1, $\pm p_i$) for $1 \leq i \leq k$ with $p_1 < p_2 < \cdots < p_k$, each n_i being a positive integer and $k \geq 0$. Because of the difficulty of factorising integers in the above form, it is important that the reduction of every matrix A over $\mathbb{Z}$ to its Smith normal form D can be carried out in a systematic way *without* using this theorem.

EXERCISES 1.2

1. (a) Determine all 2×2 matrices A over $\mathbb{Z}$ such that applying $r_1 \leftrightarrow r_2$ to A has the same effect as applying $c_1 \leftrightarrow c_2$ to A.
   Hint: Use Lemma 1.4 and $T = \left(\begin{smallmatrix} 0 & 1 \\ 1 & 0 \end{smallmatrix}\right)$.

   (b) Use the rows e_i of the identity matrix I to describe the elementary matrix P_1 corresponding to the *ero* $r_j + lr_k$ ($j \neq k$) and hence write down a proof of Lemma 1.4 in this case.

   (c) Use the columns e_i^T of I to describe the elementary matrix Q_1 corresponding to the *eco* $c_j \leftrightarrow c_k$ ($j \neq k$), and hence write down a proof of Lemma 1.4 in this case.

   (d) Show that $\equiv$ as defined in Definition 1.5 is an equivalence relation on the set of all $s \times t$ matrices over $\mathbb{Z}$, that is,
   (i) $A \equiv A$ for all such matrices A,
   (ii) $A \equiv B \Rightarrow B \equiv A$,
   (iii) $A \equiv B$ and $B \equiv C \Rightarrow A \equiv C$ where A, B, C are $s \times t$ matrices over $\mathbb{Z}$.

2. List the eight 3×3 matrices D in Smith normal form such that $\det D$ is a divisor of 12. How many $s \times s$ matrices D in Smith normal form are there with

$$\text{(i)}\quad \det D = 105 \qquad \text{(ii)}\quad \det D = 100?$$

3. The $s \times t$ matrix $A = (a_{ij})$ over $\mathbb{Z}$ is such that $a_{11} > 0$, $a_{i1} = 0$ for $2 \leq i \leq s$, $a_{11} = \gcd\{a_{11}, a_{12}, \ldots, a_{1t}\}$. List the $t - 1$ *ecos* needed to reduce A to B as in Lemma 1.9.

4. (a) Reduce the 4×4 matrix $A = \text{diag}(10, 1, 5, 25)$ to Smith normal form D using 7 elementary operations.

   (b) Reduce the 4×4 matrix $A = \text{diag}(12, 5, 50, 200)$ to Smith normal form D using 15 elementary operations.

   (c) Let $D' = \text{diag}(d_1, d_2, \ldots, d_s)$ be an $s \times s$ matrix in Smith normal form and let d be a positive integer. Show that the $(s+1) \times (s+1)$ matrix $D_1 = \text{diag}(d, d_1, d_2, \ldots, d_s)$ can be reduced to its Smith normal form $S(D_1)$ using at most $5s$ elementary operations.

(d) Show that every 2×2 diagonal matrix over $\mathbb{Z}$ can be reduced to Smith normal form using at most 5 elementary operations. Reduce $\mathrm{diag}(18, -24)$ to Smith normal form using 5 elementary operations. For $s \geq 2$ show that every $s \times s$ diagonal matrix over $\mathbb{Z}$ can be reduced to Smith normal form using at most $(5/2)s(s-1)$ elementary operations.

Hint: Use Lemma 1.10 throughout.

5. (a) Show that an invertible $s \times s$ matrix P over $\mathbb{Z}$ can be reduced to $I = S(P)$ by $s(s-1)/2$ applications of Lemma 1.7 together with at most $1 + s(s-1)/2$ *eros*.

Hint: The entries in each row of P have gcd 1.

(b) Use the method of Corollary 1.14 to reduce the following invertible matrices over $\mathbb{Z}$ to I:

$$P = \begin{pmatrix} 3 & 1 \\ 17 & 6 \end{pmatrix}, \qquad Q = \begin{pmatrix} 1 & 2 & 3 \\ 2 & 7 & 7 \\ 3 & 23 & 15 \end{pmatrix}$$

(c) Express P above as a product of elementary matrices over $\mathbb{Z}$.

(d) Specify a sequence of *eros* reducing Q to I.

(e) Specify a sequence of *ecos* reducing Q to I.

6. (a) Reduce $A = \begin{pmatrix} 390 & 780 \\ 330 & 667 \end{pmatrix}$ to Smith normal form D using the technique of Lemma 1.9.

(b) The sequence $a_0, a_1, a_2, \ldots$ of positive integers is defined by $a_0 = 1$, $a_1 = 2$, $a_{n+1} = a_n(4a_{n-1} + 1)$ for $n \geq 1$. Calculate a_n for $2 \leq n \leq 4$. Show that $\gcd\{a_{n+1}, 2a_n\} = a_n$ for $n \geq 1$.
Show that $a_{n+1} = 2(4a_{n-1} + 1)(4a_{n-2} + 1) \cdots (4a_0 + 1)$ for $n \geq 1$ and deduce $a_n/a_{n-r+1} \equiv 1 \pmod{4a_{n-r}}$ for $1 \leq r \leq n$.
Using the method of Lemma 1.9, reduce $\begin{pmatrix} a_4 & 2a_3 \\ 0 & -1 \end{pmatrix}$ to a diagonal matrix.
Let $A_n = \begin{pmatrix} a_n & 2a_{n-1} \\ 0 & -1 \end{pmatrix}$ where $n \geq 3$. Using the notation of Lemma 1.9 show that B_r or B_r^T is equal to $\begin{pmatrix} a_{n-r} & 2a_{n-r-1}(a_n/a_{n-r+1}) \\ 0 & -(a_n/a_{n-r}) \end{pmatrix}$ according as r is even or odd for $1 \leq r < n$.

Hint: Use induction on r, applying the *eco* $c_2 - 2a_{n-r}q_{n-r+1}c_1$ to B_{r-1} (r odd, $r > 1$) and the *ero* $r_2 - 2a_{n-r}q_{n-r+1}r_1$ to B_{r-1} (r even, $r > 2$) where $a_n/a_{n-r+2} = 1 + q_{n-r+1}a_{n-r+1}$.
Conclude $B_n = \mathrm{diag}(2, -a_n/2)$. Hence show that the algorithm of Theorem 1.11 requires $3n + 1$ elementary operations to reduce A_n to $S(A_n) = \mathrm{diag}(1, a_n)$ where $n \geq 3$.
Specify a sequence of four elementary operations which reduces A_n to its Smith normal form $S(A_n)$.

7. Let G denote the group of all pairs (P, Q) where P and Q are invertible $s \times s$ and $t \times t$ matrices respectively over $\mathbb{Z}$, the group operation being componentwise multiplication.

   (a) Let D be an $s \times t$ matrix over $\mathbb{Z}$. Verify that the *'centraliser'* $Z(D) = \{(P, Q) \in G : PD = DQ\}$ is a *subgroup* of G.
   Hint: Show that $Z(D)$ is closed under multiplication, closed under inversion, and contains the identity (I, I) of G.

   (b) Suppose $s = t$ and $D = \mathrm{diag}(d_1, d_2, \ldots, d_t)$ is in Smith normal form with $d_t > 0$. Write $(P, Q) = ((p_{ij}), (q_{ij})) \in G$. Show that $(P, Q) \in Z(D) \Leftrightarrow p_{ji} = (d_j/d_i)q_{ji}, q_{ij} = (d_j/d_i)p_{ij}$ for all $i \leq j$.

   (c) You and your classmate reduce the $t \times t$ matrix A to Smith normal form D in different ways. You get $P'A = DQ'$ and your classmate gets $P''A = DQ''$. Show that $(P', Q')(P'', Q'')^{-1} \in Z(D)$.

   (d) Use the two reductions of A in Example 1.1 to derive an element in $Z(\mathrm{diag}(2, 4))$ of the type $(P', Q')(P'', Q'')^{-1}$.

   (e) Modify part (b) to cover the case $d_t = 0$.

1.3 Uniqueness of the Smith Normal Form

Let A be an $s \times t$ matrix over $\mathbb{Z}$. Our task in this section is to show that A can be reduced to only one matrix D in Smith normal form. A closely related question is: what do the matrices A and PAQ^{-1} have in common, where P and Q are invertible over $\mathbb{Z}$? In other words, which properties of A are preserved when A undergoes elementary operations? The reader will know that for matrices over a field F the *rank* of a matrix is all that matters: two $s \times t$ matrices over F are equivalent if and only if their ranks are equal. The matrix A over $\mathbb{Z}$ can be regarded as a matrix over the rational field $\mathbb{Q}$; the rank of A is then the number of non-zero invariant factors d_j of A, but this single number is not enough to determine equivalence over $\mathbb{Z}$: equivalent matrices over $\mathbb{Z}$ have the same rank, but matrices over $\mathbb{Z}$ of equal rank can be inequivalent over $\mathbb{Z}$. For instance

$$I = \begin{pmatrix} 1 & 0 \\ 0 & 1 \end{pmatrix} \quad \text{and} \quad J = \begin{pmatrix} 1 & 0 \\ 0 & 2 \end{pmatrix}$$

both have rank 2 but are not equivalent over $\mathbb{Z}$, as J is not invertible over $\mathbb{Z}$ whereas all matrices equivalent to I are invertible over $\mathbb{Z}$. However, as we'll see shortly, the property which 'does the trick', that is, determines equivalence over $\mathbb{Z}$, is a combination of two concepts already known to the reader, namely gcds and determinants.

We begin the details by introducing gcds of any finite collection of integers.

Let $X = \{l_1, l_2, \ldots, l_n\}$ be a non-empty set of integers. An integer d is called a *gcd* of X if

(i) $d|l_i$ for $1 \le i \le n$

(ii) whenever an integer d' satisfies $d'|l_i$ for $1 \le i \le n$, then $d'|d$.

Condition (i) says that d is a common divisor of the integers in X. Condition (ii) says that every common divisor d' of the integers in X is a divisor of d, and so d is 'greatest' in this sense. Could X have two gcds, d_1 and d_2 say? If so, then d_1 being a common divisor and d_2 being a greatest common divisor gives $d_1|d_2$. Reversing the roles of d_1 and d_2 gives $d_2|d_1$. Therefore $d_1 = \pm d_2$, showing that X has at most one *non-negative* gcd.

The reader will know, in the case $X = \{l_1, l_2\}$, that the non-negative integer $d = \gcd\{l_1, l_2\}$ satisfies conditions (i) and (ii) above. Further, apart from the special case $l_1 = l_2 = 0$, $d = 0$, the integer d can be singled out (*characterised* is the technical term) as the smallest positive integer expressible in the form $a_1 l_1 + a_2 l_2$ where a_1, a_2 are integers. We now use this property of d to show that every finite set X of n integers has a gcd.

Consider the set K of all integer linear combinations $k = b_1 l_1 + b_2 l_2 + \cdots + b_n l_n$ of the integers in $X = \{l_1, l_2, \ldots, l_n\}$; here the l_i are arbitrary integers. It is straightforward to verify that K is an *ideal* of $\mathbb{Z}$, that is, K is a subset of $\mathbb{Z}$ satisfying:

(i) $k_1 + k_2 \in K$ for all $k_1, k_2 \in K$ (K is closed under addition),

(ii) $0 \in K$ (K contains the zero integer),

(iii) $-k \in K$ for all $k \in K$ (K is closed under negation),

(iv) $bk \in K$ for all $b \in \mathbb{Z}$, $k \in K$ (K is closed under integer multiplication).

Conditions (i), (ii), (iii) together tell us that K is an *additive subgroup* of $\mathbb{Z}$, a concept which we'll study in Chapter 2. Condition (iv) says that all integer multiples of integers in K are again in K; this condition becomes important when $\mathbb{Z}$ is replaced by a more general ring, as we will see later. We write

$$K = \langle l_1, l_2, \ldots, l_n \rangle$$

and describe K as the ideal *generated* by $l_1, l_2, \ldots, l_n$.

The set $\langle 2 \rangle$ of even integers is an ideal of $\mathbb{Z}$. More generally

$$\langle d \rangle = \{md : m \in \mathbb{Z}\} \quad \text{for } d \in \mathbb{Z}$$

is an ideal of $\mathbb{Z}$, that is, the subset $\langle d \rangle$, consisting of all integer multiples of the given integer d, satisfies (i) to (iv) above. For instance

$$\langle 6 \rangle = \{\ldots, -18, -12, -6, 0, 6, 12, 18, \ldots\} = \langle -6 \rangle.$$

Notice that the smallest positive integer in $\langle 6 \rangle$ is – you've guessed it – none other than 6; this 'obvious' observation will help our understanding of the next theorem. Notice $\langle 1 \rangle = \mathbb{Z}$ as every integer is an integer multiple of 1, and $\langle 0 \rangle = \{0\}$ as 0 is the only integer which is a multiple of 0. An ideal of the type $\langle d \rangle$ is called a *principal ideal* of $\mathbb{Z}$. The integer d is called a *generator* of the ideal $\langle d \rangle$. As $\langle d \rangle = \langle -d \rangle$ we see that $-d$ is also a generator of $\langle d \rangle$. Does $\mathbb{Z}$ have any other ideals? We show next that the answer is: No!

Theorem 1.15

Each ideal K of $\mathbb{Z}$ is principal and has a unique non-negative generator d.

Proof

Notice first that the zero ideal $K = \{0\}$ is principal with generator 0, since $\{0\} = \langle 0 \rangle$. Now suppose $K \neq \{0\}$ and so $k_0 \in K$ where k_0 is some non-zero integer. Then $-k_0 \in K$ by condition (iii). As one of $\pm k_0$ is positive, we see K contains at least one positive integer. How can we find a generator of K?

$$\text{Let } d \text{ denote the smallest positive integer in } K$$

and hope for the best! Then $md \in K$ for all $m \in \mathbb{Z}$ by condition (iv), that is, $\langle d \rangle$ is a subset of K. The proof is completed by showing that every integer $k \in K$ belongs to $\langle d \rangle$. Divide k by d obtaining $q, r \in \mathbb{Z}$ with $k = qd + r$, $0 \leq r < d$. As $k, -qd \in K$ we deduce $r = k - qd \in K$ by condition (i). But $r \in K$, $r < d$ shows that r cannot be positive, d being the smallest positive integer in K. As $0 \leq r$, we conclude $r = 0$. Hence $k = qd$, that is $k \in \langle d \rangle$ and so $K = \langle d \rangle$. So d is a positive generator of K. Therefore every ideal K of $\mathbb{Z}$ is principal with non-negative generator d.

Suppose $K = \langle d \rangle = \langle d' \rangle$ where d and d' are non-negative. Then $K = \{0\} \Rightarrow d = d' = 0$. Also $K \neq \{0\} \Rightarrow d = d'$ as the positive integers d and d' satisfy $d \mid d'$ and $d' \mid d$. $\qquad \square$

An integral domain such that all its ideals are principal is called a *Principal Ideal Domain (PID)*. So Theorem 1.15 tells us that $\mathbb{Z}$ is a PID. We now use Theorem 1.15 to show the existence of gcds.

Corollary 1.16

Let $X = \{l_1, l_2, \ldots, l_n\}$ be a set of n integers, $n \geq 1$. Then X has a unique non-negative gcd d. Further there are n integers $a_1, a_2, \ldots, a_n$ such that $d = a_1 l_1 + a_2 l_2 + \cdots + a_n l_n$ and $\langle d \rangle = \langle l_1, l_2, \ldots, l_n \rangle$.

Proof

We already know that X cannot have two non-negative gcds. To show that X does have a non-negative gcd, consider the ideal $\langle l_1, l_2, \ldots, l_n \rangle$ of $\mathbb{Z}$ consisting of all linear combinations $b_1l_1 + b_2l_2 + \cdots + b_nl_n$ of the integers l_i in X where the b_i are arbitrary integers. It's convenient to write $\langle l_1, l_2, \ldots, l_n \rangle = \langle X \rangle$. From Theorem 1.15 we know that $\langle X \rangle = \langle d \rangle$ for some non-negative integer d, which is a concise description of $\langle X \rangle$. As $d \in \langle X \rangle$ we see that d is some linear combination of the generators l_i, that is, taking $b_i = a_i$ we obtain $d = a_1l_1 + a_2l_2 + \cdots + a_nl_n$. So d is expressible as stated. The proof is completed by showing that d is indeed a gcd of X, that is, conditions (i) and (ii) for gcds are satisfied by d.

Take $b_i = 1$, $b_j = 0$ for $i \neq j$ to obtain $l_i \in \langle X \rangle$ for $1 \leq i \leq n$. So each integer l_i in X belongs to $\langle X \rangle = \langle d \rangle$, and this means $d|l_i$ for $1 \leq i \leq n$. Therefore d satisfies condition (i). To verify condition (ii) consider $d' \in \mathbb{Z}$ satisfying $d'|l_i$ for $1 \leq i \leq n$. Then $l_i = q_id'$ for some $q_i \in \mathbb{Z}$. Substituting for l_i in the above expression for d produces $d = (a_1q_1 + a_2q_2 + \cdots + a_nq_n)d'$, that is $d'|d$ as $a_1q_1 + a_2q_2 + \cdots + a_nq_n$ is an integer. So d satisfies condition (ii). The conclusion is: d is a non-negative gcd of X. □

We write gcd X for the unique non-negative gcd of the set X.

The integer $d = \gcd\{l_1, l_2, \ldots, l_n\}$ can be found by applying the Euclidean algorithm $n - 1$ times. For example let $d = \gcd X$ where $X = \{231, 385, 495\}$. Then $d = \gcd\{\gcd\{231, 385\}, 495\} = \gcd\{77, 495\} = 11$. The ideal $K = \{b_1 \times 231 + b_2 \times 385 + b_3 \times 495 : b_1, b_2, b_3 \in \mathbb{Z}\}$ generated by 231, 385 and 495 is the principal ideal generated by 11, that is, $K = \langle 11 \rangle$. Also $11 = 13 \times 77 - 2 \times 495 = 13(2 \times 231 - 385) - 2 \times 495 = 26 \times 231 - 13 \times 385 - 2 \times 495$ shows explicitly that 11 belongs to K. The job of verifying the details is left to the reader: apply the Euclidean algorithm to the pair 231, 385 to get 77 as their gcd, and then apply the Euclidean algorithm to the pair 77, 495 obtaining 11 as their gcd. Finally express 11 as an integer linear combination of the integers in X by tracing backwards the steps in these algorithms.

Let $X' = \{l'_1, l'_2, \ldots, l'_{n'}\}$ be a set of n' integers. In the application Corollary 1.17 of this theory which we have in mind, each l'_j is an integer linear combination of the integers l_i, in other words, X' is a subset of $\langle X \rangle$, that is, $X' \subseteq \langle X \rangle$.

For example let $X = \{231, 385, 495\}$ with gcd $d = 11$ as before, and let $X' = \{231 + 385, 231 + 495\} \subseteq \langle X \rangle$. So $X' = \{616, 726\}$ which has gcd $d' = 22$. Clearly d is a divisor of d'. We now deal with the general case.

Corollary 1.17

Let d and d' be the non-negative gcds of X and X' respectively. Using the above notation

$$X' \subseteq \langle X \rangle \quad \Leftrightarrow \quad \langle X' \rangle \subseteq \langle X \rangle \quad \Leftrightarrow \quad d|d'.$$

Proof

Suppose $X' \subseteq \langle X \rangle$. A typical element of $\langle X' \rangle$ is $k' = b'_1 l'_1 + b'_2 l'_2 + \cdots + b'_{n'} l_{n'}$ where the b'_j are integers. As each l'_j belongs to the ideal $\langle X \rangle$, by property (iv) of ideals we see $b'_j l'_j \in \langle X \rangle$. By property (i) of ideals and induction on n' we deduce $k' \in \langle X \rangle$ and so $\langle X' \rangle \subseteq \langle X \rangle$.

Suppose $\langle X' \rangle \subseteq \langle X \rangle$, that is, $\langle d' \rangle \subseteq \langle d \rangle$. As $d' \in \langle d' \rangle$ we see $d' \in \langle d \rangle$, which means $d' = qd$ for some integer q, that is, $d \mid d'$.

Suppose $d \mid d'$. Then $d' = qd$ with $q \in \mathbb{Z}$. For each l'_j in X' there is q'_j in $\mathbb{Z}$ with $l'_j = q'_j d'$. Hence $l'_j = q'_j q d$ showing $l'_j \in \langle d \rangle$ for $1 \leq j \leq n'$ since $q'_j q \in \mathbb{Z}$. Therefore $X' \subseteq \langle X \rangle$ which gets us back to the original assumption.   □

All we need to know about gcds is contained in Corollary 1.17 which can be expressed

$$\langle X' \rangle \subseteq \langle X \rangle \quad \Leftrightarrow \quad \gcd X \mid \gcd X'.$$

We now turn our attention to determinants. Let A be an $s \times t$ matrix over $\mathbb{Z}$ and let l be an integer in the range $1 \leq l \leq \min\{s, t\}$. Suppose l rows and l columns of A are selected. The determinant of the $l \times l$ matrix which remains on deleting the unselected $s - l$ rows and $t - l$ columns of A is called an l-*minor of* A.

So the l-minors of A are integers. The number of l-minors of A is the product $\binom{s}{l} \times \binom{t}{l}$ of binomial coefficients and we'll be interested in the non-negative gcd of the l-minors of A.

For example, the gcd of the 1-minors of $A = \left(\begin{smallmatrix} 6 & 4 & 0 \\ 8 & 8 & 4 \end{smallmatrix} \right)$ is 2, as the 1-minors are simply the entries and $\gcd\{6, 4, 0, 8, 8, 4\} = 2$. The 2-minors of A are

$$\begin{vmatrix} 6 & 4 \\ 8 & 8 \end{vmatrix} = 16, \qquad \begin{vmatrix} 6 & 0 \\ 8 & 4 \end{vmatrix} = 24, \qquad \begin{vmatrix} 4 & 0 \\ 8 & 4 \end{vmatrix} = 16$$

and $\gcd\{16, 24, 16\} = 8$. The importance of these gcds comes from the fact, proved in Theorem 1.21, that they remain unchanged when elementary operations over $\mathbb{Z}$ are applied to A. From Theorem 1.11 we know that A can be reduced to $D = \left(\begin{smallmatrix} d_1 & 0 & 0 \\ 0 & d_2 & 0 \end{smallmatrix} \right)$ in Smith normal form using elementary operations over $\mathbb{Z}$. The set $\{d_1, 0, 0, 0, d_2, 0\}$ of 1-minors of D has gcd d_1, and the set

$$\left\{ \begin{vmatrix} d_1 & 0 \\ 0 & d_2 \end{vmatrix}, \begin{vmatrix} d_1 & 0 \\ 0 & 0 \end{vmatrix}, \begin{vmatrix} 0 & 0 \\ d_2 & 0 \end{vmatrix} \right\} = \{d_1 d_2, 0, 0\}$$

of 2-minors of D has gcd $d_1 d_2$. Assuming Theorem 1.21 for the moment, we deduce $d_1 = 2$, $d_1 d_2 = 8$ on comparing these gcds, giving $D = \mathrm{diag}(2, 4)$. The conclusion is that A can be reduced to one and only one matrix in Smith normal form.

The reader will be familiar with the $(t-1)$-minors of a $t \times t$ matrix $A = (a_{ij})$ over a field F because they are, apart from sign, the entries in

the adjugate matrix adj A.

In fact, adj A has (j, i)-entry A_{ij} where $(-1)^{i+j} A_{ij}$ is the $(t-1)$-minor of A obtained by deleting row i and col j. Then adj A satisfies the matrix equation

$$A(\text{adj } A) = (\det A)I = (\text{adj } A)A$$

which includes, on comparing diagonal entries, the rules for expanding $\det A$ along any row or column, and explains why A_{ij} is known as the *cofactor* of a_{ij} in $\det A$. In fact the above equation holds for square matrices A over any commutative ring (with identity element) R and shows, together with the multiplicative property of determinants (reviewed in Theorem 1.18 below),

$$A \text{ is invertible over } R \quad \Leftrightarrow \quad \det A \text{ is an invertible element of } R.$$

Should $\det A$ be invertible over R, the above matrix equation can be divided throughout by $\det A$ to produce the familiar equations

$$AA^{-1} = I = A^{-1}A \quad \text{where } A^{-1} = (1/\det A)\,\text{adj } A.$$

Since elementary operations can be carried out by matrix multiplication Lemma 1.4, we now address the question: how are the l-minors of a matrix product BC related to the l-minors of B and the l-minors of C? The reader will know that a typical $l \times l$ submatrix of BC is formed by multiplying certain l rows of B by certain l columns of C. So we lose nothing by assuming that B is an $l \times r$ matrix and C is an $r \times l$ matrix, where r is a positive integer, in which case $\det BC$ is the unique l-minor of BC. If $r < l$ then $\det BC = 0$ (see Exercises 1.3, Question 4(d)). We now assume $r \geq l$ and let Y denote a subset of $\{1, 2, \ldots, r\}$ having l elements; there are $\binom{r}{l}$ such subsets Y. Let B_Y denote the $l \times l$ submatrix of B obtained by deleting column j for all $j \notin Y$. Similarly let $_YC$ denote the $l \times l$ submatrix of C obtained by deleting row j for all $j \notin Y$.

To help the reader through the next proof we look first at the case $l = 2, r = 3$. So

$$B = \begin{pmatrix} b_{11} & b_{12} & b_{13} \\ b_{21} & b_{22} & b_{23} \end{pmatrix} \quad \text{and} \quad C = \begin{pmatrix} c_{11} & c_{12} \\ c_{21} & c_{22} \\ c_{31} & c_{32} \end{pmatrix}.$$

The proof uses the columns of B which we denote by B_1, B_2, B_3. Then $(B_1, B_3) = B_Y$ and $\det {}_YC = \begin{vmatrix} c_{11} & c_{12} \\ c_{31} & c_{32} \end{vmatrix} = c_{11}c_{32} - c_{12}c_{31}$ where $Y = \{1, 3\}$ and so on. The determinant of the 2×2 matrix BC can be expressed as the sum of three terms $\det B_Y \det {}_YC$ where Y runs through the subsets $\{1, 2\}, \{1, 3\}, \{2, 3\}$ of $\{1, 2, 3\}$ as follows:

$$\det BC = \begin{vmatrix} b_{11}c_{11} + b_{12}c_{21} + b_{13}c_{31} & b_{11}c_{12} + b_{12}c_{22} + b_{13}c_{32} \\ b_{21}c_{11} + b_{22}c_{21} + b_{23}c_{31} & b_{21}c_{12} + b_{22}c_{22} + b_{23}c_{32} \end{vmatrix}$$

$$= |B_1 c_{11} + B_2 c_{21} + B_3 c_{31}, \quad B_1 c_{12} + B_2 c_{22} + B_3 c_{32}|$$

$$= \sum_{i,j=1}^{3} |B_i, B_j| c_{i1} c_{j2} = \sum_{i \neq j} |B_i, B_j| c_{i1} c_{j2}$$

$$= \sum_{i<j} |B_i, B_j| (c_{i1} c_{j2} - c_{i2} c_{j1}) = \sum_Y \det B_Y \det {}_Y C$$

where $Y = \{i, j\}$, $i < j$.

In the steps above we have used standard properties of the determinant function: it is *multilinear* and so $\det BC$ is a sum of 9 terms $|B_i, B_j| c_{i1} c_{j2}$. Three of these terms, those with $i = j$, are zero as $|B_i, B_i| = 0$. The 6 remaining terms occur in 3 pairs corresponding to the 3 subsets $Y = \{i, j\}$ as $|B_j, B_i| = -|B_i, B_j|$ for $i < j$. We have shown that the 2-minors of BC (there is only one, namely $\det BC$) are expressible as integer linear combinations of the 2-minors of B as well as integer linear combinations of the 2-minors of C.

We are now ready to state and prove a general theorem which was discovered independently by the French mathematicians Binet and Cauchy in 1812.

Theorem 1.18 (The Cauchy–Binet theorem over $\mathbb{Z}$)

Let B be an $l \times r$ matrix over $\mathbb{Z}$ and let C be an $r \times l$ matrix over $\mathbb{Z}$ where $r \geq l$. For each subset Y of $\{1, 2, \ldots, r\}$ having l elements, let B_Y be the $l \times l$ submatrix of B formed by deleting column j for all $j \notin Y$. Let ${}_Y C$ be the $l \times l$ submatrix of C formed by deleting row j for all $j \notin Y$. Then

$$\det BC = \sum_Y \det B_Y \det {}_Y C.$$

Proof

We first note, by way of encouragement, that two special cases of this theorem are already familiar to us: when $l = 1$ it is nothing more than the formula for an entry in a matrix product. When $l = r$ it is the *multiplicative property of determinants* $\det BC = \det B \det C$.

Write B_i for column i of B and let $C = (c_{ij})$ where $1 \leq i \leq r$, $1 \leq j \leq l$. Then

$$(\text{column } j \text{ of } BC) = \sum_i B_i c_{ij} = \sum_{ij} B_{ij} c_{ijj}$$

on replacing i by i_j where $1 \leq i_j \leq l$, as we need a different summation index for each of the l columns of BC. Using the multilinear property of the determinant function we obtain

$$\det BC = \left| \sum_{i_1} B_{i_1} c_{i_1 1}, \sum_{i_2} B_{i_2} c_{i_2 2}, \ldots, \sum_{i_l} B_{i_l} c_{i_l l} \right|$$

$$= \sum_{i_1, i_2, \ldots, i_l} \det(B_{i_1}, B_{i_2}, \ldots, B_{i_l}) c_{i_1 1} c_{i_2 2} \cdots c_{i_l l}$$

There are r^l terms in the above summation but, as a determinant with two equal columns is zero, we need consider only the $r(r-1) \cdots (r-l+1) = r!/(r-l)!$ terms with $i_1, i_2, \ldots, i_l$ distinct, that is, no two of $i_1, i_2, \ldots, i_l$ are equal. Each such term gives rise to a subset $Y = \{i_1, i_2, \ldots, i_l\}$ of $\{1, 2, \ldots, r\}$ and so these $r!/(r-l)!$ terms partition into $r!/((r-l)!l!) = \binom{r}{l}$ equivalence classes of size $l!$, two terms being equivalent if they give rise to the same subset Y. We calculate $\det BC$ by summing up the terms in each equivalence class and finally adding these sums together.

Let $Y = \{j_1, j_2, \ldots, j_l\}$ where $1 \leq j_1 < j_2 < \cdots < j_l \leq r$ and suppose also that $Y = \{i_1, i_2, \ldots, i_l\}$. Then $i_1, i_2, \ldots, i_l$ is a permutation of $j_1, j_2, \ldots, j_l$ and so

$$\left| B_{i_1}, B_{i_2}, \ldots, B_{i_l} \right| = \pm \left| B_{j_1}, B_{j_2}, \ldots, B_{j_l} \right| = \pm \det B_Y$$

according as the above permutation is even (plus sign) or odd (minus sign); remember that a permutation is even/odd according as it is the product of an even/odd number of interchanges and each interchange (of columns of B) produces a change of sign in the determinant. Adding the $l!$ terms with $\{i_1, i_2, \ldots, i_l\} = Y$ gives

$$\sum_{\{i_1, i_2, \ldots, i_l\} = Y} |B_{i_1}, B_{i_2}, \ldots, B_{i_l}| c_{i_1 1} c_{i_2 2} \cdots c_{i_l l}$$

$$= \det B_Y \left(\sum_{\{i_1, i_2, \ldots, i_l\} = Y} \pm c_{i_1 1} c_{i_2 2} \cdots c_{i_l l} \right) = \det B_Y \det {}_Y C.$$

Therefore adding these sums, one for each subset Y of l integers from among $\{1, 2, \ldots, r\}$, gives the formula of Theorem 1.18. $\qquad\square$

The theory of determinants is valid for matrices with entries from any *commutative ring R*, and the Cauchy–Binet theorem is also valid for such matrices as the above proof goes through unchanged. The formula of Theorem 1.18 is then an equality between elements of R and we refer to this equation as *the Cauchy–Binet theorem over R*.

We are ready to put gcds and determinants together.

Let $g_l(A)$ denote the non-negative gcd of the l-minors of the $s \times t$ matrix A over $\mathbb{Z}$ for $1 \leq l \leq \min\{s, t\}$.

In the case of $A = \left(\begin{smallmatrix} 6 & 4 & 0 \\ 8 & 8 & 4 \end{smallmatrix} \right)$ as we saw earlier $2 = g_1(A)$ and $8 = g_2(A)$.

Corollary 1.19

Let B be an $s \times r$ matrix over $\mathbb{Z}$ and let C be an $r \times t$ matrix over $\mathbb{Z}$. Then $g_l(B)$ and $g_l(C)$ are divisors of $g_l(BC)$ for $1 \leq l \leq \min\{r, s, t\}$.

Proof

Every $l \times l$ submatrix of BC is of the type $B'C'$, where B' is an $l \times r$ submatrix of B and C' is an $r \times l$ submatrix of C. By Theorem 1.18

$$\det B'C' = \sum_Y \det B'_Y \det {}_Y C'.$$

Let X denote the set of l-minors of C. Then $\det {}_Y C'$ belongs to X. Let X' denote the set of l-minors of BC. The above equation tells us that each integer in X' is an integer linear combination of the integers in X. Therefore $X' \subseteq \langle X \rangle$ in the notation of Corollary 1.17. Now $\langle X \rangle = \langle g_l(C) \rangle$ and $\langle X' \rangle = \langle g_l(BC) \rangle$. So $g_l(C)$ is a divisor of $g_l(BC)$ by Corollary 1.17. Since $\det B'_Y$ is an l-minor of B, we deduce in a similar way that $g_l(B)$ is also a divisor of $g_l(BC)$. $\qquad\square$

Therefore just as B and C are factors of BC, so $g_l(B)$ and $g_l(C)$ are factors (divisors) of $g_l(BC)$.

We come next to the climax of the theory!

Corollary 1.20

Let A and B be $s \times t$ matrices over $\mathbb{Z}$.

$$\text{Then } A \equiv B \quad \Leftrightarrow \quad g_l(A) = g_l(B) \text{ for } 1 \leq l \leq \min\{s, t\}.$$

Further A is equivalent to a unique matrix D in Smith normal form Definition 1.6.

Proof

Suppose $A \equiv B$. By Definition 1.5 there are invertible matrices P and Q over $\mathbb{Z}$ with $PAQ^{-1} = B$. From Corollary 1.19 we deduce $g_l(A)$ is a divisor of $g_l(PA)$

and $g_l(PA)$ is a divisor of $g_l(PAQ^{-1}) = g_l(B)$. So $g_l(A)$ is a divisor of $g_l(B)$. But pre and postmultiplying $PAQ^{-1} = B$ by P^{-1} and Q respectively gives $P^{-1}BQ = A$, showing $B \equiv A$. Hence the roles of A and B can be reversed in the first part of the proof to show that $g_l(B)$ is a divisor of $g_l(A)$. So $g_l(A) = g_l(B)$, as these integers are non-negative and each is a divisor of the other.

From Theorem 1.11 there is an $s \times t$ matrix $D = \text{diag}(d_1, d_2, \ldots, d_{\min\{s,t\}})$ in Smith normal form with $A \equiv D$. Selecting the first l rows and the first l columns of D we obtain an l-minor of D equal to $d_1 d_2 \cdots d_l$. In fact $g_l(D) = d_1 d_2 \cdots d_l$ since every l-minor of D is zero or a product $d_{j_1} d_{j_2} \cdots d_{j_l}$ where $1 \leq j_1 < j_2 < \cdots < j_l \leq \min\{s, t\}$, and $d_i | d_{j_i}$ for $1 \leq i \leq l$ as $i \leq j_i$. By the first paragraph of the proof we deduce

$$g_l(A) = d_1 d_2 \cdots d_l \quad \text{for } 1 \leq l \leq \min\{s, t\}.$$

These equations determine the invariant factors $d_1, d_2, \ldots, d_{\min\{s,t\}}$ of A in terms of the integers $g_l(A)$ using induction on l: $d_1 = g_1(A)$ and for $l > 1$ either $d_l = g_l(A)/g_{l-1}(A)$ if $g_l(A) \neq 0$ or $d_l = 0$ if $g_l(A) = 0$. The conclusion is that A is equivalent over $\mathbb{Z}$ to a *unique* matrix D in Smith normal form.

Suppose $g_l(A) = g_l(B)$ for $1 \leq l \leq \min\{s, t\}$. By Theorem 1.11 and the above paragraph, A and B are equivalent to the same matrix D in Smith normal form, that is, $A \equiv D$ and $B \equiv D$. There are invertible matrices P, P', Q, Q' over $\mathbb{Z}$ with $PAQ^{-1} = D$ and $P'B(Q')^{-1} = D$. Therefore $PAQ^{-1} = P'B(Q')^{-1}$ which rearranges to give $P''A(Q'')^{-1} = B$ where $P'' = (P')^{-1}P$ and $Q'' = (Q')^{-1}Q$. As P'' and Q'' are invertible over $\mathbb{Z}$, we conclude $A \equiv B$. □

From Theorem 1.11 and Corollary 1.20 we see that it is legitimate to refer to *the* Smith normal form $D = S(A)$ of a matrix A over $\mathbb{Z}$ and *the* invariant factors $d_i = d_i(A)$ of A.

Our next and last theorem is relevant to the theory of subgroups and quotient groups of finitely generated abelian groups which is covered in Section 3.3. How does the Smith normal form behave under matrix multiplication? Let A be an $r \times s$ matrix over $\mathbb{Z}$ where $r \leq s$. We write

$$S(A) = \text{diag}(d_1(A), d_2(A), \ldots, d_r(A)) \text{ for the Smith normal form of } A$$

and so $d_k(A)$ is the kth invariant factor of A for $1 \leq k \leq r$.

Theorem 1.21

Let A be an $r \times s$ matrix over $\mathbb{Z}$ and B an $s \times t$ matrix over $\mathbb{Z}$ where $r \leq s \leq t$. Suppose all the invariant factors of A, B and AB are positive. Then $d_k(A)$ and $d_k(B)$ are divisors of $d_k(AB)$ for $1 \leq k \leq r$.

Proof

There are invertible matrices P_1, P_2, Q_1, Q_2 over $\mathbb{Z}$ such that $P_1 A Q_1^{-1} = S(A)$ and $P_2 A B Q_2^{-1} = S(AB)$. On substituting $A = P_1^{-1} S(A) Q_1$ in the second equation and rearranging, we obtain $PS(AB) = S(A)C$ where $P_1 P_2^{-1} = P = (p_{ij})$ is an invertible $r \times r$ matrix over $\mathbb{Z}$ and $Q_1 B Q_2^{-1} = C = (c_{ij})$. Comparing (i, j)-entries in $PS(AB) = S(A)C$ gives

$$p_{ij} d_j(AB) = d_i(A) c_{ij} \qquad (\clubsuit)$$

for all i and j with $1 \leq i, j \leq r$. For k with $1 \leq k \leq r$ we restrict our attention to p_{ij} where $j \leq k \leq i \leq r$. We now perform some number-theoretic juggling to show that all these p_{ij} are divisible by a certain integer. First multiply $(\clubsuit)$ by the positive integer $d_k(AB)/d_j(AB)$ to get

$$p_{ij} d_k(AB) = d_k(A) c'_{ij} \qquad (\blacklozenge)$$

where $c'_{ij} = (d_i(A)/d_k(A)) c_{ij} (d_k(AB)/d_j(AB))$ is an integer. Next divide $(\blacklozenge)$ through by $\bar{d}_k = \gcd\{d_k(A), d_k(AB)\}$ obtaining

$$p_{ij}(d_k(AB)/\bar{d}_k) = (d_k(A)/\bar{d}_k) c'_{ij}. \qquad (\heartsuit)$$

The integers $d_k(AB)/\bar{d}_k$ and $d_k(A)/\bar{d}_k$ are *coprime* (their gcd is 1) and so $(\heartsuit)$ shows that

$$d_k(A)/\bar{d}_k \text{ is a divisor of } p_{ij} \quad \text{for } j \leq k \leq i \leq r. \qquad (\spadesuit)$$

Next we partition the $r \times r$ matrix

$$P = \left(\begin{array}{c|c} P_{11} & P_{12} \\ \hline P_{21} & P_{22} \end{array} \right)$$

where P_{11} is the leading $(k-1) \times (k-1)$ submatrix. By $(\spadesuit)$ not only are all the entries in P_{21} divisible by $d_k(A)/\bar{d}_k$ but so also are all the entries in the first column of P_{22}. We are nearly there! Working modulo $d_k(A)/\bar{d}_k$, the determinant of the invertible matrix P is

$$|P| = \left| \begin{array}{c|c} P_{11} & P_{12} \\ \hline P_{21} & P_{22} \end{array} \right| \equiv \left| \begin{array}{c|c} P_{11} & P_{12} \\ \hline 0 & P_{22} \end{array} \right| = |P_{11}| \times |P_{22}| \equiv |P_{11}| \times 0 = 0$$

showing that $|P| = \pm 1$ is divisible by $d_k(A)/\bar{d}_k$. There is only one way out of this apparent impasse: $d_k(A)/\bar{d}_k = 1$, the *only* positive integer divisor of ± 1. We conclude $d_k(A) = \bar{d}_k = \gcd\{d_k(A), d_k(AB)\}$ is a divisor of $d_k(AB)$ for $1 \leq k \leq r$.

Finally we show in a similar way that $d_k(B)$ is a divisor of $d_k(AB)$ for $1 \leq k \leq r$. There are invertible matrices P_2, P_3, Q_2, Q_3 over $\mathbb{Z}$ with $P_2 A B Q_2^{-1} = S(AB)$ and $P_3 B Q_3^{-1} = S(B)$. So $S(AB)Q = ES(B)$ where $Q = Q_2 Q_3^{-1}$ is invertible over $\mathbb{Z}$ and $E = P_2 A P_3^{-1}$. Write $Q = (q_{ij})$ and $E = (e_{ij})$. Comparing (i, j)-entries in $S(AB)Q = ES(B)$ for $1 \le i \le r$, $1 \le j \le s$ gives

$$d_i(AB)q_{ij} = e_{ij}d_j(B) \tag{$\clubsuit'$}$$

and for $1 \le i \le r$, $s < j \le t$ gives $d_i(AB)q_{ij} = 0$. So $q_{ij} = 0$ for $1 \le i \le r$, $s < j \le t$ as $d_i(AB) > 0$. Consider q_{ij} with $i \le k \le j \le s$. As in the first part of the proof ($\clubsuit'$) implies that

$$q_{ij} \text{ is divisible by } d_k(B)/\bar{\bar{d}}_k \quad \text{for } i \le k \le j \le s \tag{$\spadesuit'$}$$

where $\bar{\bar{d}}_k = \gcd\{d_k(B), d_k(AB)\}$. We partition the invertible $t \times t$ matrix

$$Q = \left(\begin{array}{c|c} Q_{11} & Q_{12} \\ \hline Q_{21} & Q_{22} \end{array} \right)$$

where Q_{11} is the leading $(k-1) \times (k-1)$ submatrix. Using ($\spadesuit'$) all entries in Q_{12} are divisible by $d_k(B)/\bar{\bar{d}}_k$ as are all entries in the first row of Q_{22}. Hence, as in the first part of the proof, $\det Q = \pm 1$ is divisible by $d_k(B)/\bar{\bar{d}}_k$. So $d_k(B)/\bar{\bar{d}}_k = 1$ giving $d_k(B) = \bar{\bar{d}}_k$ which is a divisor of $d_k(AB)$. $\square$

EXERCISES 1.3

1. (a) Use the Euclidean algorithm to find the non-negative gcd d of each of the following sets X of integers. In each case express d as an integer linear combination of the integers in X.

(i) $\{21, 75, 175\};$ (ii) $\{42, 66, 154, 231\}.$

Hint: Use the Euclidean algorithm twice for (i) and three times for (ii).

(b) Let X_1 and X_2 be finite sets of integers. Show that

$$\gcd\{\gcd X_1, \gcd X_2\} = \gcd X_1 \cup X_2.$$

(c) Let $n_1, n_2, \ldots, n_k$ be positive integers. Write $l_i = n_1 n_2 \cdots n_k / n_i$ for $1 \le i \le k$. Show that $m_0 = n_1 n_2 \cdots n_k / \gcd\{l_1, l_2, \ldots, l_k\}$ is the lcm (least common multiple) of $n_1, n_2, \ldots, n_k$, that is, show
(i) $n_i | m_0$ for $1 \le i \le k$ and
(ii) $m_0 | m$ for all integers m satisfying $n_i | m$ for all $1 \le i \le k$.
Hint: Use Corollary 1.16 for (ii).

2. (a) Find the positive generator of the ideal K of $\mathbb{Z}$ consisting of all integers of the type $k = 63b_1 + 231b_2 + 429b_3$ where $b_1, b_2, b_3 \in \mathbb{Z}$. List the integers $k \in K$ satisfying $-10 < k < 10$.

(b) Let d and d' be integers. Show that $\langle d \rangle \subseteq \langle d' \rangle \Leftrightarrow d'|d$. List the 8 ideals of $\mathbb{Z}$ which contain the ideal $\langle 30 \rangle$. Which ideals of $\mathbb{Z}$ contain the ideal $\langle p \rangle$ where p is prime? List the ideals of $\mathbb{Z}$ containing $\langle 64 \rangle$, beginning with the smallest and ending with the largest.

(c) Let K be an additive subgroup of $\mathbb{Z}$ and let $k \in K$. Prove by induction that $nk \in K$ for all positive integers n. Hence show that K is an ideal of $\mathbb{Z}$.

(d) Let K_1 and K_2 be ideals of $\mathbb{Z}$. Show that $K_1 \cap K_2$ and $K_1 + K_2 = \{k_1 + k_2 : k_1 \in K_1, k_2 \in K_2\}$ are ideals of $\mathbb{Z}$. By Theorem 1.15 there are non-negative integers d_1, d_2 with $K_1 = \langle d_1 \rangle$, $K_2 = \langle d_2 \rangle$. Show $K_1 + K_2 = \langle \gcd\{d_1, d_2\} \rangle$ and $K_1 \cap K_2 = \langle \operatorname{lcm}\{d_1, d_2\} \rangle$.
Hint: Use the first part of (b) above.

3. (a) Find the numbers of l-minors of a 5×6 matrix over $\mathbb{Z}$ for $1 \leq l \leq 5$.

(b) For each of the following $s \times t$ matrices A over $\mathbb{Z}$, calculate the gcd $g_l(A)$ of l-minors, $1 \leq l \leq \min\{s, t\}$, and find $S(A)$ without further calculation:

$$\text{(i)} \quad \begin{pmatrix} 2 & 1 & 3 \\ 4 & 6 & 8 \end{pmatrix}; \qquad \text{(ii)} \quad \begin{pmatrix} 6 & 18 & 0 \\ 24 & 9 & 27 \end{pmatrix};$$

$$\text{(iii)} \quad \begin{pmatrix} 6 & 0 & 0 \\ 0 & 10 & 0 \\ 0 & 0 & 15 \end{pmatrix}.$$

(c) Calculate $\operatorname{adj} A$ and verify $A(\operatorname{adj} A) = (\det A)I = (\operatorname{adj} A)A$ for

$$A = \begin{pmatrix} 1 & 1 & 1 \\ 2 & 5 & 2 \\ 4 & 4 & 7 \end{pmatrix}.$$

Find invertible 3×3 matrices P and Q over $\mathbb{Z}$ with $PAQ^{-1} = S(A)$ and verify that $Q(\operatorname{adj} A)P^{-1} = \operatorname{adj} S(A)$. What is the Smith normal form of $\operatorname{adj} A$?

(d) Let A be a $t \times t$ matrix over $\mathbb{Z}$ with $\det A > 0$ and let P and Q be invertible $t \times t$ matrices over $\mathbb{Z}$ with $PAQ^{-1} = S(A) = \operatorname{diag}(d_1, d_2, \ldots, d_t)$ where $t \geq 2$. Show that $Q(\operatorname{adj} A)P^{-1} = \operatorname{adj} S(A)$ and hence express the invariant factors of $\operatorname{adj} A$ in terms of the invariant factors d_i of A. How do these formulae change in the case $\det A < 0$?
Hint: Start by multiplying PAQ^{-1} and $Q(\operatorname{adj} A)P^{-1}$ together.
Suppose now that $\operatorname{rank} A = t - 1$. Show that $Q(\operatorname{adj} A)P^{-1} = \pm \operatorname{diag}(0, \ldots, 0, d_1 d_2 \cdots d_{t-1})$ and hence find $S(\operatorname{adj} A)$. What is $S(\operatorname{adj} A)$ in the case $\operatorname{rank} A \leq t - 2$?

(e) Specify the $t \times t$ matrices A over $\mathbb{Z}$ with $t \geq 2$ such that $A \equiv \operatorname{adj} A$.
   Hint: Consider first the case $t = 2$. Then try $t = 3$.

4. (a) Verify the Cauchy–Binet theorem for

$$B = \begin{pmatrix} 2 & 1 & 4 \\ 3 & 5 & 7 \end{pmatrix}, \qquad C = B^T.$$

(b) Find x and y satisfying

$$\begin{vmatrix} 14 & 2 + 2x + 3y \\ 2 + 2x + 3y & 4 + x^2 + y^2 \end{vmatrix} = 0$$

by applying the Cauchy–Binet theorem to

$$B = \begin{pmatrix} 1 & 2 & 3 \\ 2 & x & y \end{pmatrix}, \qquad C = B^T.$$

(c) Let B be an $s \times t$ matrix over $\mathbb{Z}$ where $s < t$. Show $\det BB^T = 0$ if and only if $g_s(B) = 0$.

(d) Let B be an $l \times r$ matrix over $\mathbb{Z}$ and let C be an $r \times l$ matrix over $\mathbb{Z}$ where $r < l$. Show $\det BC = 0$.
   Hint: Construct $l \times l$ matrices

$$B' = (B \mid 0) \quad \text{and} \quad C' = \begin{pmatrix} C \\ 0 \end{pmatrix}$$

and use Theorem 1.18.

5. (a) Let A be an $s \times t$ matrix over $\mathbb{Z}$ with $s \geq t$ and let Y be a subset of $\{1, 2, \ldots, t\}$. Denote by A_Y the $s \times l$ submatrix consisting of columns j of A for $j \in Y$, where $l = |Y|$. Show that $g_l((PA)_Y) = g_l(A_Y)$ for all invertible $s \times s$ matrices P over $\mathbb{Z}$ where $1 \leq l \leq t$.
   Let $S(A) = \operatorname{diag}(d_1, d_2, \ldots, d_t)$ be the Smith normal form of A. Show that A can be changed into $S(A)$ using only *eros* if and only if $g_1(A_Y) = d_j$ for $Y = \{j\}$, $1 \leq j \leq t$, and $g_l(A_Y) = d_1 d_2 \cdots d_l$ for $Y = \{1, 2, \ldots, l\}$, $2 \leq l \leq t$.
   Hint: Adapt the method of Lemma 1.9 by applying Lemma 1.7 transposed to the columns of A.

(b) Which of the following matrices A can be changed into Smith normal form using *eros* only?

$$\text{(i)} \quad \begin{pmatrix} 1 & 2 & 8 \\ 1 & 4 & 4 \\ 3 & 6 & 0 \end{pmatrix}; \qquad \text{(ii)} \quad \begin{pmatrix} 1 & 4 & 8 \\ 2 & 10 & 16 \\ 1 & 4 & 12 \end{pmatrix}.$$

(c) Let A be an $s \times t$ matrix over $\mathbb{Z}$ with $s < t$ and $g_s(A) = 1$. Show that A has Smith normal form $S(A) = (I_s \mid 0)$ where I_s is the $s \times s$ identity matrix and that A can be changed into $S(A)$ using *ecos* only. Deduce that there is an invertible $t \times t$ matrix Q over $\mathbb{Z}$ having the rows of A as its first s rows.

Hint: Consider Q where $AQ^{-1} = S(A)$.

For each of the following matrices A, find a suitable Q:

$$\text{(i)} \quad (6, 10, 15); \qquad \text{(ii)} \quad \begin{pmatrix} 10 & 9 & 10 \\ 15 & 15 & 16 \end{pmatrix}.$$

6. (a) Let A and B be $t \times t$ matrices over $\mathbb{Z}$ having non-zero *coprime* determinants. Use Theorem 1.21 to show $S(AB) = S(A)S(B)$.

(b) Suppose $S(A) = \text{diag}(2, 4)$ and $S(B) = \text{diag}(1, 3)$; what is $S(AB)$? For $S(A) = \text{diag}(2, 6)$ and $S(B) = \text{diag}(1, 4)$, find the two possible matrices $S(AB)$.

(c) Let A be a $t \times t$ matrix over $\mathbb{Z}$ with $d_t(A) \neq 0$. Let $p_1, \ldots, p_l$ be the distinct prime divisors of $d_t(A)$. Show that there are $t \times t$ matrices A_j over $\mathbb{Z}$, unique up to equivalence, such that $A = A_1 A_2 \cdots A_l$ and $d_k(A_j)$ is a power of p_j for $1 \leq j \leq l$.

Hint: Use $A = P^{-1}S(A)Q$.

(d) Express $A = A_1 A_2$ as in part (c) where $A = \begin{pmatrix} 4 & 8 \\ 12 & 10 \end{pmatrix}$. Are $A_1 A_2$ and $A_2 A_1$ necessarily equivalent?

2

Basic Theory of Additive Abelian Groups

In this chapter we discuss cyclic groups, the quotient group construction, the direct sum construction and the first isomorphism theorem, in the context of additive abelian groups; we also discuss free modules. These concepts are necessary, as well as the matrix theory of Chapter 1, for the study of finitely generated abelian groups in Chapter 3. At the same time the material provides the reader with a taster for general group theory.

2.1 Cyclic ℤ-Modules

We begin with a brief review of abelian groups. They arise in additive and multiplicative notation. Additive notation is more suited to our purpose and we'll adopt it wherever possible. The term ℤ-module is simply another name for an additive abelian group. However it signals an approach which emphasises the analogy between vector spaces and abelian groups. The structure-preserving mappings of abelian groups, their *homomorphisms* in other words, are analogous to linear mappings of vector spaces. So in a sense the reader will have seen it all before. But be careful as the analogy is by no means perfect! For instance from $\lambda v = 0$ in a vector space one can safely deduce that either $\lambda = 0$ or $v = 0$ (or both). By contrast, in a ℤ-module the equation $mg = 0$ may hold although $m \neq 0$ and $g \neq 0$ as we will see.

Next we study cyclic groups, that is, groups generated by a single element. Their theory is not too abstract and should help the reader's appreciation of the more general theorems on ℤ-modules later in Chapter 2.

C. Norman, *Finitely Generated Abelian Groups and Similarity of Matrices over a Field*,   47
Springer Undergraduate Mathematics Series,
DOI 10.1007/978-1-4471-2730-7_2, © Springer-Verlag London Limited 2012

Let $(G, +)$ denote a set G with a binary operation denoted by addition. So G is *closed under addition*, that is,

$$g_1 + g_2 \in G \quad \text{for all } g_1, g_2 \in G.$$

Then $(G, +)$ is *an (additive) abelian group* if the following laws hold:

1. *The associative law of addition*: $(g_1 + g_2) + g_3 = g_1 + (g_2 + g_3)$ for all $g_1, g_2, g_3 \in G$.
2. *The existence of a zero element*: there is an element, denoted by 0, in G satisfying $0 + g = g$ for all $g \in G$.
3. *The existence of negatives*: for each $g \in G$ there is an element, denoted by $-g$, in G satisfying $-g + g = 0$.
4. *The commutative law of addition*: $g_1 + g_2 = g_2 + g_1$ for all $g_1, g_2 \in G$.

We now drop the notation $(G, +)$ and in its place refer simply to the abelian group G, the binary operation of addition being taken for granted. Laws 1, 2, 3 are *the group axioms* in additive notation. The reader should know that the zero element 0 of a group G is unique, that is, given laws 1, 2 and 3 then there is only one element 0 as in law 2. Similarly each element g of a group G has a unique negative $-g$. Laws 3 and 4 give $g + (-g) = 0$ which tells us that g is the negative of $-g$, that is, all elements g of an additive abelian group G satisfy the unsurprising equation $-(-g) = g$.

The reader might like to simplify the expression $((-g_1) + (-g_2)) + (g_1 + g_2)$ where g_1 and g_2 are elements of an additive abelian group G, using at each step one of the above four laws (start by applying law 4 followed by law 1 (twice) and then law 3, etc.). After six steps you should get the zero element 0 of G. The conclusion is: $(-g_1) + (-g_2)$ is the negative of $g_1 + g_2$ by law 3, as the sum of these two elements of G is zero. In other words $(-g_1) + (-g_2) = -(g_1 + g_2)$ for all g_1 and g_2 in G. Luckily the manipulation of elements of an additive group need not involve the laborious application of its laws as we now explain.

Let $g_1, g_2, \ldots, g_n$ be elements of an additive group G where $n \geq 3$. These elements can be summed (added up) in order in various ways, but all ways produce the same element of G. In the case $n = 4$

$$((g_1 + g_2) + g_3) + g_4 = ((g_1 + (g_2 + g_3)) + g_4 = g_1 + ((g_2 + g_3) + g_4)$$

$$= g_1 + (g_2 + (g_3 + g_4)) = (g_1 + g_2) + (g_3 + g_4)$$

using only law 1. Omitting the brackets we see that $g_1 + g_2 + g_3 + g_4$ has an unambiguous meaning, namely any one of the above (equal) elements. Using law 1 and induction on n, it can be shown (Exercises 2.1, Question 8(a)) that brackets may be left out when adding up, in order, any finite number n of elements $g_1, g_2, \ldots, g_n$ of an additive group G to give *the generalised associative law of addition*. So the sum $g_1 + g_2 + \cdots + g_n$ of n elements of an additive abelian group G is unambiguously defined and what is more this sum is unchanged when the suffixes are permuted (*the generalised commutative* law *of addition*). In the case $n = 3$

$$g_1 + g_2 + g_3 = g_1 + g_3 + g_2 = g_3 + g_1 + g_2$$

$$= g_3 + g_2 + g_1 = g_2 + g_3 + g_1 = g_2 + g_1 + g_3.$$

Let g be an element of an additive abelian group G. The elements $2g = g + g$, $3g = g + g + g$ belong to G. More generally for every positive integer n the group element ng is obtained by adding together n elements equal g, that is,

$$ng = g + g + \cdots + g \quad (n \text{ terms}).$$

So $1g = g$ and as $n(-g) + ng = 0$ we deduce that $n(-g) = -(ng)$. Therefore it makes sense to define $(-n)g$ to be the group element $n(-g)$ and to define $0g$ to be the zero element 0 of G. It follows that $(-n)(-g) = n(-(-g)) = ng$. But more importantly we have given meaning to mg for *all* integers m (positive, negative and zero) and all elements g in G and

$$mg \in G \quad \text{for all } m \in \mathbb{Z}, \ g \in G$$

showing that G is *closed* under *integer multiplication*.

Integer multiplication on G and the group operation of addition on G are connected by the following laws:

5. *The distributive laws*:

$$m(g_1 + g_2) = mg_1 + mg_2 \quad \text{for all } m \in \mathbb{Z} \text{ and all } g_1, g_2 \in G,$$

$$(m_1 + m_2)g = m_1 g + m_2 g \quad \text{for all } m_1, m_2 \in \mathbb{Z} \text{ and all } g \in G.$$

6. *The associative law of multiplication*:

$$(m_1 m_2)g = m_1(m_2 g) \quad \text{for all } m_1, m_2 \in \mathbb{Z} \text{ and all } g \in G.$$

7. *The identity law*: $1g = g$ for all $g \in G$.

Laws 5 and 6 are the familiar laws of indices expressed in additive notation, rather than the more usual multiplicative notation; they allow elements of additive abelian groups to be manipulated with minimum fuss (see Exercises 2.1, Question 8(c)). Law 7 is frankly something of an anti-climax, but its presence will help us generalise these ideas later in a coherent way. The structure of G is expressed concisely by saying

$$G \text{ is a } \mathbb{Z}\text{-module}$$

meaning that laws 1–7 above hold. Notice the close connection between the laws of a $\mathbb{Z}$-module and the laws of a vector space: they are almost identical! Think of the elements of G as being 'vectors' and the elements of $\mathbb{Z}$ as being 'scalars'. The only thing which prevents a $\mathbb{Z}$-module from being a vector space is the fact that $\mathbb{Z}$ is not a field.

The reader will already have met many examples of additive abelian groups: for example the additive group $(\mathbb{Q}, +)$ of all rational numbers m/n $(m, n \in \mathbb{Z}, n > 0)$;

this group is obtained from the rational field $\mathbb{Q}$ by ignoring products of non-integer rational numbers – they simply don't arise in $(\mathbb{Q}, +)$. In the same way, ignoring the multiplication on any *ring R*, we obtain its *additive group* $(R, +)$. Of particular importance are the additive groups of the *residue class* rings $\mathbb{Z}_n$ (we'll shortly review their properties) as well as the additive group of the ring $\mathbb{Z}$ itself.

Let H be a *subgroup* of the additive abelian group G. So H is a sub*set* of G satisfying

(a) $h_1 + h_2 \in H$ for all $h_1, h_2 \in H$ (H is closed under addition)

(b) $0 \in H$ (H contains the zero element of G)

(c) $-h \in H$ for all $h \in H$ (H is closed under negation).

By (a) we see that H is a set with a binary operation of addition and so it makes sense to ask: is $(H, +)$ an abelian group? As $H \subseteq G$ and law 1 holds in G, we see that $(h_1 + h_2) + h_3 = h_1 + (h_2 + h_3)$ for all h_1, h_2, h_3 in H, that is, law 1 holds in H. In the same way law 4 holds in H. Also (b) and (c) ensure that law 2 and law 3 hold in H. So $(H, +)$ is an abelian group as laws 1–4 hold with G replaced by H. Hence laws 5, 6 and 7 also hold with G replaced by H, that is, H is a $\mathbb{Z}$-module. The relationship between H and G is described by saying

$$H \text{ is a submodule of the } \mathbb{Z}\text{-module } G.$$

The set $\langle 2 \rangle$ of even integers is a subgroup of the additive group $\mathbb{Z}$ and so $\langle 2 \rangle$ is a submodule of $\mathbb{Z}$. The discussion preceding Theorem 1.15 shows that $\langle 2 \rangle$ is an ideal of the ring $\mathbb{Z}$. More generally H is a *submodule* of the $\mathbb{Z}$-module $\mathbb{Z}$ if and only if H is an *ideal* of the ring $\mathbb{Z}$, and when this is the case Theorem 1.15 tells us $H = \langle d \rangle$ where d is a non-negative integer.

We now discuss in detail *cyclic* $\mathbb{Z}$-modules. They are crucial: on the one hand they and their submodules are easily described as we will soon see, and on the other hand every finitely generated $\mathbb{Z}$-module can be constructed using them as building blocks, as we show in Section 3.1.

Definition 2.1

Let G be a $\mathbb{Z}$-module containing an element g such that every element of G is of the type mg for some $m \in \mathbb{Z}$. Then G is said to be *cyclic* with *generator g* and we write $G = \langle g \rangle$.

The additive group $\mathbb{Z}$ of all integers is a cyclic $\mathbb{Z}$-module with generator 1 and so $\mathbb{Z} = \langle 1 \rangle$. As $\mathbb{Z}$ contains an infinite number of elements we say that $\mathbb{Z}$ is an *infinite cyclic group*. Let K be a subgroup of $\mathbb{Z}$ (we know that subgroups of $\mathbb{Z}$ are ideals Theorem 1.15 and so we denote them by K rather than H). As $K = \langle d \rangle$ where d is non-negative, every subgroup K of the infinite cyclic group $\mathbb{Z}$ is itself cyclic because

K has generator d. Note that $-d$ is also a generator of K as $\langle -d \rangle = \langle d \rangle$. The subgroup $\langle 2 \rangle$ of all even integers is infinite cyclic just like its 'parent' $\mathbb{Z}$ and the same is true of $\langle d \rangle$ with $d \neq 0$. Notice that $\langle 6 \rangle \subseteq \langle 2 \rangle$ as 2 is a divisor of 6 and so all integer multiples of 6 are even integers. More generally $\langle d_1 \rangle \subseteq \langle d_2 \rangle$ if and only if $d_2 | d_1$. where $d_1, d_2 \in \mathbb{Z}$.

Let n be a positive integer. The reader is assumed to have met the ring $\mathbb{Z}_n$ of integers modulo n; however, we now briefly review its construction and properties. A typical element of $\mathbb{Z}_n$ is the congruence class $\bar{r}$ of an integer r, that is, $\bar{r} = \{nq + r : q \in \mathbb{Z}\}$. So $\bar{r}$ is the subset of integers $m = nq + r$ which differ from r by an integer multiple q of n. Therefore $m - r = nq$, that is, the difference between m and r is divisible by n; this is expressed by saying that m is congruent to r modulo n and writing $m \equiv r \pmod{n}$. So $\mathbb{Z}_n$ has n elements and

$$\mathbb{Z}_n = \{\bar{0}, \bar{1}, \bar{2}, \ldots, \overline{n-1}\}$$

since the n congruence classes $\bar{r}$ correspond to the n possible remainders r on dividing an arbitrary integer m by n. You should know that $\mathbb{Z}_n$ is a commutative ring, the rules of addition and multiplication being unambiguously defined by $\bar{m} + \bar{m}' = \overline{m + m'}$, $(\bar{m})(\bar{m}') = \overline{mm'}$ for all $m, m' \in \mathbb{Z}$. The 0-element and 1-element of $\mathbb{Z}_n$ are $\bar{0}$ and $\bar{1}$ respectively. You should also know that $\mathbb{Z}_n$ is a field if and only if n is prime. The smallest field is $\mathbb{Z}_2 = \{\bar{0}, \bar{1}\}$ having two elements, namely the set $\bar{0}$ of all even integers and the set $\bar{1}$ of all odd integers.

The additive group of $\mathbb{Z}_n$ is cyclic, being generated by $\bar{1}$ as $\bar{m} = m\bar{1}$. Having n elements, $\mathbb{Z}_n$ is a cyclic group of order n.

For example, taking $n = 4$ we obtain the cyclic group $\mathbb{Z}_4 = \{\bar{0}, \bar{1}, \bar{2}, \bar{3}\}$ of order 4 with addition table:

+	$\bar{0}$	$\bar{1}$	$\bar{2}$	$\bar{3}$
$\bar{0}$	$\bar{0}$	$\bar{1}$	$\bar{2}$	$\bar{3}$
$\bar{1}$	$\bar{1}$	$\bar{2}$	$\bar{3}$	$\bar{0}$
$\bar{2}$	$\bar{2}$	$\bar{3}$	$\bar{0}$	$\bar{1}$
$\bar{3}$	$\bar{3}$	$\bar{0}$	$\bar{1}$	$\bar{2}$

The element $x + y$ appears in the table where the row headed by x meets the column headed by y, for $x, y \in \mathbb{Z}_4$. By inspection $\mathbb{Z}_4$ has three subgroups $\{\bar{0}\}$, $\{\bar{0}, \bar{2}\}$ and $\mathbb{Z}_4$ itself. The union of the congruence classes belonging to any one of these subgroups of $\mathbb{Z}_4$ is a subgroup of $\mathbb{Z}$. Thus $\bar{0} = \langle 4 \rangle$ since $\bar{0}$ consists of integers which are multiples of 4. Similarly $\bar{0} \cup \bar{2} = \langle 2 \rangle$ since $\bar{2}$ consists of integers which are multiples of 2 but not multiples of 4. Also $\bar{0} \cup \bar{1} \cup \bar{2} \cup \bar{3} = \mathbb{Z} = \langle 1 \rangle$. The three subgroups of $\mathbb{Z}_4$ correspond in this way to the three subgroups $\langle 4 \rangle, \langle 2 \rangle, \langle 1 \rangle$ of $\mathbb{Z}$ which contain $\langle 4 \rangle$. What is more, the subgroups of $\mathbb{Z}_4$ are cyclic with generators $\bar{4}, \bar{2}, \bar{1}$. We now return to $\mathbb{Z}_n$ and show that these ideas can be generalised.

Lemma 2.2

Let n be a positive integer. Each subgroup H of the additive group $\mathbb{Z}_n$ is cyclic with generator $\overline{d}$ where $d|H| = n$ and $|H|$ is the number of elements in H.

Proof

For each subgroup H of $\mathbb{Z}_n$ let $K = \{m \in \mathbb{Z} : \overline{m} \in H\}$. So K consists of those integers m which belong to a congruence class in H. We show that K is a subgroup of $\mathbb{Z}$ containing $\langle n \rangle$. Let $m_1, m_2 \in K$. Then $\overline{m}_1, \overline{m}_2 \in H$. As H is closed under addition, $\overline{m_1 + m_2} = \overline{m}_1 + \overline{m}_2 \in H$. So $m_1 + m_2 \in K$ showing that K is closed under addition. Now $\overline{0} \in H$ as H contains the zero element of $\mathbb{Z}_n$. But $\overline{0}$ consists of all integer multiples of n. So $\overline{0} = \langle n \rangle \subseteq K$ and in particular $0 \in K$. Let $m \in K$. Then $\overline{m} \in H$ and so $\overline{-m} = -\overline{m} \in H$ as H is closed under negation. Therefore $-m \in K$ showing that K is closed under negation. We have shown that K is a subgroup of $\mathbb{Z}$ containing $\langle n \rangle$. So $K = \langle d \rangle$ for some non-negative integer d by Theorem 1.15. As $\langle n \rangle \subseteq \langle d \rangle$ we conclude that $d|n$. As n is positive, d cannot be zero and so d is positive also. As $d \in K$ we see that $\overline{d} \in H$. Finally consider $\overline{m} \in H$. Then $m \in K$ and so $m = qd$ for some $q \in \mathbb{Z}$. So $\overline{m} = q\overline{d}$ showing that H is cyclic with generator $\overline{d}$.

Now $|H|$ is the *order* of H (the number of elements in H) and so K is the union of $|H|$ congruence classes (mod n). Let $m \in K$. As $K = \langle d \rangle$ there is $q \in \mathbb{Z}$ with $m = qd$. Divide q by n/d to obtain integers q', r with $q = q'(n/d) + r$ where $0 \leq r < n/d$. Then $m = (q'(n/d)+r)d = q'n+rd$, showing that K consists of the n/d congruence classes $\overline{rd}$. Hence $|H| = n/d$ and so $d|H| = n$. □

For example, by Lemma 2.2 the additive group $\mathbb{Z}_{18}$ has 6 subgroups corresponding to the 6 positive divisors 1, 2, 3, 6, 9, 18, of 18.

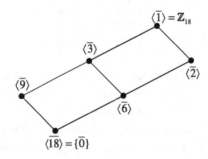

These 6 subgroups can be arranged in their lattice diagram, as shown, in which subgroup H_1 is contained in subgroup H_2 if and only if there is a sequence of upwardly sloping lines joining H_1 to H_2. For instance $\langle \overline{6} \rangle \subseteq \langle \overline{1} \rangle$ but $\langle \overline{9} \rangle \nsubseteq \langle \overline{2} \rangle$.

The proof of Lemma 2.2 shows that each subgroup H of $\mathbb{Z}_n$ corresponds to a subgroup K of $\mathbb{Z}$ which contains $\langle n \rangle$. From the last paragraph we see that each such subgroup K has a positive generator d and is made up of n/d congruence classes (mod n). These n/d elements form a subgroup H of $\mathbb{Z}_n$. So for each K there is one H and vice-versa. We show in Theorem 2.17 that bijective correspondences of this type arise in a general context. To get the idea, consider

$$\text{the natural mapping } \eta : \mathbb{Z} \to \mathbb{Z}_n$$

which maps each integer m to its congruence class $\overline{m}$ modulo n. We use $(m)\eta$ to denote the image of m by η and so $(m)\eta = \overline{m}$ for all integers m. Now η is *additive*, that is,

$$(m + m')\eta = (m)\eta + (m')\eta \quad \text{for all } m, m' \in \mathbb{Z}$$

as $\overline{m + m'} = \overline{m} + \overline{m'}$. Such additive mappings provide meaningful comparisons between additive abelian groups and surprisingly each one gives rise to a bijective correspondence as above. The *image* of η is the set of all elements $(m)\eta$ and is denoted by $\operatorname{im}\eta$. As η is surjective (onto) we see $\operatorname{im}\eta = \mathbb{Z}_n$. The *kernel* of η is the set of elements m such that $(m)\eta = \overline{0}$, the zero element of $\mathbb{Z}_n$, and is denoted by $\ker\eta$. So $\ker\eta = \langle n \rangle$. With this terminology the correspondence between H and K, where $K = \{m \in \mathbb{Z} : (m)\eta \in H\}$, is bijective from the set of subgroups H of $\operatorname{im}\eta$ to the set of subgroups K of $\mathbb{Z}$ which contain $\ker\eta$. We take up this theme in Theorem 2.17.

Definition 2.3

Let G and G' be additive abelian groups. A mapping $\theta : G \to G'$ such that $(g_1 + g_2)\theta = (g_1)\theta + (g_2)\theta$ for all $g_1, g_2 \in G$ is called *additive* or a *homomorphism*.

Such mappings θ respect the group operation and satisfy $(0)\theta = 0'$, $(-g)\theta = -(g)\theta$ for $g \in G$, that is, θ maps the zero of G to the zero of G' and θ respects negation (see Exercises 2.1, Question 4(c)). With $g_1 = g_2 = g$ in Definition 2.3 we obtain $(2g)\theta = 2((g)\theta)$. Using induction the additive mapping θ satisfies

$$(mg)\theta = m((g)\theta)$$

for all $m \in \mathbb{Z}$, $g \in G$. We describe θ as being $\mathbb{Z}$-*linear*, meaning that θ is additive and satisfies the above equation, that is, θ maps each integer multiple mg of each element g of G to m times the element $(g)\theta$ of G'.

The natural mapping $\eta : \mathbb{Z} \to \mathbb{Z}_n$ is $\mathbb{Z}$-linear. The 'doubling' mapping $\theta : \mathbb{Z} \to \mathbb{Z}$, where $(m)\theta = 2m$ for all $m \in \mathbb{Z}$, is also $\mathbb{Z}$-linear.

How can we tell whether two $\mathbb{Z}$-modules are really different or essentially the same? The next definition provides the terminology to tackle this problem.

Definition 2.4

A bijective (one-to-one and onto) $\mathbb{Z}$-linear mapping is called an *isomorphism*. The $\mathbb{Z}$-modules G and G' are called *isomorphic* if there is an isomorphism $\theta : G \to G'$ in which case we write $\theta : G \cong G'$ or simply $G \cong G'$. An isomorphism $\theta : G \cong G$ of G to itself is called an *automorphism* of G.

For example $\theta : \mathbb{Z} \cong \langle 2 \rangle$, where $(m)\theta = 2m$ for all $m \in \mathbb{Z}$, is an isomorphism showing that the $\mathbb{Z}$-module of all integers is isomorphic to the $\mathbb{Z}$-module of all even integers. In the same way $\mathbb{Z} \cong \langle d \rangle$ for every non-zero integer d. Isomorphic $\mathbb{Z}$-modules are abstractly identical and differ at most in notation.

The inverse of an isomorphism $\theta : G \cong G'$ is an isomorphism $\theta^{-1} : G' \cong G$ and the composition of compatible isomorphisms $\theta : G \cong G'$ and $\theta' : G' \cong G''$ is an isomorphism $\theta\theta' : G \cong G''$ (Exercises 2.1, Question 4(d)).

The additive cyclic group $\mathbb{Z}$ is generated by 1 and also by -1. As $\mathbb{Z}$ has no other generators there is just one non-identity automorphism of $\mathbb{Z}$, namely $\tau : \mathbb{Z} \to \mathbb{Z}$ defined by $(m)\tau = -m$ for all $m \in \mathbb{Z}$. Notice that $(1)\tau = -1$ and $(-1)\tau = 1$. More generally, every automorphism of a cyclic group permutes the generators amongst themselves.

Let g be an element of the additive group G. The *smallest* positive integer n such that $ng = 0$ is called the *order* of g; if there is no such integer n then g is said to have *infinite* order. We now reformulate this concept in a more convenient manner – it will enable finite and infinite cyclic groups to be dealt with in a unified way.

Let $K = \{m \in \mathbb{Z} : mg = 0\}$, that is, K consists of those integers m such that mg is the zero element of G. It's routine to show that K is an ideal of $\mathbb{Z}$. So K is a principal ideal of $\mathbb{Z}$ with non-negative generator n, that is, $K = \langle n \rangle$ by Theorem 1.15. Then $n = 0$ means that g has infinite order whereas $n > 0$ means that g has finite order n. The ideal $K = \langle n \rangle$ is called the *order ideal* of g. Notice

$$mg = 0 \quad \Leftrightarrow \quad m \in K \quad \Leftrightarrow \quad n | m$$

which is a useful criterion for finding the order of a group element in particular cases. For instance suppose $36g = 0$, $18g \neq 0$, $12g \neq 0$. Then g has order n such that n is a divisor of 36 but not a divisor of either $18 = 36/2$ or $12 = 36/3$. There is only one such positive integer n, namely 36. So g has order 36. More generally

$$g \text{ has finite order } n \quad \Leftrightarrow \quad ng = 0 \quad \text{and} \quad (n/p)g \neq 0 \quad \text{for all prime divisors } p \text{ of } n.$$

The $\Leftarrow$ implication is valid because every positive divisor d of n with $d < n$ satisfies $d | (n/p)$ for some prime divisor p of n; so d cannot be the order of g, and hence n is the order of g.

Be careful! From $24g = 0$ and $12g \neq 0$ one cannot deduce that g has order 24, as g could have order 8.

Once again let g be an element of the additive group G. Then $\langle g \rangle = \{mg : m \in \mathbb{Z}\}$ is a subgroup of G. As $\langle g \rangle$ is cyclic with generator g it is reasonable to call $\langle g \rangle$ *the cyclic subgroup of G generated by g*.

We now explain how the order ideal K of a group element g determines the *isomorphism type* Definition 2.6 of the cyclic group $\langle g \rangle$.

Theorem 2.5

Every cyclic group G is isomorphic either to the additive group $\mathbb{Z}$ or to the additive group $\mathbb{Z}_n$ for some positive integer n.

Proof

Let g generate G and so $G = \langle g \rangle$. Consider $\theta : \mathbb{Z} \to G$ defined by $(m)\theta = mg$ for all integers m. Then θ is $\mathbb{Z}$-linear by laws 5 and 6 of a $\mathbb{Z}$-module. Now θ is surjective (onto) since every element of G is of the form $(m)\theta$ for some $m \in \mathbb{Z}$, that is, $\mathrm{im}\,\theta = G$ meaning that G is the image of θ (we'll state the general definition of image and kernel in Section 2.3). The kernel of θ is $\ker\theta = \{m \in \mathbb{Z} : (m)\theta = 0\} = \{m \in \mathbb{Z} : mg = 0\} = K$ which is the order ideal of g. By Theorem 1.15 there is a non-negative integer n with $\ker\theta = \langle n \rangle$.

Suppose $n = 0$. Then θ is injective (one-to-one) because suppose $(m)\theta = (m')\theta$. Then $(m - m')\theta = (m)\theta - (m')\theta = 0$ showing that $m - m'$ belongs to $\ker\theta = \langle 0 \rangle = \{0\}$. So $m - m' = 0$, that is, $m = m'$. Therefore θ is bijective and so $\theta : \mathbb{Z} \cong G$, that is,

all infinite cyclic groups are isomorphic to the additive group $\mathbb{Z}$ of integers.

Suppose $n > 0$. As above we suppose $(m)\theta = (m')\theta$. This means $m - m' \in \ker\theta = K = \langle n \rangle$ and so $m - m'$ is an integer multiple of n, that is, $m \equiv m' \pmod n$. The steps can be reversed to show that $m \equiv m' \pmod n$ implies $(m)\theta = (m')\theta$. So θ has the same effect on integers m and m' which are congruent $\pmod n$. In other words θ has the same effect on all the integers of each congruence class $\overline{m}$, and it makes sense to introduce the mapping $\tilde{\theta} : \mathbb{Z}_n \to G$ defined by $(\overline{m})\tilde{\theta} = (m)\theta$ for all $m \in \mathbb{Z}$. As θ is additive and surjective, the same is true of $\tilde{\theta}$. As θ has different effects on different congruence classes $\pmod n$, we see that $\tilde{\theta}$ is injective. Therefore $\tilde{\theta} : \mathbb{Z}_n \cong G$ which shows:

every cyclic group of finite order n is isomorphic to the additive group $\mathbb{Z}_n$. □

We will see in Theorem 2.16 that every $\mathbb{Z}$-linear mapping θ gives rise to an isomorphism $\tilde{\theta}$ as in the above proof. To illustrate Theorem 2.5 let $g = \overline{18} \in \mathbb{Z}_{60}$. The order of g is the smallest positive integer n satisfying $18n \equiv 0 \pmod{60}$. Dividing through by $\gcd\{18, 60\} = 6$ we obtain $3n \equiv 0 \pmod{10}$ and so $n = 10$. Therefore g has order 10 and hence generates a subgroup $\langle g \rangle$ of $\mathbb{Z}_{60}$ which is isomorphic to the additive group $\mathbb{Z}_{10}$ by Theorem 2.5. The reader can check

$$\langle g \rangle = \{\overline{0}, \overline{18}, \overline{36}, \overline{54}, \overline{12}, \overline{30}, \overline{48}, \overline{6}, \overline{24}, \overline{42}\} \subseteq \mathbb{Z}_{60}$$

and $\tilde{\theta} : \mathbb{Z}_{10} \cong \langle g \rangle$ where $(\overline{m})\tilde{\theta} = \overline{18m}$ for $\overline{m} \in \mathbb{Z}_{10}$.

From the proof of Theorem 2.5 we see

$$g \text{ has finite order } n \quad \Leftrightarrow \quad |\langle g \rangle| = n$$

In other words, each element g of finite order n generates a cyclic subgroup $\langle g \rangle$ of order n.

Definition 2.6

Let n be a positive integer. A cyclic $\mathbb{Z}$-module G is said to be of *isomorphism type C_n* or C_0 according as G is isomorphic to the additive group $\mathbb{Z}_n$ or the additive group $\mathbb{Z}$.

So for $n > 0$, groups of isomorphism type C_n are cyclic groups of order n. Groups of type C_0 are infinite cyclic groups. Groups of type C_1 are *trivial* because they contain only one element, namely their zero element. Notice that for all non-negative integers n

> $G = \langle g \rangle$ has isomorphism type C_n where $\langle n \rangle$ is the order ideal of g

How is the order of mg related to the order of g? Should g have infinite order then mg also has infinite order for $m \neq 0$ and order 1 for $m = 0$.

Lemma 2.7

Let the $\mathbb{Z}$-module element g have finite order n. Then mg has finite order $n/\gcd\{m, n\}$ for $m \in \mathbb{Z}$.

Proof

The order ideal of g is $\langle n \rangle$. Let $\langle n' \rangle$ be the order ideal of mg where $n' \geq 0$. Write $d = \gcd\{m, n\}$. Then $(n/d)mg = (m/d)ng = (m/d)0 = 0$ showing that n/d annihilates mg, that is, n/d belongs to the order ideal $\langle n' \rangle$ of mg. Hence n' is a divisor of n/d and also $n' > 0$. On the other hand $n'mg = 0$ shows that $n'm$ belongs to the order ideal $\langle n \rangle$ of g. So $n'm = qn$ for some integer q. Hence n/d is a divisor of n' (m/d). But m/d

and n/d are *coprime integers* meaning $\gcd\{m/d, n/d\} = 1$. Hence n/d is a divisor of n'. As n/d and n' are both positive and each is a divisor of the other we conclude that $n' = n/d$. So mg has order $n' = n/d = n/\gcd\{m, n\}$. □

For example the element $\overline{1}$ of the additive group $\mathbb{Z}_{60}$ has order 60. So $\overline{18} = 18(\overline{1})$ in $\mathbb{Z}_{60}$ has order $60/\gcd\{18, 60\} = 60/6 = 10$. As we saw earlier, $\overline{18}$ generates a cyclic subgroup of $\mathbb{Z}_{60}$ with $|\langle \overline{18} \rangle| = 10$. We will see in Section 3.1 that Lemma 2.7 plays an crucial part in the theory of finite abelian groups.

Finally we work through an application of these ideas which involves an abelian group in *multiplicative* notation. This abelian group is familiar to the reader yet has a hint of mystery!

Example 2.8

Let G denote the multiplicative group $\mathbb{Z}_{43}^*$ of non-zero elements of the field $\mathbb{Z}_{43}$. In Corollary 3.17 it is shown that the multiplicative group F^* of every finite field F is cyclic. So $\mathbb{Z}_p^*$ is cyclic for all primes p. In particular $\mathbb{Z}_{43}^*$ is cyclic, and we set out to find a generator by 'trial and error'. As $|G| = 42 = 2 \times 3 \times 7$ each generator g has order 42, that is, $g^{42} = \overline{1}$ the identity element of $\mathbb{Z}_{43}^*$, and $g^{42/7} = g^6 \neq \overline{1}$, $g^{42/3} = g^{14} \neq \overline{1}$, $g^{42/2} = g^{21} \neq \overline{1}$.

We first try $g = \overline{2}$. Then $g^6 = \overline{64} = \overline{21} \neq \overline{1}$, and so $g^7 = g g^6 = \overline{2} \times \overline{21} = \overline{42} = -\overline{1}$. Squaring the last equation gives $g^{14} = g^7 g^7 = (-\overline{1})^2 = \overline{1}$ showing that $\overline{2}$ is not a generator of $\mathbb{Z}_{43}^*$. In fact $\overline{2}$ has order 14.

Now try $g = \overline{3}$. Then $g^4 = \overline{81} = -\overline{5}$ and so $g^6 = g^4 g^2 = (-\overline{5}) \times \overline{9} = -\overline{45} = -\overline{2} \neq \overline{1}$. Hence $g^7 = g^6 g = (-\overline{2}) \times \overline{3} = -\overline{6}$. Squaring gives $g^{14} = g^7 g^7 = (-\overline{6})^2 = \overline{36} = -\overline{7} \neq \overline{1}$. Then $g^{21} = g^7 g^{14} = (-\overline{6}) \times (-\overline{7}) = \overline{42} = -\overline{1} \neq \overline{1}$. Squaring now gives $g^{42} = g^{21} g^{21} = (-\overline{1})^2 = \overline{1}$. So $g^{42} = \overline{1}$ and $g^6 \neq \overline{1}$, $g^{14} \neq \overline{1}$, $g^{21} \neq \overline{1}$ which show that g has order 42. Therefore the integer powers of $g = \overline{3}$ are the elements of a cyclic subgroup H of order 42. As $H \subseteq \mathbb{Z}_{43}^*$ and both H and $\mathbb{Z}_{43}^*$ have exactly 42 elements, we conclude $H = \mathbb{Z}_{43}^*$. So $\mathbb{Z}_{43}^*$ is cyclic with generator $\overline{3}$.

Having found one generator of $\mathbb{Z}_{43}^*$ we use Lemma 2.7 to find all g with $\langle g \rangle = \mathbb{Z}_{43}^*$. There is an integer m with $g = (\overline{3})^m$ and $1 \leq m \leq 42$. Comparing orders gives $\langle g \rangle = \mathbb{Z}_{43}^* \Leftrightarrow 42 = 42/\gcd\{m, 42\} \Leftrightarrow \gcd\{m, 42\} = 1$ by Lemma 2.7. So $\mathbb{Z}_{43}^*$ has 12 generators $g = (\overline{3})^m$ where $m \in \{1, 5, 11, 13, 17, 19, 23, 25, 29, 31, 37, 41\}$. For instance $(\overline{3})^5 = \overline{81} \times \overline{3} = -\overline{5} \times \overline{3} = -\overline{15} = \overline{28}$ generates $\mathbb{Z}_{43}^*$. In Section 2.2 we will meet the *Euler ϕ-function* and see that $12 = \phi(42)$.

By *Fermat's little theorem*, which is proved at the beginning of Section 2.2, every element $g = \overline{r}$ of $G = \mathbb{Z}_p^*$, p prime, satisfies $g^{p-1} = \overline{1}$. So

g is a generator of $\mathbb{Z}_p^*$ $\Leftrightarrow$ $g^{(p-1)/p'} \neq \overline{1}$ for all prime divisors p' of $p - 1$.

EXERCISES 2.1

1. (a) Write out the addition table of the additive group $\mathbb{Z}_5$. Does $\overline{4} \in \mathbb{Z}_5$ generate $\mathbb{Z}_5$? Which elements of $\mathbb{Z}_5$ are generators? Specify the (two) subgroups of $\mathbb{Z}_5$.

    (b) Write out the addition table of the $\mathbb{Z}$-module $\mathbb{Z}_6$. Express the elements $27(\overline{2})$, $-17(\overline{4})$, $15(\overline{3}) + 13(\overline{4})$ in the form $\overline{r}$, $0 \leq r < 6$. List the elements in each of the four submodules H of $\mathbb{Z}_6$ and express the corresponding submodules K of $\mathbb{Z}$ in the form $\langle d \rangle$, $d \geq 0$. Specify a generator of each H. Which elements generate $\mathbb{Z}_6$?
    Hint: Use Lemma 2.2.

    (c) List the elements in the submodule of the $\mathbb{Z}$-module $\mathbb{Z}_{21}$ generated by

    $$\text{(i)} \quad \overline{14}; \qquad \text{(ii)} \quad \overline{15}.$$

    What are the orders of $\overline{14}$ and $\overline{15}$ in $\mathbb{Z}_{21}$? What common property do the 12 elements of $\mathbb{Z}_{21}$ not in either of these submodules have?
    Hint: Use Definition 2.1.

2. (a) Calculate $\gcd\{91, 289\}$. Does $\overline{91}$ generate the $\mathbb{Z}$-module $\mathbb{Z}_{289}$? Does $\overline{51}$ generate this $\mathbb{Z}$-module?
    Hint: Use Lemma 2.7.

    (b) Use Lemma 2.7 to show that $\overline{m} \in \mathbb{Z}_n$ is a generator of the $\mathbb{Z}$-module $\mathbb{Z}_n$ if and only if $\gcd\{m, n\} = 1$.

    (c) List the elements of $\mathbb{Z}_{25}$ which are *not* generators of the $\mathbb{Z}$-module $\mathbb{Z}_{25}$. Do these elements form a submodule of $\mathbb{Z}_{25}$? How many generators does the $\mathbb{Z}$-module $\mathbb{Z}_{125}$ have?

    (d) Let p be prime. Find the number of generators in each of the following $\mathbb{Z}$-modules:

    $$\text{(i)} \quad \mathbb{Z}_p; \qquad \text{(ii)} \quad \mathbb{Z}_{p^2}; \qquad \text{(iii)} \quad \mathbb{Z}_{p^3}; \qquad \text{(iv)} \quad \mathbb{Z}_{p^l}.$$

3. (a) Show that $\overline{2} \in \mathbb{Z}_{13}$ satisfies $(\overline{2})^4 = \overline{3}$, $(\overline{2})^6 = \overline{-1}$. Deduce that $\overline{2}$ has order 12. Express each power $(\overline{2})^l$ for $1 \leq l \leq 12$ in the form $\overline{r}$ where $1 \leq r \leq 12$. Does $\overline{2}$ generate the multiplicative group $\mathbb{Z}_{13}^*$ of non-zero elements of $\mathbb{Z}_{13}$? Use Lemma 2.7 to find the elements $\overline{r}$ which generate $\mathbb{Z}_{13}^*$.

    (b) Find a generator g of the multiplicative group $\mathbb{Z}_{17}^*$ by 'trial and error'. Specify the 5 subgroups of $\mathbb{Z}_{17}^*$ (each is cyclic with generator a power of g). How many generators does $\mathbb{Z}_{17}^*$ have?

    (c) Verify that $2^8 \equiv -3 \pmod{37}$. Hence show that $\overline{2}$ generates $\mathbb{Z}_{37}^*$ (it's not enough to show that $(\overline{2})^{36} = \overline{1}$). Arrange the 9 subgroups of $\mathbb{Z}_{37}^*$ in their lattice diagram. How many generators does $\mathbb{Z}_{37}^*$ have?

(d) Find the orders of each of the elements $\bar{2}, \bar{3}, \bar{4}, \bar{5}$ of $\mathbb{Z}_{41}^*$. Find a generator of $\mathbb{Z}_{41}^*$.

4. (a) Let G be a ℤ-module and c an integer. Show that $\theta : G \to G$, given by $(g)\theta = cg$ for all $g \in G$, is ℤ-linear.

Let $\theta : G \to G$ be ℤ-linear and G cyclic with generator g_0. Show that there is an integer c as above. Let $\langle n \rangle$ be the order ideal of g_0. Show that c is unique modulo n. Show further that θ is an automorphism of G if and only if $\gcd\{c, n\} = 1$.

Deduce that the additive group ℤ has exactly 2 automorphisms.

Show that every ℤ-linear mapping $\theta : \mathbb{Z}_n \to \mathbb{Z}_n$ for $n > 0$ is of the form $(\bar{m})\theta = \bar{c}\,\bar{m}$ for all $\bar{m} \in \mathbb{Z}_n$ and some $\bar{c} \in \mathbb{Z}_n$. Show also that $\theta : \mathbb{Z}_n \cong \mathbb{Z}_n$ if and only if $\gcd\{c, n\} = 1$. How many automorphisms does the additive group $\mathbb{Z}_9$ have? Are all of these automorphisms powers of $\theta_2 : \mathbb{Z}_9 \to \mathbb{Z}_9$ defined by $(\bar{m})\theta_2 = 2\bar{m}$ for all $\bar{m} \in \mathbb{Z}_9$?

(b) Let G and G' be ℤ-modules and let $\varphi : G \to G'$ be a ℤ-linear mapping. For g_0 in G let $\langle n \rangle$ and $\langle n' \rangle$ be the order ideals of g_0 and $(g_0)\varphi$ respectively. Show that $n' | n$.

Suppose now that $G = \langle g_0 \rangle$ and let g_0' in G' have order ideal $\langle d \rangle$ where $d | n$. Show that there is a unique ℤ-linear mapping $\theta : G \to G'$ with $(g_0)\theta = g_0'$.

How many ℤ-linear mappings $\theta : \mathbb{Z} \to \mathbb{Z}_{12}$ are there? How many of these mappings are surjective?

Show that $\bar{r} \in \mathbb{Z}_n$ has order ideal $\langle n/\gcd\{r, n\} \rangle$ for $n > 0$. Show that the number of ℤ-linear mappings $\theta : \mathbb{Z}_m \to \mathbb{Z}_n$ is $\gcd\{m, n\}$. Specify explicitly the five ℤ-linear mappings $\mathbb{Z}_{10} \to \mathbb{Z}_{15}$ and the five ℤ-linear mappings $\mathbb{Z}_{15} \to \mathbb{Z}_{10}$.

(c) Let $\theta : G \to G'$ be a homomorphism Definition 2.3 where G and G' are abelian groups with zero elements 0 and $0'$ respectively. Show that $(0)\theta = 0'$ and $(-g)\theta = -(g)\theta$ for all $g \in G$.

(d) Let G, G', G'' be ℤ-modules and let $\theta : G \to G'$, $\theta' : G' \to G''$ be ℤ-linear mappings. Show that $\theta\theta' : G \to G''$ is ℤ-linear where $(m)\theta\theta' = ((m)\theta)\theta' \; \forall m \in \mathbb{Z}$. For bijective θ show that $\theta^{-1} : G' \to G$ is ℤ-linear. Deduce that the automorphisms of G are the elements of a multiplicative group $\mathrm{Aut}\, G$, the group operation being composition of mappings. Is $\mathrm{Aut}\,\mathbb{Z}_9$ cyclic? Is $\mathrm{Aut}\,\mathbb{Z}_8$ cyclic?

5. (a) Let G be a ℤ-module. Show that $H = \{2g : g \in G\}$ and $K = \{g \in G : 2g = 0\}$ are submodules of G. Find examples of G with

(i) $H \subset K$;   (ii) $H = K$;   (iii) $K \subset H$;

(iv) $H \not\subseteq K$, $K \not\subseteq H$.

Hint: Consider $G = \mathbb{Z}_n$.

(b) Show that the submodule K in (a) above has the structure of a vector space over $\mathbb{Z}_2$. Find dim K where $G = \mathbb{Z}_n$.

6. (a) Let q_1 and q_2 be rational numbers. Show that the set $\langle q_1, q_2 \rangle$ of all rationals of the form $m_1 q_1 + m_2 q_2$ $(m_1, m_2 \in \mathbb{Z})$ is a submodule of the $\mathbb{Z}$-module $(\mathbb{Q}, +)$ of all rational numbers.

(b) Using the above notation show $1/6 \in \langle 3/2, 2/3 \rangle$. Show that $\langle 1/6 \rangle = \langle 3/2, 2/3 \rangle$.

(c) Write $q_i = a_i / b_i \neq 0$ where $a_i, b_i \in \mathbb{Z}$, $\gcd\{a_i, b_i\} = 1$, $b_i > 0$ for $i = 1, 2$. Let $a_i' = a_i / \gcd\{a_1, a_2\}$, $b_i' = b_i / \gcd\{b_1, b_2\}$. Show that $\gcd\{a_1' b_2', a_2' b_1'\} = 1$ and deduce $q_0 = \gcd\{a_1, a_2\} / \operatorname{lcm}\{b_1, b_2\} \in \langle q_1, q_2 \rangle$. Conclude that q_0 generates $\langle q_1, q_2 \rangle$.

Hint: Remember $\operatorname{lcm}\{b_1, b_2\} = b_1 b_2 / \gcd\{b_1, b_2\}$ and $\gcd\{a, b\} = \gcd\{a, c\} = 1 \Rightarrow \gcd\{a, bc\} = 1$ for $a, b, c \in \mathbb{Z}$.

Is $\langle q_1, q_2, q_3 \rangle = \{m_1 q_1 + m_2 q_2 + m_3 q_3 : m_1, m_2, m_3 \in \mathbb{Z}\}$, where $q_1, q_2, q_3 \in \mathbb{Q}$, necessarily a cyclic submodule of $\mathbb{Q}$?

(d) Find a generator of $\langle 6/35, 75/56 \rangle$ and a generator of $\langle 6/35, 75/56, 8/15 \rangle$. Do either of these submodules contain $\mathbb{Z}$?

7. (a) Let H_1 and H_2 be subgroups of the additive abelian group G. Show
    (i) the intersection $H_1 \cap H_2$ is a subgroup of G,
    (ii) the sum $H_1 + H_2 = \{h_1 + h_2 : h_1 \in H_1, h_2 \in H_2\}$ is a subgroup of G,
    (iii) the union $H_1 \cup H_2$ is a subgroup of G $\Leftrightarrow$ either $H_1 \subseteq H_2$ or $H_2 \subseteq H_1$.

    Hint: Show $\Rightarrow$ by contradiction.

(b) Find generators of $H_1 \cap H_2$ and $H_1 + H_2$ in the case of $G = \mathbb{Z}$, $H_1 = \langle 30 \rangle$, $H_2 = \langle 100 \rangle$. Generalise your answer to cover the case $G = \mathbb{Z}$, $H_1 = \langle m_1 \rangle$, $H_2 = \langle m_2 \rangle$.

8. (a) Let $g_1, g_2, \ldots, g_n$ $(n \geq 3)$ be elements of an additive group G. The elements s_i of G are defined inductively by $s_1 = g_1$, $s_2 = s_1 + g_2$, $s_3 = s_2 + g_3$, $\ldots$, $s_n = s_{n-1} + g_n$ $(1 \leq i \leq n)$. Use the associative law of addition and induction to show that all ways of summing $g_1, g_2, \ldots, g_n$ in order give s_n.

Hint: Show first that each summation of $g_1, g_2, \ldots, g_n$ decomposes as $s_i + s_{n-i}'$ where s_{n-i}' is a summation of $g_{i+1}, g_{i+2}, \ldots, g_n$ for some i with $1 \leq i < n$.

Deduce the generalised associative law of addition: brackets may be omitted in any sum of n elements of G.

(b) Let $g_1, g_2, \ldots, g_n$ $(n \geq 2)$ be elements of an additive abelian group G. Use the associative and commutative laws of addition, and induction, to show that all ways of summing $g_1, g_2, \ldots, g_n$ in any order give s_n as defined in (a) above.

(c) Let g, g_1, g_2 be elements of an additive abelian group G. Use (b) above to verify laws 5 and 6 of a $\mathbb{Z}$-module namely: $m(g_1 + g_2) = mg_1 + mg_2$, $(m_1 + m_2)g = m_1g + m_2g$, $(m_1m_2)g = m_1(m_2g)$ for all integers m, m_1, m_2.

Hint: Suppose first that m, m_1, m_2 are positive.

2.2 Quotient Groups and the Direct Sum Construction

Two ways of obtaining new abelian groups from old are discussed: the *quotient group* construction and the *direct sum* construction. Particular cases of both constructions are already known to the reader, the most significant being $\mathbb{Z}_n$ which is the quotient of $\mathbb{Z}$ by its subgroup $\langle n \rangle$, that is, $\mathbb{Z}_n = \mathbb{Z}/\langle n \rangle$ for all positive integers n. Keep this familiar example in mind as the theory unfolds.

Let G be an additive abelian group having a subgroup K. We construct the quotient group G/K which can be thought of informally as G modulo K. Formally the elements of G/K are subsets of G of the type

$$K + g_0 = \{k + g_0 : k \in K\} \quad \text{for } g_0 \in G.$$

Subsets of this kind are called *cosets of K in G*. The elements of $K + g_0$ are sums $k + g_0$ where k runs through K and g_0 is a given element of G. We write $\overline{g_0} = K + g_0$ to emphasise the close analogy between cosets and congruence classes of integers. Notice that $g \in \overline{g_0}$ means $g = k + g_0$ for some $k \in K$, that is, $g - g_0 \in K$ showing that g differs from g_0 by an element of K. The condition $g - g_0 \in K$ is expressed by writing $g \equiv g_0 \pmod{K}$ and saying that g is *congruent* to g_0 modulo K. Our next lemma deals with the set-theoretic properties of cosets. Notice that each coset has as many aliases (alternative names) as it has elements!

Lemma 2.9

Let K be a subgroup of the additive abelian group G. Using the above notation $\overline{g} = \overline{g_0}$ if and only if $g \equiv g_0 \pmod{K}$. Congruence modulo K is an equivalence relation on G. Each element of G belongs to exactly one coset of K in G.

Proof

The subgroup K contains the zero element 0 of G. Hence $g = 0 + g \in \overline{g}$, showing that each g in G belongs to the coset $\overline{g}$. Suppose $\overline{g} = \overline{g_0}$ which means that the sets $\overline{g}$ and $\overline{g_0}$ consist of exactly the same elements. As $g \in \overline{g}$ we see $g \in \overline{g_0}$ and so, as above, we conclude that $g \equiv g_0 \pmod{K}$.

Now suppose $g \equiv g_0 \pmod{K}$. Then $g - g_0 = k_0 \in K$. We first show $\overline{g} \subseteq \overline{g_0}$. Consider $x \in \overline{g}$. Then $x = k + g$ for $k \in K$. Hence $x = k + k_0 + g_0$ which belongs to $\overline{g_0}$ since $k + k_0 \in K$ as K is closed under addition. So we have shown $\overline{g} \subseteq \overline{g_0}$. Now K is closed under negation and so $g_0 - g = -k_0 \in K$ showing $g_0 \equiv g \pmod{K}$. Interchanging the roles of g and g_0 in the argument, we see that $\overline{g_0} \subseteq \overline{g}$. The sets $\overline{g}$ and $\overline{g_0}$ are such that each is a subset of the other, that is, $\overline{g} = \overline{g_0}$.

We now use the 'if and only if' condition $\overline{g} = \overline{g_0} \Leftrightarrow g \equiv g_0 \pmod{K}$ to prove that congruence modulo K satisfies the three laws of an equivalence relation. As $\overline{g} = \overline{g}$ we see that $g \equiv g \pmod{K}$ for all $g \in G$, that is, congruence modulo K is reflexive. Suppose $g_1 \equiv g_2 \pmod{K}$ for some $g_1, g_2 \in G$; then $\overline{g_1} = \overline{g_2}$ and so $\overline{g_2} = \overline{g_1}$ which means $g_2 \equiv g_1 \pmod{K}$, that is, congruence modulo K is symmetric. Suppose $g_1 \equiv g_2 \pmod{K}$ and $g_2 \equiv g_3 \pmod{K}$ where $g_1, g_2, g_3 \in G$; then $\overline{g_1} = \overline{g_2}$ and $\overline{g_2} = \overline{g_3}$ and so $\overline{g_1} = \overline{g_3}$ (it really is that easy!) which gives $g_1 \equiv g_3 \pmod{K}$, that is, congruence modulo K is transitive. Congruence modulo K satisfies the reflexive, symmetric and transitive laws and so is an equivalence relation on G.

The proof is finished by showing that no element g in G can belong to two different cosets of K in G. We know $g \in \overline{g}$ as $0 \in K$. Suppose $g \in \overline{g_0}$ for some $g_0 \in G$. The preliminary discussion shows $g \equiv g_0 \pmod{K}$ and hence $\overline{g} = \overline{g_0}$. So g belongs to $\overline{g}$ and to no other coset of K in G. $\square$

The cosets of K in G *partition* the set G, that is, these cosets are non-empty, non-overlapping subsets having G as their union. In other words, each element of G belongs to a *unique* coset of K in G.

For example, let $G = \mathbb{Z}$ and $K = \langle 3 \rangle$. Then $m \equiv m' \pmod{K}$ means $m \equiv m' \pmod{3}$ for $m, m' \in \mathbb{Z}$. There are three cosets of K in G, namely $\overline{0} = K + 0 = K = \{\ldots, -9, -6, -3, 0, 3, 6, 9, \ldots\}$, $\overline{1} = K + 1 = \{\ldots, -8, -5, -2, 1, 4, 7, 10, \ldots\}$, $\overline{2} = K + 2 = \{\ldots, -7, -4, -1, 2, 5, 8, 11, \ldots\}$ that is, the congruence classes of integers modulo 3 and these cosets partition $\mathbb{Z}$. So $\mathbb{Z}_3 = \{\overline{0}, \overline{1}, \overline{2}\} = G/K$ in this case.

Now suppose $G = \mathbb{Z}_{12}$ and $K = \langle \overline{4} \rangle$. There are four cosets of K in G and these are $K + \overline{0} = K = \{\overline{0}, \overline{4}, \overline{8}\}$, $K + \overline{1} = \{\overline{1}, \overline{5}, \overline{9}\}$, $K + \overline{2} = \{\overline{2}, \overline{6}, \overline{10}\}$, $K + \overline{3} = \{\overline{3}, \overline{7}, \overline{11}\}$.

These cosets partition $\mathbb{Z}_{12}$ and we will see shortly that they are the elements of a cyclic group of order 4.

The number $|G/K|$ of cosets of K in G is called *the index* of the subgroup K in its parent group G. The index is either a positive integer or infinite. Let G be a *finite* abelian group, that is, the number $|G|$ of elements in G is a positive integer. In this case $|G|$ is called *the order* of G. The index of the subgroup K in G is the positive integer $|G/K|$. Every coset $K + g_0$ has exactly $|K|$ elements. As these $|G/K|$ cosets partition G, we obtain the equation $|G| = |G/K||K|$ and so

$$|G/K| = |G|/|K|$$

on counting the elements in G coset by coset. So $|K|$ is a divisor of $|G|$, that is,

> the order $|K|$ of every subgroup K of a finite abelian group G is a divisor
> of the order $|G|$ of G

which is known as *Lagrange's theorem* for finite abelian groups.

Each element g of G generates a cyclic subgroup $\langle g \rangle$. Suppose again that G is finite. Writing $K = \langle g \rangle$, from Theorem 2.5 we see that g has finite order $|K|$, and so $|K|g = 0$. Hence $|G|g = |G/K||K|g = |G/K| \times 0 = 0$. We have proved:

> $|G|g = 0$ for all elements g of the finite abelian group G.

We call this useful fact the $|G|$-*lemma*. For instance every element g of an additive abelian group of order 27 satisfies $27g = 0$. For each prime p the multiplicative group $\mathbb{Z}_p^*$ is abelian of order $p - 1$, and so using multiplicative notation for a moment we obtain $(\bar{r})^{p-1} = \bar{1}$ for all $\bar{r} \in \mathbb{Z}_p^*$, that is, $r^{p-1} \equiv 1 \pmod{p}$ for all integers r and primes p with $\gcd\{r, p\} = 1$. On multiplying through by r we get

$$r^p \equiv r \pmod{p} \text{ for all integers } r \text{ and primes } p$$

which is known as *Fermat's 'little' theorem.*

Returning to the general case of an additive abelian group G with subgroup K, let $\eta : G \to G/K$ be the *natural mapping* defined by $(g)\eta = \bar{g}$ for all $g \in G$. So η maps each element g to the coset $\bar{g}$. Each coset of K in G is of the form $\bar{g}$ for some $g \in G$ and so η is surjective. Can addition of cosets be introduced in such a way that G/K is an abelian group and η is an additive mapping as in Definition 2.3? There is only one possible way in which this can be done, because $(g_1 + g_2)\eta = (g_1)\eta + (g_2)\eta$, that is,

$$\overline{g_1 + g_2} = \overline{g_1} + \overline{g_2} \tag{♣}$$

tells us that the sum $\bar{g}_1 + \bar{g}_2$ of cosets must be the coset containing $g_1 + g_2$. The following lemma assures us that this rule of coset addition is unambiguous: it does not depend on the particular aliases used for $\overline{g_1}$ and $\overline{g_2}$, and it does the job of turning the set G/K into an abelian group.

Lemma 2.10

Let G be an additive abelian group with subgroup K. Let g_1, g_1', g_2, g_2' be elements of G such that $g_1 \equiv g_1' \pmod{K}$, $g_2 \equiv g_2' \pmod{K}$. Then $g_1 + g_2 \equiv g_1' + g_2' \pmod{K}$. The above rule (♣) of coset addition is unambiguous and G/K, with this addition, is an abelian group.

Proof

By hypothesis there are k_1, k_2 in K with $g_1 = k_1 + g_1'$, $g_2 = k_2 + g_2'$. Adding these equations and rearranging the terms, which is allowed as G is abelian, we obtain $g_1 + g_2 = (k_1 + g_1') + (k_2 + g_2') = (k_1 + k_2) + (g_1' + g_2')$ showing that $g_1 + g_2 \equiv g_1' + g_2' \pmod{K}$ as $k_1 + k_2 \in K$. So it is legitimate to add congruences modulo K. In terms of cosets, starting with $\overline{g_1} = \overline{g_1'}$, $\overline{g_2} = \overline{g_2'}$, we have shown $\overline{g_1 + g_2} = \overline{g_1' + g_2'}$. Therefore coset addition is indeed unambiguously defined by

$$\overline{g_1} + \overline{g_2} = \overline{g_1 + g_2} \quad \text{for all } g_1, g_2 \in G$$

as the right-hand side is unchanged when the representatives of the cosets on the left-hand side are changed from g_i to g_i', $i = 1, 2$.

We now verify that coset addition satisfies the laws of an abelian group. The associative law is satisfied as

$$(\overline{g_1} + \overline{g_2}) + \overline{g_3} = \overline{(g_1 + g_2)} + \overline{g_3} = \overline{(g_1 + g_2) + g_3}$$

$$= \overline{g_1 + (g_2 + g_3)} = \overline{g_1} + \overline{(g_2 + g_3)} = \overline{g_1} + (\overline{g_2} + \overline{g_3})$$

for all $g_1, g_2, g_3 \in G$. The coset $\overline{0} = \{k + 0 : k \in K\} = \{k : k \in K\} = K$ is the zero element of G/K since $\overline{0} + \overline{g} = \overline{0 + g} = \overline{g}$ for all $g \in G$. The coset $\overline{g}$ has negative $\overline{-g}$ since $\overline{-g} + \overline{g} = \overline{-g + g} = \overline{0}$ and so $-\overline{g} = \overline{(-g)}$ for all $g \in G$. Finally, the commutative law holds in G/K since $\overline{g_1} + \overline{g_2} = \overline{g_1 + g_2} = \overline{g_2 + g_1} = \overline{g_2} + \overline{g_1}$ for all $g_1, g_2 \in G$.   □

The group G/K is called the *quotient* (or *factor*) *group* of G by K and $\eta : G \to G/K$ is called the *natural homomorphism*.

The additive group $\mathbb{R}$ of real numbers contains the subgroup $\mathbb{Z}$ of integers. A typical element of $\mathbb{R}/\mathbb{Z}$ is the coset $\overline{x} = \{\ldots, x - 2, x - 1, x, x + 1, x + 2, \ldots\}$ consisting of all those numbers which differ from the real number x by an integer. For instance $\overline{1/3} = \{\ldots, -5/3, -2/3, 1/3, 4/3, 7/3, \ldots\}$. Every x is uniquely expressible $x = \lfloor x \rfloor + r$ where $\lfloor x \rfloor$ is an integer called the *integer part* of x, and r is a real number called the *fractional part* of x with $0 \le r < 1$. For instance $\lfloor \pi \rfloor = 3$ and $\pi - \lfloor \pi \rfloor = 0.14159\ldots$. In the group $\mathbb{R}/\mathbb{Z}$ the integer part plays no role, but the fractional part is all-important as $\overline{x} = \overline{y}$ if and only if x and y have the same fractional part. So every element of $\mathbb{R}/\mathbb{Z}$ can be expressed uniquely as $\overline{r}$ where $0 \le r < 1$. Suppose $0 \le r_1, r_2 < 1$. Then coset addition in terms of fractional parts is given by

$$\overline{r_1} + \overline{r_2} = \begin{cases} \overline{r_1 + r_2} & \text{for } r_1 + r_2 < 1 \\ \overline{r_1 + r_2 - 1} & \text{for } r_1 + r_2 \ge 1. \end{cases}$$

For instance $\overline{1/2} + \overline{2/3} = \overline{1/6}$. The group $\mathbb{R}/\mathbb{Z}$ is known as the *reals mod one* and we'll see in Section 2.3 that it's isomorphic to the multiplicative group of complex numbers of modulus 1.

The reader knows already that the additive group $\mathbb{Z}_n$ is the particular case of the above construction with $G = \mathbb{Z}$, $K = \langle n \rangle$, that is, $\mathbb{Z}_n = \mathbb{Z}/\langle n \rangle$ is the standard example of a cyclic group of type C_n as defined in Definition 2.6 for $n \geq 0$. Note that $\mathbb{Z}_0 \cong \mathbb{Z}$ as the elements of $\mathbb{Z}_0$ are singletons (sets with exactly one element) $\{m\}$ for $m \in \mathbb{Z}$ and $\{m\} \to m$ is an isomorphism. At the other end of the scale the singleton $\mathbb{Z}_1 = \{\mathbb{Z}\}$ is the standard example of a trivial abelian group. Both $\mathbb{Z}$ and its subgroups $\langle n \rangle$ are examples of *free* $\mathbb{Z}$-modules, that is, $\mathbb{Z}$-modules having $\mathbb{Z}$-*bases*. This concept will be discussed in Section 2.3. For the moment let's note that $\mathbb{Z}$ has $\mathbb{Z}$-basis 1 (or -1) and $\langle n \rangle$ has $\mathbb{Z}$-basis n (or $-n$) for $n > 0$; the trivial $\mathbb{Z}$-module $\langle 0 \rangle$ has $\mathbb{Z}$-basis the empty set $\emptyset$. The equation $\mathbb{Z}_n = \mathbb{Z}/\langle n \rangle$ expresses the additive group of $\mathbb{Z}_n$ as a quotient of $\mathbb{Z}$ (which is free) by its subgroup $\langle n \rangle$ (which is also free). It turns out that all f.g. abelian groups are best thought of as quotients of free $\mathbb{Z}$-modules by free subgroups, as we'll see in Section 3.1.

We now discuss the direct sum construction. You will be used to the formula $(x_1, y_1) + (x_2, y_2) = (x_1 + x_2, y_1 + y_2)$ for the sum of vectors. Carrying out addition in this componentwise way is the distinguishing feature of a direct sum.

Let G_1 and G_2 be additive abelian groups. The elements of $G_1 \oplus G_2$ are ordered pairs (g_1, g_2) where $g_1 \in G_1$, $g_2 \in G_2$. The rule of addition in $G_1 \oplus G_2$ is

$$(g_1, g_2) + (g_1', g_2') = (g_1 + g_1', g_2 + g_2') \quad \text{for } g_1, g_1' \in G_1 \text{ and } g_2, g_2' \in G_2.$$

It is straightforward to show that $G_1 \oplus G_2$ is itself an additive abelian group, and we confidently leave this job to the reader (Exercises 2.2, Question 4(f)). Suffice it to say that $(0_1, 0_2)$ is the zero element of $G_1 \oplus G_2$ where 0_i is the zero of G_i for $i = 1, 2$ and $-(g_1, g_2) = (-g_1, -g_2)$ showing that negation, like addition, is carried out component by component. The abelian group $G_1 \oplus G_2$ is called *the external direct sum* of G_1 and G_2.

This construction, which is easier to grasp than the quotient group construction, produces 'at a stroke' a vast number of abelian groups. For instance

$$\mathbb{Z}_2 \oplus \mathbb{Z}_2, \qquad \mathbb{Z}_3 \oplus \mathbb{Z}, \qquad (\mathbb{Z}_1 \oplus \mathbb{Z}_5) \oplus \mathbb{Z}, \qquad (\mathbb{Z}_2 \oplus \mathbb{Z}_4) \oplus (\mathbb{Z}_4 \oplus \mathbb{Z}_7)$$

these groups being built up using cyclic groups and the direct sum construction. It turns out that all groups constructed in this way are abelian and finitely generated. Can *every* finitely generated abelian group be built up in this way? We will see in the next chapter that the answer is: Yes! What is more, the Smith normal form will help us decide which pairs of these groups are isomorphic.

We now look in detail at the group $G = \mathbb{Z}_2 \oplus \mathbb{Z}_2$. Write $0 = (\overline{0}, \overline{0})$, $u = (\overline{1}, \overline{0})$,

$v = (\overline{0}, \overline{1})$, $w = (\overline{1}, \overline{1})$. Then $G = \{0, u, v, w\}$ has addition table

+	0	u	v	w
0	0	u	v	w
u	u	0	w	v
v	v	w	0	u
w	w	v	u	0

Notice that the sum of any two of u, v, w is the other one, $v + w = u$ etc., and each element is equal to its negative, as $v + v = 0$ means $v = -v$ for instance. This group has five subgroups namely $\langle 0 \rangle$, $\langle u \rangle$, $\langle v \rangle$, $\langle w \rangle$ and G itself. Now G is not cyclic and we write $G = \langle u, v \rangle$ meaning that each element of G is of the form $lu + mv$ for some integers l and m. In fact G is the smallest non-cyclic group. Any group isomorphic to G is called a *Klein 4-group* after the German mathematician Felix Klein. Being the direct sum of two cyclic groups of order 2, G is said to be of *isomorphism type* $C_2 \oplus C_2$ (see Definition 2.13). You may have already met this group in the context of vector spaces because $\mathbb{Z}_2 \oplus \mathbb{Z}_2$ is the standard example of a 2-dimensional vector space over the field $\mathbb{Z}_2$. The elements $0, u, v, w$ of G are the vectors and the elements $\overline{0}, \overline{1}$ of $\mathbb{Z}_2$ are the scalars. The ordered pair u, v is a basis of this vector space. The subgroups of G are precisely the subspaces and the automorphisms of G are precisely the invertible linear mappings θ of this vector space. For example $\theta : G \cong G$ such that $(0)\theta = 0$, $(u)\theta = v$, $(v)\theta = w$, $(w)\theta = u$ is an automorphism of G.

Using the approach outlined in the introduction, we now give the reader a glimpse ahead to Chapter 3. Just as the natural mapping $\eta : \mathbb{Z} \to \mathbb{Z}_2$ encapsulates the relationship between $\mathbb{Z}$ and $\mathbb{Z}_2$, so the $\mathbb{Z}$-linear mapping $\theta : \mathbb{Z} \oplus \mathbb{Z} \to G$, defined by $(l, m)\theta = lu + mv$ for all $l, m \in \mathbb{Z}$, tells us all there is to know about $G = \mathbb{Z}_2 \oplus \mathbb{Z}_2$ in terms of the more tractable module $\mathbb{Z} \oplus \mathbb{Z}$; in fact $\mathbb{Z} \oplus \mathbb{Z}$ is a free $\mathbb{Z}$-module of *rank 2* because $e_1 = (1, 0)$, $e_2 = (0, 1)$ is a $\mathbb{Z}$-basis having two 'vectors' (the term *rank* rather than *dimension* is used in this context). Note that $(e_1)\theta = 1u + 0v = u$ and similarly $(e_2)\theta = v$, $(e_1 + e_2)\theta = w$. So θ is surjective, that is, $\operatorname{im}\theta = G$. Which pairs (l, m) of integers belong to the kernel of θ? In other words, which pairs (l, m) of integers are mapped by θ to the zero element of G? The answer is: l and m are both even, because this is the condition for the equation $lu + mv = 0$ to be true. So $\ker\theta = \langle 2e_1, 2e_2 \rangle$, that is, $\ker\theta$ consists of all integer linear combinations of $2e_1 = (2, 0)$ and $2e_2 = (0, 2)$. In fact $2e_1, 2e_2$ is a $\mathbb{Z}$-*basis* of $\ker\theta$ which is therefore a free subgroup of $\mathbb{Z} \oplus \mathbb{Z}$. Notice $2 = \operatorname{rank}\ker\theta$. The $\mathbb{Z}$-bases of $\mathbb{Z} \oplus \mathbb{Z}$ and $\ker\theta$ are related by

$$D = \begin{pmatrix} 2 & 0 \\ 0 & 2 \end{pmatrix}.$$

We will see in Section 3.1 that f.g. abelian groups are not usually as prettily presented as this one; here the matrix D is already in Smith normal form. There are four cosets of $K = \ker \theta$ in $\mathbb{Z} \oplus \mathbb{Z}$ depending on the parity of the integers l and m, that is,

$$(\mathbb{Z} \oplus \mathbb{Z})/K = \{K, K + e_1, K + e_2, K + e_1 + e_2\}$$

For instance, the elements of $K + e_1$ are the pairs (l, m) of integers with l odd, m even. These cosets correspond, using θ, to the elements $0, u, v, w$ respectively of $\operatorname{im} \theta = G$, that is, $\tilde{\theta} : (\mathbb{Z} \oplus \mathbb{Z})/\ker \theta \cong G$ where

$$(K)\tilde{\theta} = (0)\theta = 0, \qquad (K + e_1)\tilde{\theta} = (e_1)\theta = u, \qquad (K + e_2)\tilde{\theta} = (e_2)\theta = v,$$

$$(K + e_1 + e_2)\tilde{\theta} = (e_1 + e_2)\theta = w.$$

Kernels and images are defined at the start of Section 2.3. The isomorphism $\tilde{\theta}$, which is a particular case of Theorem 2.16, shows that the Klein 4-group is isomorphic to $(\mathbb{Z} \oplus \mathbb{Z})/\ker \theta$, a quotient of two free $\mathbb{Z}$-modules. The point is: every f.g. abelian group can be analysed in this way as we'll see in Theorem 3.4.

Frequently occurring examples of the direct sum construction are provided by the *Chinese remainder theorem*. This theorem which we now discuss plays an important role in the decomposition of rings and abelian groups. Let $\bar{r}_n$ denote the congruence class of the integer r in $\mathbb{Z}_n$ for all positive integers n. Let m and n be given positive integers and consider the mapping

$$\alpha : \mathbb{Z}_{mn} \to \mathbb{Z}_m \oplus \mathbb{Z}_n \quad \text{defined by } (\bar{r}_{mn})\alpha = (\bar{r}_m, \bar{r}_n) \quad \text{for all } \bar{r}_{mn} \in \mathbb{Z}_{mn}.$$

For instance with $m = 5$, $n = 7$ and $r = 24$ we have $(\overline{24})\alpha = (\bar{4}, \bar{3})$ since $24 \equiv 4 \pmod 5$ and $24 \equiv 3 \pmod 7$. Then α is unambiguously defined and respects addition since

$$(\bar{s}_{mn} + \bar{t}_{mn})\alpha = (\overline{(s+t)}_{mn})\alpha = (\overline{(s+t)}_m, \overline{(s+t)}_n) = (\bar{s}_m + \bar{t}_m, \bar{s}_n + \bar{t}_n)$$

$$= (\bar{s}_m, \bar{s}_n) + (\bar{t}_m, \bar{t}_n) = (\bar{s}_{mn})\alpha + (\bar{t}_{mn})\alpha \quad \text{for all integers } s, t.$$

The group $\mathbb{Z}_m \oplus \mathbb{Z}_n$ becomes a commutative ring (the direct sum of the rings $\mathbb{Z}_m$ and $\mathbb{Z}_n$) provided multiplication is carried out, like addition, component by component, that is, $(x, y)(x', y') = (xx', yy')$ for all $x, x' \in \mathbb{Z}_m$ and $y, y' \in \mathbb{Z}_n$. Replacing each '$+$' in the above equations by the product symbol '$\cdot$' produces

$$(\bar{s}_{mn} \cdot \bar{t}_{mn})\alpha = (\bar{s}_{mn})\alpha \cdot (\bar{t}_{mn})\alpha \quad \text{for all integers } s, t$$

showing that α respects multiplication. Also $\bar{1}_{mn}$ is the 1-element of $\mathbb{Z}_{mn}$ and $(\bar{1}_{mn})\alpha = (\bar{1}_m, \bar{1}_n)$ is the 1-element of $\mathbb{Z}_m \oplus \mathbb{Z}_n$. Therefore

$$\alpha \text{ is a } \textit{ring homomorphism}$$

meaning that α is a mapping of rings which respects addition, multiplication and 1-elements.

Theorem 2.11 (The Chinese remainder theorem)

Let m and n be positive integers with $\gcd\{m, n\} = 1$. Then $\alpha : \mathbb{Z}_{mn} \cong \mathbb{Z}_m \oplus \mathbb{Z}_n$ is a ring isomorphism.

Proof

Using the above theory, α is a ring homomorphism and so it is enough to show that α is bijective. As $\mathbb{Z}_{mn}$ and $\mathbb{Z}_m \oplus \mathbb{Z}_n$ both contain exactly mn elements it is enough to show that α is surjective. Consider a typical element $(\bar{s}_m, \bar{t}_n)$ of $\mathbb{Z}_m \oplus \mathbb{Z}_n$. We may assume $0 \leq s < m$ and $0 \leq t < n$. Can an integer r be found which leaves remainder s on division by m and remainder t on division by n? (Special cases of this problem were solved in ancient China – hence the name of the theorem.) The answer is: Yes! Let $r = atm + bsn$ where a, b are integers with $am + bn = 1$. Then $r \equiv bsn \pmod{m}$ and $bsn = s - sam \equiv s \pmod{m}$. So $r \equiv s \pmod{m}$. Similarly $r \equiv t \pmod{n}$ and so r leaves remainders s, t on division by m, n respectively. Therefore $(\bar{r}_{mn})\alpha = (\bar{r}_m, \bar{r}_n) = (\bar{s}_m, \bar{t}_n)$ showing that α is indeed surjective. $\qquad\square$

Let R be a ring with 1-element e. An element u of R is *a unit (invertible element) of R* if there is an element v of R with $uv = e = vu$. It is straightforward to verify that the product uu' of units of R is itself a unit of R, and together with this product the set of units of R is a multiplicative group $U(R)$. Note that $U(F) = F^*$ for every field F, as every non-zero element of F is a unit of F. The groups $U(\mathbb{Z}_n)$ are studied in Section 3.3. We now use Theorem 2.11 to determine the order $|U(\mathbb{Z}_n)|$ of $U(\mathbb{Z}_n)$ in terms of the prime factorisation of the positive integer n. The reader will know that $\bar{r}$ is a unit of $\mathbb{Z}_n$ if and only if $\gcd\{r, n\} = 1$. It is convenient to assume (as we may) that $1 \leq r \leq n$. The reader may also have met the *Euler ϕ-function* defined by

$$\phi(n) = |\{r : 1 \leq r \leq n, \gcd\{r, n\} = 1|$$

that is, $\phi(n)$ is the *number* of integers r between 1 and n which are coprime to the positive integer n, and so $\phi(n) = |U(\mathbb{Z}_n)|$. One sees directly that $\phi(1) = 1$ and $\phi(p) = p - 1$ for all primes p. The closed interval $[1, p^l]$ contains p^l integers and the p^{l-1} multiples of p in this interval are exactly those which are *not* coprime to p since $\gcd\{r, p^l\} \neq 1 \Leftrightarrow p|r$; hence $\phi(p^l) = p^l - p^{l-1}$. In particular $\phi(7) = 7 - 1 = 6$, $\phi(8) = 8 - 4 = 4$, $\phi(9) = 9 - 3 = 6$.

Corollary 2.12

The Euler ϕ-function is multiplicative, that is, $\phi(mn) = \phi(m)\phi(n)$ where m, n are coprime positive integers. Let $n = p_1^{l_1} p_2^{l_2} \cdots p_k^{l_k}$ where $p_1, p_2, \ldots, p_k$ are different primes. Then $\phi(n) = (p_1^{l_1} - p_1^{l_1 - 1})(p_2^{l_2} - p_2^{l_2 - 1}) \ldots (p_k^{l_k} - p_k^{l_k - 1})$.

Proof

As multiplication in $\mathbb{Z}_m \oplus \mathbb{Z}_n$ is carried out componentwise, $(\bar{s}_m, \bar{t}_n)$ is a unit of the ring $\mathbb{Z}_m \oplus \mathbb{Z}_n$ if and only if $\bar{s}_m$ is a unit of the ring $\mathbb{Z}_m$ and $\bar{t}_n$ is a unit of the ring $\mathbb{Z}_n$. Therefore $U(\mathbb{Z}_m \oplus \mathbb{Z}_n) = U(\mathbb{Z}_m) \times U(\mathbb{Z}_n)$ where $\times$ denotes the Cartesian product (see Exercises 2.3, Question 4(d)). Comparing sizes of these sets gives $|U(\mathbb{Z}_m \oplus \mathbb{Z}_n)| = |U(\mathbb{Z}_m)||U(\mathbb{Z}_n)| = \phi(m)\phi(n)$. Suppose $\gcd\{m, n\} = 1$. As isomorphic rings have isomorphic groups of units, or specifically in our case, $\bar{r}_{mn}$ is a unit of $\mathbb{Z}_{mn}$ if and only if $(\bar{r}_{mn})\alpha$ is a unit of $\mathbb{Z}_m \oplus \mathbb{Z}_n$ by Theorem 2.11, we deduce $\phi(mn) = |U(\mathbb{Z}_{mn})| = |U(\mathbb{Z}_m \oplus \mathbb{Z}_n)|$. So $\phi(mn) = \phi(m)\phi(n)$ where $\gcd\{m, n\} = 1$.

We use induction on the number k of distinct prime divisors of n. As $\phi(1) = 1$ we take $k > 0$ and assume $\phi(p_2^{l_2} \cdots p_k^{l_k}) = (p_2^{l_2} - p_2^{l_2-1}) \cdots (p_k^{l_k} - p_k^{l_k-1})$. As $\gcd\{p_1^{l_1}, p_2^{l_2}m, \ldots, p_k^{l_k}\} = 1$, the multiplicative property of ϕ gives

$$\phi(n) = \phi(p_1^{l_1}(p_2^{l_2} \cdots p_k^{l_k})) = \phi(p_1^{l_1})\phi(p_2^{l_2} \cdots p_k^{l_k})$$

$$= (p_1^{l_1} - p_1^{l_1-1})(p_2^{l_2} - p_2^{l_2-1}) \cdots (p_k^{l_k} - p_k^{l_k-1})$$

as in Corollary 2.12. By induction, the formula for $\phi(n)$ is as stated. $\qquad\square$

For example $\phi(500) = \phi(2^2 5^3) = \phi(2^2)\phi(5^3) = (2^2 - 2)(5^3 - 5^2) = 200$.

Let n be a positive integer. Which elements $\bar{r} \in \mathbb{Z}_n$ satisfy $\langle \bar{r} \rangle = \mathbb{Z}_n$? In other words, which elements of the additive abelian group $\mathbb{Z}_n$ have order n? We may assume $1 \leq r \leq n$. As $\bar{1}_n$ has order n and $\bar{r} = r\bar{1}_n$, by Lemma 2.7 we see that $\bar{r}$ has order $n/\gcd\{r, n\}$. So $\bar{r}$ has order n if and only if $\gcd\{r, n\} = 1$. Hence

each finite cyclic group of order n has $\phi(n)$ generators.

For instance $\mathbb{Z}_{10}$ has $\phi(10) = (2 - 1)(5 - 1) = 4$ generators and $\mathbb{Z}_{10} = \langle \bar{1} \rangle = \langle \bar{3} \rangle = \langle \bar{7} \rangle = \langle \bar{9} \rangle$.

The direct sum construction can be extended to any finite number t of $\mathbb{Z}$-modules. Let $G_1, G_2, \ldots, G_t$ be $\mathbb{Z}$-modules. Their *external direct sum* $G_1 \oplus G_2 \oplus \cdots \oplus G_t$ is the $\mathbb{Z}$-module having all ordered t-tuples $(g_1, g_2, \ldots, g_t)$ where $g_i \in G_i$ $(1 \leq i \leq t)$ as its elements, addition and integer multiplication being carried out componentwise. So $(g_1, g_2, \ldots, g_t) + (g_1', g_2', \ldots, g_t') = (g_1 + g_1', g_2 + g_2', \ldots, g_t + g_t')$ for $g_i, g_i' \in G_i$ and $m(g_1, g_2, \ldots, g_t) = (mg_1, mg_2, \ldots, mg_t)$ for $m \in \mathbb{Z}$, $g_i \in G_i$ where $1 \leq i \leq t$. We now generalise Definition 2.6.

Definition 2.13

Suppose the $\mathbb{Z}$-module G_i is cyclic of isomorphism type C_{d_i} for $1 \leq i \leq t$. Any $\mathbb{Z}$-module G isomorphic to $G_1 \oplus G_2 \oplus \cdots \oplus G_t$ is said to be of *isomorphism type* $C_{d_1} \oplus C_{d_2} \oplus \cdots \oplus C_{d_t}$.

For instance the additive group G of the ring $\mathbb{Z}_2 \oplus \mathbb{Z}_3$ has isomorphism type $C_2 \oplus C_3$. By Theorem 2.11 we know that G is cyclic of isomorphism type C_6 and so we write $C_2 \oplus C_3 = C_6$ since the isomorphism class of $\mathbb{Z}$-modules of type $C_2 \oplus C_3$ coincides with the isomorphism class of $\mathbb{Z}$-modules of type C_6. Also $C_2 \oplus C_3 = C_3 \oplus C_2$ as $G_1 \oplus G_2 \cong G_2 \oplus G_1$ for all $\mathbb{Z}$-modules G_1 and G_2. More generally for positive integers m and n we have

$$C_m \oplus C_n = C_n \oplus C_m \quad \text{and} \quad C_m \oplus C_n = C_{mn} \quad \text{in case } \gcd\{m, n\} = 1$$

by Theorem 2.11. We will use these rules in Chapter 3 to manipulate the isomorphism type symbols and show Theorem 3.4 that every finitely generated $\mathbb{Z}$-module G is of isomorphism type $C_{d_1} \oplus C_{d_2} \oplus \cdots \oplus C_{d_t}$ where the non-negative integers d_i are successive divisors, that is, $d_i | d_{i+1}$ for $1 \leq i < t$.

Next we generalise the Chinese remainder theorem. Using Theorem 2.11 and induction on k we obtain the ring isomorphism

$$\alpha : \mathbb{Z}_n \cong \mathbb{Z}_{q_1} \oplus \mathbb{Z}_{q_2} \oplus \cdots \oplus \mathbb{Z}_{q_k} \quad \text{given by } (r_n)\alpha = (r_{q_1}, r_{q_2}, \ldots, r_{q_k})$$

for all $r \in \mathbb{Z}$, where $n = q_1 q_2 \cdots q_k$ and $q_1, q_2, \ldots, q_k$ are powers of distinct primes $p_1, p_2, \ldots, p_k$. For example consider $\alpha : \mathbb{Z}_{60} \cong \mathbb{Z}_4 \oplus \mathbb{Z}_3 \oplus \mathbb{Z}_5$. As 11 leaves remainders 3, 2, 1 on division by 4, 3, 5 respectively we see (suppressing the subscripts) that $(\overline{11})\alpha = (\overline{3}, \overline{2}, \overline{1})$. Doubling gives $(\overline{22})\alpha = (\overline{6}, \overline{4}, \overline{2}) = (\overline{2}, \overline{1}, \overline{2})$ and negating gives $(\overline{49})\alpha = (-\overline{11})\alpha = (-\overline{3}, -\overline{2}, -1) = (\overline{1}, \overline{1}, \overline{4})$. Squaring gives $((\overline{22})^2)\alpha = (\overline{2}, \overline{1}, \overline{2})^2 = (\overline{2}^2, \overline{1}^2, \overline{2}^2) = (\overline{0}, \overline{1}, \overline{4}) = (\overline{4})\alpha$ and so $(\overline{22})^2 = \overline{4}$ in $\mathbb{Z}_{60}$. Similarly $((\overline{49})^2\alpha = (\overline{1}, \overline{1}, \overline{1}) = (\overline{1})\alpha$ showing that $\overline{49}$ is a self-inverse element of $\mathbb{Z}_{60}$ as $(\overline{49})^2 = \overline{1}$, that is, $(\overline{49})^{-1} = \overline{49}$ in $\mathbb{Z}_{60}$. It is an amazing fact that the 60 triples in $\mathbb{Z}_4 \oplus \mathbb{Z}_3 \oplus \mathbb{Z}_5$ add and multiply in exactly the same way as the 60 elements of $\mathbb{Z}_{60}$.

We now look at the direct sum construction from the opposite point of view. Under which circumstances is the $\mathbb{Z}$-module G isomorphic to a direct sum $G_1 \oplus G_2$ of $\mathbb{Z}$-modules? We will see shortly that the submodules of G hold the answer to this question. Let 0_i denote the zero element of the $\mathbb{Z}$-module G_i for $i = 1, 2$. Then $G_1 \oplus G_2$ has submodules $G'_1 = \{(g_1, 0_2) : g_1 \in G_1\}$ and $G'_2 = \{(0_1, g_2) : g_2 \in G_2\}$ which are isomorphic to G_1 and G_2 respectively. Also each element (g_1, g_2) of $G_1 \oplus G_2$ is *uniquely* expressible as a sum $g'_1 + g'_2$, where $g'_1 \in G'_1$, $g'_2 \in G'_2$, since $(g_1, g_2) = g'_1 + g'_2$ if and only if $g'_1 = (g_1, 0_2)$, $g'_2 = (0_1, g_2)$. Consider an isomorphism $\alpha : G \cong G_1 \oplus G_2$ and let $H_i = \{h_i \in G : (h_i)\alpha \in G'_i\}$ for $i = 1, 2$. So H_1 and H_2 are the submodules of G which correspond under α to G'_1 and G'_2. We write

$$G = H_1 \oplus H_2 \text{ and call } G \text{ the } \textit{internal direct sum} \text{ of its submodules } H_1 \text{ and } H_2$$

as each element g of G is *uniquely* expressible as $g = h_1 + h_2$ where $h_1 \in H_1$ and $h_2 \in H_2$, since α is an isomorphism.

For example $\alpha : \mathbb{Z}_6 \rightarrow \mathbb{Z}_2 \dot{\oplus} \mathbb{Z}_3$ as above leads to the submodules $H_1 = \{\bar{0}, \bar{3}\}$ and $H_2 = \{\bar{0}, \bar{2}, \bar{4}\}$. The six elements of $\mathbb{Z}_6$ are

$$\bar{0} = \bar{0} + \bar{0}, \qquad \bar{1} = \bar{3} + \bar{4}, \qquad \bar{2} = \bar{0} + \bar{2}, \qquad \bar{3} = \bar{3} + \bar{0}, \qquad \bar{4} = \bar{0} + \bar{4}, \qquad \bar{5} = \bar{3} + \bar{2}$$

and they coincide as shown with the six elements $h_1 + h_2$ where $h_1 \in H_1$, $h_2 \in H_2$. So $\mathbb{Z}_6 = H_1 \oplus H_2$ is the internal direct sum of its submodules H_1 and H_2. Of course it is equally true that $\mathbb{Z}_6 = H_2 \oplus H_1$. More generally the order in which the H_i (the *summands*) appear in any internal direct sum is not important. The Klein 4-group $G = \{0, u, v, w\}$ can be *decomposed* (expressed as an internal direct sum) as

$$G = \langle u \rangle \oplus \langle v \rangle = \langle v \rangle \oplus \langle w \rangle = \langle w \rangle \oplus \langle u \rangle$$

which tells us (three times!) that G, being the internal direct sum of two cyclic subgroups of order 2, has isomorphism type $C_2 \oplus C_2$.

The argument of the paragraph above can be extended. Let G be a $\mathbb{Z}$-module with submodules $H_1, H_2, \ldots, H_t$ such that for each element g in G there are *unique* elements h_i in H_i $(1 \leq i \leq t)$ with $g = h_1 + h_2 + \cdots + h_t$. It is straightforward to check that $\alpha : G \cong H_1 \oplus H_2 \oplus \cdots \oplus H_t$, defined by $(g)\alpha = (h_1, h_2, \ldots, h_t)$, is an isomorphism; so G is isomorphic to the external direct sum of the $\mathbb{Z}$-modules $H_1, H_2, \ldots, H_t$. Generalising the above paragraph it is usual to write $G = H_1 \oplus H_2 \oplus \cdots \oplus H_t$ and call G the *internal direct sum* of its submodules $H_1, H_2, \ldots, H_t$.

Confused? We've shown that the internal direct sum of the H_i, when it exists, is isomorphic to the external direct sum of the H_i. Nevertheless we'll usually tell the reader, when it occurs in the theory ahead, which version of direct sum we have in mind.

As we have already seen $\mathbb{Z}_6 = \langle \bar{3} \rangle \oplus \langle \bar{4} \rangle$ as $\mathbb{Z}_6$ is the internal direct sum of $H_1 = \langle \bar{3} \rangle$ and $H_2 = \langle \bar{4} \rangle$; note that $(\bar{3})\alpha = (\bar{1}, \bar{0})$ and $(\bar{4})\alpha = (\bar{0}, \bar{1})$. In the same way let us look at $\alpha : \mathbb{Z}_{60} \cong \mathbb{Z}_4 \oplus \mathbb{Z}_3 \oplus \mathbb{Z}_5$. How can we quickly find $\bar{r}$ in $\mathbb{Z}_{60}$ with $(\bar{r})\alpha = (\bar{1}, \bar{0}, \bar{0})$ and $1 \leq r \leq 60$? As r is divisible by 3 and 5 there are only four possibilities: 15, 30, 45 and 60. So $r = 45$ as $r \equiv 1 \pmod{4}$. The reader can check that $(\overline{40})\alpha = (\bar{0}, \bar{1}, \bar{0})$, $(\overline{36})\alpha = (\bar{0}, \bar{0}, \bar{1})$. So $\mathbb{Z}_{60} = \langle \overline{45} \rangle \oplus \langle \overline{40} \rangle \oplus \langle \overline{36} \rangle$ shows how $\mathbb{Z}_{60}$ decomposes as an internal direct sum.

Let $H_1, H_2, \ldots, H_t$ be submodules of the $\mathbb{Z}$-module G. It is straightforward to verify that their *sum*

$$H_1 + H_2 + \cdots + H_t = \{h_1 + h_2 + \cdots + h_t : h_i \in H_i \quad \text{for all } 1 \leq i \leq t\}$$

is a submodule of G. For example take $G = \mathbb{Z}_{60}$, $H_1 = \langle \bar{6} \rangle$, $H_2 = \langle \overline{10} \rangle$, $H_3 = \langle \overline{15} \rangle$. As $\bar{1} = \bar{6} + \overline{10} - \overline{15}$ and so $\bar{r} = \overline{6r} + \overline{10r} + \overline{(-15r)} \in H_1 + H_2 + H_3$ for $\bar{r} \in \mathbb{Z}_{60}$, we see $G = H_1 + H_2 + H_3$. However $\bar{1} = \bar{6} - 2 \times \overline{10} + \overline{15}$ showing that $\bar{1}$ can be expressed in at least two ways as a sum $h_1 + h_2 + h_3$ with $h_i \in H_i$. Conclusion: G is *not* the internal direct sum of H_1, H_2 and H_3 as there's no such thing! The uniqueness condition of the internal direct sum is violated. The next lemma tells us the best way of checking whether or not a sum of submodules is direct.

Definition 2.14

The submodules H_i ($1 \leq i \leq t$) of the $\mathbb{Z}$-module G are called *independent* if the equation $h_1 + h_2 + \cdots + h_t = 0$, where $h_i \in H_i$ for all i with $1 \leq i \leq t$, holds only in the case $h_1 = h_2 = \cdots = h_t = 0$.

The reader should note the similarity between Definition 2.14 and linear independence of vectors. We show next that the internal direct sum of independent submodules always exists.

Lemma 2.15

Let $H_1, H_2, \ldots, H_t$ be independent submodules of the $\mathbb{Z}$-module G such that $G = H_1 + H_2 + \cdots + H_t$. Then $G = H_1 \oplus H_2 \oplus \cdots \oplus H_t$.

Proof

Consider $g \in G$. As $G = H_1 + H_2 + \cdots + H_t$ there are $h_i \in H_i$ ($1 \leq i \leq t$) with $g = h_1 + h_2 + \cdots + h_t$. Suppose $g = h'_1 + h'_2 + \cdots + h'_t$ where $h'_i \in H_i$ ($1 \leq i \leq t$). Subtracting produces $0 = g - g = (h_1 - h'_1) + (h_2 - h'_2) + \cdots + (h_t - h'_t)$. As $h_i - h'_i \in H_i$ we deduce $h_i - h'_i = 0$ ($1 \leq i \leq t$) using the independence of $H_1, H_2, \ldots, H_t$. Hence $h_i = h'_i$ for $1 \leq i \leq t$ showing that g is uniquely expressible as a sum of elements, one from each H_i. Therefore $G = H_1 \oplus H_2 \oplus \cdots \oplus H_t$. $\square$

Remember that the external direct sum $G_1 \oplus G_2$ makes sense for all $\mathbb{Z}$-modules G_1 and G_2. But the internal direct sum of submodules exists only in the special case of *independent* submodules detailed above. Nevertheless we shall see in Chapter 3 that this special case frequently occurs.

EXERCISES 2.2

1. (a) Let $G = \mathbb{Z}_8$ and $K = \{\overline{0}, \overline{4}\}$. List the 4 cosets of K in G. Show that G/K is cyclic and state its isomorphism type.
   (b) Let $G = \mathbb{Z}_{12}$ and $K = \{\overline{0}, \overline{3}, \overline{6}, \overline{9}\} = \langle \overline{3} \rangle$. List the 3 cosets of K in G. Show that $K + \overline{1}$ generates G/K. State the isomorphism type of G/K.

(c) Let $G = \mathbb{Z}_{24}$ and $K = \langle \overline{18} \rangle$. Show $|K| = 4$. What is the order of G/K? Show that G/K is cyclic and state its isomorphism type.

(d) Write $G = \mathbb{Z}_n$, $K = \langle \overline{m} \rangle$ where $\overline{m} \in \mathbb{Z}_n$. Use Lemma 2.7 to determine $|K|$ and $|G/K|$. Show that G/K is cyclic and state its isomorphism type.

2. (a) Let d be a positive divisor of the positive integer n. Use Lemma 2.2 to show that $\mathbb{Z}_n$ has a unique subgroup K of index d.

(b) Let d be a positive integer. Use Theorem 1.15 to show that $\mathbb{Z}$ has a unique subgroup K of index d. Does every subgroup of $\mathbb{Z}$ have finite index?

(c) Let G be a cyclic group with subgroup K. Show that G/K is cyclic.
    Hint: Use a generator of G to find a generator of G/K.

3. (a) Let $\mathbb{Q}$ denote the additive group of rational numbers. In the group $\mathbb{Q}/\mathbb{Z}$ of *rationals mod one* find the orders of $\mathbb{Z} + 1/3$ and $\mathbb{Z} + 5/8$. Show that every element of $\mathbb{Q}/\mathbb{Z}$ has finite order.

(b) Let K be a subgroup of $\mathbb{Q}/\mathbb{Z}$ and suppose $\mathbb{Z} + m/n$, $\mathbb{Z} + m'/n'$ belong to K where $\gcd\{m, n\} = 1$, $\gcd\{m', n'\} = 1$. Use integers a, b with $am + bn = 1$ to show that $\mathbb{Z} + 1/n \in K$. Show also $\mathbb{Z} + d/nn' \in K$ where $d = \gcd\{n, n'\}$. Suppose K is finite. Show that K is cyclic.
    Hint: Consider $\mathbb{Z} + 1/n \in K$ with n maximum.

(c) Let n be a positive integer. Show that $\mathbb{Q}/\mathbb{Z}$ has a unique subgroup of order n.
    Hint: Use (b) above.

(d) Let $K = \{\mathbb{Z} + 1/2^s : l, s \in \mathbb{Z}, s \geq 0\}$. Show that K is a subgroup of $\mathbb{Q}/\mathbb{Z}$ having infinite order. List the finite subgroups of K. Show that K has a unique infinite subgroup. (The group K is denoted by $\mathbb{Z}(2^\infty)$.)

4. (a) Verify that $\mathbb{Z}_3 \oplus \mathbb{Z}_4$ has generator $g = (\overline{1}_3, \overline{1}_4)$ by listing the elements $g, 2g, 3g, 4g, \ldots$ in the form $(\overline{s}, \overline{t})$ where $0 \leq s < 3$, $0 \leq t < 4$. State its isomorphism type.

(b) Find the orders of the 8 non-zero elements of $G = \mathbb{Z}_3 \oplus \mathbb{Z}_3$. Specify generators of the 4 subgroups of order 3. Express G in six ways as the internal direct sum of subgroups of order 3. State the isomorphism type of G.
    Hint: G is a vector space over $\mathbb{Z}_3$ and the subgroups are subspaces.

(c) Let the elements g_i of the $\mathbb{Z}$-module G_i have finite order n_i ($i = 1, 2$). Show that the element (g_1, g_2) of the external direct sum $G_1 \oplus G_2$ has order $l = \text{lcm}\{n_1, n_2\} = n_1 n_2 / \gcd\{n_1, n_2\}$.
    Hint: Start by showing that (g_1, g_2) has finite order n say, where $n | l$. Then show $n_i | n$.

(d) Let m and n be coprime positive integers. Use Lemma 2.7 and part (c) above to show that $(\overline{s}, \overline{t})$ in $\mathbb{Z}_m \oplus \mathbb{Z}_n$ has order mn if and only if

$\gcd\{s, m\} = 1$, $\gcd\{t, n\} = 1$. How many generators does the cyclic group $\mathbb{Z}_7 \oplus \mathbb{Z}_8$ have?

(e) Let g and h be elements of an additive abelian group having orders m and n where $\gcd\{m, n\} = 1$. Show that $g + h$ has order mn.

The additive abelian group G has a cyclic subgroup K of order m such that G/K is cyclic of order n where $\gcd\{m, n\} = 1$. Show that G is cyclic of order mn.

Hint: Let $K + h_0$ generate G/K. Deduce from Exercises 2.1, Question 4(b) that n is a divisor of the order s of h_0. Now use g and h where $\langle g \rangle = K$ and $h = (s/n)h_0$.

(f) Let G_1 and G_2 be additive abelian groups. Show that their external direct sum $G_1 \oplus G_2$ is an additive abelian group. Show that $G_1 \oplus G_2$ and $G_2 \oplus G_1$ are isomorphic.

5. (a) Find $\bar{r}$ in $\mathbb{Z}_{143}$ such that r leaves remainders 7 and 6 on division by 11 and 13 respectively.

(b) Find the 4 elements $x \in \mathbb{Z}_{143}$ satisfying $x^2 = x$.

Hint: First solve $x^2 = x$ for $x \in \mathbb{Z}_{11}$ and secondly for $x \in \mathbb{Z}_{13}$. Then use the Chinese remainder theorem.

(c) Find the 4 elements $x \in \mathbb{Z}_{143}$ satisfying $x^2 = \bar{1}$, and the 9 elements $x \in \mathbb{Z}_{143}$ satisfying $x^3 = x$.

Hint: Use the method of (b) above.

(d) How many $\mathbb{Z}$-linear mappings $\theta : \mathbb{Z}_3 \oplus \mathbb{Z}_5 \to \mathbb{Z}_{15}$ are there? How many of these mappings are (i) group isomorphisms, (ii) ring isomorphisms?

Hint: Consider $\bar{r} \in \mathbb{Z}_{15}$ where $(\bar{1}_3, \bar{1}_5)\theta = \bar{r}$.

6. (a) Let $m_1, m_2, \ldots, m_t$ be integers. Show

$$\mathbb{Z} = \langle m_1 \rangle + \langle m_2 \rangle + \cdots + \langle m_t \rangle \quad \Leftrightarrow \quad \gcd\{m_1, m_2, \ldots, m_t\} = 1.$$

Is $\mathbb{Z} = \langle 15 \rangle + \langle 36 \rangle + \langle 243 \rangle$? Is $\mathbb{Z} = \langle 15 \rangle + \langle 36 \rangle + \langle 80 \rangle$?

(b) Suppose $\mathbb{Z} = H_1 \oplus H_2$ (internal direct sum of subgroups). Use Theorem 1.15 to show that $\mathbb{Z}$ is *indecomposable*, that is, either H_1 or H_2 is trivial.

(c) Let H_1 and H_2 be submodules of the $\mathbb{Z}$-module G such that $H_1 \cap H_2 = \{0\}$. Show that H_1, H_2 are independent. More generally, let $H_1, H_2, \ldots, H_{t-1}, H_t$ be submodules of G such that $H_1, H_2, \ldots, H_{t-1}$ are independent and $(H_1 + H_2 + \cdots + H_{t-1}) \cap H_t = \{0\}$. Show that $H_1, H_2, \ldots, H_{t-1}, H_t$ are independent. What is the order of $H_1 \oplus H_2 \oplus \cdots \oplus H_t$ given that each H_i is finite?

(d) Write $G = \mathbb{Z}_3 \oplus \mathbb{Z}_9$ (external direct sum of abelian groups). Use Question 4(d) above to show that G has 18 elements of order 9 and 8 elements of order 3. Deduce that G has 3 cyclic subgroups of order 9 and

4 cyclic subgroups of order 3. (Remember that a cyclic group of order n has $\phi(n)$ generators.) Specify generators of these 7 cyclic subgroups of G. Find the number of pairs of cyclic subgroups H_1, H_2 of G with $|H_1| = 3$, $|H_2| = 9$ such that $G = H_1 \oplus H_2$.

Hint: Choose H_2 first and then H_1 with $H_1 \cap H_2 = \{0\}$.

(e) Let $H_1, H_2, \ldots, H_t$ be independent submodules of a $\mathbb{Z}$-module G and let K_i be a submodule of H_i for $1 \le i \le t$. Show that $K_1, K_2, \ldots, K_t$ are independent. Write

$$H = H_1 \oplus H_2 \oplus \cdots \oplus H_t \quad \text{and} \quad K = K_1 \oplus K_2 \oplus \cdots \oplus K_t$$

(internal direct sums). Show

$$H/K \cong (H_1/K_1) \oplus (H_2/K_2) \oplus \cdots \oplus (H_t/K_t)$$

(external direct sum).

Hint: Consider α defined by

$$(K + h)\alpha = (K_1 + h_1, K_2 + h_2, \ldots, K_t + h_t)$$

where $h = h_1 + h_2 + \cdots + h_t$, $h_i \in H_i$ for $1 \le i \le t$. Show first that α is unambiguously defined.

2.3 The First Isomorphism Theorem and Free Modules

In this section we introduce the last two topics required for our onslaught on f.g. abelian groups. First we explain how each homomorphism of abelian groups gives rise to an isomorphism; this is the first isomorphism theorem and it plays a vital role in expressing every f.g. abelian group as a quotient group $\mathbb{Z}^t/K$, both $\mathbb{Z}^t$ (the external direct sum of t copies of $\mathbb{Z}$) and its subgroup K being free $\mathbb{Z}$-modules. Secondly we discuss bases of free modules. Some of the theorems are analogous to those familiar to the reader in the context of finite-dimensional vector spaces – it's nice to know that two bases of the same free module are guaranteed to have the same number of elements (this number is called the *rank* of the free module and is analogous to *dimension* of a vector space). Also the rows of invertible $t \times t$ matrices over $\mathbb{Z}$ are, as one might expect, precisely the $\mathbb{Z}$-bases of $\mathbb{Z}^t$. So far so good, but the analogy has its shortcomings. For example only certain $\mathbb{Z}$-independent subsets of $\mathbb{Z}^t$ can be extended to $\mathbb{Z}$-bases of $\mathbb{Z}^t$ (see Exercises 1.3, Question 5(c)). Dually, there are subsets of $\mathbb{Z}^t$ which generate $\mathbb{Z}^t$ but which don't contain a $\mathbb{Z}$-basis of $\mathbb{Z}^t$; in fact $\{2, 3\}$ is such a subset of $\mathbb{Z} = \mathbb{Z}^1$ as $\langle 2, 3 \rangle = \mathbb{Z}$, $\langle 2 \rangle \ne \mathbb{Z}$, $\langle 3 \rangle \ne \mathbb{Z}$. The message is: take nothing for granted!

Let G and G' be $\mathbb{Z}$-modules and let $\theta : G \to G'$ be a $\mathbb{Z}$-linear mapping. As we've seen in previous discussions, there are two important submodules associated with θ. The first is the kernel of θ and consists of those elements of G which θ maps to the zero element $0'$ of G'. Therefore

$$\boxed{\ker \theta = \{g \in G : (g)\theta = 0'\}}$$

It is routine to show that $\ker \theta$ is a submodule of G.

The second is the image of θ and consists of those elements of G' which are images by θ of elements in G. Therefore

$$\boxed{\operatorname{im} \theta = \{(g)\theta : g \in G\}}$$

Again it is routine to show that $\operatorname{im} \theta$ is a submodule of G' (see Exercises 2.3, Question 1(a)).

Next we show how θ gives rise to an isomorphism $\tilde{\theta}$.

Theorem 2.16 (The first isomorphism theorem for $\mathbb{Z}$-modules)

Let G and G' be $\mathbb{Z}$-modules and let $\theta : G \to G'$ be a $\mathbb{Z}$-linear mapping. Write $K = \ker \theta$. Then $\tilde{\theta}$, defined by $(K + g)\tilde{\theta} = (g)\theta$ for all $g \in G$, is an isomorphism

$$\tilde{\theta} : G/K \cong \operatorname{im} \theta.$$

Proof

All the elements in the coset $K + g$ are mapped by θ to $(g)\theta$ because $(k + g)\theta = (k)\theta + (g)\theta = 0' + (g)\theta = (g)\theta$ for all $k \in K$. So the above definition of $\tilde{\theta}$ makes sense and produces the mapping $\tilde{\theta} : G/K \to \operatorname{im} \theta$. Suppose $(g_1)\theta = (g_2)\theta$ for $g_1, g_2 \in G$. Then $(g_1 - g_2)\theta = (g_1)\theta - (g_2)\theta = 0'$ showing that $g_1 - g_2 = k \in K$, that is, $g_1 = k + g_2$ and so g_1 and g_2 belong to the same coset of $\ker \theta$ in G. Therefore θ has a different effect on the elements of different cosets, in other words, $\tilde{\theta}$ is injective. As θ is additive, so also is $\tilde{\theta}$ because

$$((K + g_1) + (K + g_2))\tilde{\theta} = (K + (g_1 + g_2))\tilde{\theta} = (g_1 + g_2)\theta = (g_1)\theta + (g_2)\theta$$

$$= (K + g_1)\tilde{\theta} + (K + g_2)\tilde{\theta} \quad \text{for all } g_1, g_2 \in G.$$

Finally $\operatorname{im} \tilde{\theta} = \operatorname{im} \theta$ and so $\tilde{\theta}$ is surjective. Therefore $\tilde{\theta}$ is an isomorphism being bijective and additive. $\qquad \square$

The isomorphism $\tilde{\theta}$ is said to be *induced* by the homomorphism θ. So every homomorphism $\theta : G \to G'$ induces (gives rise to) an isomorphism $\tilde{\theta}$ as in Theorem 2.16 between the quotient group $G/\ker \theta$ and the subgroup $\operatorname{im} \theta$ of G'.

We've met particular cases of Theorem 2.16 already in our discussion of cyclic groups. Let's briefly recapitulate Theorem 2.5. Suppose that G is a cyclic group with generator g and let $\theta : \mathbb{Z} \to G$ be the $\mathbb{Z}$-linear mapping defined by $(m)\theta = mg$ for all $m \in \mathbb{Z}$. Then $G = \langle g \rangle = \mathrm{im}\,\theta$ and $K = \ker\theta = \langle n \rangle$ is the order ideal of g where the non-negative integer n is unique. Applying Theorem 2.16 we obtain the isomorphism $\tilde{\theta} : \mathbb{Z}/\langle n \rangle \cong G$ where $(\overline{m})\tilde{\theta} = (K + m)\tilde{\theta} = (m)\theta = mg$ for all $\overline{m} \in \mathbb{Z}/\langle n \rangle$. Finally C_n is the isomorphism type of G. So the isomorphism types C_n of cyclic groups G correspond bijectively to the non-negative integers n. From the classification point of view this is all there is to know about cyclic groups!

Applying Theorem 2.16 to the natural homomorphism $\eta : \mathbb{Z} \to \mathbb{Z}_n$ produces the isomorphism $\tilde{\eta} : \mathbb{Z}/\langle n \rangle \cong \mathbb{Z}_n$ given by $(\overline{m})\tilde{\eta} = \overline{m}$ for all $m \in \mathbb{Z}$ as $\ker\eta = \langle n \rangle$. So in fact $\mathbb{Z}/\langle n \rangle = \mathbb{Z}_n$ and $\tilde{\eta}$ is the identity mapping. More generally let K be a submodule of the $\mathbb{Z}$-module G. Remember that the natural homomorphism $\eta : G \to G/K$ is defined by $(g)\eta = \overline{g} = K + g$ for all $g \in G$. Also remember that the 0-element of G/K is the coset $K + 0 = K$. Therefore

$$\ker\eta = \{g \in G : (g)\eta = K\} = \{g \in G : K + g = K\} = \{g \in G : g \in K\} = K.$$

We've shown that the natural homomorphism $\eta : G \to G/K$ has K as its kernel. A typical element of G/K is $K + g = (g)\eta$ and so η is surjective, that is, $\mathrm{im}\,\eta = G/K$. We're ready to apply Theorem 2.16 to η and the outcome is something of an anti-climax because $\tilde{\eta} : G/K \cong G/K$ is nothing more than the identity mapping as $(\overline{g})\tilde{\eta} = \overline{g}$ for all $g \in G$.

The mapping $\theta : \mathbb{Z} \oplus \mathbb{Z} \to \mathbb{Z}$, defined by $(l, m)\theta = l - m$ for all $l, m \in \mathbb{Z}$, is additive and surjective. In this case $\mathrm{im}\,\theta = \mathbb{Z}$ and $\ker\theta = \{(l, m) : (l, m)\theta = l - m = 0\} = \{(l, l) : l \in \mathbb{Z}\} = \langle (1, 1) \rangle$. From Theorem 2.16 we conclude that $(\mathbb{Z} \oplus \mathbb{Z})/\langle (1, 1) \rangle$ is an infinite cyclic group as $\tilde{\theta} : (\mathbb{Z} \oplus \mathbb{Z})/\langle (1, 1) \rangle \cong \mathbb{Z}$. Each coset of $\langle (1, 1) \rangle$ in $\mathbb{Z} \oplus \mathbb{Z}$ can be expressed as $\overline{(l, 0)}$ for a unique integer l and $(\overline{(l, 0)})\tilde{\theta} = l - 0 = l$.

Isomorphisms occur between abelian groups in additive notation and abelian groups in multiplicative notation. Let $\mathbb{C}^*$ denote the multiplicative group of non-zero complex numbers and let $\theta : \mathbb{R} \to \mathbb{C}^*$ be the mapping defined by $(x)\theta = \cos 2\pi x + i \sin 2\pi x$ for all $x \in \mathbb{R}$. Therefore $(x)\theta$ is the complex number of modulus 1 and argument $2\pi x$. The reader will certainly know that multiplication of complex numbers is carried out by multiplying moduli and adding arguments and so

$$(x + x')\theta = (x)\theta \cdot (x')\theta \quad \text{for all } x, x' \in \mathbb{R}.$$

Therefore θ is a homomorphism from the additive group $\mathbb{R}$ to the multiplicative group $\mathbb{C}^*$. In this context $\ker\theta$ consists of the real numbers x which θ maps to the identity element 1 of $\mathbb{C}^*$. Now $(x)\theta = 1$ if and only if x is a whole number, that is, $\ker\theta = \mathbb{Z}$. From Theorem 2.16 we deduce that $\tilde{\theta} : \mathbb{R}/\mathbb{Z} \cong \mathrm{im}\,\theta$, and so $\mathbb{R}/\mathbb{Z}$ (the reals modulo 1) is isomorphic to the group $\mathrm{im}\,\theta$ of complex numbers of modulus 1.

Let G be an abelian group. Any group isomorphic to a quotient group G/K, where K is a subgroup of G, is called a *homomorphic image of G*. The reason for this terminology is as follows. Let $\theta : G \rightarrow G'$ be a homomorphism from G to an abelian group G'. Then $(G)\theta = \mathrm{im}\,\theta \cong G/K$ where $K = \ker\theta$ by Theorem 2.16. So every homomorphic image $(G)\theta$ of G is isomorphic to a quotient group G/K. On the other hand, every quotient group G/K is a homomorphic image of G since $G/K = (G)\eta = \mathrm{im}\,\eta$ where $\eta : G \rightarrow G/K$ is the natural homomorphism. The preceding paragraph shows that the multiplicative group of complex numbers of modulus 1 is a homomorphic image of the additive group of real numbers.

We next generalise the discussion following Lemma 2.2 by showing that every $\mathbb{Z}$-linear mapping gives rise to a bijective correspondence between two sets of submodules.

Theorem 2.17

Let G and G' be $\mathbb{Z}$-modules and let $\theta : G \rightarrow G'$ be a $\mathbb{Z}$-linear mapping. Let $\mathbb{L}$ be the set of submodules H of G with $\ker\theta \subseteq H$. Let $\mathbb{L}'$ be the set of submodules H' of G' with $H' \subseteq \mathrm{im}\,\theta$. Then $(H)\theta = \{(h)\theta : h \in H\}$ is in $\mathbb{L}'$ for all H in $\mathbb{L}$. The mapping $H \rightarrow (H)\theta$ is a bijection from $\mathbb{L}$ to $\mathbb{L}'$ and satisfies

$$H_1 \subseteq H_2 \quad \Leftrightarrow \quad (H_1)\theta \subseteq (H_2)\theta \quad \text{for all } H_1, H_2 \in \mathbb{L}.$$

Proof

For each submodule H of G it is routine to verify that $(H)\theta$ is a submodule of $\mathrm{im}\,\theta$, and so $(H)\theta$ belongs to $\mathbb{L}'$ for H in $\mathbb{L}$. For each submodule H' of $\mathrm{im}\,\theta$ let $(H')\varphi = \{h \in G : (h)\theta \in H'\}$. Again it is routine to verify that $(H')\varphi$ is a submodule of G. As the zero $0'$ of G' belongs to H' we see that $\ker\theta \subseteq (H')\varphi$, that is, $(H')\varphi$ belongs to $\mathbb{L}$ for all H' in $\mathbb{L}'$. The proof is completed by showing that the mapping $\mathbb{L} \rightarrow \mathbb{L}'$ given by $H \rightarrow (H)\theta$ for all $H \in \mathbb{L}$ and the mapping $\mathbb{L}' \rightarrow \mathbb{L}$ given by $H' \rightarrow (H')\varphi$ for all $H' \in \mathbb{L}'$ are inverses of each other. (The reader is reminded that only bijective mappings have inverses and often the best way (as here) of showing that a mapping is a bijection amounts to 'conjuring up' another mapping which turns out to be its inverse.) Notice $H \subseteq (H)\theta\varphi$ as $(h)\theta \in (H)\theta$ for all $h \in H$. Now consider $g \in (H)\theta\varphi = ((H)\theta)\varphi$. Then $(g)\theta \in (H)\theta$. So $(g)\theta = (h)\theta$ for some $h \in H$. However $g - h = k \in \ker\theta$ since $(g - h)\theta = (g)\theta - (h)\theta = 0'$. So $g = h + k \in H$ since $\ker\theta \subseteq H$ and H is closed under addition. So $(H)\theta\varphi \subseteq H$. Therefore $H = (H)\theta\varphi$ for all H in $\mathbb{L}$.

In a similar way $(H')\varphi\theta = ((H')\varphi)\theta \subseteq H'$ since $(g)\theta \in H'$ for all $g \in (H')\varphi$. Let $h' \in H'$. Then $h' = (g)\theta$ for $g \in G$ since $H' \subseteq \mathrm{im}\,\theta$. But $(g)\theta \in H'$ means $g \in (H')\varphi$ and so $h' = (g)\theta \in ((H')\varphi)\theta = (H')\varphi\theta$. We've shown $H' \subseteq (H')\varphi\theta$ and so $H' =$

$(H')\varphi\theta$ for all H' in $\mathbb{L}'$. The mapping $\mathbb{L} \to \mathbb{L}'$, in which $H \to (H)\theta$, has an inverse and so this mapping is bijective. Finally it's straightforward to show that $H_1 \subseteq H_2 \Rightarrow (H_1)\theta \subseteq (H_2)\theta$ for $H_1, H_2 \in \mathbb{L}$. Now suppose $(H_1)\theta \subseteq (H_2)\theta$ for $H_1, H_2 \in \mathbb{L}$. Then $H_1 = (H_1)\theta\varphi \subseteq (H_2)\theta\varphi = H_2$ and therefore

$$H_1 \subseteq H_2 \quad \Leftrightarrow \quad (H_1)\theta \subseteq (H_2)\theta. \qquad \qquad \square$$

Each pair of submodules H_1 and H_2 in $\mathbb{L}$ gives rise to submodules $H_1 \cap H_2$ and $H_1 + H_2$ in $\mathbb{L}$. The set $\mathbb{L}$, partially ordered by inclusion, is therefore a *lattice* as is $\mathbb{L}'$. We shall not have much to say about lattices *per se*, but it is often illuminating to draw their diagrams as below.

We return to the $\mathbb{Z}$-linear mapping $\theta : \mathbb{Z} \oplus \mathbb{Z} \to \mathbb{Z}_2 \oplus \mathbb{Z}_2$ mentioned before Theorem 2.11. So $(l, m)\theta = lu + mv = (\bar{l}, \bar{m}) \in \mathbb{Z}_2 \oplus \mathbb{Z}_2$ for all $l, m \in \mathbb{Z}$, where $u = (\bar{1}, \bar{0})$, $v = (\bar{0}, \bar{1}) \in \mathbb{Z}_2 \oplus \mathbb{Z}_2$. The Klein 4-group $\mathbb{Z}_2 \oplus \mathbb{Z}_2 = \langle u, v \rangle = \operatorname{im} \theta$ has five subgroups $\{\bar{0}\}, \langle u \rangle, \langle v \rangle, \langle u + v \rangle, \langle u, v \rangle$. These are the subgroups H' of Theorem 2.17 shown in their lattice diagram:

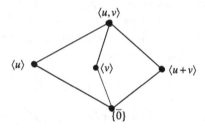

The five corresponding subgroups $H = (H')\varphi = \{(l, m) \in \mathbb{Z} \oplus \mathbb{Z} : (\bar{l}, \bar{m}) \in H'\}$ of $\mathbb{Z} \oplus \mathbb{Z}$ are $\langle 2e_1, 2e_2 \rangle, \langle e_1, 2e_2 \rangle, \langle 2e_1, e_2 \rangle, \langle e_1 + e_2, 2e_2 \rangle, \langle e_1, e_2 \rangle$ respectively and by Theorem 2.17 they fit together in the same way:

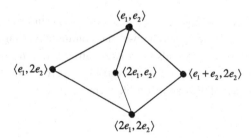

Notice that $\langle e_1 + e_2, 2e_2 \rangle = \langle e_1 + e_2, 2e_1 \rangle = \{(l, m) : \text{parity } l = \text{parity } m\}$. Each of these subgroups H has a $\mathbb{Z}$-basis as shown, that is, each element of H is uniquely expressible as an integer linear combination of the $\mathbb{Z}$-basis elements which themselves belong to H. We show in Theorem 3.1 that *all* subgroups H of $\mathbb{Z} \oplus \mathbb{Z}$ have $\mathbb{Z}$-bases. This fact allows the abstract theory to be expressed using matrices over $\mathbb{Z}$: for each subgroup H we construct a matrix A over $\mathbb{Z}$ having as its rows a set of generators of H. So the above five subgroups H of $\mathbb{Z} \oplus \mathbb{Z}$ give rise to the following five 2×2 matrices A over $\mathbb{Z}$:

$$\begin{pmatrix} 2 & 0 \\ 0 & 2 \end{pmatrix}, \quad \begin{pmatrix} 1 & 0 \\ 0 & 2 \end{pmatrix}, \quad \begin{pmatrix} 2 & 0 \\ 0 & 1 \end{pmatrix}, \quad \begin{pmatrix} 1 & 1 \\ 0 & 2 \end{pmatrix}, \quad \begin{pmatrix} 1 & 0 \\ 0 & 1 \end{pmatrix}.$$

We develop this idea in Chapter 3. For the moment notice that the invariant factors of these matrices present the isomorphism types of the corresponding quotient groups $\mathbb{Z} \oplus \mathbb{Z}/H \cong \mathbb{Z}_2 \oplus \mathbb{Z}_2/H'$ on a plate! Thus $\mathbb{Z} \oplus \mathbb{Z}/\langle 2e_1, 2e_2 \rangle \cong \mathbb{Z}/\langle 2 \rangle \oplus \mathbb{Z}/\langle 2 \rangle$ has isomorphism type $C_2 \oplus C_2$ being the direct sum of two cyclic groups of type C_2. Similarly $\mathbb{Z} \oplus \mathbb{Z}/\langle e_1, 2e_2 \rangle \cong \mathbb{Z}/\langle 1 \rangle \oplus \mathbb{Z}/\langle 2 \rangle$ has isomorphism type $C_1 \oplus C_2 = C_2$ as the trivial C_1 term can be omitted. The second, third and fourth of the above matrices are equivalent over $\mathbb{Z}$, the corresponding quotient groups being isomorphic. Also $\mathbb{Z} \oplus \mathbb{Z}/\langle e_1, e_2 \rangle \cong \mathbb{Z}/\langle 1 \rangle \oplus \mathbb{Z}/\langle 1 \rangle$ has isomorphism type $C_1 \oplus C_1 = C_1$.

Much of the abstract theory of subgroups and quotient groups developed here applies, with some minor changes, to non-abelian groups G, which are usually expressed in multiplicative notation. For such groups the quotient group G/K makes sense only when K is a *normal subgroup of* G, that is,

$$Kg = gK \quad \text{for all } g \text{ in } G.$$

The above equation means each element kg for $k \in K$, $g \in G$ can be expressed as gk' for some $k' \in K$, and conversely each element gk' for $g \in G$, $k' \in K$ can be expressed as kg for some $k \in K$. The kernel of every homomorphism $\theta : G \to G'$ between groups is a normal subgroup of G. Lagrange's theorem and the conclusions of Theorems 2.16 and 2.17 are valid for groups in general (see Exercises 2.3, Question 4).

We now discuss (finitely generated) free $\mathbb{Z}$-modules. In fact the following theory 'works' when $\mathbb{Z}$ is replaced by any non-trivial commutative ring R with 1-element. The theory of determinants extends to square matrices over R as was pointed out in Section 1.3. So a $t \times t$ matrix P over R is invertible over R if and only if $\det P \in U(R)$, that is, the determinant of P is an invertible element of R.

Lemma 2.18

Let P and Q be $t \times t$ matrices over a non-trivial commutative ring R such that $PQ = I$ where I is the $t \times t$ identity matrix over R. Then $QP = I$.

Proof

Comparing determinants in the matrix equation $PQ = I$ gives $\det PQ = \det I = 1$ the 1-element of R. Using the multiplicative property Theorem 1.18 of determinants we obtain $(\det P)(\det Q) = 1$ and so $\det P$ is an invertible element of R. The matrix $P^{-1} = (1/\det P) \operatorname{adj} P$ over R satisfies $P^{-1}P = I = PP^{-1}$. Hence $P^{-1} = P^{-1}I = P^{-1}PQ = IQ = Q$ and so $QP = P^{-1}P = I$. $\square$

Interchanging the roles of P and Q we see that $QP = I \Rightarrow PQ = I$. So from the single equation $PQ = I$ we can deduce that P and Q are both invertible over R and each is the inverse of the other: $Q = P^{-1}$ and $P = Q^{-1}$. We will need this fact in the proof of the next theorem.

The set of $t \times t$ matrices over the ring R is closed under matrix addition and matrix multiplication and is itself a ring $\mathfrak{M}_t(R)$. The invertible elements of $\mathfrak{M}_t(R)$ form the *general linear group $GL_t(R)$ of degree t over R*. We will study certain aspects of $\mathfrak{M}_t(F)$, where F is a field, in the second half of the book. Let us suppose that F is a *finite* field with q elements. You should be aware that q must be a power of a prime (Exercises 2.3, Question 5(a)). Then $|\mathfrak{M}_t(F)| = q^{t^2}$, there being q choices for each of the t^2 entries in a $t \times t$ matrix over F. How can we find the number $|GL_t(F)|$ of invertible $t \times t$ matrices P over F? The reader will know that P is invertible over F if and only if the rows of P form a basis of the vector space F^t of all t-tuples over F. What is more each basis $v_1, v_2, \ldots, v_t$ of F^t can be built up, vector by vector, ensuring linear independence at each stage as we now explain. There are $q^t - 1$ choices for v_1 (any of the $|F^t| = q^t$ vectors in F^t except the zero vector). Suppose i linearly independent vectors $v_1, v_2, \ldots, v_i$ have been chosen where $1 \le i < t$. Then $v_1, v_2, \ldots, v_i, v_{i+1}$ are linearly independent $\Leftrightarrow v_{i+1} \notin \langle v_1, v_2, \ldots, v_i \rangle$. So there are $q^t - q^i$ choices for v_{i+1} (any of the $|F^t| = q^t$ vectors in F^t except for the $q^i = |\langle v_1, v_2, \ldots, v_i \rangle|$ vectors in $\langle v_1, v_2, \ldots, v_i \rangle$). Hence

$$|GL_t(F)| = (q^t - 1)(q^t - q)(q^t - q^2) \cdots (q^t - q^{t-1})$$

there being $q^t - q^i$ remaining choices for row $i + 1$ of a matrix P in $GL_t(F)$, the previous i rows of P having already been chosen. In particular the number of 3×3 matrices over $\mathbb{Z}_2$ is $2^9 = 512$ and $(2^3 - 1)(2^3 - 2)(2^3 - 4) = 7 \times 6 \times 4 = 168$ of these are invertible. So $|\mathfrak{M}_3(\mathbb{Z}_2)| = 512$ and $|GL_3(\mathbb{Z}_2)| = 168$.

The Chinese remainder theorem generalises to decompose $\mathfrak{M}_t(\mathbb{Z}_{mn})$ where $\gcd\{m, n\} = 1$ using the ring isomorphism $\alpha : \mathbb{Z}_{mn} \cong \mathbb{Z}_m \oplus \mathbb{Z}_n$ of Theorem 2.11 as follows: let $\bar{r}_{mn}$ denote the (i, j)-entry in the $t \times t$ matrix A over $\mathbb{Z}_{mn}$ for $1 \le i, j \le t$. Write $(A)\alpha = (B, C)$ where B is the $t \times t$ matrix over $\mathbb{Z}_m$ with (i, j)-entry $\bar{r}_m$ and C is the $t \times t$ matrix over $\mathbb{Z}_n$ with (i, j)-entry $\bar{r}_n$, that is, B and C are obtained by reducing each entry in A modulo m and modulo n respectively. Then

$$\alpha : \mathfrak{M}_t(\mathbb{Z}_{mn}) \cong \mathfrak{M}_t(\mathbb{Z}_m) \oplus \mathfrak{M}_t(\mathbb{Z}_n) \quad \text{for } \gcd\{m, n\} = 1$$

as α is a ring isomorphism. So α maps invertible elements to invertible elements and hence we obtain the group isomorphism

$$\alpha| : GL_t(\mathbb{Z}_{mn}) \cong GL_t(\mathbb{Z}_m) \times GL_t(\mathbb{Z}_n) \quad \text{for } \gcd\{m, n\} = 1$$

where the right-hand side denotes the *external direct product* of the indicated groups (see Exercises 2.3, Question 4(d)) and $\alpha|$ denotes the restriction of α to $GL_t(\mathbb{Z}_{mn})$.

For example take $t = 2$, $m = 7$, $n = 8$ and

$$A = \begin{pmatrix} \overline{27} & \overline{17} \\ \overline{44} & \overline{51} \end{pmatrix} \in \mathfrak{M}_2(\mathbb{Z}_{56}).$$

Then

$$(A)\alpha = (B, C) = \left(\begin{pmatrix} \overline{6} & \overline{3} \\ \overline{2} & \overline{2} \end{pmatrix}, \begin{pmatrix} \overline{3} & \overline{1} \\ \overline{4} & \overline{3} \end{pmatrix} \right) \in \mathfrak{M}_2(\mathbb{Z}_7) \oplus \mathfrak{M}_2(\mathbb{Z}_8).$$

In fact A, B and C are invertible and

$$(A^{-1})\alpha = (B^{-1}, C^{-1}) = \left(\begin{pmatrix} \overline{5} & \overline{3} \\ \overline{2} & \overline{1} \end{pmatrix}, \begin{pmatrix} \overline{7} & \overline{3} \\ \overline{4} & \overline{7} \end{pmatrix} \right).$$

Hence

$$A^{-1} = \begin{pmatrix} \overline{47} & \overline{3} \\ \overline{44} & \overline{15} \end{pmatrix}$$

on applying Theorem 2.11 to each entry. Note $|\mathfrak{M}_2(\mathbb{Z}_{56})| = 56^4$ and

$$|GL_2(\mathbb{Z}_{56})| = |GL_2(\mathbb{Z}_7)| \times |GL_2(\mathbb{Z}_8)| = (7^2 - 1)(7^2 - 7) \times 4^4 \times 6 = 3096576$$

as

$$|GL_2(\mathbb{Z}_8)| = 4^4 \times |GL_2(\mathbb{Z}_2)|.$$

More generally let $\det A = \overline{m}$ where $A \in \mathfrak{M}_t(\mathbb{Z}_n)$. As $\overline{m}$ is an invertible element of $\mathbb{Z}_n \Leftrightarrow \gcd\{m, n\} = 1$, we see that $A \in GL_t(\mathbb{Z}_n) \Leftrightarrow \gcd\{m, n\} = 1$. Also $|GL_t(\mathbb{Z}_q)| = p^{t^2(s-1)}|GL_t(\mathbb{Z}_p)|$ where $q = p^s$, p prime (see Exercises 2.3, Question 5(b)).

In Chapter 5 we use the concept of an $F[x]$-module M in order to discuss the theory of *similarity* of square matrices over the field F. Here $F[x]$ is the ring of all polynomials $a_0 + a_1x + a_2x^2 + \cdots + a_nx^n$ over F. There is a close analogy between the theory of similarity and the theory of finite abelian groups as we'll come to realise. Both theories involve R-modules where R is a principal ideal domain. However, the aspect of the theory which we deal with next 'works' in the general context of R being merely a non-trivial commutative ring (with 1-element). So we assume in the following theory that R is such a ring.

Let M be a set closed under a binary operation called 'addition' and denoted in the familiar way by '$+$'. Suppose that $(M, +)$ satisfies laws 1, 2, 3 and 4 of an abelian group as introduced at the beginning of Section 2.1. Suppose also that it makes sense to multiply elements r of a commutative ring R and elements v of M together, the result always being an element of M, that is,

$$rv \in M \quad \text{for all } r \in R, \ v \in M.$$

Then M is called an R-*module* if the above product rv satisfies:

5. $r(v_1 + v_2) = rv_1 + rv_2$ for all $r \in R$ and all $v_1, v_2 \in M$,
   $(r_1 + r_2)v = r_1v + r_2v$ for all $r_1, r_2 \in R$ and all $v \in M$,
6. $(r_1r_2)v = r_1(r_2v)$ for all $r_1, r_2 \in R$ and all $v \in M$,
7. $1v = v$ for all $v \in M$ where 1 denotes the 1-element of R.

There are no surprises here! We have simply mimicked the $\mathbb{Z}$-module definition at the start of Section 2.1. Should the ring R happen to be a field F then laws 1–7 above are the laws of a vector space, that is,

> the concepts F-module and vector space over F are the same.

Definition 2.19

Let M be an R-module containing $v_1, v_2, \ldots, v_t$.

(i) The elements $v_1, v_2, \ldots, v_t$ *generate* M if each element of M can be expressed $r_1v_1 + r_2v_2 + \cdots + r_tv_t$ for some $r_1, r_2, \ldots, r_t \in R$ in which case we write

$$M = \langle v_1, v_2, \ldots, v_t \rangle.$$

(ii) The elements $v_1, v_2, \ldots, v_t$ are R-*independent* if the equation

$$r_1v_1 + r_2v_2 + \cdots + r_tv_t = 0$$

holds only in the case $r_1 = r_2 = \cdots = r_t = 0$.

(iii) The ordered set $v_1, v_2, \ldots, v_t$ is an R-*basis* of M if $v_1, v_2, \ldots, v_t$ generate M and are R-independent.

The above definitions are modelled on the corresponding vector space concepts which will be well-known to the reader. You are used to regarding the bases v_1, v_2 and v_2, v_1 of a 2-dimensional vector space V as being different – the order in which the vectors appear is important and the same goes for R-bases.

Let the R-module M have R-basis $v_1, v_2, \ldots, v_t$ and let $v \in M$. As $v_1, v_2, \ldots, v_t$ generate M there are ring elements $r_1, r_2, \ldots, r_t$ with $v = r_1v_1 + r_2v_2 + \cdots + r_tv_t$.

In fact $r_1, r_2, \ldots, r_t$ are unique because suppose $v = r_1' v_1 + r_2' v_2 + \cdots + r_t' v_t$ where $r_1', r_2', \ldots, r_t' \in R$. Subtracting we obtain

$$0 = v - v = (r_1 - r_1')v_1 + (r_2 - r_2')v_2 + \cdots + (r_t - r_t')v_t$$

and so from the R-independence of $v_1, v_2, \ldots, v_t$ we deduce $r_1 - r_1' = 0, r_2 - r_2' = 0,$ $\ldots, r_t - r_t' = 0$; therefore $r_i = r_i'$ for $1 \leq i \leq t$, showing that each v in M can be expressed in one and only one way as an R-linear combination of $v_1, v_2, \ldots, v_t$. In particular (as in the next proof) from $v_i = r_1 v_1 + r_2 v_2 + \cdots + r_t v_t$ we deduce $r_i = 1$ and $r_k = 0$ for $k \neq i$.

It is not encouraging that some generating sets of a $\mathbb{Z}$-module G do not contain any $\mathbb{Z}$-basis of G, that some $\mathbb{Z}$-independent subsets of G are not contained in any $\mathbb{Z}$-basis of G, and that quite possibly G has no $\mathbb{Z}$-basis at all. However, as a consequence of the next theorem, should an R-module M have an R-basis consisting of exactly t (a nonnegative integer) elements then every R-basis of M has t elements also, in which case M is said to be a *free R-module of rank t*.

Theorem 2.20

Let R be a non-trivial commutative ring and suppose that M is an R-module with R-basis $v_1, v_2, \ldots, v_t$. Suppose also that M contains elements $u_1, u_2, \ldots, u_s$ which generate M. Then $s \geq t$.

Proof

Each v_i is a linear combination of $u_1, u_2, \ldots, u_s$ and so there are ring elements $p_{ij} \in R$ with $v_i = \sum_{j=1}^{s} p_{ij} u_j$ for $1 \leq i \leq t$. Let $P = (p_{ij})$ denote the $t \times s$ matrix over R with (i, j)-entry p_{ij}. In the same way each module element u_j is expressible as a linear combination of $v_1, v_2, \ldots, v_t$ and so there are ring elements $q_{jk} \in R$ with $u_j = \sum_{k=1}^{t} q_{jk} v_k$ for $1 \leq j \leq s$. Let $Q = (q_{jk})$ be the $s \times t$ matrix over R with (j, k)-entry q_{jk}. We've chosen the symbols i, j, k so that the (i, k)-entry $\sum_{j=1}^{s} p_{ij} q_{jk}$ in the $t \times t$ matrix PQ appears in the familiar notation. Substituting for u_j we obtain

$$v_i = \sum_{j=1}^{s} p_{ij} u_j = \sum_{j=1}^{s} p_{ij} \left(\sum_{k=1}^{t} q_{jk} v_k \right) = \sum_{k=1}^{t} \left(\sum_{j=1}^{s} p_{ij} q_{jk} \right) v_k \quad \text{for } 1 \leq i \leq t$$

which must in fact be no more than the unsurprising equation $v_i = v_i$ as $v_1, v_2, \ldots, v_t$ are R-independent. Looking at the last term above we see that the (i, k)-entry in PQ is 1 or 0 according as $i = k$ or $i \neq k$, that is,

$$PQ = I_t \quad \text{the } t \times t \text{ identity matrix over } R. \tag{$\blacklozenge$}$$

Suppose $s < t$. We'll shortly discover a contradiction to this supposition and that will complete the proof. We can't use Lemma 2.18 and leap to the conclusion that P and Q are inverses of each other as neither P nor Q is a square matrix. But the reader should have the feeling that something is wrong: the condition $s < t$ means that P is 'long and thin' and Q is 'short and fat', but nevertheless their product PQ is the large 'virile' identity matrix I_t. We clinch the matter by partitioning $P = \binom{P_1}{P_2}$ and $Q = (Q_1 \; Q_2)$ where P_1 and Q_1 are $s \times s$ matrices, and so P_2 is $(t - s) \times s$ and Q_2 is $s \times (t - s)$. Then ($\blacklozenge$) gives

$$PQ = \binom{P_1}{P_2}(Q_1 \mid Q_2) = \left(\begin{array}{c|c} P_1Q_1 & P_1Q_2 \\ \hline P_2Q_1 & P_2Q_2 \end{array}\right) = I_t = \left(\begin{array}{c|c} I_s & 0 \\ \hline 0 & I_{t-s} \end{array}\right)$$

and so $P_1Q_1 = I_s$ on comparing leading entries. Now Lemma 2.18 can be used to give $Q_1P_1 = I_s$ as the $s \times s$ matrices P_1 and Q_1 are inverses of each other. Comparing $(1, 2)$-entries in the above partitioned matrices gives $P_1Q_2 = 0$ and hence $Q_2 = I_sQ_2 = (Q_1P_1)Q_2 = Q_1(P_1Q_2) = Q_10 = 0$. The 1-element of R cannot be zero (for if so then $R = \{0\}$). Comparing $(2, 2)$-entries above now gives $P_2Q_2 = I_{t-s} \neq 0$ whereas $P_2Q_2 = P_20 = 0$. We have found the contradiction to $s < t$ we are looking for as P_2Q_2 cannot be both non-zero and zero! Therefore $s \geq t$. $\qquad\square$

Corollary 2.21

Let M be an R-module with R-basis $v_1, v_2, \ldots, v_t$. Then every R-basis of M has exactly t elements. Let $u_1, u_2, \ldots, u_t$ be elements of M and let $Q = (q_{jk})$ be the $t \times t$ matrix over R such that $u_j = \sum_{k=1}^{t} q_{jk}v_k$ for $1 \leq j \leq t$. Then $u_1, u_2, \ldots, u_t$ is an R-basis of M if and only if Q is invertible over R.

Proof

Let $u_1, u_2, \ldots, u_s$ be an R-basis of M. As $u_1, u_2, \ldots, u_s$ generate M we deduce $t \leq s$ from Theorem 2.20. As $u_1, u_2, \ldots, u_s$ is an R-basis of M and $v_1, v_2, \ldots, v_t$ generate M, interchanging the roles of the u's and v's, we obtain $s \leq t$ from Theorem 2.20. So $s = t$. Using ($\blacklozenge$) above and Lemma 2.18 we see that the $t \times t$ matrix Q is invertible over R.

Now suppose that $u_1, u_2, \ldots, u_t$ are elements of M such that Q is invertible over R. Write $Q^{-1} = (p_{ij})$. Multiplying $u_j = \sum_{k=1}^{t} q_{jk}v_k$ by p_{ij} and summing over j gives

$$\sum_{j=1}^{t} p_{ij}u_j = \sum_{j=1}^{t} p_{ij}\left(\sum_{k=1}^{t} q_{jk}v_k\right) = \sum_{k=1}^{t}\left(\sum_{j=1}^{t} p_{ij}q_{jk}\right)v_k = v_i \quad \text{for } 1 \leq i \leq t \quad (\heartsuit)$$

which shows that each v_i is an R-linear combination of $u_1, u_2, \ldots, u_t$. Consider $v \in M$. As $M = \langle v_1, v_2, \ldots, v_t \rangle$ there are elements $r_1, r_2, \ldots, r_t \in R$ with $v = \sum_{i=1}^{t} r_i v_i$. Using ($\heartsuit$) we see

$$v = \sum_{i=1}^{t} r_i \left(\sum_{j=1}^{t} p_{ij} u_j \right) = \sum_{j=1}^{t} \left(\sum_{i=1}^{t} r_i p_{ij} \right) u_j = \sum_{j=1}^{t} r'_j u_j$$

where $r'_j = \sum_{j=1}^{t} r_i p_{ij}$ for $1 \leq j \leq t$, that is, $(r'_1, r'_2, \ldots, r'_t) = (r_1, r_2, \ldots, r_t) Q^{-1}$. So $u_1, u_2, \ldots, u_t$ generate M.

Finally we show that $u_1, u_2, \ldots, u_t$ are R-independent. Suppose $\sum_{j=1}^{t} r'_j u_j = 0$ where $r'_1, r'_2, \ldots, r'_t \in R$. On multiplying $u_j = \sum_{k=1}^{t} q_{jk} v_k$ by r'_j and summing over j we obtain

$$0 = \sum_{j=1}^{t} r'_j u_j = \sum_{j=1}^{t} r'_j \left(\sum_{k=1}^{t} q_{jk} v_k \right) = \sum_{k=1}^{t} \left(\sum_{j=1}^{t} r'_j q_{jk} \right) v_k = \sum_{k=1}^{t} r_k v_k$$

where $r_k = \sum_{j=1}^{t} r'_j q_{jk}$ for $1 \leq k \leq t$, that is, $(r_1, r_2, \ldots, r_t) = (r'_1, r'_2, \ldots, r'_t) Q$. As $v_1, v_2, \ldots, v_t$ are R-independent we see $r_1 = r_2 = \cdots = r_t = 0$. Hence $(r'_1, r'_2, \ldots, r'_t) = (r_1, r_2, \ldots, r_t) Q^{-1} = 0 \times Q^{-1} = 0$ showing $r'_1 = r'_2 = \cdots = r'_t = 0$. So $u_1, u_2, \ldots, u_t$ are R-independent and hence they form an R-basis of M.   □

Definition 2.22

Let R be a commutative ring. An R-module M having an R-basis is called *free*. The number t of elements in any R-basis of a free R-module M is called the *rank* of M.

So the concept 'rank of a module' applies only to free modules. This concept makes sense for R-modules by Corollary 2.21 and generalises the familiar idea of dimension of a finite-dimensional vector space. We'll use rank as defined in Definition 2.22 to establish the important *Invariance Theorem* 3.7 concerning f.g. $\mathbb{Z}$-modules in Section 3.1.

The set R^t of t-tuples $(r_1, r_2, \ldots, r_t)$, where each r_i belongs to the non-trivial commutative ring R, is itself an R-module, the module operations being carried out componentwise. It should come as no surprise to the reader that R^t has an R-basis, namely

$$e_1 = (1, 0, 0, \ldots, 0), \qquad e_2 = (0, 1, 0, \ldots, 0), \qquad \ldots, \qquad e_t = (0, 0, 0, \ldots, 1)$$

which is known as the *standard basis* of R^t. So R^t is free of rank t.

Our next corollary tells us how to recognise R-bases of R^t: they are nothing more than the rows of invertible $t \times t$ matrices over R.

Corollary 2.23

Let $\rho_1, \rho_2, \ldots, \rho_t$ denote the rows of the $t \times t$ matrix Q over a non-trivial commutative ring R. Then $\rho_1, \rho_2, \ldots, \rho_t$ is an R-basis of R^t if and only if Q is invertible over R.

Proof

Write $Q = (q_{jk})$. Then $\rho_j = \sum_{k=1}^{t} q_{jk} e_k$ for $1 \leq j \leq t$. On applying Corollary 2.21 with $M = R^t$, $u_j = \rho_j$ and $v_k = e_k$ we see that $\rho_1, \rho_2, \ldots, \rho_t$ is an R-basis of R^t if and only if Q is invertible over R. $\qquad\qquad\square$

We'll use the case $R = \mathbb{Z}$ of Corollary 2.23 in Section 3.1. As an illustration consider $\rho_1 = (4, 5)$, $\rho_2 = (5, 6)$. Then ρ_1, ρ_2 is a $\mathbb{Z}$-basis of $\mathbb{Z}^2$ as $P = \left(\begin{smallmatrix} \rho_1 \\ \rho_2 \end{smallmatrix}\right) = \left(\begin{smallmatrix} 4 & 5 \\ 5 & 6 \end{smallmatrix}\right)$ is invertible over $\mathbb{Z}$ since $\det P = -1$ is an invertible element of $\mathbb{Z}$. The rows of $P^{-1} = \left(\begin{smallmatrix} -6 & 5 \\ 5 & -4 \end{smallmatrix}\right)$ tell us how the elements e_1, e_2 of the standard $\mathbb{Z}$-basis of $\mathbb{Z}^2$ are expressible as $\mathbb{Z}$-linear combinations of ρ_1, ρ_2 because

$$\begin{pmatrix} e_1 \\ e_2 \end{pmatrix} = I = P^{-1}P = \begin{pmatrix} -6 & 5 \\ 5 & -4 \end{pmatrix} \begin{pmatrix} \rho_1 \\ \rho_2 \end{pmatrix} = \begin{pmatrix} -6\rho_1 + 5\rho_2 \\ 5\rho_1 - 4\rho_2 \end{pmatrix},$$

that is, $e_1 = -6\rho_1 + 5\rho_2$, $e_2 = 5\rho_1 - 4\rho_2$ on equating rows. Hence $(m_1, m_2) = (-6m_1 + 5m_2)\rho_1 + (5m_1 - 4m_2)\rho_2$ for all $(m_1, m_2) \in \mathbb{Z}^2$, showing explicitly that ρ_1, ρ_2 generate $\mathbb{Z}^2$, that is, $\langle \rho_1, \rho_2 \rangle = \mathbb{Z}^2$.

Definition 2.24

Let M and M' be R-modules. A mapping $\theta : M \to M'$ is called R-*linear* if $(v_1 + v_2)\theta = (v_1)\theta + (v_2)\theta$ for all $v_1, v_2 \in M$ and $(rv)\theta = r((v)\theta)$ for all $r \in R$, $v \in M$. A bijective R-linear mapping θ is called an *isomorphism*. If there is an isomorphism $\theta : M \to M'$, then the R-modules M and M' are called *isomorphic* and we write $\theta : M \cong M'$.

The above definition mimics Definitions 2.3 and 2.4 replacing $\mathbb{Z}$ by the commutative ring R. The following lemma will be used in Section 3.1.

Lemma 2.25

Let M be a free R-module of rank t and let M' be an R-module which is isomorphic to M. Then M' is also free of rank t.

Proof

There is an isomorphism $\theta : M \cong M'$. The free R-module M has an R-basis $v_1, v_2, \ldots, v_t$. Write $v_i' = (v_i)\theta$ for $1 \le i \le t$. It is straightforward to show that $v_1', v_2', \ldots, v_t'$ is an R-basis of M' (see Exercises 2.3, Question 7(a)). Hence M' is free of rank t by Definition 2.22. □

One advantage of expressing abelian groups in the language of $\mathbb{Z}$-modules is that some theorems we have already met painlessly generalise to R-modules. In particular this is true of Lemma 2.10, Theorems 2.16 and 2.17 as we now outline.

Definition 2.26

Let N be a subset of an R-module M. Suppose that N is a subgroup of the additive group of M and $ru \in N$ for all $r \in R$ and all $u \in N$. Then N is called a *submodule* of the R-module M.

So submodules of the R-module M are subgroups N of the abelian group $(M, +)$ which are closed under multiplication by elements of the ring R. It is important to realise that submodules of R-modules are themselves R-modules: laws 1–7 of an R-module hold with M replaced by N throughout. The reader will be familiar with this type of thing as subspaces of vector spaces are vector spaces 'in their own right'. Indeed should the ring R be a field F, then Definition 2.26 tells us that submodules of the F-module M are exactly subspaces of the vector space M.

Let N be a submodule of the R-module M. As N is a subgroup of the additive group M the quotient group M/N can be constructed as in Lemma 2.10 (here M and N replace G and K respectively). The elements of M/N are cosets $N + v$ for $v \in M$ where (unsurprisingly) $N + v = \{u + v : u \in N\}$. Can the abelian group M/N be given the extra structure of an R-module? The answer is: Yes!

Lemma 2.27

Let N be a submodule of the R-module M. Write $r(N + v) = N + rv$ for $r \in R$ and $v \in M$. This product is unambiguously defined and using it M/N is an R-module. The natural mapping $\eta : M \to M/N$ is R-linear where $(v)\eta = N + v$ for all $v \in N$.

Proof

Suppose $N + v = N + v'$ where $v, v' \in M$. Then $v - v' \in N$ as in Lemma 2.9. So $rv - rv' = r(v - v') \in N$ as N is a submodule of the R-module M. But $rv - rv' \in N$

gives $N + rv = N + rv'$ by Lemma 2.9 and shows that the given definition of $r(N + v)$ is unambiguous.

By Lemma 2.10 coset addition in M/N obeys laws 1, 2, 3, 4 of a $\mathbb{Z}$-module. We should check that laws 5, 6 and 7 are obeyed by the product $r(N + v)$ defined above. Consider $r \in R$ and $v, v_1, v_2 \in M$. Then

$$r((N + v_1) + (N + v_2)) = r(N + (v_1 + v_2)) = N + r(v_1 + v_2)$$

$$= N + (rv_1 + rv_2) = (N + rv_1) + (N + rv_2)$$

$$= r(N + v_1) + r(N + v_2)$$

which shows that the first part of law 5 is obeyed. The remaining parts can be checked in a similar way (see Exercises 2.3, Question 7(d)) showing that M/N is an R-module.

As $(rv)\eta = N + rv = r(N + v) = r((v)\eta)$ for all $r \in R$, $v \in M$ we see that η is R-linear. □

The reader should verify that kernels and images of R-linear mappings are submodules (see Exercises 2.3, Question 7(c)).

We are now ready to generalise Theorem 2.16 and 2.17.

Corollary 2.28 (The first isomorphism theorem for R-modules)

Let M and M' be R-modules and let $\theta : M \to M'$ be an R-linear mapping. Write $K = \ker \theta$. Then $\tilde{\theta} : M/K \cong \operatorname{im} \theta$ is an isomorphism of R-modules where $\tilde{\theta}$ is defined by $(K + v)\tilde{\theta} = (v)\theta$ for all $v \in M$.

Proof

By Theorem 2.16 we know that $\tilde{\theta}$ is an isomorphism of $\mathbb{Z}$-modules. So it is enough to check that $\tilde{\theta}$ is R-linear: $(r(K + v))\tilde{\theta} = (K + rv)\tilde{\theta} = (rv)\theta = r((v)\theta) = r((K + v)\tilde{\theta})$ for $r \in R$, $v \in M$ using the R-linearity of θ and the definition of $\tilde{\theta}$. □

Corollary 2.29

Let M and M' be R-modules and let $\theta : M \to M'$ be an R-linear mapping. Let $\mathbb{L}$ be the set of submodules N of M with $\ker \theta \subseteq N$. Let $\mathbb{L}'$ be the set of submodules N' of M' with $N' \subseteq \operatorname{im} \theta$. Then $(N)\theta = \{(u)\theta : u \in N\}$ is in $\mathbb{L}'$ for all N in $\mathbb{L}$. The mapping $N \to (N)\theta$ is a bijection from $\mathbb{L}$ to $\mathbb{L}'$ and satisfies $N_1 \subseteq N_2 \Leftrightarrow (N_1)\theta \subseteq (N_2)\theta$ for all $N_1, N_2 \in \mathbb{L}$.

Proof

In view of Theorem 2.17 there is not a great deal left to prove and what's left is routine. We know that $(N)\theta$ is a subgroup of $(M', +)$. Consider $r \in R$ and $u \in N$. Then $ru \in N$ as N is a submodule of M. So $r((u)\theta) = (ru)\theta \in (N)\theta$ showing that $(N)\theta$ is a submodule of M'. Therefore $N \rightarrow (N)\theta$ is a mapping from $\mathbb{L}$ to $\mathbb{L}'$. Following the proof of Theorem 2.17 for each submodule N' of M' write $(N')\varphi = \{v \in M : (v)\theta \in N'\}$. The diligent reader will have checked that $(N')\varphi$ is a subgroup of $(M, +)$. Consider $r \in R$ and $v \in (N')\varphi$. Is $rv \in (N')\varphi$? Yes it is, as $v' = (v)\theta \in N'$ and so $(rv)\theta = r((v)\theta) = rv' \in N'$ as N' is a submodule of M'. The conclusion is: $(N')\varphi$ is a submodule of M and $N' \rightarrow (N')\varphi$ is a mapping from $\mathbb{L}'$ to $\mathbb{L}$. As before these mappings are inverses of each other and are inclusion-preserving. Therefore $N \rightarrow (N)\theta$ is a bijection from $\mathbb{L}$ to $\mathbb{L}'$ satisfying $N_1 \subseteq N_2 \Leftrightarrow (N_1)\theta \subseteq (N_2)\theta$ for all $N_1, N_2 \in \mathbb{L}$. $\square$

EXERCISES 2.3

1. (a) Let G and G' be $\mathbb{Z}$-modules and let $\theta : G \rightarrow G'$ be a $\mathbb{Z}$-linear mapping.

   (i) Show that $\ker \theta$ is a submodule of G. Show that $\ker \theta = \{0\} \Leftrightarrow \theta$ is injective.

   (ii) Show that $\operatorname{im} \theta$ is a submodule of G'. Is it true that $\operatorname{im} \theta = G' \Leftrightarrow \theta$ is surjective?

   (b) The $\mathbb{Z}$-linear mapping $\theta : \mathbb{Z} \oplus \mathbb{Z} \rightarrow \mathbb{Z}$ is given by $(l, m)\theta = 4l - 2m$ for all $l, m \in \mathbb{Z}$. Verify that $(1, 2) \in \ker \theta$. Do $(-1, -2)$ and $(2, 4)$ belong to $\ker \theta$? Show that $\ker \theta = \langle (1, 2) \rangle$. Which integers belong to $\operatorname{im} \theta$? Is $\operatorname{im} \theta$ infinite cyclic? Using the notation of Theorem 2.16, are the integers $(\ker \theta + (12, 20))\tilde{\theta}$ and $(\ker \theta + (17, 30))\tilde{\theta}$ equal? Show that $\mathbb{Z} \oplus \mathbb{Z} / \ker \theta$ is infinite cyclic and specify a generator.

   (c) Specify the subgroups of each of the following groups and hence determine the isomorphism types of their homomorphic images.

   (i) $\mathbb{Z}_8$;   (ii) $\mathbb{Z}_{12}$;   (iii) $\mathbb{Z}_n$ $(n > 0)$;

   (iv) $\mathbb{Z}$;   (v) $\mathbb{Z}_2 \oplus \mathbb{Z}_2$.

   (d) Let G_1 and G_2 be additive abelian groups. By applying Theorem 2.16 to suitable homomorphisms, establish the isomorphisms: $G_1/\{0\} \cong G_1$, $G_1/G_1 \cong \{0\}$. Show also that G_1 and G_2 are homomorphic images of $G_1 \oplus G_2$. Are the groups G_1 and $(G_1 \oplus G_1)/K$ isomorphic where $K = \{(g_1, g_1) : g_1 \in G_1\}$?

   (e) Let G be a $\mathbb{Z}$-module and $\theta : G \rightarrow G$ a $\mathbb{Z}$-linear mapping which is *idempotent* (i.e. $\theta^2 = \theta$). Use the equation $g = (g - (g)\theta) + (g)\theta$ for

$g \in G$ to show $G = \ker\theta \oplus \operatorname{im}\theta$. Let $G = \mathbb{Z}_2 \oplus \mathbb{Z}_4$ and let $(\overline{l}, \overline{m})\theta = (\overline{m}, \overline{2l - m})$ for all $l, m \in \mathbb{Z}$. Show that $\theta : G \to G$ is idempotent and find generators of $\ker\theta$ and $\operatorname{im}\theta$.

2. (a) Let G and G' be $\mathbb{Z}$-modules and $\theta : G \to G'$ a $\mathbb{Z}$-linear mapping.
     (i) For each subgroup H of G show that $H' = \{(h)\theta : h \in H\}$ is a subgroup of G contained in $\operatorname{im}\theta$. Show $H/(\ker\theta \cap H) \cong H'$ by applying Theorem 2.16 to the *restriction* of θ to H (i.e. the $\mathbb{Z}$-linear mapping $\theta|_H : H \to G'$ defined by $(h)\theta|_H = (h)\theta$ for all $h \in H$).
     (ii) For each subgroup H' of G', show that $H = \{h \in G : (h)\theta \in H'\}$ is a subgroup of G containing $\ker\theta$. Show $H/\ker\theta \cong H' \cap \operatorname{im}\theta$ by applying Theorem 2.16 to $\theta|_H$.
    (b) The $\mathbb{Z}$-linear mapping $\theta : \mathbb{Z} \oplus \mathbb{Z} \to \mathbb{Z}_2$ is defined by $(l, m)\theta = \overline{l} - m$ for all $l, m \in \mathbb{Z}$. Verify that $(1, 1)$ and $(2, 0)$ belong to $\ker\theta$. Show that $(1, 1), (2, 0)$ is a $\mathbb{Z}$-basis of $\ker\theta$. Show $\operatorname{im}\theta = \mathbb{Z}_2$. Using Theorem 2.16 determine the isomorphism type of $\mathbb{Z} \oplus \mathbb{Z}/\ker\theta$. Use Theorem 2.17 to show that $\ker\theta$ is a *maximal* subgroup of $\mathbb{Z} \oplus \mathbb{Z}$ (i.e. $\ker\theta \neq \mathbb{Z} \oplus \mathbb{Z}$ and there are no subgroups H of $\mathbb{Z} \oplus \mathbb{Z}$ with $\ker\theta \subset H \subset \mathbb{Z} \oplus \mathbb{Z}$).
    (c) The $\mathbb{Z}$-linear mapping $\theta : \mathbb{Z} \oplus \mathbb{Z} \to \mathbb{Z}_4$ is defined by $(l, m)\theta = \overline{l} + m$ for all $l, m \in \mathbb{Z}$. Show that $\ker\theta = \langle (1, -1), (4, 0) \rangle$. Verify that $\operatorname{im}\theta = \mathbb{Z}_4$ and hence find the isomorphism type of $(\mathbb{Z} \oplus \mathbb{Z})/\ker\theta$. List the subgroups H' of $\mathbb{Z}_4$ and the corresponding subgroups H of $\mathbb{Z} \oplus \mathbb{Z}$ with $\ker\theta \subseteq H$ as in Theorem 2.17. Taking $H' = \langle \overline{2} \rangle$ specify a $\mathbb{Z}$-basis of the corresponding H.
    (d) The $\mathbb{Z}$-linear mapping $\theta : \mathbb{Z} \oplus \mathbb{Z} \to \mathbb{Z}_2 \oplus \mathbb{Z}_4 = G'$ is defined by $(l, m)\theta = (\overline{l}, \overline{m})$ for all $l, m \in \mathbb{Z}$. Verify that $2e_1, 4e_2$ is a $\mathbb{Z}$-basis of $\ker\theta$. For each of the following subgroups H' of G' specify a $\mathbb{Z}$-basis ρ_1, ρ_2 of $H = (H')\varphi = \{h \in \mathbb{Z} \oplus \mathbb{Z} : (h)\theta \in H'\}$:

$$\langle (\overline{1}, \overline{0}) \rangle, \qquad \langle (\overline{0}, \overline{2}) \rangle, \qquad \langle (\overline{1}, \overline{1}) \rangle.$$

In each case find the Smith normal form $\operatorname{diag}(d_1, d_2)$ of the 2×2 matrix $A = \left(\begin{smallmatrix} \rho_1 \\ \rho_2 \end{smallmatrix} \right)$ and check that G'/H' has isomorphism type $C_{d_1} \oplus C_{d_2}$.
    (e) Let G and G' be $\mathbb{Z}$-modules and $\theta : G \to G'$ a surjective $\mathbb{Z}$-linear mapping. Let H' be a subgroup of G' and $H = \{h \in G : (h)\theta \in H'\}$. By applying Theorem 2.16 to $\theta\eta$, where $\eta : G' \to G'/H'$ is the natural homomorphism, show $G/H \cong G'/H'$.

3. (a) Let R be a ring with 1-element e. A subgroup K of the additive group of R is called *an ideal of R* if rk and kr belong to K for all $r \in R$, $k \in K$. Show that multiplication of cosets is unambiguously defined by $(K + r_1)(K + r_2) = K + r_1 r_2$ for all $r_1, r_2 \in R$ (i.e. show that if $K + r_1 = K + r_1'$ and $K + r_2 = K + r_2'$ then $K + r_1 r_2 = K + r_1' r_2'$).

Using the notation $K + r = \bar{r}$ and Lemma 2.10, show that the set $R/K = \{\bar{r} : r \in R\}$ of cosets is a ring (*the quotient ring* of R by K). Show further that $\eta : R \to R/K$, given by $(r)\eta = \bar{r}$ for all $r \in R$, is a ring homomorphism (the *natural homomorphism* from R to R/K). What are im η and ker η?

(b) Let R and R' be rings and let $\theta : R \to R'$ be a ring homomorphism. Show that im θ is a *subring* of R' (i.e. im θ is a subgroup of the additive group of R', im θ is closed under multiplication and im θ contains the 1-element e' of R').

Show that $K = \ker \theta = \{k \in R : (k)\theta = 0'\}$ is an ideal of R where $0'$ is the 0-element of R'. Prove *the first isomorphism theorem for rings* namely $\tilde{\theta} : R/K \cong \mathrm{im}\,\theta$ (i.e. $\tilde{\theta}$ defined by $(K + r)\tilde{\theta} = (r)\theta$ for all $r \in R$ is a ring isomorphism).

(c) Let $\theta : \mathbb{Z} \to R$ be a ring homomorphism from the ring $\mathbb{Z}$ of integers to a ring R. Use part (b) above and Theorem 1.15 to show that there is a non-negative integer d with $\mathbb{Z}/\langle d \rangle \cong \mathrm{im}\,\theta$, that is, the rings $\mathbb{Z}_d$ are, up to isomorphism, the (ring) homomorphic images of $\mathbb{Z}$.

(d) Let R, R', R'' be rings and let $\theta : R \to R'$, $\theta' : R' \to R''$ be ring homomorphisms. Show that $\theta\theta' : R \to R''$ is a ring homomorphism. Suppose θ is a ring isomorphism. Show that θ^{-1} is also a ring isomorphism. Deduce that the automorphisms θ of R (the ring isomorphisms $\theta : R \to R$) form a group Aut R, the group operation being mapping composition.

(e) Let R_1 and R_2 be rings. Show that $R_1 \oplus R_2 = \{(r_1, r_2) : r_1 \in R_1, r_2 \in R_2\}$, with addition and multiplication of ordered pairs defined by $(r_1, r_2) + (r_1', r_2') = (r_1 + r_1', r_2 + r_2')$, $(r_1, r_2)(r_1', r_2') = (r_1 r_1', r_2 r_2')$ for all $r_1, r_1' \in R_1$ and all $r_2, r_2' \in R_2$, is itself a ring (*the direct sum* of R_1 and R_2).

(f) Let K and L be ideals of a ring R (see Question 3(a) above). Show that $K \cap L$ and $K + L = \{k + l : k \in K, l \in L\}$ are ideals of R. Establish the *generalised Chinese remainder theorem* which states: suppose $K + L = R$; then $\tilde{\alpha} : R/(K \cap L) \cong R/K \oplus R/L$ is a ring isomorphism where $(r + K \cap L)\tilde{\alpha} = (r + K, r + L)$ for all $r \in R$.

Hint: Consider $\alpha : R \to R/K \oplus R/L$ defined by $(r)\alpha = (r + K, r + L)$ for all $r \in R$.

4. (a) Let G be a multiplicative group with subgroup K. Then K is called *normal in* G if $g^{-1}kg \in K$ for all $k \in K$, $g \in G$. Write $Kg = \{kg : k \in K\}$ and $gK = \{gk : k \in K\}$. Show that K is normal in G if and only if $Kg = gK$ for all $g \in G$.

The group S_3 consisting of the 6 bijections (permutations) of $\{1, 2, 3\}$ to $\{1, 2, 3\}$ contains the elements σ and τ where $(1)\sigma = 2$, $(2)\sigma = 3$,

$(3)\sigma = 1$ and $(1)\tau = 2$, $(2)\tau = 1$, $(3)\tau = 3$. Show that $\langle\sigma\rangle = \{\sigma, \sigma^2, \sigma^3\}$ is normal in S_3 but $\langle\tau\rangle = \{\tau, \tau^2\}$ is not normal in S_3.
Hint: $\langle\sigma\rangle$ is the subgroup of even permutations in S_3.

Suppose that K is normal in G. Show that the product of cosets is unambiguously defined by $(Kg_1)(Kg_2) = K(g_1g_2)$ for $g_1, g_2 \in G$. Hence show that the set G/K of all cosets $Kg(g \in G)$ is a group, *the quotient group of G by K* – it's the multiplicative version of Lemma 2.10.

(b) Let $G = U(\mathbb{Z}_{15})$ the multiplicative group of invertible elements in the ring $\mathbb{Z}_{15}$. List the 8 elements in G. For each of the following subgroups K, partition G into cosets of K, construct the multiplication table of G/K and state the isomorphism type of G/K:

$$K = \{\bar{1}, \bar{4}\}, \qquad K = \{\bar{1}, \overline{14}\}, \qquad K = \{\bar{1}, \bar{4}, \overline{11}, \overline{14}\},$$

$$K = \{\bar{1}, \bar{2}, \bar{4}, \bar{8}\}.$$

From your results decide whether or not (i) $K_1 \cong K_2$ implies $G/K_1 \cong G/K_2$, (ii) $G/K_1 \cong G/K_2$ implies $K_1 \cong K_2$, where K_1 and K_2 are normal subgroups of G.

(c) Let G and G' be multiplicative groups. Suppose the mapping $\theta : G \to G'$ satisfies $(g_1g_2)\theta = (g_1)\theta(g_2)\theta$ for all $g_1, g_2 \in G$. Then θ is called a group homomorphism. Show that $(e)\theta = e'$ where e, e' are the identity elements of G, G' by considering $(e^2)\theta$. Deduce that $(g^{-1})\theta = ((g)\theta)^{-1}$ for all $g \in G$.
Show that $K = \ker\theta = \{k \in G : (k)\theta = e'\}$ is a normal subgroup of G. Show that $\operatorname{im}\theta = \{(g)\theta : g \in G\}$ is a subgroup of G'. Prove *the first isomorphism theorem for groups* namely $\tilde{\theta} : G/K \cong \operatorname{im}\theta$, i.e. $\tilde{\theta}$ defined by $(Kg)\tilde{\theta} = (g)\theta$ for all $g \in G$ is a *group isomorphism* (a bijective group homomorphism) – it's the multiplicative version of Theorem 2.16.
Let R be a non-trivial commutative ring and let t be a positive integer. Use Theorem 1.18 to show that $\theta : GL_t(R) \to U(R)$ given by $(A)\theta = \det A$ for all $A \in GL_t(R)$ is a group homomorphism. Show $\operatorname{im}\theta = U(R)$ and write $\ker\theta = SL_t(R)$ *the special linear group of degree t over R*. Find a formula for $|SL_t(\mathbb{Z}_p)|$ where p is prime.

(d) Let G_1 and G_2 be multiplicative groups. Show that the Cartesian product $G_1 \times G_2$ (the set of all ordered pairs (g_1, g_2) where $g_1 \in G_1$, $g_2 \in G_2$) with componentwise multiplication is a group, *the external direct product of G_1 and G_2* – it's the multiplicative version of the external direct sum (Exercises 2.2, Question 4(e)).
The projection homomorphisms $\pi_i : G_1 \times G_2 \to G_i$ are defined by $(g_1, g_2)\pi_i = g_i$ for all $(g_1, g_2) \in G_1 \times G_2$ where $i = 1, 2$. Show that

$G_1 \times G_2$ has normal subgroups K_1 and K_2, isomorphic to G_1 and G_2, such that $K_1 \cap K_2$ is trivial and $K_1 K_2 = \{k_1 k_2 : k_i \in K_i\} = G_1 \times G_2$. *Hint*: Use $\ker \pi_i$.

5. (a) Let F be a field with 1-element e. Show that the mapping $\chi : \mathbb{Z} \to F$, defined by $(m)\chi = me$ for all $m \in \mathbb{Z}$, is a ring homomorphism. The non-negative integer d with $\ker \chi = \langle d \rangle$ is called *the characteristic of F* (d exists by Theorem 1.15). Show that $\tilde{\chi} : \mathbb{Z}_d \cong \operatorname{im} \chi$ (i.e. $\operatorname{im} \chi$ is a subring of F which is isomorphic to $\mathbb{Z}_d$). Using the fact that F has no divisors of zero, deduce that either $d = 0$ or $d = p$ (prime). It is customary to write $d = \chi(F)$.

   Let F be a finite field. Explain why $\mathbb{Z}_p \cong \operatorname{im} \chi$ in this case; $\operatorname{im} \chi = F_0$ is called *the prime subfield of F*. Explain how F has the structure of a vector space over F_0. Why is this vector space finitely generated? Show $|F| = p^s$ where s is the dimension of F over F_0. Use the binomial theorem to show $(a + b)^p = a^p + b^p$ for $a, b \in F$. Hence show that $\theta : F \to F$ defined by $(a)\theta = a^p$ for all $a \in F$ is an automorphism of F (*the Frobenius automorphism*).

   (b) Let d and n be positive integers with $d \mid n$. Using the notation of Theorem 2.11, let $\delta_1 : \mathbb{Z}_n \to \mathbb{Z}_d$ be the surjective ring homomorphism given by $(\overline{m}_n)\delta_1 = \overline{m}_d$ for all $m \in \mathbb{Z}$. Let $A = (a_{ij})$ be the $t \times t$ matrix over $\mathbb{Z}_n$ with (i, j)-entry a_{ij}. Write $(A)\delta_t = ((a_{ij})\delta_1)$, i.e. $(A)\delta_t$ is the $t \times t$ matrix over $\mathbb{Z}_d$ with (i, j)-entry $(a_{ij})\delta_1$. Show that $\delta_t : \mathfrak{M}_t(\mathbb{Z}_n) \to \mathfrak{M}_t(\mathbb{Z}_d)$ is a surjective ring homomorphism. Describe the elements of $\ker \delta_t$ and show $|\ker \delta_t| = (n/d)^{t^2}$.

   Take $d = p$ (prime), $n = p^s$ where $s > 0$ and write $\det A = \overline{m}_n$. Show that

   $$A \in GL_t(\mathbb{Z}_n) \quad \Leftrightarrow \quad \gcd\{m, p^s\} \quad \Leftrightarrow \quad \gcd\{m, p\} = 1$$

   $$\Leftrightarrow \quad (A)\delta_t \in GL_t(\mathbb{Z}_d).$$

   Deduce that the restriction $\delta_t | : GL_t(\mathbb{Z}_n) \to GL_t(\mathbb{Z}_p)$ is a surjective homomorphism of multiplicative groups having kernel the coset $\ker \delta_t + I$, where I is the $t \times t$ identity matrix over $\mathbb{Z}_n$. Hence show

   $$|GL_t(\mathbb{Z}_{p^s})| = p^{(s-1)t^2}(p^t - 1)(p^t - p) \cdots (p^t - p^{t-1}).$$

   Calculate $|GL_3(\mathbb{Z}_4)|$ and $|GL_2(\mathbb{Z}_{17})|$. Does $GL_2(\mathbb{Z}_{125})$ have fewer elements than $GL_2(\mathbb{Z}_{128})$? Taking $p = s = t = 2$, list the 16 matrices in the normal subgroup

   $$\ker \delta_2 | = \ker \delta_2 + \begin{pmatrix} \overline{1} & \overline{0} \\ \overline{0} & \overline{1} \end{pmatrix} \quad \text{of } GL_2(\mathbb{Z}_4).$$

(c) Let $\alpha : \mathfrak{M}_2(\mathbb{Z}_{72}) \cong \mathfrak{M}_2(\mathbb{Z}_8) \oplus \mathfrak{M}_2(\mathbb{Z}_9)$ be the generalised Chinese remainder theorem isomorphism. Find A_1 and A_2 in $\mathfrak{M}_2(\mathbb{Z}_{72})$ with

$$(A_1)\alpha = \left(\begin{pmatrix} \bar{1} & \bar{2} \\ \bar{3} & \bar{4} \end{pmatrix}, \begin{pmatrix} \bar{5} & \bar{6} \\ \bar{7} & \bar{8} \end{pmatrix} \right),$$

$$(A_2)\alpha = \left(\begin{pmatrix} \bar{1} & \bar{0} \\ \bar{0} & \bar{1} \end{pmatrix}, \begin{pmatrix} \bar{0} & \bar{1} \\ \bar{1} & \bar{0} \end{pmatrix} \right).$$

Calculate $A_1 + A_2$, $A_1 A_2$ and check

$$(A_1 + A_2)\alpha = (A_1)\alpha + (A_2)\alpha, \qquad (A_1 A_2)\alpha = (A_1)\alpha(A_2)\alpha.$$

Let $n = p_1^{s_1} p_2^{s_2} \cdots p_k^{s_k}$ where $p_1, p_2, \ldots, p_k$ are distinct primes. Write down a formula for $|GL_t(\mathbb{Z}_n)|$ in terms of t, p_i and $q_i = p_i^{s_i}$ for $1 \leq i \leq k$.

6. (a) Let $\rho_1, \rho_2, \ldots, \rho_s$ denote the rows of the $s \times t$ matrix A over $\mathbb{Z}$ where $s \leq t$. Show that $\rho_1, \rho_2, \ldots, \rho_s$ are $\mathbb{Z}$-independent elements of $\mathbb{Z}^t$ if and only if all the invariant factors $d_1, d_2, \ldots, d_s$ of A are positive. Use Exercises 1.3, Question 5(c) to show that there is a $\mathbb{Z}$-basis of $\mathbb{Z}^t$ beginning with $\rho_1, \rho_2, \ldots, \rho_s$ if and only if $d_1 = d_2 = \cdots = d_s = 1$.

(b) Let $\rho_1, \rho_2, \ldots, \rho_s$ denote the rows of the $s \times t$ matrix A over $\mathbb{Z}$ where $s \geq t$. Let $d_1, d_2, \ldots, d_t$ denote the invariant factors of A. Show that $\rho_1, \rho_2, \ldots, \rho_s$ generate $\mathbb{Z}^t$ if and only if $d_1 = d_2 = \cdots = d_t = 1$.
Hint: Use Corollaries 1.13 and 1.19.
Suppose $\langle \rho_1, \rho_2, \ldots, \rho_s \rangle = \mathbb{Z}^t$ and let $\rho_1', \rho_2', \ldots, \rho_t'$ denote the rows of A^T. Show that a $\mathbb{Z}$-basis of $\mathbb{Z}^t$ can be selected from $\rho_1, \rho_2, \ldots, \rho_s$ if and only if there are $s - t$ rows of the $s \times s$ identity matrix which together with $\rho_1', \rho_2', \ldots, \rho_t'$ form a $\mathbb{Z}$-basis of $\mathbb{Z}^s$.
Hint: It is possible to select a $\mathbb{Z}$-basis of $\mathbb{Z}^t$ from $\rho_1, \rho_2, \ldots, \rho_s$ if and only if A has a t-minor equal to ± 1.

(c) Test each of the following sets for $\mathbb{Z}$-independence. Which of them is contained in a $\mathbb{Z}$-basis of $\mathbb{Z}^3$?

(i) $(1, 3, 2), (4, 6, 5), (7, 9, 8)$;   (ii) $(1, 3, 7), (3, 5, 9)$;

(iii) $(1, 2, 3), (4, 3, 5)$.

Which of the following sets generate $\mathbb{Z}^3$? Which of them contains a $\mathbb{Z}$-basis of $\mathbb{Z}^3$?

(iv) $(1, 2, 3), (3, 1, 2), (1, 1, 4), (4, 1, 1)$;

(v) $(1, 1, 1), (1, 2, 3), (1, 3, 4,), (1, 5, 6)$;

(vi) $(1, 1, 1), (1, 1, 2), (1, 3, 1), (4, 1, 1)$.

7. (a) Let M and M' be R-modules and let $\theta : M \to M'$ be an R-linear
       mapping. Suppose that M is free with R-basis $v_1, v_2, \ldots, v_t$. Let
       $v_i' = (v_i)\theta$ for $1 \leq i \leq t$. Show that $v_1', v_2', \ldots, v_t'$ generate $M' \Leftrightarrow \theta$
       is surjective. Show that $v_1', v_2', \ldots, v_t'$ are R-independent $\Leftrightarrow \theta$ is in-
       jective.

       Let M and M' be free R-modules of rank t and t' respectively. Show
       that M and M' are isomorphic if and only if $t = t'$.

   (b) Let M be a free R-module of rank t. Suppose $u_1, u_2, \ldots, u_t$ gener-
       ate M. Use Lemma 2.18 and Corollary 2.21 to show that $u_1, u_2, \ldots, u_t$
       form an R-basis of M.

   (c) Let M and M' be R-modules and let $\theta : M \to M'$ be an R-linear
       mapping. Show that $\ker \theta = \{v \in M : (v)\theta = 0\}$ is a submodule of M.
       Show that $\operatorname{im} \theta = \{(v)\theta : v \in M\}$ is a submodule of M'. Suppose θ is
       bijective; show that θ^{-1} is R-linear. Is the inverse of an isomorphism
       of R-modules itself such an isomorphism?

   (d) Let N be a submodule of an R-module M. Complete the proof
       Lemma 2.27 that M/N is an R-module.

   (e) A non-empty subset N of an R-module M is closed under addition
       and $ru \in N$ for all $r \in R$, $u \in N$. Is N a submodule of M? Justify
       your answer.

       Let N_1 and N_2 be submodules of an R-module M. Show that
       $N_1 + N_2 = \{u_1 + u_2 : u_1 \in N_1, \ u_2 \in N_2\}$ is a submodule of M. Show
       that $N_1 \cap N_2$ is a submodule of M.

   (f) Let M_1 and M_2 be R-modules. Using Exercises 2.2, Question 4(f)
       show that the Cartesian product

       $$M_1 \times M_2 = \{(v_1, v_2) : v_1 \in M_1, \ v_2 \in M_2\}$$

       is an R-module (the external direct sum $M_1 \oplus M_2$ of M_1 and M_2)
       on defining $(v_1, v_2) + (v_1', v_2') = (v_1 + v_1', v_2 + v_2')$ and $r(v_1, v_2) = (rv_1, rv_2)$ for all $v_1, v_1' \in M_1$, $v_2, v_2' \in M_2$, $r \in R$.

       Use (e) above to show that $N_1 = \{(v_1, 0) \in M_1 \oplus M_2\}$ and
       $N_2 = \{(0, v_2) \in M_1 \oplus M_2\}$ are submodules of $M_1 \oplus M_2$ satisfying
       $N_1 + N_2 = M_1 \oplus M_2$, $N_1 \cap N_2 = \{(0, 0)\}$. Show that there are R-linear
       isomorphisms $\alpha_1 : N_1 \cong M_1$, and $\alpha_2 : N_2 \cong M_2$.

       What is the connection between the external direct sum $M_1 \oplus M_2$ and
       the internal direct sum $N_1 \oplus N_2$? (The answer is *very* short!)

3

Decomposition of Finitely Generated
$\mathbb{Z}$-Modules

In the first part of this chapter the theory of f.g. abelian groups G is completed: each G gives rise to a unique sequence $(d_1, d_2, \ldots, d_{t'})$ of non-negative integers, *the invariant factor sequence of* G, such that $d_i | d_{i+1}$ for $1 \le i < t'$ where $d_1 \ne 1$ and $t' \ge 0$. Further two f.g. abelian groups are isomorphic if and only if their invariant factor sequences are identical. One could not wish for a more concise conclusion! Notice that 1 cannot be an invariant factor of G and any zero invariant factors occur at the end of the sequence. For example $(2, 6, 6, 24, 96, 96, 0, 0, 0)$ and $(6, 6, 18, 18, 18, 36, 72)$ are invariant factor sequences.

Nevertheless some questions remain unanswered, for instance: what is the number of isomorphism classes of abelian groups of order n? The answer depends on the prime factorisation of n as we show in Section 3.2. Indeed, just as n can be resolved into a product of prime powers, so every finite abelian group can be decomposed into a direct sum of groups of prime power order, its *primary decomposition*.

Two applications of the theory are discussed in Section 3.3. First the abelian multiplicative groups of units (invertible elements) of finite fields and of the residue class rings $\mathbb{Z}_n$ are analysed. Secondly the isomorphism types of subgroups and quotient groups of a given f.g. abelian group are determined.

C. Norman, *Finitely Generated Abelian Groups and Similarity of Matrices over a Field*,
Springer Undergraduate Mathematics Series,
DOI 10.1007/978-1-4471-2730-7_3, © Springer-Verlag London Limited 2012

3.1 The Invariant Factor Decomposition

Here the theory of matrices over $\mathbb{Z}$ is combined with the theory of abelian groups (aka $\mathbb{Z}$-modules). The main ingredients are Theorem 1.11 on Smith normal form and the first isomorphism theorem 2.16. At the centre of attention is the free $\mathbb{Z}$-module of rank t

$$\mathbb{Z}^t = \mathbb{Z} \oplus \mathbb{Z} \oplus \cdots \oplus \mathbb{Z}$$

made up of t copies of the infinite cyclic group $\mathbb{Z}$. The elements of $\mathbb{Z}^t$ are t-tuples $(m_1, m_2, \ldots, m_t)$ of integers, that is, row vectors with t integer entries. We saw in Section 2.3 that in some ways the theory of $\mathbb{Z}^t$ is analogous to the theory of t-dimensional vector spaces. Our first job here is to show that the rank of submodules of $\mathbb{Z}^t$ behaves as one might expect. Further every finitely generated $\mathbb{Z}$-module G is a homomorphic image of $\mathbb{Z}^t$ for some positive integer t, and so G is isomorphic, by Theorem 2.16, to a quotient group $\mathbb{Z}^t/K$ where K is a submodule of $\mathbb{Z}^t$. The matrix theory of Chapter 1 is then applied to lick the concrete group $\mathbb{Z}^t/K$ into a recognisable shape. In particular the *minimum number* t' of elements needed to generate G can be read off. The conclusion is: G contains t' non-trivial cyclic submodules H_i such that G is the internal direct sum

$$G = H_1 \oplus H_2 \oplus \cdots \oplus H_{t'}$$

where H_i is of type C_{d_i} and $d_i | d_{i+1}$ for $1 \leq i < t'$, $0 \leq t' \leq t$. Although there are in general many ways of decomposing a given f.g. abelian group G as above, Corollary 3.5, Lemma 3.6 and Theorem 3.7 together prove that the sequence $d_1, d_2, \ldots, d_{t'}$ is *unique*.

We first study submodules K of the $\mathbb{Z}$-module $\mathbb{Z}^t$. You should glance back at Theorem 1.15 because it is crucial for our next proof that every ideal of $\mathbb{Z}$ is principal.

Theorem 3.1

Let K be a submodule of $\mathbb{Z}^t$. Then K has a $\mathbb{Z}$-basis $z_1, z_2, \ldots, z_s$ where $0 \leq s \leq t$, that is, K is a free $\mathbb{Z}$-module of rank s.

Proof

The proof is by induction on t. First consider $t = 1$. Submodules K of $\mathbb{Z}^1 = \mathbb{Z}$ are precisely ideals K of $\mathbb{Z}$. By Theorem 1.15 there is d in $\mathbb{Z}$ with $K = \langle d \rangle$. By convention the empty set $\emptyset$ is regarded as being a $\mathbb{Z}$-basis of $K = \langle 0 \rangle$, that is, $K = \{0\}$ is free of rank $s = 0$. For $K \neq \{0\}$ the single non-zero integer $d = z_1$ is a $\mathbb{Z}$-basis of K and so $s = 1$.

Now suppose $t > 1$. To help the induction we regard $\mathbb{Z}^{t-1}$ as being the submodule of $\mathbb{Z}^t$ consisting of t-tuples having last entry zero, that is, $\mathbb{Z}^{t-1} = \{(l_1, l_2, \ldots, l_t) \in \mathbb{Z}^t : l_t = 0\}$. Let K be a submodule of $\mathbb{Z}^t$ and consider $K' = \{l_t : (l_1, l_2, \ldots, l_t) \in K\}$, that is, K' consists of those integers l_t which occur in the last place of t-tuples in K. Then K' is an ideal of $\mathbb{Z}$ (we leave the reader to check this fact). By Theorem 1.15 there is d in $\mathbb{Z}$ with $K' = \langle d \rangle$. The intersection $K \cap \mathbb{Z}^{t-1}$ is a submodule of $\mathbb{Z}^{t-1}$ and so by inductive hypothesis $K \cap \mathbb{Z}^{t-1}$ has a $\mathbb{Z}$-basis $z_1, z_2, \ldots, z_{s-1}$ where $s \leq t$. If $d = 0$ then $K \subseteq \mathbb{Z}^{t-1}$ and $K = K \cap \mathbb{Z}^{t-1}$ has $\mathbb{Z}$-basis as above; so K is free of rank $s - 1 < t$. If $d \neq 0$ then $d \in K'$, that is, there is a t-tuple z_s in K having last entry d. We finish the proof by showing that $z_1, z_2, \ldots, z_{s-1}, z_s$ is a $\mathbb{Z}$-basis of K.

Let $k \in K$. The last entry in the t-tuple k belongs to K' and so is qd for some q in $\mathbb{Z}$. Hence $k - qz_s$ has last entry zero. As z_s belongs to K so also does $k - qz_s$. Therefore $k - qz_s \in K \cap \mathbb{Z}^{t-1} = \langle z_1, z_2, \ldots, z_{s-1} \rangle$. Hence there are integers $l_1, l_2, \ldots, l_{s-1}$ such that $k = l_1 z_1 + l_2 z_2 + \cdots + l_{s-1} z_{s-1} + q z_s$. So $K = \langle z_1, z_2, \ldots, z_{s-1}, z_s \rangle$, that is, $z_1, z_2, \ldots, z_{s-1}, z_s$ generate K according to Definition 2.19(i).

We show that $z_1, z_2, \ldots, z_{s-1}, z_s$ are $\mathbb{Z}$-independent. Suppose there are integers $l_1, l_2, \ldots, l_{s-1}, l_s$ with $l_1 z_1 + l_2 z_2 + \cdots + l_{s-1} z_{s-1} + l_s z_s = 0$ which is an equality between t-tuples of integers. Comparing last entries gives $l_s d = 0$ as the last entry in each of $z_1, z_2, \ldots, z_{s-1}$ is zero and z_s has last entry d. Since $d \neq 0$ we deduce $l_s = 0$. This leaves $l_1 z_1 + l_2 z_2 + \cdots + l_{s-1} z_{s-1} = 0$. As $z_1, z_2, \ldots, z_{s-1}$ form a $\mathbb{Z}$-basis of $K \cap \mathbb{Z}^{t-1}$ they are $\mathbb{Z}$-independent Definition 2.19 and so $l_1 = l_2 = \cdots = l_{s-1} = 0$. Hence $z_1, z_2, \ldots, z_{s-1}, z_s$ are indeed $\mathbb{Z}$-independent and so they form a $\mathbb{Z}$-basis of K. Therefore rank $K = s \leq t$ and the induction is complete. $\qquad \square$

By Theorem 3.1 every submodule K of the free $\mathbb{Z}$-module $\mathbb{Z}^t$ is itself free and $K \cong \mathbb{Z}^s$ for $1 \leq s \leq t$ or $K = \{0\}$ by Exercises 2.3, Question 7(a). So when K is taken out of context its structure is seen to be almost dull. However the interest lies in the relationship between K and its parent $\mathbb{Z}^t$. Suppose $K \neq \{0\}$. Then K has $\mathbb{Z}$-basis $z_1, z_2, \ldots, z_s$.

Let A denote the $s \times t$ matrix over $\mathbb{Z}$ having row i equal to z_i for $1 \leq i \leq s$.

Regarding A as a matrix over the rational field $\mathbb{Q}$ we see that $s = \operatorname{rank} A$ as A has s linearly independent rows (the reader should think through the implication: $z_1, z_2, \ldots, z_s$ linearly dependent over $\mathbb{Q}$ implies $z_1, z_2, \ldots, z_s$ linearly dependent over $\mathbb{Z}$). Now $A = (a_{ij})$ relates the $\mathbb{Z}$-basis $z_1, z_2, \ldots, z_s$ of K to the standard $\mathbb{Z}$-basis $e_1, e_2, \ldots, e_t$ of $\mathbb{Z}^t$ by the s row equations

$$z_i = \sum_{j=1}^{t} a_{ij} e_j \quad \text{for } 1 \leq i \leq s.$$

How does A change when the $\mathbb{Z}$-bases of K and $\mathbb{Z}^t$ are changed? The answer should come as a pleasant surprise! Let $z'_1, z'_2, \ldots, z'_s$ be a $\mathbb{Z}$-basis of K. Form the $s \times t$ matrix

$A' = (a'_{ij})$ having z'_i as row i. By Corollary 2.21 there is an invertible $s \times s$ matrix $P = (p_{ij})$ over $\mathbb{Z}$ such that $z'_i = \sum_{j=1}^{t} p_{ij} z_j$ for $1 \le i \le t$, that is, $A' = PA$. Let $\rho_1, \rho_2, \ldots, \rho_t$ be a $\mathbb{Z}$-basis of $\mathbb{Z}^t$. By Corollary 2.23 the $t \times t$ matrix Q with ρ_j as row j ($1 \le j \le t$) is invertible over $\mathbb{Z}$. Let $B = (b_{ij})$ be the $s \times t$ matrix relating the $\mathbb{Z}$-basis $z'_1, z'_2, \ldots, z'_s$ of K to the $\mathbb{Z}$-basis $\rho_1, \rho_2, \ldots, \rho_t$ of $\mathbb{Z}^t$, that is,

$$z'_i = \sum_{j=1}^{t} b_{ij} \rho_j \quad \text{for } 1 \le i \le s.$$

The above s row equations amount to the single matrix equation $A' = BQ$. Therefore $PA = BQ$ showing that

$$A \text{ changes to the equivalent matrix } B = PAQ^{-1}$$

on changing the $\mathbb{Z}$-bases of K and $\mathbb{Z}^t$. We have already seen the above matrix equation in Definition 1.5! The matrices P and Q, both invertible over $\mathbb{Z}$, are at our disposal – we can choose them to suit our purpose. Using Theorem 1.11 we choose P and Q such that $PAQ^{-1} = D = \text{diag}(d_1, d_2, \ldots, d_s)$ is in Smith normal form, because this choice will reveal the structure of the quotient group $\mathbb{Z}^t/K$. Notice that the d_i are non-zero as rank $A = s$. In fact $B = D$ gives $z'_i = d_i \rho_i$ for $1 \le i \le s$, showing that K has $\mathbb{Z}$-basis $d_1 \rho_1, d_2 \rho_2, \ldots, d_s \rho_s$ made up of integer multiples of the first s elements of the $\mathbb{Z}$-basis $\rho_1, \rho_2, \ldots, \rho_t$ of $\mathbb{Z}^t$.

Let G be a finitely generated $\mathbb{Z}$-module. Then G contains a finite number t of elements $g_1, g_2, \ldots, g_t$ such that $G = \langle g_1, g_2, \ldots, g_t \rangle$, that is, every element g of G is expressible $g = m_1 g_1 + m_2 g_2 + \cdots + m_t g_t$ where $m_1, m_2, \ldots, m_t$ are integers. The next step should come almost as second nature to the reader: consider the $\mathbb{Z}$-linear mapping

$$\theta : \mathbb{Z}^t \to G \quad \text{defined by } (m_1, m_2, \ldots, m_t)\theta = m_1 g_1 + m_2 g_2 + \cdots + m_t g_t$$

for all integers $m_1, m_2, \ldots, m_t$. As we have seen already in special cases the homomorphism θ is the key which unlocks the structure of G. Notice that $(e_j)\theta = g_j$ showing that θ maps the jth element e_j of the standard $\mathbb{Z}$-basis of $\mathbb{Z}^t$ to the jth element g_j of the given generators of G for $1 \le j \le t$. Also θ is surjective as $g = m_1 g_1 + m_2 g_2 + \cdots + m_t g_t$ can be expressed $g = (m_1, m_2, \ldots, m_t)\theta$, that is, $G = \text{im}\,\theta$. Let $K = \ker\theta$ and then

$$\tilde{\theta} : \mathbb{Z}^t/K \cong G \quad \text{where } (K + z)\tilde{\theta} = (z)\theta \quad \text{for all } z = (m_1, m_2, \ldots, m_t) \in \mathbb{Z}^t$$

by Theorem 2.16. The concrete quotient group $\mathbb{Z}^t/K$ is therefore an isomorphic replica of the abstract f.g. abelian group G.

Before tackling the decomposition in general, we study two particular examples.

Example 3.2

Let G be a $\mathbb{Z}$-module generated by g_1, g_2, g_3 subject to the relations

$$3g_1 + 5g_2 + 3g_3 = 0, \qquad 3g_1 + 3g_2 + 5g_3 = 0, \qquad 7g_1 + 3g_2 + 7g_3 = 0.$$

This description is called *a presentation of G*, meaning that the abstract group G is the homomorphic image of a concrete group, the kernel of the homomorphism being specified by a set of generators. In our case $G = \langle g_1, g_2, g_3 \rangle = \operatorname{im} \theta$ where $\theta : \mathbb{Z}^3 \to G$ is given by $(m_1, m_2, m_3)\theta = m_1 g_1 + m_2 g_2 + m_3 g_3$. Write $z_1 = (3, 5, 3)$, $z_2 = (3, 3, 5)$, $z_3 = (7, 3, 7)$. Then the above equations can be expressed as $(z_1)\theta = 0$, $(z_2)\theta = 0$, $(z_3)\theta = 0$ and so z_1, z_2, z_3 belong to $\ker \theta$. The above phrase 'subject to the relations...' means, by convention, that all relations between g_1, g_2, g_3 are $\mathbb{Z}$-linear combinations of the three given relations. In other words $K = \ker \theta$ is generated by z_1, z_2, z_3, that is, $K = \langle z_1, z_2, z_3 \rangle$ and so the rows of

$$A = \begin{pmatrix} z_1 \\ z_2 \\ z_3 \end{pmatrix} = \begin{pmatrix} 3 & 5 & 3 \\ 3 & 3 & 5 \\ 7 & 3 & 7 \end{pmatrix}$$

generate K. In this case z_1, z_2, z_3 is a $\mathbb{Z}$-basis of K as will become clear shortly. Using the method of Chapter 1, the sequence of elementary operations:

$$c_2 - c_1, \qquad c_1 - c_2, \qquad c_2 - 2c_1, \qquad c_3 - 3c_1, \qquad r_2 - 3r_1,$$

$$r_3 - 11r_1, \qquad -r_2, \qquad -r_3, \qquad c_2 - c_3, \qquad c_3 - 2c_2$$

reduces A to $D = \operatorname{diag}(1, 2, 26)$. So there are invertible 3×3 matrices P and Q over $\mathbb{Z}$ satisfying $PA = DQ$. The rows of the diagonal matrix D are non-zero and so these rows are $\mathbb{Z}$-independent. Hence the rows of $A = P^{-1}DQ$ are $\mathbb{Z}$-independent also, that is, z_1, z_2, z_3 is a $\mathbb{Z}$-basis of K. Now $D = PAQ^{-1}$ and P and Q can be found as in Chapter 1. Applying, in order, the *eros* in the above sequence, that is, $r_2 - 3r_1$, $r_3 - 11r_1$, $-r_2$, $-r_3$ to the 3×3 identity matrix I produces

$$P = \begin{pmatrix} 1 & 0 & 0 \\ 3 & -1 & 0 \\ 11 & 0 & -1 \end{pmatrix}.$$

Applying, in order, the *eros* $r_1 + r_2$, $r_2 + r_1$, $r_1 + 2r_2$, $r_1 + 3r_3$, $r_3 + r_2$, $r_2 + 2r_3$, that is, the *conjugates* of the *ecos* in the above sequence, to the 3×3 identity matrix I produces the important matrix

$$Q = \begin{pmatrix} \rho_1 \\ \rho_2 \\ \rho_3 \end{pmatrix} = \begin{pmatrix} 3 & 5 & 3 \\ 3 & 6 & 2 \\ 1 & 2 & 1 \end{pmatrix}.$$

So the rows $\rho_1 = (3, 5, 3)$, $\rho_2 = (3, 6, 2)$, $\rho_3 = (1, 2, 1)$ of Q form a $\mathbb{Z}$-basis of $\mathbb{Z}^3$. As the rows of A form a $\mathbb{Z}$-basis of K, the rows of PA also form a $\mathbb{Z}$-basis of K, that is, the rows ρ_1, $2\rho_2$, $26\rho_3$ of DQ form a $\mathbb{Z}$-basis of K. Hence K is free of rank 3. So we've arrived at the desirable situation where $\mathbb{Z}^3 = \langle \rho_1, \rho_2, \rho_3 \rangle$ and $K = \langle \rho_1, 2\rho_2, 26\rho_3 \rangle$, the $\mathbb{Z}$-basis of K consisting of integer multiples of the $\mathbb{Z}$-basis of $\mathbb{Z}^3$. At this point we abandon the standard $\mathbb{Z}$-basis e_1, e_2, e_3 of $\mathbb{Z}^3$ in favour of ρ_1, ρ_2, ρ_3 which is tailor-made for the analysis of G.

Consider the cyclic subgroups $\langle (\rho_1)\theta \rangle$, $\langle (\rho_2)\theta \rangle$, $\langle (\rho_3)\theta \rangle$ of G. As $\rho_1 \in K = \ker \theta$ we see $\langle (\rho_1)\theta \rangle = \langle 0 \rangle$ and this trivial group is promptly discarded. Write $H_1 = \langle (\rho_2)\theta \rangle$ and $H_2 = \langle (\rho_3)\theta \rangle$. We show next that $G = H_1 \oplus H_2$ (internal direct sum) and that H_1 and H_2 are cyclic of orders 2 and 26 respectively.

As θ is surjective, for each $g \in G$ there is $z \in \mathbb{Z}^3$ with $g = (z)\theta$. As ρ_1, ρ_2, ρ_3 form a $\mathbb{Z}$-basis of $\mathbb{Z}^3$, there are integers l_1, l_2, l_3 with $z = l_1\rho_1 + l_2\rho_2 + l_3\rho_3$. As θ is $\mathbb{Z}$-linear $g = (l_1\rho_1 + l_2\rho_2 + l_3\rho_3)\theta = l_1(\rho_1)\theta + l_2(\rho_2)\theta + l_3(\rho_3)\theta \in H_1 + H_2$ since $(\rho_1)\theta = 0$. Therefore $G = H_1 + H_2$.

We now show that H_1 and H_2 are independent subgroups of G. Suppose $h_1 + h_2 = 0$ where $h_1 \in H_1$, $h_2 \in H_2$. There are $l_2, l_3 \in \mathbb{Z}$ with $h_1 = l_2(\rho_2)\theta$ and $h_2 = l_3(\rho_3)\theta$. So $l_2(\rho_2)\theta + l_3(\rho_3)\theta = 0$. As θ is $\mathbb{Z}$-linear we deduce $(l_2\rho_2 + l_3\rho_3)\theta = 0$ which gives $l_2\rho_2 + l_3\rho_3 \in \ker \theta = K$. As $K = \langle \rho_1, 2\rho_2, 26\rho_3 \rangle$ there are integers l'_1, l'_2, l'_3 with $l_2\rho_2 + l_3\rho_3 = l'_1\rho_1 + 2l'_2\rho_2 + 26l'_3\rho_3$. Now ρ_1, ρ_2, ρ_3 are $\mathbb{Z}$-independent and so 'comparing coefficients' of ρ_1, ρ_2, ρ_3, as explained after Definition 2.19, gives

$$0 = l'_1, \qquad l_2 = 2l'_2, \qquad l_3 = 26l'_3. \qquad (\clubsuit)$$

Two consequences follow from ($\clubsuit$). First

$$h_1 = l_2(\rho_2)\theta = 2l'_2(\rho_2)\theta = l'_2(2\rho_2)\theta = 0,$$

$$h_2 = l_3(\rho_3)\theta = 26l'_3(\rho_3)\theta = l'_3(26\rho_3)\theta = 0$$

since $2\rho_2$, $26\rho_3 \in K = \ker \theta$. Therefore $h_1 = h_2 = 0$ and so

$$G = \langle (\rho_2)\theta \rangle \oplus \langle (\rho_3)\theta \rangle = H_1 \oplus H_2$$

by Lemma 2.15 as we set out to prove. Secondly the orders of the generators $(\rho_i)\theta$ for $i = 2, 3$ can be directly deduced from ($\clubsuit$): suppose $l_2(\rho_2)\theta = 0$ for $l_2 \in \mathbb{Z}$. Taking $l_3 = 0$ we obtain the equation $l_2(\rho_2)\theta + l_3(\rho_3)\theta = 0$ and so conclude $l_2 = 2l'_2$ from ($\clubsuit$). Hence $(\rho_2)\theta$ has order 2 since $(\rho_2)\theta \neq 0$ but $2(\rho_2)\theta = (2\rho_2)\theta = 0$. In the same way suppose $l_3(\rho_3)\theta = 0$ for $l_3 \in \mathbb{Z}$. Setting $l_2 = 0$ we again obtain $l_2(\rho_2)\theta + l_3(\rho_3)\theta = 0$. From ($\clubsuit$) we obtain $l_3 = 26l'_3$. So $r(\rho_3)\theta \neq 0$ for $1 \leq r < 26$, and as $26(\rho_3)\theta = (26\rho_3)\theta = 0$ the order of $(\rho_3)\theta$ is 26. We've shown that $H_1 = \langle (\rho_2)\theta \rangle$ and $H_2 = \langle (\rho_3)\theta \rangle$ are cyclic of isomorphism types C_2 and C_{26} respectively and so G has isomorphism type $C_2 \oplus C_{26}$. The *invariant factor sequence* of G is $(2, 26)$ by

Definition 3.8. Finally rows 2 and 3 of Q express the generators of H_1 and H_2 as integer linear combinations of the original generators g_1, g_2, g_3 of G, that is,

$$(\rho_2)\theta = (3, 6, 2)\theta = 3g_1 + 6g_2 + 2g_3,$$

$$(\rho_3)\theta = (1, 2, 1)\theta = g_1 + 2g_2 + g_3$$

and so

$$G = H_1 \oplus H_2 = \langle 3g_1 + 6g_2 + 2g_3 \rangle \oplus \langle g_1 + 2g_2 + g_3 \rangle.$$

Example 3.3

Let $G = \langle g_1, g_2, g_3 \rangle$ where

$$2g_1 + 4g_2 + 6g_3 = 0,$$

$$8g_1 + 10g_2 + 12g_3 = 0,$$

$$14g_1 + 16g_2 + 18g_3 = 0.$$

This presentation means first that $G = \operatorname{im}\theta$ where $\theta : \mathbb{Z}^3 \to G$ is the $\mathbb{Z}$-linear mapping with $(e_i)\theta = g_i$ for $i = 1, 2, 3$. Secondly $K = \ker\theta$ is the subgroup of $\mathbb{Z}^3$ generated by the rows of

$$A = \begin{pmatrix} 2 & 4 & 6 \\ 8 & 10 & 12 \\ 14 & 16 & 18 \end{pmatrix}$$

and so G is isomorphic to $\mathbb{Z}^3/K$; in fact $\tilde{\theta} : \mathbb{Z}^3/K \cong G$ by Theorem 2.16. We know from Example 1.2 that A has Smith normal form $D = \operatorname{diag}(2, 6, 0)$ and that there are invertible 3×3 matrices P and Q over $\mathbb{Z}$ with $PA = DQ$. The zero entry in the diagonal of D shows that the rows of $A = P^{-1}DQ$ are $\mathbb{Z}$-linearly dependent and so in this case the rows of A do *not* form a $\mathbb{Z}$-basis of K. The rows of

$$Q = \begin{pmatrix} \rho_1 \\ \rho_2 \\ \rho_3 \end{pmatrix} = \begin{pmatrix} 1 & 2 & 3 \\ 0 & 1 & 2 \\ 0 & 0 & 1 \end{pmatrix}$$

form a $\mathbb{Z}$-basis ρ_1, ρ_2, ρ_3 of $\mathbb{Z}^3$ and the *non-zero* rows $2\rho_1, 6\rho_2$ of $PA = DQ$ generate K and are $\mathbb{Z}$-independent, that is, K has $\mathbb{Z}$-basis $2\rho_1, 6\rho_2$. So in this case K is free of rank 2. As in Example 3.2, and in all cases as we'll see in Theorem 3.4, the rows of Q provide us with a suitable $\mathbb{Z}$-basis of $\mathbb{Z}^t$ for decomposing $\mathbb{Z}^t/K \cong G$; this is because the non-zero rows of $PA = DQ$ form a $\mathbb{Z}$-basis of K consisting of integer multiples of the rows of Q. What is more these integer multiples are the non-zero diagonal entries in D and so divide each other successively. Let H_i denote the cyclic submodule of G

generated by $(\rho_i)\theta$, that is, $H_i = \langle (\rho_i)\theta \rangle$ for $i = 1, 2, 3$. As in Example 3.2 we obtain the internal direct sum decomposition:

$$G = H_1 \oplus H_2 \oplus H_3 = \langle (\rho_1)\theta \rangle \oplus \langle (\rho_2)\theta \rangle \oplus \langle (\rho_3)\theta \rangle$$

$$= \langle g_1 + 2g_2 + g_3 \rangle \oplus \langle g_1 + g_2 \rangle \oplus \langle g_3 \rangle.$$

Comparing coefficients of ρ_1, ρ_2, ρ_3, as in Example 3.2 we see that the order ideals of $(\rho_1)\theta$, $(\rho_2)\theta$, $(\rho_3)\theta$ are $\langle 2 \rangle$, $\langle 6 \rangle$, $\langle 0 \rangle$ respectively. So H_1, H_2, H_3 are of isomorphism types C_2, C_6, C_0 being isomorphic to $\mathbb{Z}/\langle 2 \rangle$, $\mathbb{Z}/\langle 6 \rangle$, $\mathbb{Z}/\langle 0 \rangle$ respectively. Finally G has isomorphism type $C_2 \oplus C_6 \oplus C_0$ and so has (as we will see in Definition 3.8) *invariant factor sequence* $(2, 6, 0)$.

We are ready for the first important theorem of this section.

Theorem 3.4 (The invariant factor decomposition of f.g. $\mathbb{Z}$-modules)

Let G be a finitely generated $\mathbb{Z}$-module. Then G contains $t' \geq 0$ non-trivial cyclic submodules H_i $(1 \leq i \leq t')$ such that $G = H_1 \oplus H_2 \oplus \cdots \oplus H_{t'}$ is their internal direct sum, where H_i is of isomorphism type C_{d_i} with $d_i | d_{i+1}$ for $1 \leq i < t'$. Therefore G is of isomorphism type $C_{d_1} \oplus C_{d_2} \oplus \cdots \oplus C_{d_{t'}}$.

Proof

Let $g_1, g_2, \ldots, g_t$ generate G and let $\theta : \mathbb{Z}^t \to G$ be the surjective $\mathbb{Z}$-linear mapping defined, as usual, by $(m_1, m_2, \ldots, m_t)\theta = m_1 g_1 + m_2 g_2 + \cdots + m_t g_t$ for all integers $m_1, m_2, \ldots, m_t$. So $G = \langle g_1, g_2, \ldots, g_t \rangle = \operatorname{im} \theta$. What about $K = \ker \theta$? Suppose first $K = \{0\}$. Then $\theta : \mathbb{Z}^t \cong G$ and so the decomposition $\mathbb{Z}^t = \langle e_1 \rangle \oplus \langle e_2 \rangle \oplus \cdots \oplus \langle e_t \rangle$ gives $G = H_1 \oplus H_2 \oplus \cdots \oplus H_t$ where $H_i = \langle g_i \rangle = \langle (e_i)\theta \rangle$ is of isomorphism type C_0 for $1 \leq i \leq t$. In this case G is free of rank t, $d_i = 0$ for $1 \leq i \leq t$ and $t' = t$.

Suppose now $K \neq \{0\}$. From Theorem 3.1 we know K has a $\mathbb{Z}$-basis $z_1, z_2, \ldots, z_s$ where $1 \leq s \leq t$. Let A denote the $s \times t$ matrix over $\mathbb{Z}$ having z_i as row i and so $e_i A = z_i$ $(1 \leq i \leq s)$. The invariant factors of A consist of 1 (r times) followed by d_i for $1 \leq i \leq s - r$. Let $s' = s - r$ (we allow $r = 0$ but insist that $d_1 \neq 1$). By Corollary 1.13 there are invertible matrices P and Q over $\mathbb{Z}$ such that $PA = DQ$ where $D = \operatorname{diag}(1, 1, \ldots, 1, d_1, d_2, \ldots, d_{s'})$ is in Smith normal form. As $\operatorname{rank} D = \operatorname{rank} A = s$ the integers $d_1, d_2, \ldots, d_{s'}$ are non-zero. By Corollary 2.23 the rows $\rho_1, \rho_2, \ldots, \rho_t$ of Q form a $\mathbb{Z}$-basis of $\mathbb{Z}^t$. As the rows of A form a $\mathbb{Z}$-basis of K we see from Corollary 2.21 that the rows of PA also form a $\mathbb{Z}$-basis of K. Therefore the rows of DQ, that is,

$$\rho_1, \ldots, \rho_r, d_1 \rho_{r+1}, \ldots, d_{s'} \rho_s \quad \text{form a } \mathbb{Z}\text{-basis of } K.$$

As $\rho_1, \rho_2, \ldots, \rho_t$ generate $\mathbb{Z}^t$ and θ is surjective, it follows that $(\rho_1)\theta, (\rho_2)\theta, \ldots, (\rho_t)\theta$ generate $\mathrm{im}\,\theta = G$. But $(\rho_1)\theta = (\rho_2)\theta = \cdots = (\rho_r)\theta = 0$ since $\rho_j \in \ker\theta$ for $1 \le j \le r$. Discarding these r zero generators we are left with $(\rho_{r+1})\theta, (\rho_{r+2})\theta, \ldots, (\rho_t)\theta$ which generate G. Let $t' = t - r$ and let $H_1 = \langle(\rho_{r+1})\theta\rangle$, $H_2 = \langle(\rho_{r+2})\theta\rangle$, $\ldots, H_{t'} = \langle(\rho_t)\theta\rangle$. So H_i is a cyclic submodule of G for $1 \le i \le t'$ and $G = H_1 + H_2 + \cdots + H_{t'}$. We now use Lemma 2.15 to show that G is the internal direct sum of $H_1, H_2, \ldots, H_{t'}$. At the same time we will discover the isomorphism type of each H_i.

Suppose $h_1 + h_2 + \cdots + h_{t'} = 0$ where $h_i \in H_i$ for $1 \le i \le t'$. For each such i there is an integer l_{r+i} with $h_i = l_{r+i}(\rho_{r+i})\theta$ as H_i is cyclic with generator $(\rho_{r+i})\theta$. Hence $l_{r+1}(\rho_{r+1})\theta + l_{r+2}(\rho_{r+2})\theta + \cdots + l_t(\rho_t)\theta = 0$ on substituting for each h_i. Since θ is $\mathbb{Z}$-linear we deduce $(l_{r+1}\rho_{r+1} + l_{r+2}\rho_{r+2} + \cdots + l_t\rho_t)\theta = 0$, that is, $l_{r+1}\rho_{r+1} + l_{r+2}\rho_{r+2} + \cdots + l_t\rho_t \in \ker\theta = K$. As $\rho_1, \ldots, \rho_r, d_1\rho_{r+1}, \ldots, d_{s'}\rho_s$ is a $\mathbb{Z}$-basis of K there are unique integers $l'_1, l'_2, \ldots, l'_s$ satisfying

$$l_{r+1}\rho_{r+1} + l_{r+2}\rho_{r+2} + \cdots + l_t\rho_t = l'_1\rho_1 + \cdots + l'_r\rho_r + l'_{r+1}d_1\rho_{r+1} + \cdots + l'_s d_{s'}\rho_s. \quad (\blacklozenge)$$

As $\rho_1, \rho_2, \ldots, \rho_t$ are $\mathbb{Z}$-independent we can legitimately 'compare coefficients' in equation $(\blacklozenge)$. The first r elements $\rho_1, \rho_2, \ldots, \rho_r$ do not appear on the left-hand side of $(\blacklozenge)$ and so $l'_1 = l'_2 = \cdots = l'_r = 0$. The next s' elements $\rho_{r+1}, \rho_{r+2}, \ldots, \rho_s$ occur on both sides of $(\blacklozenge)$ and so $l_{r+i} = l'_{r+i}d_i$ for $1 \le i \le s'$. Hence $h_i = l_{r+i}(\rho_{r+i})\theta = l'_{r+i}d_i(\rho_{r+i})\theta = l'_{r+i}(d_i\rho_{r+i})\theta = l'_{r+i} \times 0 = 0$ as $d_i\rho_{r+i} \in \ker\theta$ for $1 \le i \le s'$. The last $t - s$ elements $\rho_{s+1}, \rho_{s+2}, \ldots, \rho_t$ do not appear on the right-hand side of $(\blacklozenge)$ and so $l_{s+1} = l_{s+2} = \cdots = l_t = 0$. Hence $h_i = l_{r+i}(\rho_{r+i})\theta = 0(\rho_{r+i})\theta = 0$ for $s' < i \le t'$. Therefore $h_i = 0$ for $1 \le i \le t'$ showing that $H_1, H_2, \ldots, H_{t'}$ are independent submodules of G. By Lemma 2.15

$$G = H_1 \oplus H_2 \oplus \cdots \oplus H_{t'}.$$

Every f.g. abelian group is an internal direct sum of cyclic subgroups.

This is an important part of Theorem 3.4, but there is still more to prove! What are the isomorphism types Definition 2.6 of the above subgroups H_i? Suppose first that $1 \le i \le s'$ and let l_{r+i} belong to the order ideal of the generator $(\rho_{r+i})\theta$ of H_i. Then $l_{r+i}(\rho_{r+i})\theta = 0$ and hence $(l_{r+i}\rho_{r+i})\theta = 0$, that is, $l_{r+i}\rho_{r+i} \in \ker\theta$. Equation $(\blacklozenge)$ holds with $l_{r+j} = 0$ for $1 \le j \le t'$, $j \ne i$. From $(\blacklozenge)$ we deduce $d_i | l_{r+i}$, that is, l_{r+i} is an integer multiple of d_i. On the other hand $d_i(\rho_{r+i})\theta = (d_i\rho_{r+i})\theta = 0$ as $d_i\rho_{r+i}$ belongs to K (it is present in a $\mathbb{Z}$-basis of K). We conclude that $\langle d_i\rangle$ is the order ideal of $(\rho_{r+i})\theta$ and so $H_i = \langle(\rho_{r+i})\theta\rangle$ has isomorphism type C_{d_i} for $1 \le i \le s'$.

Secondly suppose $s' < i \le t'$ and let l_{r+i} belong to the order ideal of the generator $(\rho_{r+i})\theta$ of H_i. Then $l_{r+i}(\rho_{r+i})\theta = 0$ and hence $(l_{r+i}\rho_{r+i})\theta = 0$, that is, $l_{r+i}\rho_{r+i} \in \ker\theta$ as before. Also equation $(\blacklozenge)$ holds with $l_{r+j} = 0$ for $1 \le j \le t'$, $j \ne i$. The conclusion is $l_{r+i} = 0$ in this case and $\langle 0 \rangle$ is the order ideal of $(\rho_{r+i})\theta$. So $H_i = \langle (\rho_{r+i})\theta \rangle$ has isomorphism type C_0 for $s' < i \leq t'$ and

the number of infinite cyclic H_i is $t' - s' = t - s = \operatorname{rank} \mathbb{Z}^t - \operatorname{rank} K$.

Let us define $d_i = 0$ for $s' < i \leq t'$. Then the two preceding paragraphs show that H_i is of isomorphism type C_{d_i} for $1 \leq i \leq t'$. As $D = \operatorname{diag}(1, 1, \ldots, 1, d_1, d_2, \ldots, d_{s'})$ is in Smith normal form we know $d_i | d_{i+1}$ for $1 \leq i < s'$. Therefor $d_i | d_{i+1}$ for $1 \leq i < t'$ as $d_{s'} | 0$ and $0 | 0$. So G is of isomorphism type $C_{d_1} \oplus C_{d_2} \oplus \cdots \oplus C_{d_{t'}}$ by Definition 2.13 and the proof is complete. $\square$

The decomposition theorem 3.4 is a milestone, and you know that it has taken some effort to get this far. However it is an existence theorem – every finitely generated ℤ-module G can be decomposed in a certain way – and as such raises a number of related questions. We mentioned earlier that the cyclic submodules H_i appearing in the decomposition Theorem 3.4 are not unique. Are the non-negative integers d_i, which specify the isomorphism types of the H_i, unique? Our immediate aim is to convince you that the answer here is: Yes! As a consequence the additive abelian groups

$$\mathbb{Z}_2 \oplus \mathbb{Z}_6 \oplus \mathbb{Z}_{24} \oplus \mathbb{Z}_{48} \quad \text{and} \quad \mathbb{Z}_2 \oplus \mathbb{Z}_{12} \oplus \mathbb{Z}_{12} \oplus \mathbb{Z}_{48}$$

are not isomorphic although they have many properties in common: both have order $2^9 \times 3^3$, both have *exponent* 48 (both contain an element of order 48 but no element of higher order), both are generated by 4 of their elements, both have 15 elements of order 2 and 26 elements of order 3.

The proof of the uniqueness of the integers d_i lies in the study of *invariants of* ℤ-*modules*, that is, those properties of ℤ-modules which are preserved by isomorphisms. Anticipating Definition 3.8 the sequence $(d_1, d_2, \ldots, d_{t'})$ as in Theorem 3.4 is uniquely determined by the f.g. ℤ-module G and is called *the sequence of invariant factors* of G. From the abstract point of view this tells us all there is to know about finitely generated ℤ-modules: two such modules are isomorphic if and only if their invariant factor sequences are identical. Note that the integer 1 never occurs as an invariant factor: because $d_1 \neq 1$ and so either $d_1 = 0$ or $d_1 > 1$; as $d_1 | d_i$ we see $d_i \neq 1$ for $1 \leq i \leq t'$. The empty set $\emptyset$ is the invariant factor sequence of every trivial ℤ-module as $t' = 0$ in this case. We now present the details leading up to Definition 3.8.

Let G be a ℤ-module. An element g of G has *finite order* if there is a non-zero integer l with $lg = 0$. From $l_1 g_1 = 0$, $l_2 g_2 = 0$, $l_1 \neq 0$, $l_2 \neq 0$ we deduce $l_1 l_2 (g_1 + g_2) = l_2 l_1 g_1 + l_1 l_2 g_2 = 0 + 0 = 0$, showing that the sum $g_1 + g_2$ of elements $g_1, g_2 \in G$ of finite order is itself an element of finite order since $l_1 l_2 \neq 0$. Let T denote the set of all elements g in G having finite order. It is routine to check that

T is a submodule of G. It is customary to call T the *torsion subgroup* or *torsion submodule of* G. For example the torsion subgroup T of $G = \mathbb{Z}_2 \oplus \mathbb{Z}_6 \oplus \mathbb{Z}$ consists of the 12 triples $(\bar{r}, \bar{s}, 0)$ where $\bar{r} \in \mathbb{Z}_2$, $\bar{s} \in \mathbb{Z}_6$, and so $T \cong \mathbb{Z}_2 \oplus \mathbb{Z}_6$.

When the f.g. $\mathbb{Z}$-module G is decomposed as in Theorem 3.4, it is a relatively simple matter to locate its torsion submodule T as we show next. What is more we'll see that the quotient G/T is free and its rank is a useful invariant.

Corollary 3.5

Let $G = H_1 \oplus H_2 \oplus \cdots \oplus H_{t'}$ be an internal direct sum decomposition as in Theorem 3.4 and so H_i is of isomorphism type C_{d_i} for $1 \le i \le t'$ and $d_i | d_{i+1}$ for $1 \le i < t'$. Let s' be the integer in the range $0 \le s' \le t'$ such that $d_i > 1$ for $1 \le i \le s'$ and $d_i = 0$ for $t' \ge i > s'$. Then G has torsion subgroup $T = H_1 \oplus H_2 \oplus \cdots \oplus H_{s'}$ and $|T| = d_1 d_2 \cdots d_{s'}$. The quotient module G/T is free of rank $t' - s'$.

Proof

Each element g in G can be expressed $g = h_1 + h_2 + \cdots + h_{t'}$ where $h_i \in H_i$. By the uniqueness property of the internal direct sum, we see that $lg = 0$ if and only if $lh_i = 0$ for each i with $1 \le i \le t'$. So g has finite order if and only if each h_i has finite order. For $1 \le i \le s'$ each h_i has finite order since it belongs to the finite cyclic group H_i. For $s' < i \le t'$ the cyclic group H_i is of type C_0 as $d_i = 0$; in other words H_i is infinite cyclic and the only element h_i in such a group having finite order is $h_i = 0$. So T consists of elements $g = h_1 + h_2 + \cdots + h_{s'}$ and hence $T = H_1 \oplus H_2 \oplus \cdots \oplus H_{s'}$. As $|H_i| = d_i$ for $1 \le i \le s'$ we see that T is a finite abelian group of order $d_1 d_2 \cdots d_{s'}$.

The mapping $\pi : G \to G$ defined by $(g)\pi = h_{s'+1} + h_{s'+2} + \cdots + h_{t'}$ where $g = h_1 + h_2 + \cdots + h_{t'}$ is $\mathbb{Z}$-linear. In fact π is the *projection* of G onto its *direct summand* $\operatorname{im} \pi = H_{s'+1} \oplus H_{s'+2} \oplus \cdots \oplus H_{t'}$. Now $(g)\pi = 0$ if and only if $h_i = 0$ for $s' < i \le t'$, that is $\ker \pi = H_1 \oplus H_2 \oplus \cdots \oplus H_{s'} = T$. By Theorem 2.16 we know $G/\ker \pi \cong \operatorname{im} \pi$ and so $G/T \cong \operatorname{im} \pi$. However $\operatorname{im} \pi$ has $\mathbb{Z}$-basis $(\rho_{s+1})\theta, (\rho_{s+2})\theta, \ldots, (\rho_t)\theta$ using the notation of Theorem 3.4, as H_i has isomorphism type C_0 if and only if $s' < i \le t'$. So $\operatorname{im} \pi$ is free of rank $t - s = t' - s'$ and the same is true of G/T by Lemma 2.25.   □

From Theorem 3.4 and Corollary 3.5 it follows that every f.g. $\mathbb{Z}$-module G has an internal direct sum decomposition

$$G = T \oplus M$$

where T is the torsion submodule of G and M is a free submodule of G, because $M = \operatorname{im} \pi$ satisfies these conditions. However M is not usually unique: in the case

$G = \mathbb{Z}_2 \oplus \mathbb{Z}$ then $T = \{(\bar{0}, 0), (\bar{1}, 0)\}$ and both $M_0 = \langle (\bar{0}, 1) \rangle$, $M_1 = \langle (\bar{1}, 1) \rangle$ are free submodules of rank 1 satisfying $G = T \oplus M_0 = T \oplus M_1$.

We show next that isomorphic modules have isomorphic torsion submodules and also that the corresponding quotient modules are isomorphic.

Lemma 3.6

Let $\alpha : G \cong G'$ be an isomorphism between the ℤ-modules G and G'. Let T and T' be the torsion subgroups of G and G' respectively. Then $(T)\alpha = T'$ and so $T \cong T'$. Also $\tilde{\alpha} : G/T \cong G'/T'$ where $(T + g)\tilde{\alpha} = T' + (g)\alpha$ for all $g \in G$.

Proof

Consider $g_0 \in T$. Then $lg_0 = 0$ for some non-zero integer l. Hence $l(g)\alpha = 0$ by the ℤ-linearity of α. So $(g_0)\alpha \in T'$ which means that $(T)\alpha \subseteq T'$. Now let $g_1, g_2 \in G$ be such that $g_1 \equiv g_2 \pmod{T}$. Then $g_1 - g_2 \in T$ and applying α gives $(g_1 - g_2)\alpha \in T'$, that is, $(g_1)\alpha - (g_2)\alpha \in T'$ and so $(g_1)\alpha \equiv (g_2)\alpha \pmod{T'}$. We have shown that $T + g_1 = T + g_2$ implies $T' + (g_1)\alpha = T' + (g_2)\alpha$. So it makes sense to define $\tilde{\alpha}$ as above. Since α is ℤ-linear, so also is $\tilde{\alpha} : G/T \to G'/T'$.

The bijective property of α now comes into play. The mapping $\alpha^{-1} : G' \to G$ is ℤ-linear by Exercises 2.1, Question 4(d), and so α^{-1} is a ℤ-module isomorphism. Replacing α by $\beta = \alpha^{-1}$ in the above paragraph we obtain $(T')\beta \subseteq T$. Applying α to this set containment gives $T' = (T')\beta\alpha \subseteq (T)\alpha$. Therefore $(T)\alpha = T'$ and so $T \cong T'$ as α restricted to T is an isomorphism between T and T'. Also $\tilde{\beta} : G'/T' \to G/T$, defined by $(T' + g')\tilde{\beta} = T + (g')\beta$ for all $g' \in G$, is ℤ-linear. As α, β are an inverse pair of isomorphisms, the same is true of $\tilde{\alpha}$, $\tilde{\beta}$. Therefore $\tilde{\alpha} : G/T \cong G'/T'$. □

We apply Lemma 3.6 to the finitely generated ℤ-module G with decomposition $G = H_1 \oplus H_2 \oplus \cdots \oplus H_{t'}$ as in Theorem 3.4 and the isomorphic ℤ-module G' with an analogous decomposition $G' = H_1' \oplus H_2' \oplus \cdots \oplus H_{t''}'$ where H_j' is cyclic of type $C_{d_j'}$ with $d_1' \neq 1$, $d_j' | d_{j+1}'$ for $1 \leq j \leq t''$. Notice that the case $G = G'$ is included here, that is, amongst other things, we are considering two decompositions of the same ℤ-module. What must these decompositions have in common? Let us suppose that the integer s'' with $0 \leq s'' \leq t''$ is such that $d_j' > 1$ for $1 \leq j \leq s''$ and $d_j' = 0$ for $s'' < j \leq t''$. Then $T' = H_1' \oplus H_2' \oplus \cdots \oplus H_{s''}'$ is the torsion submodule of G' and G'/T' is free of rank $t'' - s''$, on applying Corollary 3.5 to G'. By Lemma 3.6 the free ℤ-modules G/T and G'/T' are isomorphic and so have equal rank by Lemma 2.25, that is,

$$t' - s' = t'' - s''$$

showing that the numbers of infinite cyclic summands in the two decompositions are equal. Therefore this number, which is called the *torsion-free rank of* G, is unambiguously defined and is an invariant of f.g. $\mathbb{Z}$-modules.

We now compare the torsion submodules T and T' of G and G'. From Lemma 3.6 we know that T and T' are isomorphic. Our aim is to show $s' = s''$ and $d_i = d_i'$ for $1 \leq i \leq s'$. This is achieved by studying certain corresponding submodules of T and T' as we now explain. For each integer n and $\mathbb{Z}$-module G, let nG denote the submodule of G consisting of all elements ng for $g \in G$. In other words $nG = \operatorname{im} \mu_n$ where the $\mathbb{Z}$-linear mapping $\mu_n : G \to G$ is given by $(g)\mu_n = ng$ for all $g \in G$. Write $G_{(n)} = \{g \in G : ng = 0\}$ and so $G_{(n)} = \ker \mu_n$ is a submodule of G. For $\mathbb{Z}$-modules $G_1, G_2, \ldots, G_s$ it is straightforward to verify

$$n(G_1 \oplus G_2 \oplus \cdots \oplus G_s) = nG_1 \oplus nG_2 \oplus \cdots \oplus nG_s \quad \text{and}$$

$$(G_1 \oplus G_2 \oplus \cdots \oplus G_s)_{(n)} = (G_1)_{(n)} \oplus (G_2)_{(n)} \oplus \cdots \oplus (G_s)_{(n)}$$

(see Exercises 3.1, Question 5(c)). Suppose the $\mathbb{Z}$-module element g has positive order d. Then ng has order $d/\gcd\{n,d\}$ by Lemma 2.7. Therefore

$$n\mathbb{Z}_d \cong \mathbb{Z}_{d/\gcd\{n,d\}}$$

on taking $g = \bar{1} \in \mathbb{Z}_d$. In particular $n\mathbb{Z}_d$ is trivial if and only if $d/\gcd\{n,d\} = 1$, that is, if and only if $d|n$. The $\mathbb{Z}$-module element $g' = (d/\gcd\{n,d\})g$ has order $d/\gcd\{d/\gcd\{n,d\}, d\} = d/(d/\gcd\{n,d\}) = \gcd\{n,d\}$ by Lemma 2.7. With $g = \bar{1} \in \mathbb{Z}_d$ as above we obtain

$$(\mathbb{Z}_d)_{(n)} \cong \mathbb{Z}_{\gcd\{n,d\}}$$

as $g' = (d/\gcd\{n,d\})\bar{1}$ generates $(\mathbb{Z}_d)_{(n)}$ (see Exercises 3.1, Question 5(d)).

There is just one more thing to point out: for $n \geq 2$ the order of each element of $G_{(n)}$ is a divisor of n and so the $\mathbb{Z}$-module status of $G_{(n)}$ can be upgraded to that of a $\mathbb{Z}_n$-module at no extra cost! Specifically the product $\bar{i}g = (\langle n \rangle + i)g = ig$ for $\bar{i} \in \mathbb{Z}_n$, $g \in G_{(n)}$ is unambiguously defined and gives $G_{(n)}$ the structure of a $\mathbb{Z}_n$-module.

As an illustration consider $T = \mathbb{Z}_6 \oplus \mathbb{Z}_6 \oplus \mathbb{Z}_{12} \oplus \mathbb{Z}_{36}$. Then

$$2T \cong \mathbb{Z}_3 \oplus \mathbb{Z}_3 \oplus \mathbb{Z}_6 \oplus \mathbb{Z}_{18}, \qquad 3T \cong \mathbb{Z}_2 \oplus \mathbb{Z}_2 \oplus \mathbb{Z}_4 \oplus \mathbb{Z}_{12},$$

$$4T \cong \mathbb{Z}_3 \oplus \mathbb{Z}_3 \oplus \mathbb{Z}_3 \oplus \mathbb{Z}_9, \qquad 5T \cong \mathbb{Z}_6 \oplus \mathbb{Z}_6 \oplus \mathbb{Z}_{12} \oplus \mathbb{Z}_{36},$$

$$6T \cong \mathbb{Z}_1 \oplus \mathbb{Z}_1 \oplus \mathbb{Z}_2 \oplus \mathbb{Z}_6, \qquad 8T \cong \mathbb{Z}_3 \oplus \mathbb{Z}_3 \oplus \mathbb{Z}_3 \oplus \mathbb{Z}_9,$$

$$T_{(2)} \cong \mathbb{Z}_2 \oplus \mathbb{Z}_2 \oplus \mathbb{Z}_2 \oplus \mathbb{Z}_2, \qquad T_{(3)} \cong \mathbb{Z}_3 \oplus \mathbb{Z}_3 \oplus \mathbb{Z}_3 \oplus \mathbb{Z}_3,$$

$$T_{(4)} \cong \mathbb{Z}_2 \oplus \mathbb{Z}_2 \oplus \mathbb{Z}_4 \oplus \mathbb{Z}_4, \qquad T_{(5)} \cong \mathbb{Z}_1 \oplus \mathbb{Z}_1 \oplus \mathbb{Z}_1 \oplus \mathbb{Z}_1,$$

$$T_{(6)} \cong \mathbb{Z}_6 \oplus \mathbb{Z}_6 \oplus \mathbb{Z}_6 \oplus \mathbb{Z}_6, \qquad T_{(8)} \cong \mathbb{Z}_2 \oplus \mathbb{Z}_2 \oplus \mathbb{Z}_4 \oplus \mathbb{Z}_4.$$

The decompositions $6T$ and $T_{(6)}$ are the most useful: the following proof is by induction on the number of non-isomorphic non-trivial summands (terms) present which is 3 for T but only 2 for $6T$. Also $T_{(6)} \cong \mathbb{Z}_6^4$ is a free $\mathbb{Z}_6$-module of rank 4 and this shows, as we will see, that every decomposition of T as in Theorem 3.4 has exactly 4 non-trivial summands. We remind the reader that every finite cyclic group of order d is isomorphic to (the additive group of the ring) $\mathbb{Z}_d$ (see Theorem 2.5) and this fact is used in the next proof.

We are now ready for the final theorem of Section 3.1.

Theorem 3.7 (The invariance theorem for finite $\mathbb{Z}$-modules)

Let T and T' be isomorphic finite $\mathbb{Z}$-modules. By Corollary 3.5 there are positive integers $d_1, d_2, \ldots, d_{s'}$ such that $T \cong \mathbb{Z}_{d_1} \oplus \mathbb{Z}_{d_2} \oplus \cdots \oplus \mathbb{Z}_{d_{s'}}$ where $d_1 > 1$ and $d_i | d_{i+1}$ for $1 \le i < s'$. In the same way there are positive integers $d'_1, d'_2, \ldots, d'_{s''}$ such that $T' \cong \mathbb{Z}_{d'_1} \oplus \mathbb{Z}_{d'_2} \oplus \cdots \oplus \mathbb{Z}_{d'_{s''}}$ where $d'_1 > 1$ and $d'_i | d'_{i+1}$ for $1 \le i < s''$. Then $s' = s''$ and $d_i = d'_i$ for $1 \le i \le s'$.

Proof

By hypothesis there is an isomorphism $\alpha : T \cong T'$. Let n be an integer. The $\mathbb{Z}$-linearity of α gives $\mu_n \alpha = \alpha \mu_n$ as $(ng)\alpha = n(g)\alpha$ for all $g \in T$. Therefore $(nT)\alpha = \{(ng)\alpha : g \in T\} = \{n(g)\alpha : g \in T\} \subseteq nT'$ as $(g)\alpha \in T'$ for $g \in T$. Replacing α by α^{-1} we see $(nT')\alpha^{-1} \subseteq nT$. Applying α to this inclusion gives $nT' \subseteq (nT)\alpha$. Therefore $(nT)\alpha = nT'$ showing that the submodules nT and nT' correspond under α and so are isomorphic. We write $\alpha| : nT \cong nT'$ as the restriction $\alpha|$ of α to nT is an isomorphism between nT and nT'.

In the same way for $n \in \mathbb{Z}$ and $g \in T_{(n)}$ we have $(g)\mu_n = 0$ and so $0 = (g)\mu_n \alpha = (g)\alpha \mu_n$ showing $(g)\alpha \in T'_{(n)}$. We've shown $(T_{(n)})\alpha \subseteq T'_{(n)}$. Replacing α by α^{-1} gives $(T'_{(n)})\alpha^{-1} \subseteq T_{(n)}$ and so, on applying α, we obtain $T'_{(n)} \subseteq (T_{(n)})\alpha$. Therefore $(T_{(n)})\alpha = T'_{(n)}$ showing that the submodules $T_{(n)}$ and $T'_{(n)}$ correspond under α and so are isomorphic. As above $\alpha| : T_{(n)} \cong T'_{(n)}$.

Take $n = d_1$ and focus on the isomorphism $\alpha| : T_{(d_1)} \cong T'_{(d_1)}$. Using the preliminary theory $(\mathbb{Z}_{d_i})_{(d_1)} \cong \mathbb{Z}_{d_1}$ since $\gcd\{d_1, d_i\} = d_1$ for $1 \le i \le s'$. From $T \cong \mathbb{Z}_{d_1} \oplus \mathbb{Z}_{d_2} \oplus \cdots \oplus \mathbb{Z}_{d_{s'}}$ we deduce

$$T_{(d_1)} \cong (\mathbb{Z}_{d_1})_{(d_1)} \oplus (\mathbb{Z}_{d_2})_{(d_1)} \oplus \cdots \oplus (\mathbb{Z}_{d_{s'}})_{(d_1)}$$

$$\cong \mathbb{Z}_{d_1} \oplus \mathbb{Z}_{d_1} \oplus \cdots \oplus \mathbb{Z}_{d_1} = (\mathbb{Z}_{d_1})^{s'}.$$

The $\mathbb{Z}_{d_1}$-module $T_{(d_1)}$ is therefore isomorphic to the free $\mathbb{Z}_{d_1}$-module $(\mathbb{Z}_{d_1})^{s'}$ of rank s'. By Lemma 2.25 both $T_{(d_1)}$ and $T'_{(d_1)}$ are free $\mathbb{Z}_{d_1}$-modules of rank s'. Combining $(\mathbb{Z}_{d_i'})_{(d_1)} \cong \mathbb{Z}_{\gcd\{d_1,d_i'\}}$ for $1 \le i \le s''$ and $T' \cong \mathbb{Z}_{d_1'} \oplus \mathbb{Z}_{d_2'} \oplus \cdots \oplus \mathbb{Z}_{d_{s''}'}$ gives

$$T'_{(d_1)} \cong \mathbb{Z}_{\gcd\{d_1,d_1'\}} \oplus \mathbb{Z}_{\gcd\{d_1,d_2'\}} \oplus \cdots \oplus \mathbb{Z}_{\gcd\{d_1,d_{s''}'\}}$$

showing that $T'_{(d_1)}$ is the direct sum of s'' cyclic submodules and so is generated by s'' of its elements. From Theorem 2.20 we deduce $s'' \ge s'$. We have reached the tipping point in the proof! Using the isomorphism $\alpha^{-1} : T' \cong T$ the preceding theory 'works' on interchanging T and T'. Using the isomorphism $\alpha^{-1}| : T'_{(d_1')} \cong T_{(d_1')}$ we see that the $\mathbb{Z}_{d_1'}$-module $T'_{(d_1')}$ is isomorphic to the free $\mathbb{Z}_{d_1'}$-module $(\mathbb{Z}_{d_1'})^{s''}$ of rank s''. Hence

$$T_{(d_1')} \cong \mathbb{Z}_{\gcd\{d_1',d_1\}} \oplus \mathbb{Z}_{\gcd\{d_1',d_2\}} \oplus \cdots \oplus \mathbb{Z}_{\gcd\{d_1',d_{s'}\}}$$

is a free $\mathbb{Z}_{d_1'}$-module of rank s'' by Lemma 2.25 and is generated by s' of its elements. Therefore $s' \ge s''$ by Theorem 2.20 and so $s' = s''$. From Lemma 2.18 and Corollary 2.21 the s' generators of $T'_{(d_1)}$ form a $\mathbb{Z}_{d_1}$-basis of the $\mathbb{Z}_{d_1}$-module $T'_{(d_1)}$ (see Exercises 2.3, Question 7(b)). Therefore each of these s' generators has order ideal $\{\overline{0}\} = \{\langle d_1 \rangle\}$ in the $\mathbb{Z}_{d_1}$-module $T'_{(d_1)}$ and so has order d_1 in the $\mathbb{Z}$-module $T'_{(d_1)}$. From the first of these generators we deduce $\gcd\{d_1, d_1'\} = d_1$ showing $d_1|d_1'$. Interchanging the roles of T and T' gives $d_1'|d_1$ and so $d_1 = d_1'$.

Let m_1 denote the number of i with $d_i = d_1$ and let m_1' denote the number of i with $d_i' = d_1$. As $d_1\mathbb{Z}_{d_i} \cong \mathbb{Z}_{d_i/\gcd\{d_1,d_i\}} = \mathbb{Z}_{d_i/d_1}$, from the preliminary discussion, we obtain $d_1 T \cong \sum_{m_1 < i \le s'} \oplus \mathbb{Z}_{d_i/d_1}$. So $d_1 T$ is the direct sum of $s' - m_1$ non-trivial cyclic submodules and, since $d_i/d_1 | d_j/d_1$ for $m_1 < i \le j \le s'$, this decomposition is again as in Corollary 3.5. In the same way $d_1 T' \cong \sum_{m_1' < i \le s'} \oplus \mathbb{Z}_{d_i'/d_1}$ which is a decomposition of $d_1 T'$ into $s' - m_1'$ non-trivial cyclic submodules as in Corollary 3.5. As $\alpha| : d_1 T \cong d_1 T'$ the proof can be completed by induction on the number, r say, of *different* integers among $d_1, d_2, \ldots, d_{s'}$. Take $r = 1$. Then $m_1 = s'$ and $d_1 T$ is trivial. So $d_1 T'$ is also trivial and $m_1' = s'$. Therefore $d_i = d_1 = d_i'$ for $1 \le i \le s'$. Now take $r > 1$. There are $r - 1$ different integers among $d_{m_1+1}/d_1, d_{m_1+2}/d_1, \ldots, d_{s'}/d_1$ and so the conclusion of Theorem 3.7 holds on replacing $\alpha : T \cong T'$ by $\alpha| : d_1 T \cong d_1 T'$, that is, $s' - m_1 = s' - m_1'$ (showing $m_1 = m_1'$) and also $d_i/d_1 = d_i'/d_1$ for $m_1 < i \le s'$. We now have $s' = s''$, $d_i = d_1 = d_i'$ for $1 \le i \le m_1 = m_1'$ and $d_i = d_i'$ for $m_1 < i \le s'$ on multiplying by d_1. The induction is therefore complete. $\qquad\square$

We have finally arrived and all we need to do is pull ourselves together!

Definition 3.8

Let G be a finitely generated $\mathbb{Z}$-module. Then G decomposes $G = H_1 \oplus H_2 \oplus \cdots \oplus H_{t'}$ as the internal direct sum of t' non-trivial cyclic subgroups H_i $(1 \le i \le t')$ where H_i is of isomorphism type C_{d_i} and $d_i | d_{i+1}$ $(1 \leq i < t')$ by Theorem 3.4. By Corollary 3.5, Lemma 3.6, Theorem 3.7 the integers d_i are unique and so it is legitimate to say:

$$(d_1, d_2, \ldots, d_{t'})$$

is *the invariant factor sequence of G.*

Let s' be as in Corollary 3.5. Then $(d_1, d_2, \ldots, d_{t'}) = (d_1, d_2, \ldots, d_{s'}, 0, 0, \ldots, 0)$ showing that the invariant factor sequence of G terminates in $t' - s'$ zeros.

For example suppose $(2, 6, 42, 84, 0, 0, 0)$ is the invariant factor sequence of G. Then G has torsion-free rank 3 (the number of zero invariant factors) and the free $\mathbb{Z}$-module G/T has invariant factor sequence $(0, 0, 0)$. Also $(2, 6, 42, 84)$ is the invariant factor sequence of the torsion subgroup T of G and so $|T| = 2 \times 6 \times 42 \times 84 = 2^5 \times 3^3 \times 7^2$.

Our next corollary 'sows up' the theory of isomorphism classes of f.g. $\mathbb{Z}$-modules: there is just one class for each invariant factor sequence.

Corollary 3.9 (Classification of finitely generated $\mathbb{Z}$-modules)

Let G and G' be finitely generated $\mathbb{Z}$-modules. Then G and G' are isomorphic if and only if their invariant factor sequences $(d_1, d_2, \ldots, d_{t'})$ and $(d'_1, d'_2, \ldots, d'_{t''})$ are equal, that is, $t' = t''$ and $d_i = d'_i$ for $1 \leq i \leq t'$.

Proof

Suppose first that G and G' are isomorphic $\mathbb{Z}$-modules. So there is an isomorphism $\alpha : G \cong G'$. By Lemma 3.6 the torsion subgroups T and T' of G and G' are isomorphic. We deduce $d_i = d'_i$ for $1 \leq i \leq s'$ by Corollary 3.5 and Theorem 3.7, accounting for all non-zero invariant factors of G and G'. So $(d_1, d_2, \ldots, d_{t'})$ terminates with $t' - s'$ zeros and $(d'_1, d'_2, \ldots, d'_{t''})$ terminates with $t'' - s'$ zeros. By Corollary 3.5 and Lemma 3.6 the quotient $\mathbb{Z}$-modules G/T and G'/T' are isomorphic and free of ranks $t' - s'$ and $t'' - s'$ respectively. Hence $t' - s' = t'' - s'$ by Lemma 2.25 showing that G and G' have the same torsion-free rank. Therefore $t' = t''$ and $d_i = d'_i$ for $1 \leq i \leq t'$.

Conversely suppose $(d_1, d_2, \ldots, d_{t'}) = (d'_1, d'_2, \ldots, d'_{t''})$, that is, $t' = t''$ and $d_i = d'_i$ for $1 \leq i \leq t'$. Then G and G' have internal direct sum decompositions $G = H_1 \oplus H_2 \oplus \cdots \oplus H_{t'}$ and $G' = H'_1 \oplus H'_2 \oplus \cdots \oplus H'_{t'}$ where H_i is a cyclic subgroup of G and H'_i is a cyclic subgroup of G'; also H_i and H'_i are of the same isomorphism type C_{d_i} for $1 \leq i \leq t'$. By Theorem 2.5 there are isomorphisms $\alpha_i : H_i \cong H'_i$ for $1 \leq i \leq t'$. For each g in G there are unique elements $h_i \in H_i$ with $g = h_1 + h_2 + \cdots + h_{t'}$; we define $\alpha : G \to G'$ by $(g)\alpha = (h_1)\alpha_1 + (h_2)\alpha_2 + \cdots + (h_{t'})\alpha_{t'} \in G'$. As each α_i is $\mathbb{Z}$-linear so also is α. As each α_i is bijective so also is α. We conclude $\alpha : G \cong G'$ and so G and G' are isomorphic $\mathbb{Z}$-modules. $\square$

The sequence $(d_1, d_2, \ldots, d_{t'})$ of t' non-negative integers d_i is said to satisfy the *invariant factor condition* if $d_1 \neq 1$ and $d_i | d_{i+1}$ for $1 \leq i < t'$ where $t' \geq 0$. Do *all* such sequences arise from f.g. $\mathbb{Z}$-modules as in Definition 3.8? The answer is: Yes! The additive group $\mathbb{Z}/\langle n \rangle$ has invariant factor sequence (n) for all non-negative integers n $(n \neq 1)$. More generally suppose $(d_1, d_2, \ldots, d_{t'})$ satisfies the invariant factor condition; then the external direct sum $\mathbb{Z}/\langle d_1 \rangle \oplus \mathbb{Z}/\langle d_2 \rangle \oplus \cdots \oplus \mathbb{Z}/\langle d_{t'} \rangle$ is a $\mathbb{Z}$-module with invariant factor sequence $(d_1, d_2, \ldots, d_{t'})$.

There are six isomorphism classes of abelian groups G of order 72, that is, with $|G| = 72$ because there are six sequences $(d_1, d_2, \ldots, d_{t'})$ of non-negative integers with $d_1 d_2 \cdots d_{t'} = 72$ satisfying the invariant factor condition namely $(2, 2, 18)$, $(2, 6, 6)$, $(2, 36)$, $(3, 24)$, $(6, 12)$, (72). We will discover in Section 3.2 that these sequences are best found using the prime factorisation of $72 = 2^3 \times 3^2$.

It follows from Theorem 3.4 and Definition 3.8 that the finitely generated $\mathbb{Z}$-module G cannot be generated by less than t' of its elements. So the invariant factor decomposition $G = H_1 \oplus H_2 \oplus \cdots \oplus H_{t'}$ is the best one can achieve in the sense that the number t' of cyclic direct summands H_i is a minimum. In Section 3.2 we discuss a way of decomposing a finite abelian group into a direct sum of non-trivial cyclic subgroups where the number of summands is maximum. This new method of decomposition is in some ways more revealing than the invariant factor decomposition.

EXERCISES 3.1

1. Throughout this question the $\mathbb{Z}$-module G is generated by its elements g_1, g_2 and $\theta : \mathbb{Z}^2 \rightarrow G$ is the $\mathbb{Z}$-linear mapping defined by $(m_1, m_2)\theta = m_1 g_1 + m_2 g_2$ for all $m_1, m_2 \in \mathbb{Z}$.

   (a) The generators g_1, g_2 of G are subject to the relations $4g_1 + 2g_2 = 0$, $10g_1 + 2g_2 = 0$. Calculate the Smith normal form D of $A = \left(\begin{smallmatrix} 4 & 2 \\ 10 & 2 \end{smallmatrix} \right)$. Find invertible matrices P and Q over $\mathbb{Z}$ with $PA = DQ$. Do the rows z_1, z_2 of A form a $\mathbb{Z}$-basis of $K = \ker \theta$? Select a $\mathbb{Z}$-basis of $\ker \theta$ from the rows of DQ. What is the value of rank K? Express $(\rho_1)\theta$ and $(\rho_2)\theta$ as integer linear combinations of g_1, g_2 where ρ_1, ρ_2 denote the rows of Q. Is $G = \langle (\rho_1)\theta \rangle \oplus \langle (\rho_2)\theta \rangle$? (Yes/No). State the orders of $(\rho_1)\theta$ and $(\rho_2)\theta$. State the invariant factor sequence and isomorphism type of G.

   (b) The generators g_1, g_2 of G are subject to the relations $3g_1 + 5g_2 = 0$, $7g_1 + 9g_2 = 0$. Answer Question 1(a) above in the case of this $\mathbb{Z}$-module G, that is, calculate the Smith normal form D of $A = \left(\begin{smallmatrix} 3 & 5 \\ 7 & 9 \end{smallmatrix} \right)$ etc. Is G cyclic?

(c) The generators g_1, g_2 of G are subject to the relations $8g_1 + 9g_2 = 0$, $7g_1 + 8g_2 = 0$. Answer Question 1(a) above in the case of this ℤ-module G. Is G trivial?

(d) The generators g_1, g_2 of G are subject to the relations $2g_1 + 4g_2 = 0$, $4g_1 + 8g_2 = 0$. Answer Question 1(a) above in the case of this ℤ-module G. What is the torsion-free rank of G and the order of its torsion submodule?

(e) The generators g_1, g_2 of G are subject to the relations $3g_1 + 4g_2 = 0$, $6g_1 + 8g_2 = 0$. Answer Question 1(a) above in the case of this ℤ-module G. Is G infinite cyclic?

(f) The generators g_1, g_2 of G are subject to the relations $14g_1 + 36g_2 = 0$, $12g_1 + 28g_2 = 0$, $28g_1 + 12g_2 = 0$. Find the Smith normal form of

$$A = \begin{pmatrix} 14 & 36 \\ 12 & 28 \\ 28 & 12 \end{pmatrix}$$

and hence answer Question 1(a) above in the case of this ℤ-module G.

(g) The generators g_1, g_2 of G are subject to the relations $n_1 g_1 = 0$, $n_2 g_2 = 0$ where n_1, n_2 are positive integers. Use Lemma 1.10 to find the invariant factors of G. Under what condition on n_1, n_2 is G cyclic? *Hint:* The number of invariant factors of G is 0, 1 or 2.

2. Throughout this question the ℤ-module G is generated by its elements g_1, g_2, g_3 and $\theta : \mathbb{Z}^3 \to G$ is the ℤ-linear mapping defined by $(m_1, m_2, m_3)\theta = m_1 g_1 + m_2 g_2 + m_3 g_3$ for all $m_1, m_2, m_3 \in \mathbb{Z}$.

(a) The generators g_1, g_2, g_3 of G are subject to the relations

$$2g_1 + 4g_2 + 4g_3 = 0,$$

$$4g_1 + 8g_2 + 6g_3 = 0,$$

$$2g_1 + 6g_2 + 6g_3 = 0.$$

Find the Smith normal form D of the coefficient matrix

$$A = \begin{pmatrix} 2 & 4 & 4 \\ 4 & 8 & 6 \\ 2 & 6 & 6 \end{pmatrix}$$

of these relations. Find also invertible matrices P and Q over ℤ satisfying $PA = DQ$. Use the rows of Q and the non-zero rows of DQ to specify ℤ-bases of $\mathbb{Z}^3$ and $K = \ker \theta$. Hence decompose G into an internal direct sum of cyclic submodules expressing their generators as integer linear combinations of g_1, g_2, g_3. State the isomorphism type and the sequence of invariant factors of G.

(b) The generators g_1, g_2, g_3 of G are subject to the relations

$$3g_1 + 2g_2 + 5g_3 = 0,$$
$$2g_1 + 4g_2 + 8g_3 = 0,$$
$$3g_1 + 4g_2 + 7g_3 = 0.$$

Answer Question 2(a) above in the case of this $\mathbb{Z}$-module, i.e. find the Smith normal form D of the coefficient matrix

$$A = \begin{pmatrix} 3 & 2 & 5 \\ 2 & 4 & 8 \\ 3 & 4 & 7 \end{pmatrix}$$

of these relations, etc. Is this $\mathbb{Z}$-module isomorphic to the $\mathbb{Z}$-module G of Question 1(a) above?

(c) The generators g_1, g_2, g_3 of G are subject to the relations

$$2g_1 + 4g_2 + 2g_3 = 0,$$
$$4g_1 + 6g_2 + 2g_3 = 0,$$
$$2g_1 + 6g_2 + 4g_3 = 0.$$

Answer Question 2(a) above in the case of this $\mathbb{Z}$-module. State the invariant factors of the torsion submodule T of G. State the torsion-free rank of G. Is G/T cyclic?

(d) The generators g_1, g_2, g_3 of G are subject to the single relation $8g_1 + 12g_2 + 18g_3 = 0$. Answer Question 2(a) above in the case of this $\mathbb{Z}$-module. State the invariant factor sequence of the torsion submodule T of G. State the torsion-free rank of G and the invariant factor sequence of G/T.

3. (a) Let $K = \{(m_1, m_2) \in \mathbb{Z}^2 : \text{parity } m_1 = \text{parity } m_2\}$. Show that K is a submodule of $\mathbb{Z}^2$. Verify that K has $\mathbb{Z}$-basis $z_1 = (1, 1)$, $z_2 = (0, 2)$. Find the Smith normal form of the 2×2 matrix A having z_i as row i for $i = 1, 2$. Hence find the isomorphism type of $\mathbb{Z}^2/K$.

(b) Let $K = \{(m_1, m_2, m_3) \in \mathbb{Z}^3 : m_1 \equiv m_2 \equiv m_3 \pmod{2}\}$. Find a $\mathbb{Z}$-basis z_1, z_2, z_3 of the submodule K of $\mathbb{Z}^3$. Find the Smith normal form of the 3×3 matrix A having z_i as row i for $i = 1, 2, 3$. State the isomorphism type of $\mathbb{Z}^3/K$.

(c) Let t and n be given positive integers. Let $K = \{(m_1, m_2, \ldots, m_t) \in \mathbb{Z}^t : m_1 \equiv m_j \pmod{n} \text{ for all } j \text{ with } 2 \leq j \leq t\}$. Find a $\mathbb{Z}$-basis $z_1, z_2, \ldots, z_t$ of the submodule K of $\mathbb{Z}^t$. Find the Smith normal form of the $t \times t$ matrix A having z_i as row i for $1 \leq i \leq t$. Find the isomorphism type of $\mathbb{Z}^t/K$.

4. (a) Let G denote the additive group $\mathbb{Z}_8 \oplus \mathbb{Z}_{12}$. Show that $g = (\bar{1}, \bar{1}) \in G$ has order $24 = \text{lcm}\{8, 12\}$.

Hint: Start by showing $24g = 0$.

Find the orders of the elements $(\bar{2}, \bar{1})$ and $(\bar{1}, \bar{2})$ of G.

(b) Let n_1, n_2 be positive integers. Show that the element $g = (\bar{1}, \bar{1})$ of the additive group $\mathbb{Z}_{n_1} \oplus \mathbb{Z}_{n_2}$ has order $\mathrm{lcm}\{n_1, n_2\}$.

The $\mathbb{Z}$-module G has submodules $H_1, H_2, \ldots, H_t$ such that $G = H_1 \oplus H_2 \oplus \cdots \oplus H_t$. Let $h_i \in H_i$ be an element of finite order n_i $(1 \leq i \leq t)$. Show that $g = h_1 + h_2 + \cdots + h_t$ has finite order $m = \mathrm{lcm}\{n_1, n_2, \ldots, n_t\}$.

Hint: Show first that g has finite order m' where $m' | m$, and then show $n_i | m'$ $(1 \leq i \leq t)$. Finally use Exercises 1.3, Question 1(c).

(c) Suppose that the $\mathbb{Z}$-module G is the internal direct sum of cyclic submodules $H_i = \langle h_i \rangle$ $(1 \leq i \leq t)$ and so each g in G is expressible as $g = m_1 h_1 + m_2 h_2 + \cdots + m_t h_t$ where $m_1, m_2, \ldots, m_t \in \mathbb{Z}$. Let h_i have order ideal $\langle n_i \rangle$ for $1 \leq i \leq t$. Use Lemma 2.7 and (b) above to show that the order of g is

$$\mathrm{lcm}\{n_1 / \gcd\{m_1, n_1\}, n_2 / \gcd\{m_2, n_2\}, \ldots, n_t / \gcd\{m_t, n_t\}\}$$

in case each $n_i > 0$. What is the order of g if there is i with $m_i \neq 0$, $n_i = 0$?

(d) Use the formula of (c) above to calculate the orders of g_1, g_2 in the module G of Question 1(f) above.

Hint: The relevant integers m_j appear in the rows of Q^{-1}, and this matrix can be calculated directly by applying to I, in order, the *ecos* appearing in the reduction of A to D.

(e) Use the formula of (c) above to calculate the orders of g_1, g_2, g_3 in the module G of Question 2(b) above.

5. (a) Let G be a $\mathbb{Z}$-module and let $\theta : \mathbb{Z}^t \to G$ be a surjective $\mathbb{Z}$-linear mapping. Let $z_1, z_2, \ldots, z_s$ generate $\ker \theta$ and let A be the $s \times t$ matrix over $\mathbb{Z}$ having z_i as row i for $1 \leq i \leq s$. Let D be the Smith normal form of A and let P and Q be invertible matrices over $\mathbb{Z}$ satisfying $PA = DQ$. Show that the non-zero rows of DQ form a $\mathbb{Z}$-basis of $\ker \theta$.

(b) Let H be a subgroup of the f.g. $\mathbb{Z}$-module G. Use Theorem 3.1 and a surjective $\mathbb{Z}$-linear mapping $\theta : \mathbb{Z}^t \to G$ to show that H is also finitely generated.

Hint: Consider $K' = \{k \in \mathbb{Z}^t : (k)\theta \in H\}$.

(c) Let $G_1, G_2, \ldots, G_s$ be $\mathbb{Z}$-modules. Using the notation introduced before Theorem 3.7 verify

$$n(G_1 \oplus G_2 \oplus \cdots \oplus G_s) = nG_1 \oplus nG_2 \oplus \cdots \oplus nG_s \quad \text{and}$$

$$(G_1 \oplus G_2 \oplus \cdots \oplus G_s)_{(n)} = (G_1)_{(n)} \oplus (G_2)_{(n)} \oplus \cdots \oplus (G_s)_{(n)}$$

for $n \in \mathbb{Z}$.

(d) Let n and d be integers with $d \geq 1$. Show that $g' = (d/\gcd\{n,d\})\bar{1}$ generates $(\mathbb{Z}_d)_{(n)}$ where $\bar{1}$ is the 1-element of $\mathbb{Z}_d$.

(e) Let $d_1, d_2, \ldots, d_r$ be r distinct positive integers with $d_1 \geq 2$ satisfying $d_i | d_j$ for $1 \leq i \leq j \leq r$ and let $m_1, m_2, \ldots, m_r$ be positive integers. Suppose $G \cong (\mathbb{Z}_{d_1})^{m_1} \oplus (\mathbb{Z}_{d_2})^{m_2} \oplus \cdots \oplus (\mathbb{Z}_{d_r})^{m_r}$ and so d_i occurs m_i times in the invariant factor sequence Definition 3.8 of G for $1 \leq i \leq r$. Express the submodules $d_i G$ and $G_{(d_i)}$ in the same way (see the proof of Theorem 3.7) for $1 \leq i \leq r$. Is $(d_i G)_{(d_{i+1}/d_i)}$ a free R-module for $1 \leq i < r$? If so find the ring R and state the rank of this R-module.

6. (a) Let G be an additive abelian group. A homomorphism $\chi : G \to \mathbb{Q}/\mathbb{Z}$ is called *a character of* G. (Here $\mathbb{Q}/\mathbb{Z}$ denotes the additive group of rationals modulo 1 consisting of cosets $\mathbb{Z} + q$ for $q \in \mathbb{Q}$, see Exercises 2.2, Question 3.) We use, as is customary, the functional notation $\chi(g)$ for the image of $g \in G$ by χ. Let χ_1 and χ_2 be characters of G. Show that their sum $\chi_1 + \chi_2$, defined by $(\chi_1 + \chi_2)(g) = \chi_1(g) + \chi_2(g)$ for all $g \in G$, is a character of G. Show also that the set of all characters of G, together with addition as defined above, is an abelian group G^*. The group G^* is called *the character group of* G.

Let H be a subgroup of G. Show that $H^o = \{\chi \in G^* : \chi(h) = \mathbb{Z}$ (the zero element of $\mathbb{Q}/\mathbb{Z})$ for all $h \in H\}$ is a subgroup of G^*. For $\chi \in (G/H)^*$ let $(\chi)\alpha : G \to \mathbb{Q}/\mathbb{Z}$ be defined by $((\chi)\alpha)(g) = \chi(H + g)$ for all $g \in G$. Show $(\chi)\alpha \in H^o$ and $\alpha : (G/H)^* \cong H^o$.

(b) Let G be a finite abelian group. By Theorem 3.4 there are elements h_i of order d_i in G $(1 \leq i \leq t)$ such that $G = \langle h_1 \rangle \oplus \langle h_2 \rangle \oplus \cdots \oplus \langle h_t \rangle$ where t is a positive integer (there's no need here to insist that $d_1 \neq 1$ or $d_i | d_{i+1}$). For each i with $1 \leq i \leq t$ show that G has a unique character χ_i such that $\chi_i(h_i) = \mathbb{Z} + 1/d_i$, $\chi_i(h_j) = \mathbb{Z} + 0$ for all $j \neq i$. *Hint*: $\mathbb{Z} + 1/d_i$ has order d_i in $\mathbb{Q}/\mathbb{Z}$.

Show that χ_i has order d_i in G^* and $G^* = \langle \chi_1 \rangle \oplus \langle \chi_2 \rangle \oplus \cdots \oplus \langle \chi_t \rangle$. Deduce that there is a unique isomorphism $\beta : G^* \cong G$ with $(\chi_i)\beta = h_i$ for $1 \leq i \leq t$.

For each subgroup H of G write $(H)\pi = (H^o)\beta$. Let $\mathbb{L}(G)$ denote the set of all subgroups H of G. Show that $\pi : \mathbb{L}(G) \to \mathbb{L}(G)$ is a *polarity*, that is,

  (i) $(H)\pi^2 = H$ for all $H \in \mathbb{L}(G)$,

  (ii) $H_1 \subseteq H_2 \Leftrightarrow (H_1)\pi \supseteq (H_2)\pi$ where $H_1, H_2 \in \mathbb{L}(G)$ (π is inclusion-reversing).

Hint: For (i) show first $(H)\beta^{-1} \subseteq ((H^o)\beta)^o$.

7. (a) (i) Let R be a non-trivial commutative ring and let $U(R)$ denote its multiplicative group of invertible elements. For $a, b \in R$ write $a \equiv b$ if there is $u \in U(R)$ with $a = bu$. Show that $\equiv$ is an equivalence relation on R (the equivalence classes are called *associate classes*). Are $\{0\}$ and $U(R)$ associate classes? Which commutative rings partition into two associate classes? Partition $\mathbb{Z}_{12}$ into associate classes.

(ii) Let R be an integral domain. For $a, b \in R$ write $b|a$ if there is $c \in R$ with $a = bc$. Show $b|a \Leftrightarrow \langle a \rangle \subseteq \langle b \rangle$. Show also $a \equiv b$ if and only if $a|b$ and $b|a$. Deduce $\langle a \rangle = \langle b \rangle \Leftrightarrow a \equiv b$.

(b) Let R be a *principal ideal domain* (PID), that is, R is a non-trivial commutative ring with no zero-divisors such that each ideal K of R is expressible as $K = \langle d \rangle = \{rd : r \in R\}$ for some $d \in R$.

For $a, b \in R$ write $\langle a \rangle + \langle b \rangle = \langle d \rangle$ and $\langle a \rangle \cap \langle b \rangle = \langle m \rangle$. Show that d and m have the divisor properties of $\gcd\{a, b\}$ and $\text{lcm}\{a, b\}$ respectively, that is, $d|a$, $d|b$ and $d'|a$, $d'|b \Rightarrow d'|d$ for $d' \in R$ and also $a|m$, $b|m$ and $a|m'$, $b|m' \Rightarrow m|m'$ for $m' \in R$. Show $d = a'a + b'b$ for $a', b' \in R$. Also show $ab \equiv dm$.

Hint: Show ab/d has the properties of $\text{lcm}\{a, b\}$ in the case $d \neq 0$.

Let K_i be an ideal of R for each positive integer i with $K_i \subseteq K_{i+1}$. Show that $K = \bigcup_{i \geq 1} K_i$ is an ideal of R. Deduce the existence of a positive integer l with $K_i = K_l$ for $i \geq l$. Let $b_1, b_2, \ldots, b_i, \ldots$ be a sequence of elements of R such that $b_{i+1}|b_i$ for $i \geq 1$. Show that there is an integer l with $b_i \equiv b_l$ for $i \geq l$.

(c) Let R be a PID. A diagonal $s \times t$ matrix D with (i, i)-entry d_i ($1 \leq i \leq \min\{s, t\}$) over R is said to be in *Smith normal form* (Snf) if $d_i|d_j$ for $1 \leq i \leq j \leq \min\{s, t\}$. Is every 1×2 diagonal matrix over R in Snf?

(i) Let A be a 1×2 matrix over R. Adapt the proof of Lemma 1.8 to show that $A = DQ$ where the 1×2 matrix D over R is in Snf and the 2×2 matrix Q over R satisfies $\det Q = e$ the 1-element of R.

(ii) Let A be a diagonal 2×2 matrix over R. Generalise Lemma 1.10 to show that A can be reduced to D in Snf using at most five elementary operations of type (iii).

(d) Let A be an $s \times t$ matrix over R where R is a PID and $s \geq 2$. Let $P' = \left(\begin{smallmatrix} p_{11} & p_{12} \\ p_{21} & p_{22} \end{smallmatrix} \right)$ be a non-elementary matrix over R with $\det P' = e$ and let (i, j) be an ordered pair of distinct integers with $1 \leq i, j \leq s$. The operation of replacing $e_i A$ and $e_j A$ (rows i and j of A) by $p_{11} e_i A + p_{12} e_j A$ and $p_{21} e_i A + p_{22} e_j A$ respectively is called a *non-elementary row operation* (*nero*) over R. For each *nero* show that there is an $s \times s$ matrix P over R with $\det P = e$ such that PA is the result of applying this *nero* to A.

Let $t \geq 2$. A *non-elementary column operation* (*neco*) consists of replacing Ae_i^T (column i of A) by $p_{11}Ae_i^T + p_{21}Ae_j^T$ and Ae_j^T (column j of A) by $p_{12}Ae_i^T + p_{22}Ae_j^T$ all other columns of A remaining unchanged. For each *neco* show that there is a $t \times t$ matrix Q over R with $\det Q = e$ such that AQ is the matrix which results on applying this *neco* to A.

Does Lemma 1.4 apply without change on replacing $\mathbb{Z}$ by a commutative ring R? (Yes/No). (Note that multiplication of a row (or column) by any element of $U(R)$ is allowed as an *ero* (or *eco*) over R.)

(e) Let A be an $s \times t$ matrix over R where R is a PID. Outline a method of obtaining an invertible $s \times s$ matrix P over R and an invertible $t \times t$ matrix Q over R such that $PAQ^{-1} = D$ is in Snf.

Hint: Reduce A to D as in Lemma 1.9 and Theorem 1.11 but using *neros* and *necos* in place of the Euclidean algorithm Lemma 1.7.

(f) Let $D = \text{diag}(d_1, d_2, \ldots, d_{\min\{s,t\}})$ and $D' = \text{diag}(d'_1, d'_2, \ldots, d'_{\min\{s,t\}})$ be $s \times t$ matrices over a PID R which are both in Snf. Suppose $D \equiv D'$, that is, there is an invertible $s \times s$ matrix P over R and an invertible $t \times t$ matrix Q over R such that $PDQ^{-1} = D'$. Show $d_i \equiv d'_i$ for $1 \leq i \leq \min\{s, t\}$.

Hint: The theory of Corollaries 1.19 and 1.20 remains essentially unchanged on replacing $\mathbb{Z}$ by R.

8. (a) Let M be an R-module where R is an integral domain (R is a nontrivial commutative ring having no divisors of zero). Write $T(M) = \{v \in M : \text{there is } r \in R \text{ with } rv = 0, \ r \neq 0\}$. Show that $T(M)$ is a submodule Definition 2.26 of M. $T(M)$ is called *the torsion submodule* of M. Let M' be an R-module and suppose $M \cong M'$. Generalise Lemma 3.6 to show $T(M) \cong T(M')$ and $M/T(M) \cong M'/T(M')$. Describe the torsion submodules of $T(M)$ and $M/T(M)$.

(b) Let M be a free R-module of (finite) rank t where R is a PID (see Question 7(a)(ii) above). Let N be a submodule of M. Adapt the proof of Theorem 3.1 to show that N is free of rank s where $s \leq t$.

(c) Let v be an element of an R-module M where R is a PID. Show that $K = \{r \in R : rv = 0\}$ is an ideal of R. Let $K = \langle d \rangle$; then v is said to have *order d in M* (also the associative class (see Question 7(a)(i)) of d is called *the order* of v in M). Let M be finitely generated. Adapt the proof of Theorem 3.4 using Question 7(e) above to show $M = N_1 \oplus N_2 \oplus \cdots \oplus N_{t'}$ where N_j is a non-zero cyclic submodule of M having generator v_j of order d_j in M such that $d_i | d_j$ for $1 \leq i \leq j \leq t'$. Deduce $M = T(M) \oplus N_0$ where N_0 is a free submodule.

(d) Let R be a PID. Let M and M' be cyclic R-modules with generators v and v' of orders d and d' respectively. Show $M \cong M'$ if and only if $d \equiv d'$.

(e) Let M and M' be finitely generated R-modules where R is a PID.
Suppose $M = N_1 \oplus N_2 \oplus \cdots \oplus N_{t'}$ where N_j is a non-trivial cyclic
submodule of M having generator v_j of order d_j in M such that $d_i | d_j$
for $1 \leq i \leq j \leq t'$. Suppose $M' = N_1' \oplus N_2' \oplus \cdots \oplus N_{t''}'$ where N_j' is
a non-trivial cyclic submodule of M' having generator v_j' of order d_j'
in M' such that $d_i' | d_j'$ for $1 \leq i \leq j \leq t''$. Show $M \cong M'$ if and only if
$t' = t''$ and $d_j \equiv d_j'$ for $1 \leq j \leq t'$.
Hint: Use (a), (b) and (c) above and generalise Theorem 3.7.
The sequence $(\langle d_1 \rangle, \langle d_2 \rangle, \ldots, \langle d_{t'} \rangle)$ of ideals of the PID R is called
the invariant factor sequence of the f.g. R-module M.

3.2 Primary Decomposition of Finite Abelian Groups

Let G be a finite abelian group. From Theorem 3.7 we know that there is a unique se-
quence $(d_1, d_2, \ldots, d_{s'})$ of s' positive integers with $d_1 \neq 1$, $d_i | d_{i+1}$ $(1 \leq i < s')$ such
that $G = H_1 \oplus H_2 \oplus \cdots \oplus H_{s'}$ where H_i is a cyclic subgroup of G having order d_i
$(1 \leq i \leq s')$. Although invariant factor decompositions, as above, score top marks for
elegance, they do have some disadvantages: for one thing, the subgroups H_i are not
usually unique. Here we discuss the *primary decomposition* of G, which is analogous
to resolving the positive integer $|G|$ into prime powers (the fundamental theorem of
arithmetic is stated at the end of Section 1.2), and does have an important uniqueness
property. However there is a 'down-side' to this approach: there is no practical algorithm known for obtaining the prime factorisation of a general positive integer n. Here
we dodge the issue by assuming that the factorisation of $|G|$ has already been done!

Consider first an additive abelian group G of order $144 = 16 \times 9$. Then G has
subgroups $G_2 = \{g \in G : 16g = 0\}$ and $G_3 = \{g \in G : 9g = 0\}$. We'll see shortly that
G_2 and G_3 are the unique subgroups of G having orders 16 and 9 respectively. Further
G has *primary decomposition* $G = G_2 \oplus G_3$ showing that G is completely specified
by its *primary components* G_2 and G_3. Isomorphisms respect decompositions of this
type and so the analysis of G is reduced to that of G_2 and G_3. It turns out that there are
five isomorphism classes of abelian groups G_2 with $|G_2| = 16$ and two isomorphism
classes of abelian groups G_3 with $|G_3| = 9$. Hence there are $5 \times 2 = 10$ isomorphism
classes of abelian groups G of order 144.

The primary components of G are not cyclic in general. Our aim here is to obtain
an invariant factor decomposition of each primary component. We will do this by
using the primary decomposition of each cyclic subgroup H_i in an invariant factor
decomposition, as above, of G. The ultimate outcome is a decomposition of G into
the internal direct sum of a number of cyclic subgroups of prime power order, and all
subgroups of this type are *indecomposable* – they cannot themselves be expressed as
a direct sum in a non-trivial way.

Let $|G| = p_1^{n_1} p_2^{n_2} \cdots p_k^{n_k}$ be the factorisation of the order $|G|$ of G into positive powers of distinct primes $p_1, p_2, \ldots, p_k$. *The p_j-component* of G is

$$G_{p_j} = \{g \in G : p_j^{n_j} g = 0\} \quad \text{for } 1 \leq j \leq k.$$

We know that $|G|g = 0$ for all g in G by the $|G|$-lemma of Section 2.2. So G_{p_j} consists of those elements of G having orders which are powers of the prime p_j. It is straightforward to verify that G_{p_j} is a subgroup of G. Collectively $G_{p_1}, G_{p_2}, \ldots, G_{p_k}$ are called *the primary components of G*.

For example $G = \mathbb{Z}_6 \oplus \mathbb{Z}_{20}$ has order $|G| = 6 \times 20 = 2^3 \times 3 \times 5$. The primary components of G are

$$G_2 = \langle (\overline{3}, \overline{0}), (\overline{0}, \overline{5}) \rangle \cong \mathbb{Z}_2 \oplus \mathbb{Z}_4, \qquad G_3 = \langle (\overline{2}, \overline{0}) \rangle \cong \mathbb{Z}_3, \qquad G_5 = \langle (\overline{0}, \overline{4}) \rangle \cong \mathbb{Z}_5.$$

Consider an isomorphism $\alpha : G \cong G'$ between the finite abelian groups G and G'. Then $|G| = |G'|$ and $G'_{p_j} = \{g' \in G' : p_j^{n_j} g' = 0\}$ is the p_j-component of G'. For $g \in G_{p_j}$ we see $p_j^{n_j}(g)\alpha = (p_j^{n_j} g)\alpha = (0)\alpha = 0$ using the $\mathbb{Z}$-linearity of α. So $(G_{p_j})\alpha \subseteq G'_{p_j}$ showing that α maps G_{p_j} to G'_{p_j}. In the same way $(G'_{p_j})\alpha^{-1} \subseteq G_{p_j}$. So α, restricted to G_{p_j}, is an isomorphism $\alpha| : G_{p_j} \cong G'_{p_j}$. We have shown:

Isomorphic finite abelian groups have isomorphic primary components.

Taking $G = G'$ gives $(G_{p_j})\alpha = G_{p_j}$ for all automorphisms α of G. Therefore isomorphisms and automorphisms respect primary components.

We now show that every finite abelian group is the internal direct sum of its primary components.

Theorem 3.10 (The primary decomposition of finite abelian groups)

Let G be a finite abelian group and suppose $|G| = p_1^{n_1} p_2^{n_2} \cdots p_k^{n_k}$ where $p_1, p_2, \ldots, p_k$ are distinct primes. Then

$$G = G_{p_1} \oplus G_{p_2} \oplus \cdots \oplus G_{p_k}$$

where $G_{p_1}, G_{p_2}, \ldots, G_{p_k}$ are the primary components of G.

Proof

Write $m_j = |G|/p_j^{n_j}$. So m_j is the product of the $k - 1$ prime powers $p_i^{n_i}$ where $i \neq j$. The k positive integers $m_1, m_2, \ldots, m_k$ are *coprime*, meaning $\gcd\{m_1, m_2, \ldots, m_k\} = 1$, as there is no common prime divisor of $m_1, m_2, \ldots, m_k$. By Corollary 1.16 there are integers $a_1, a_2, \ldots, a_k$ such that $a_1 m_1 + a_2 m_2 + \cdots + a_k m_k = 1$. For each g in G we have

$$g = 1g = (a_1 m_1 + a_2 m_2 + \cdots + a_k m_k)g$$

$$= a_1 m_1 g + a_2 m_2 g + \cdots + a_k m_k g$$

$$= g_1 + g_2 + \cdots + g_k$$

where $g_j = a_j m_j g$ $(1 \le j \le k)$. Now $p_j^{n_j} m_j = |G|$ and hence $p_j^{n_j} g_j = a_j |G| g = 0$ as $|G| g = 0$ by the $|G|$-lemma. So $g_j \in G_{p_j}$ for $1 \le j \le k$. We have shown $G = G_{p_1} + G_{p_2} + \cdots + G_{p_k}$ as each element of G is a sum of k elements one from each of the k primary components G_{p_j}.

To show that the sum of the primary components is direct, suppose

$$g_1 + g_2 + \cdots + g_k = 0 \quad \text{where } g_j \in G_{p_j}.$$

We fix our attention on one particular term g_j. The positive integer m_j has factor $p_i^{n_i}$ and so $m_j g_i = 0$ as $p_i^{n_i} g_i = 0$ for $i \ne j$. Inserting $k - 1$ zero terms $m_j g_i$ gives

$$m_j g_j = m_j g_j + \sum_{i=1, i \ne j}^{k} m_j g_i = \sum_{i=1}^{k} m_j g_i = m_j \left(\sum_{i=1}^{k} g_i \right) = m_j \times 0 = 0.$$

We've now know $m_j g_j = 0$ and $p_j^{n_j} g_j = 0$ with $\gcd\{m_j, p_j^{n_j}\} = 1$ and so g_j doesn't stand a chance! The positive integer m_i has factor $p_j^{n_j}$ for $i \ne j$. So the integer $1 - a_j m_j = \sum_{i-1, i \ne j}^{k} a_i m_i$ has factor $p_j^{n_j}$ and hence $a_j' = (1 - a_j m_j)/p_j^{n_j}$ is an integer. Therefore

$$g_j = 1 g_j = (a_j m_j + ((1 - a_j m_j)/p_j^{n_j}) p_j^{n_j}) g_j$$

$$= a_j m_j g_j + a_j' p_j^{n_j} g_j = a_j \times 0 + a_j' \times 0 = 0$$

for $1 \le j \le k$. The equation $g_1 + g_2 + \cdots + g_k = 0$ where $g_j \in G_{p_j}$ for $1 \le j \le k$ holds only in the case $g_1 = g_2 = \cdots = g_k = 0$. By Definition 2.14 the primary components of G are independent. By Lemma 2.15 the sum of the primary components of G is direct, that is, $G = G_{p_1} \oplus G_{p_2} \oplus \cdots \oplus G_{p_k}$. $\qquad\square$

As an illustration we look at the cyclic group $G = \mathbb{Z}_{360}$. Then $|G| = 2^3 \times 3^2 \times 5$ and with $p_1 = 2$, $p_2 = 3$, $p_3 = 5$ and $n_1 = 3$, $n_2 = 2$, $n_3 = 1$ we obtain $m_1 = 45$, $m_2 = 40$, $m_3 = 72$. The element $\overline{1}$ of G has order 360. By Lemma 2.7 the element $\overline{45} = 45(\overline{1})$ has order $360/\gcd\{45, 360\} = 360/45 = 8$, and so $\langle \overline{45} \rangle = G_2$ is the 2-component of G. Similarly $\overline{40}$ has order 9 and $\overline{72}$ has order 5. So $\langle \overline{40} \rangle = G_3$ is the 3-component of G and $\langle \overline{72} \rangle = G_5$ is the 5-component of G. The primary components are cyclic, as they must be by Lemma 2.2, and we've expressed their generators as multiples of the generator $\overline{1}$ of G. Therefore $G = \langle \overline{45} \rangle \oplus \langle \overline{40} \rangle \oplus \langle \overline{72} \rangle$ is the primary decomposition Theorem 3.10 of G.

From Theorem 3.10 we deduce, as in Exercises 3.2, Question 5(b):

> Any two finite abelian groups of equal order having isomorphic primary components are isomorphic.

Definition 3.11

The *exponent* of a finite abelian group G is the smallest positive integer n with $nG = \{0\}$, that is, $ng = 0$ for all $g \in G$.

The exponent of $G = \mathbb{Z}_6 \oplus \mathbb{Z}_{20}$ is 60: on the one hand $60G$ is trivial and on the other hand the element $(\bar{1}, \bar{1})$ of G has order 60 showing $n'G \neq \{0\}$ for all integers n' with $1 \leq n' < 60$. The invariant factor sequence of G is $(2, 60)$ and, as we prove next, the largest invariant factor of every non-trivial finite abelian group is its exponent.

Corollary 3.12

Let $(d_1, d_2, \ldots, d_{s'})$ be the invariant factor sequence of the non-trivial finite abelian group G. Then $d_{s'}$ is the exponent of G and G has an element of order $d_{s'}$. Further $d_{s'}$ is a divisor of $|G|$ and $|G|$ is a divisor of $(d_{s'})^{s'}$. The order and the exponent of G have the same prime divisors.

Proof

Let $K = \{m \in \mathbb{Z} : mG = \{0\}\}$ where $mG = \{mg : g \in G\}$. We know from Section 3.1 that mG is a subgroup of G for each integer m. It is routine to check that K is an ideal of $\mathbb{Z}$. By the $|G|$-lemma of Section 2.2 we see $|G| \in K$. So K is non-zero and hence $K = \langle n \rangle$ by Theorem 1.15 where n is a positive integer. As $n \in K$ but $n' \notin K$ for all integers n' with $1 \leq n' < n$, the exponent of G is n. As $|G| \in \langle n \rangle$ we deduce that n is a divisor of $|G|$.

Now G has subgroups $H_1, H_2, \ldots, H_{s'}$ such that $G = H_1 \oplus H_2 \oplus \cdots \oplus H_{s'}$ and H_i is of isomorphism type C_{d_i} for $1 \leq i \leq s'$ by Theorem 3.4. Also $d_i | d_{i+1}$ and so there is a positive integer q_i with $d_i q_i = d_{i+1}$ for $1 \leq i < s'$. Hence $d_i | d_{s'}$ for all i with $1 \leq i \leq s'$ since $d_i(q_i q_{i+1} \cdots q_{s'-1}) = d_{s'}$. As $d_i H_i = \{0\}$ (in fact H_i has exponent d_i) we see that $d_{s'} H_i = \{0\}$. Hence $d_{s'} G = d_{s'} H_1 \oplus d_{s'} H_2 \oplus \cdots \oplus d_{s'} H_{s'} = \{0\}$ and so $d_{s'} \in K$. Therefore $n | d_{s'}$. Let $h_{s'}$ generate the cyclic subgroup $H_{s'}$. Then $H_{s'} = \langle h_{s'} \rangle$ and $h_{s'}$ has order $d_{s'}$. As $nh_{s'} = 0$ we see $d_{s'} | n$ and so $n = d_{s'}$. Therefore $d_{s'}$ is the exponent of G.

From Corollary 3.5 we know $|G| = d_1 d_2 \cdots d_{s'}$. Multiplying together the s' equations $d_i(q_i q_{i+1} \cdots q_{s'+1}) = d_{s'}$ for $1 \le i \le s'$ gives

$$|G| q_1 q_2^2 q_3^3 \cdots q_{s'-1}^{s'-1} = (d_{s'})^{s'}$$

from which we see that $|G|$ is a divisor of $(d_{s'})^{s'}$. Let p be a prime divisor of $|G|$. Then $p | (d_{s'})^{s'}$ and so $p | d_{s'}$ as a prime divisor of a product must be a divisor of at least one of its factors. Conversely all divisors of $d_{s'}$ are divisors of $|G|$. □

Let G be a finite abelian group and suppose $|G| = p_1^{n_1} p_2^{n_2} \cdots p_k^{n_k}$ as in Theorem 3.10. We apply Corollary 3.12 to the primary component G_{p_j} of G: as $p_j^{n_j} G_{p_j} = \{0\}$ the exponent of G_{p_j} is a divisor of $p_j^{n_j}$ and so is a power of p_j. Therefore $|G_{p_j}|$ is also a power of p_j by Corollary 3.12. From Theorem 3.10 we deduce

$$|G| = |G_{p_1}| \times |G_{p_2}| \times \cdots \times |G_{p_k}| \quad \text{and so} \quad |G_{p_j}| = p_j^{n_j} \quad \text{for } 1 \le j \le k$$

on comparing powers of p_j. Therefore the decomposition Theorem 3.10 of G into primary components corresponds to the factorisation of $|G|$ into prime powers.

Let G be a finite abelian group of prime exponent p. Such a group is called is called an *elementary abelian p-group*. Groups of this kind and their automorphisms will be analysed in Section 3.3. By Corollary 3.12 the order of G is a power of p, that is, $|G| = p^{s'}$, the invariant factor sequence of G being $(p, p, \ldots, p)$, that is, the s' invariant factors of G are all equal to p. The one essential fact (as we will see) is that G is the additive group of an s'-dimensional vector space over the field $\mathbb{Z}_p$. In particular the Klein 4-group has exponent 2 and its elements are the vectors of a 2-dimensional vector space over $\mathbb{Z}_2$.

Now $G = \mathbb{Z}_6 \oplus \mathbb{Z}_{20}$ has primary components $G_2 \cong \mathbb{Z}_2 \oplus \mathbb{Z}_4$, $G_3 \cong \mathbb{Z}_3$, $G_5 \cong \mathbb{Z}_5$. By Theorem 3.10 we obtain $G = G_2 \oplus G_3 \oplus G_5$, that is, $\mathbb{Z}_6 \oplus \mathbb{Z}_{20} \cong \mathbb{Z}_2 \oplus \mathbb{Z}_4 \oplus \mathbb{Z}_3 \oplus \mathbb{Z}_5$, which amounts to two applications of the Chinese remainder theorem, namely $\mathbb{Z}_6 \cong \mathbb{Z}_2 \oplus \mathbb{Z}_3$ and $\mathbb{Z}_{20} \cong \mathbb{Z}_4 \oplus \mathbb{Z}_5$, followed by a rearrangement of the summands (the terms in the direct sum). In fact the derivation of a decomposition of G into the internal direct sum of cyclic subgroups of prime power order, starting from an invariant factor decomposition of G, which we carry out shortly, is nothing more than a systematic application of the Chinese remainder theorem 2.11.

Let G be a cyclic group of prime power order p^n. The positive divisors of p^n are $1, p, p^2, \ldots, p^n$, and they form a *chain*, meaning that every two integers a, b in this list are such that either $a | b$ or $b | a$. By Lemma 2.2 every pair H, H' of subgroups of G satisfy either $H \subseteq H'$ or $H' \subseteq H$. Let's suppose $H \subseteq H'$ and ask: is it possible for $G = H \oplus H'$? So each element g in G is uniquely expressible as $g = h + h'$ where $h \in H, h' \in H'$. Suppose, just for a moment, that there is a non-zero element h in H.

Then $-h \in H$ and so $-h \in H'$. Hence the zero element 0 of G is expressible as a sum of two elements, one from H and one from H' in two different ways: $0 = h + (-h)$, $0 = 0 + 0$. By the uniqueness property of $H \oplus H'$ we deduce $H = \{0\}$. The conclusion is:

$$\text{Cyclic groups of prime power order are } \textit{indecomposable}$$

that is, G cannot be expressed in the form $G = H \oplus H'$ where both H and H' are non-trivial.

Now consider an arbitrary finite abelian group G with invariant factor sequence $(d_1, d_2, \ldots d_{s'})$ and let $G = H_1 \oplus H_2 \oplus \cdots \oplus H_{s'}$ where $H_i = \langle h_i \rangle$ is of isomorphism type C_{d_i} for $1 \leq i \leq s'$. As before let $|G| = p_1^{n_1} p_2^{n_2} \cdots p_k^{n_k}$ where $p_1, p_2, \ldots, p_k$ are different primes and each $n_j \geq 1$ for $1 \leq j \leq k$. Each invariant factor d_i has factorisation $d_i = p_1^{t_{i1}} p_2^{t_{i2}} \cdots p_k^{t_{ik}}$ where $t_{ij} \geq 0$. As $|G| = d_1 d_2 \cdots d_{s'}$ we obtain $n_j = t_{1j} + t_{2j} + \cdots + t_{s'j}$ on comparing powers of p_j. For each i let k_i denote the number of positive exponents t_{ij}. As $d_i | d_{i+1}$ for $1 \leq i < s'$ we see $k_1 \leq k_2 \leq \cdots \leq k_{s'}$. Write $m_{ij} = d_i / p_j^{t_{ij}}$. The generator h_i of H_i has order d_i and so $m_{ij} h_i$ has order $d_i / \gcd\{m_{ij}, d_i\} = p_j^{t_{ij}}$ by Lemma 2.7. Let

$$H_{i,p_j} = \langle m_{ij} h_i \rangle \quad \text{in the case } t_{ij} > 0,$$

that is, $H_{i,p_j} = (H_i)_{p_j}$ is the p_j-component of H_i. By Theorem 3.10 the primary decomposition of H_i is

$$H_i = \sum_j \oplus H_{i,p_j}$$

as H_i is the internal direct sum of its k_i non-trivial primary components H_{i,p_j}.

The prime powers $p_j^{t_{ij}} = |H_{i,p_j}| > 1$ are called the *elementary divisors* of G.

The elementary divisors of G are the primes to the highest power in the factorisations of the invariant factors of G and so they are also invariants.

For example

$$\mathbb{Z}_{10} \oplus \mathbb{Z}_{100} \cong (\mathbb{Z}_2 \oplus \mathbb{Z}_5) \oplus (\mathbb{Z}_4 \oplus \mathbb{Z}_{25}) \cong (\mathbb{Z}_2 \oplus \mathbb{Z}_4) \oplus (\mathbb{Z}_5 \oplus \mathbb{Z}_{25})$$

has elementary divisors $2, 4; 5, 25$, whereas

$$\mathbb{Z}_4 \oplus \mathbb{Z}_{100} \cong \mathbb{Z}_4 \oplus (\mathbb{Z}_4 \oplus \mathbb{Z}_{25}) \cong (\mathbb{Z}_4 \oplus \mathbb{Z}_4) \oplus \mathbb{Z}_{25}$$

has elementary divisors $4, 4; 25$.

In the general case, replacing each H_i by its primary decomposition gives

$$G = \sum_{i,j} H_{i,p_j}$$

which expresses G as the internal direct sum of $k_1 + k_2 + \cdots + k_{s'}$ non-trivial cyclic subgroups H_{i,p_j} of prime power order. This decomposition of G is 'best' in the sense that the number of non-trivial cyclic summands is as *large* as possible (Exercises 3.2, Question 6(b)). From it we now directly deduce the structure of the primary components of G. For each j the direct sum of the subgroups H_{i,p_j} is a subgroup of order $p_j^{t_{1j}} p_j^{t_{2j}} \cdots p_j^{t_{s'j}} = p_j^{n_j}$ and so this direct sum is the p_j-component of G, that is,

$$G_{p_j} = \sum_i \oplus H_{i,p_j} \qquad\qquad (\clubsuit\clubsuit)$$

Now $t_{1j} \le t_{2j} \le \cdots \le t_{s'j}$ since $d_i | d_{i+1}$ for $1 \le i < s'$. Hence, on omitting the first so many trivial summands H_{i,p_j} with $t_{ij} = 0$, $(\clubsuit\clubsuit)$ above becomes an invariant factor decomposition, as in Theorem 3.7, of G_{p_j} for $1 \le j \le k$. In other words:

> The elementary divisors of each finite abelian group are the invariant factors
> of its primary components.

We have now completed the analysis of an individual finite abelian group G. Our final task is to determine the number of isomorphism classes of abelian groups G having a specified order $|G| = p_1^{n_1} p_2^{n_2} \cdots p_k^{n_k}$. We will see that this number depends on the powers $n_1, n_2, \ldots, n_k$ of the distinct prime divisors of $|G|$ and not on the prime divisors themselves. You already know one special case of this phenomenon: any two groups of prime order p are isomorphic, that is, for each prime p there is just one isomorphism class of groups of order p, each of these groups being cyclic. This example can be generalised: consider an abelian group G such that $|G| = p_1 p_2 \cdots p_k$ (a product of distinct primes). What could the sequence $(d_1, d_2, \ldots, d_{s'})$ of invariant factors of G be? By Corollary 3.12 we have $d_{s'} | p_1 p_2 \cdots p_k$ and also $p_j | d_{s'}$ for $1 \le j \le k$. Therefore $p_1 p_2 \cdots p_k | d_{s'}$ and so $d_{s'} = p_1 p_2 \cdots p_k$. As $|G|$ is the product of all the invariant factors of G we see that $d_{s'}$ is the only invariant factor of G, that is, $s' = 1$ and G is cyclic. In particular, all abelian groups of order $105 = 3 \times 5 \times 7$ are cyclic and hence any two are isomorphic.

The following table illustrates the relationship between the invariant factors of a finite abelian group G and its elementary divisors.

$\cong G$	p_1	p_2	$\cdots$	p_k
d_1	t_{11}	t_{12}	$\cdots$	t_{1k}
d_2	t_{21}	t_{22}	$\cdots$	t_{2k}
$\vdots$	$\vdots$	$\vdots$		$\vdots$
$d_{s'}$	$t_{s'1}$	$t_{s'2}$	$\cdots$	$t_{s'k}$

The rows correspond to the invariant factors d_i of G and each contains k_i non-zero entries. The columns correspond to the prime divisors p_j of $|G|$, and the exponent t_{ij}

appears in row i and column j. We suppose $p_j < p_{j+1}$ for $1 \le j < k$. Then isomorphic groups have identical tables and non-isomorphic groups have different tables and so it is reasonable to refer to *the table of the isomorphism class of the finite abelian group* G. We use $\cong G$ to denote the isomorphism class of G.

The table of the isomorphism class of abelian groups G with invariant factor sequence $(2, 6, 60, 600)$ is shown below.

$\cong G$	2	3	5
2	1	0	0
6	1	1	0
60	2	1	1
600	3	1	2

The elementary divisors of G are $2, 2, 4, 8; 3, 3, 3; 5, 25$. The connection between elementary divisors and invariant factors is evident from the rows in the table. Starting at the bottom row: $8 \times 3 \times 25 = 600$, that is, the product of the largest elementary divisors for each prime divisor of $|G|$ is the largest invariant factor of G. From the next-to-last row we read off the product of the next-to-largest elementary divisors which is the next-to-largest invariant factor of G, that is, $4 \times 3 \times 5 = 60$ and so on. Now $|G| = 2^7 \times 3^3 \times 5^3 = 432000$. How many isomorphism classes of abelian groups of order 432000 are there? Equivalently, how many tables are there such that the non-zero entries in the 2-column are non-decreasing and have sum 7, the non-zero entries in the 3-column are non-decreasing and have sum 3, and the non-zero entries in the 5-column are non-decreasing and have sum 3? The answer is 135; read on to find out why!

We return to the general table and concentrate on the p_j-column: any zero entries occur at the top and the remaining positive entries form a non-decreasing sequence (reading downwards) with sum n_j.

Definition 3.13

Let n be a non-negative integer. A *partition of* n is a non-decreasing sequence $(t_1, t_2, \ldots, t_s)$ of positive integers with $t_1 + t_2 + \cdots + t_s = n$. The integers t_i $(1 \le i \le s)$ are called the *parts* of the partition.

The number of partitions of n is denoted by $p(n)$.

The partitions of 4 are $(1, 1, 1, 1)$, $(1, 1, 2)$, $(1, 3)$, $(2, 2)$, (4), and so $p(4) = 5$. You can check $p(1) = 1$, $p(2) = 2$, $p(3) = 3$. It is convenient to allow $s = 0$ in Definition 3.13, that is, the empty sequence $\emptyset$ is a partition of 0 and so $p(0) = 1$.

The *partition function* $p(n)$ has been extensively studied, notably by the eighteenth century Swiss mathematician Euler.

We now describe a way of calculating $p(n)$ akin to the Pascal triangle method of computing binomial coefficients. Let

$p(n, j)$ denote the number of partitions of n having no part less than j
$$(n \geq 0,\ j \geq 1).$$

Directly from Definition 3.13 we see that $p(n, 1) = p(n)$. Note $p(0, j) = 1$ for all $j \geq 1$ as the partition $\emptyset$ of 0 has no parts, but $p(n, j) = 0$ for $1 \leq n < j$. Each partition $(t_1, t_2, \ldots, t_s)$ of n with $s \geq 1$ consists of a first part t_1 and a partition $(t_2, \ldots, t_s)$ of $n - t_1$ having no part less than t_1. So there are $p(n - t_1, t_1)$ partitions of n having first part t_1. Therefore

$$p(n) = p(n - 1, 1) + p(n - 2, 2) + \cdots + p(0, n) \tag{$\blacklozenge$}$$

on counting up the partitions of n according as their first part is $1, 2, \ldots, n$. In the same way $p(n, j) = p(n - j, j) + p(n - j - 1, j + 1) + \cdots + p(0, n)$ on counting up the partitions of n having first part $j, j + 1, \ldots, n$ $(1 \leq j \leq n)$. Therefore

$$p(n, j + 1) = p(n, j) - p(n - j, j) \quad \text{for } 1 \leq j < n \tag{$\blacktriangledown$}$$

as $p(n, j+1) = p(n-j-1, j+1) + \cdots + p(0, n)$. Using ($\blacklozenge$) and ($\blacktriangledown$) the array having $p(n, j)$ in row n and column j can be constructed row by row. The first few rows are shown in the following table. Suppose rows $0, 1, 2, \ldots, 9$ have been completed. From ($\blacklozenge$) with $n = 10$ we obtain

$$p(10, 1) = p(10) = p(9, 1) + p(8, 2) + \cdots + p(0, 10)$$

$$= 30 + 7 + 2 + 1 + 1 + 0 + 0 + 0 + 0 + 1,$$

that is, $p(10) = 42$. The remaining non-zero entries in row 10 can be found using ($\blacktriangledown$) with $n = 10$ putting $j = 1, 2, \ldots, 9$ successively. So

$$p(10, 2) = p(10, 1) - p(9, 1) = 42 - 30 = 12,$$

$$p(10, 3) = p(10, 2) - p(8, 2) = 12 - 7 = 5$$

and so on to complete row 10.

n	$p(n)$	$p(n,2)$	$p(n,3)$	$p(n,4)$	$p(n,5)$	$p(n,6)$	$p(n,7)$	$p(n,8)$	$p(n,9)$	$p(n,10)$	$\cdots$
0	1	1	1	1	1	1	1	1	1	1	$\cdots$
1	1	0	0	0	0	0	0	0	0	0	$\cdots$
2	2	1	0	0	0	0	0	0	0	0	$\cdots$
3	3	1	1	0	0	0	0	0	0	0	$\cdots$
4	5	2	1	1	0	0	0	0	0	0	$\cdots$
5	7	2	1	1	1	0	0	0	0	0	$\cdots$
6	11	4	2	1	1	1	0	0	0	0	$\cdots$
7	15	4	2	1	1	1	1	0	0	0	$\cdots$
8	22	7	3	2	1	1	1	1	0	0	$\cdots$
9	30	8	4	2	1	1	1	1	1	0	$\cdots$
10	42	12	5	3	2	1	1	1	1	1	$\cdots$

A readable account of a more efficient method of calculating $p(n)$ is in Chapter 19 of Norman Biggs, *Discrete Mathematics*, OUP, 1985.

Let G be an abelian group of order p^n, where p is prime, having invariant factor sequence $(d_1, d_2, \ldots, d_{s'})$. We know from Corollary 3.5 that $d_1 d_2 \cdots d_{s'} = p^n$ and so $d_i = p^{t_i}$ for $1 \leq i \leq s'$. Further as $d_i | d_{i+1}$, on comparing powers of p it follows that $(t_1, t_2, \ldots, t_{s'})$ is a partition Definition 3.13 of n. Conversely every partition of n arises from an invariant factor sequence of G in this way and so:

There are $p(n)$ isomorphism classes of abelian groups of order p^n.

In particular there are five isomorphism classes of abelian groups of order 16, their isomorphism types being $C_2 \oplus C_2 \oplus C_2 \oplus C_2$, $C_2 \oplus C_2 \oplus C_4$, $C_2 \oplus C_8$, $C_4 \oplus C_4$, C_{16} corresponding to the five partitions of 4.

Now consider the general case of an abelian group G of order $|G| = p_1^{n_1} p_2^{n_2} \cdots p_k^{n_k}$. By the above reasoning the primary component G_{p_j} of G belongs to one of the $p(n_j)$ isomorphism classes of abelian groups of order $p_j^{n_j}$. So the non-zero entries in the p_j-column of the table of the isomorphism class of G are precisely the parts of a partition of n_j $(1 \leq j \leq k)$. Hence using Theorem 3.10 we deduce:

There are $p(n_1)p(n_2) \cdots p(n_k)$ isomorphism classes
of abelian groups of order $p_1^{n_1} p_2^{n_2} \cdots p_k^{n_k}$

as the primary components are independent of each other.

For example consider isomorphism classes of abelian groups of order $144 = 2^4 \times 3^2$. Such a class corresponds to a row in the following table:

Primary decomposition	Invariant factor decomposition
2-component and 3-component	Isomorphism type
$\mathbb{Z}_2 \oplus \mathbb{Z}_2 \oplus \mathbb{Z}_2 \oplus \mathbb{Z}_2 \oplus \mathbb{Z}_3 \oplus \mathbb{Z}_3$	$C_2 \oplus C_2 \oplus C_6 \oplus C_6$
$\mathbb{Z}_2 \oplus \mathbb{Z}_2 \oplus \mathbb{Z}_2 \oplus \mathbb{Z}_2 \oplus \mathbb{Z}_9$	$C_2 \oplus C_2 \oplus C_2 \oplus C_{18}$
$\mathbb{Z}_2 \oplus \mathbb{Z}_2 \oplus \mathbb{Z}_4 \oplus \mathbb{Z}_3 \oplus \mathbb{Z}_3$	$C_2 \oplus C_6 \oplus C_{12}$
$\mathbb{Z}_2 \oplus \mathbb{Z}_2 \oplus \mathbb{Z}_4 \oplus \mathbb{Z}_9$	$C_2 \oplus C_2 \oplus C_{36}$
$\mathbb{Z}_2 \oplus \mathbb{Z}_8 \oplus \mathbb{Z}_3 \oplus \mathbb{Z}_3$	$C_6 \oplus C_{24}$
$\mathbb{Z}_2 \oplus \mathbb{Z}_8 \oplus \mathbb{Z}_9$	$C_2 \oplus C_{72}$
$\mathbb{Z}_4 \oplus \mathbb{Z}_4 \oplus \mathbb{Z}_3 \oplus \mathbb{Z}_3$	$C_{12} \oplus C_{12}$
$\mathbb{Z}_4 \oplus \mathbb{Z}_4 \oplus \mathbb{Z}_9$	$C_4 \oplus C_{36}$
$\mathbb{Z}_{16} \oplus \mathbb{Z}_3 \oplus \mathbb{Z}_3$	$C_3 \oplus C_{48}$
$\mathbb{Z}_{16} \oplus \mathbb{Z}_9$	C_{144}

As $144 = 2^4 \times 3^2$ the number of isomorphism classes of abelian groups of order 144 is $p(4) \times p(2) = 5 \times 2 = 10$. Each abelian group of order 144 is isomorphic to a group in one of the rows of the above table.

There are $p(7)p(3)p(3) = 15 \times 3 \times 3 = 135$ isomorphism classes of abelian groups G of order $2^7 \times 3^3 \times 5^3$. Each such class corresponds to a triple of partitions (of 7, 3 and 3) arising from the three columns of its table and the number s' of rows in this table is the largest number of parts in any one of these partitions.

The reader should have gained the impression that finitely generated abelian groups are 'manageable'. This fact is exploited in other branches of mathematics: the homology of finite simplicial complexes in algebraic topology is a case in point. In Chapters 5 and 6 the analogous theory of $F[x]$-modules is developed, and this culminates in the canonical forms of square matrices over a field F.

EXERCISES 3.2

1. (a) Let G denote the additive group of $\mathbb{Z}_{10}$. List the elements in the primary components G_2 and G_5. Verify $G = G_2 + G_5$, $G_2 \cap G_5 = \{\bar{0}\}$ and deduce $G = G_2 \oplus G_5$. Let μ_3 be the automorphism of G defined by $(g)\mu_3 = 3g$ for all $g \in G$. Verify $(G_2)\mu_3 = G_2$ and $(G_5)\mu_3 = G_5$. Is Aut G cyclic?

    Hint: Use Exercises 2.1, Question 4(a).

   (b) Let G denote the additive group of $\mathbb{Z}_{12}$. List the elements in the primary components G_2 and G_3. Verify $G = G_2 + G_3$, $G_2 \cap G_3 = \{\bar{0}\}$ and deduce $G = G_2 \oplus G_3$. Does G have elementary divisors 2, 2, 3 or 4, 3? Verify $(G_2)\mu_5 = G_2$ and $(G_3)\mu_5 = G_3$ where $(g)\mu_5 = 5g$ for all $g \in G$. Is Aut G cyclic?

(c) The cyclic group G has order $|G| = p_1^{n_1} p_2^{n_2} \ldots p_k^{n_k}$ where $p_1, p_2,$ $\ldots, p_k$ are distinct primes. List the elementary divisors of G.

(d) The order of the finite abelian group G is $|G| = p^n m$ where $\gcd\{p, m\} = 1$. Show that the p-component $G_p = \{g \in G : p^n g = 0\}$ is a subgroup of G. Show $mG = G_p$. Is $mG_p = G_p$?

2. (a) Let G denote the additive group of $\mathbb{Z}_{10} \oplus \mathbb{Z}_{12}$. State the invariant factors of G and those of its 2-component G_2. List the elementary divisors of G. List the elementary divisors and invariant factors of all abelian groups of order $|G| = 120$.

(b) Show that there are six isomorphism classes of abelian groups of order 200. List the elementary divisors of these classes and their invariant factor sequences. Do any two have the same exponent?

(c) List the isomorphism types of the $p(4) = 5$ isomorphism classes of abelian groups of order 16. Do any two have the same exponent? List the elementary divisors and invariant factor sequences of the eight isomorphism classes of abelian groups of order $(900)^2$ having exponent 900.

(d) List the elementary divisors and the invariant factor sequences of the ten isomorphism classes of abelian groups of order 400. How many isomorphism classes of abelian groups of order 144 are there?

(e) Find three abelian groups having the same order (less than 100) and the same exponent but no two of which are isomorphic.

3. (a) List the $p(5) = 7$ partitions of 5. List the $p(9, 2) = 8$ partitions of 9 having all parts ≥ 2.

(b) Calculate the entries for the row $n = 11$ in the table in the text following Definition 3.13. Hence show $p(12) = 77$. Find $p(13)$ and $p(14)$.

(c) Show from the definition that $p(2n, n) = 2$ and find a formula for $p(3n, n)$.

(d) Find, directly from Definition 3.13, a formula for the number of partitions of n having all parts ≤ 2. Hence find the number of isomorphism classes of abelian groups of order p^n having exponent p^2.

(e) Let j, k, n be integers with $j \geq 1$, $n \geq 1$ and $0 \leq k \leq \lfloor n/j \rfloor$. Explain why the number of partitions of n having all parts $\geq j$ and exactly k parts equal to j is $p(n - jk, j + 1)$. Hence show that $p(n, j) = \sum_{k=0}^{\lfloor n/j \rfloor} p(n - jk, j + 1)$.
Use the table following Definition 3.13 to check the above equation in the cases $n = 10$, $k = 1, 2, 3, 4$.

4. (a) Let G be the additive group of $\mathbb{Z}_5 \oplus \mathbb{Z}_5$. How many elements of order 5 does G have? Is G an elementary abelian 5-group? (Yes/No). Show that G has 6 subgroups of order 5 and specify a generator of each. How many (ordered) pairs of subgroups H_1, H_2 of G are there with $|H_1| = |H_2| = 5$ and $G = H_1 \oplus H_2$?

(b) Let G be the additive group of $\mathbb{Z}_p \oplus \mathbb{Z}_p$ where p is prime. Is G an elementary p-group? Show that G has $p^2 - 1$ elements of order p and $p + 1$ subgroups H of order p. Show that H can be expressed either $H = \langle (\bar{1}, \bar{i}) \rangle$ where $\bar{i} \in \mathbb{Z}_p$ or $H = \langle (\bar{0}, \bar{1}) \rangle$. How many (ordered) pairs of subgroups H_1, H_2 of G are there with $|H_1| = |H_2| = p$ and $G = H_1 \oplus H_2$?

(c) Let G be the additive group of $\mathbb{Z}_9 \oplus \mathbb{Z}_{27}$ and let $H = \{h \in G : 3h = 0\}$. List the elements of H. Is H a subgroup of G? (Yes/No). If so is H an elementary abelian 3-group? State the invariant factor sequences of G, H and G/H.

(d) Let G be an additive cyclic group of order p^t where p is prime and $t \geq 1$. Let $H = \{h \in G : ph = 0\}$. Show $H = p^{t-1}G$ using Lemma 2.7. State the orders of H and G/H.

(e) Let $(t_1, t_2, \ldots, t_s)$ be a partition of $n(> 0)$ and suppose $G = H_1 \oplus H_2 \oplus \cdots \oplus H_s$ (internal direct sum) where H_i is cyclic of order p^{t_i} for $1 \leq i \leq s$ and p is prime. Write $H = \{h \in G : ph = 0\}$. Use (d) above to show $H = p^{t_1-1}H_1 \oplus p^{t_2-1}H_2 \oplus \cdots \oplus p^{t_s-1}H_s$. Deduce the order of H. State the invariant factor sequences of H and G/H. Under what condition on t_s is G/H an elementary abelian p-group?

(f) The abelian group G has order $|G| = p_1^{n_1} p_2^{n_2} \ldots p_k^{n_k}$ where $p_1, p_2, \ldots, p_k$ are distinct primes and $n_j > 0$ for $1 \leq j \leq k$. Use Theorem 3.10 and Corollary 3.12 to show that there are $n_1 n_2 \cdots n_k$ possibilities for the exponent of G.

Hint: Treat the case $k = 1$ first.

Let $H = \{h \in G : p_1 p_2 \cdots p_k h = 0\}$. Is H necessarily cyclic? Does H cyclic $\Rightarrow G$ cyclic? Justify your answer.

5. (a) The additive abelian group G is the direct sum of its subgroups H_1 and H_2. Let $\alpha_1 \in \operatorname{Aut} H_1$ and $\alpha_2 \in \operatorname{Aut} H_2$. Show that $\alpha : G \to G$, defined by $(h_1 + h_2)\alpha = (h_1)\alpha_1 + (h_2)\alpha_2$ for all $h_1 \in H_1$, $h_2 \in H_2$, is an automorphism of G. Show also that $(H_1)\alpha = H_1$ and $(H_2)\alpha = H_2$. Write $\alpha = \alpha_1 \oplus \alpha_2$.

Let $\beta \in \operatorname{Aut} G$ satisfy $(H_1)\beta = H_1$ and $(H_2)\beta = H_2$. Show that there are $\beta_1 \in \operatorname{Aut} H_1$ and $\beta_2 \in \operatorname{Aut} H_2$ with $\beta = \beta_1 \oplus \beta_2$. Is $L = \{\beta \in \operatorname{Aut} G : (H_1)\beta = H_1, (H_2)\beta = H_2\}$ a subgroup of $\operatorname{Aut} G$? (Yes/No) Is $L \cong \operatorname{Aut} H_1 \times \operatorname{Aut} H_2$? (Yes/No)

(b) Let G be a finite additive abelian group with $|G| = p_1^{n_1} p_2^{n_2} \ldots p_k^{n_k}$ where $n_1, n_2, \ldots, n_k$ are positive integers and $p_1, p_2, \ldots, p_k$ are distinct primes. Let $\alpha_j \in \operatorname{Aut} G_{p_j}$ for $1 \leq j \leq k$. Using Theorem 3.10 and the notation of Question 5(a) above, show that $\alpha = \alpha_1 \oplus \alpha_2 \oplus \cdots \oplus \alpha_k$ is an automorphism of G; here $(g)\alpha = \sum_{j=1}^{k}(g_j)\alpha_j$ with $g = \sum_{j=1}^{k} g_j$ and $g_j \in G_{p_j}$ for $1 \leq j \leq k$. Show also that each automorphism $\beta \in \operatorname{Aut} G$ can be expressed as $\beta = \beta_1 \oplus \beta_2 \oplus \cdots \oplus \beta_k$ for unique $\beta_j \in \operatorname{Aut} G_{p_j}$. Deduce that

$$\operatorname{Aut} G \cong \operatorname{Aut} G_{p_1} \times \operatorname{Aut} G_{p_2} \times \cdots \times \operatorname{Aut} G_{p_k},$$

that is, $\operatorname{Aut} G$ is isomorphic to the external direct product (Exercises 2.3, Question 4(d)) of the automorphism groups of the primary components of G.

Hint: Consider $\beta \leftrightarrow (\beta_1, \beta_2, \ldots, \beta_k)$ where $\beta = \beta_1 \oplus \beta_2 \oplus \cdots \oplus \beta_k$.

6. (a) Show that a non-trivial finitely generated $\mathbb{Z}$-module H is indecomposable if and only if H has isomorphism type either C_0 or C_{p^n} where p is prime.

   Hint: Use Exercises 2.2, Question 6(b).

   (b) Let G be a finitely generated $\mathbb{Z}$-module with torsion-free rank r and torsion subgroup T. Let l denote the number of elementary divisors of T. Suppose $G = H_1 \oplus H_2 \oplus \cdots \oplus H_m$ where H_i is a non-trivial submodule of G for $1 \le i \le m$. Show $m \le l + r$.

   Hint: Decompose each H_i into a direct sum of non-trivial indecomposable submodules using Exercises 3.1, Question 5(b), Theorem 3.4, Corollary 3.5 and Theorem 3.10.

   Deduce $m = l + r$ if and only if H_i is indecomposable for $1 \le i \le m$.

3.3 Endomorphism Rings and Isomorphism Classes of Subgroups and Quotient Groups

Let G be an abelian group. An additive mapping $\alpha : G \to G$ is called an *endomorphism of* G (see Lemma 2.9; we have used α in place of θ here, reserving θ for an important job in Theorem 3.15). So an endomorphism of the $\mathbb{Z}$-module G is a $\mathbb{Z}$-linear mapping of G to itself. Let α and α' be endomorphisms of G. It is straightforward to verify that

their *sum* $\alpha + \alpha'$, defined by $(g)(\alpha + \alpha') = (g)\alpha + (g)\alpha'$ for all $g \in G$,

is an endomorphism of G. The composition of α followed by α', that is,

their *product* $\alpha\alpha'$, defined by $(g)\alpha\alpha' = ((g)\alpha)\alpha'$ for all $g \in G$,

is also an endomorphism of G (Exercises 2.1, Question 4(d)). The set of all endomorphisms of G, together with the above binary operations of sum and product, is denoted by $\operatorname{End} G$.

Lemma 3.14

Let G be an abelian group. Then $\text{End}\,G$ is a ring and $\text{Aut}\,G = U(\text{End}\,G)$.

Proof

Consider α and α' in $\text{End}\,G$. Then $\alpha + \alpha'$ and $\alpha\alpha'$ also belong to $\text{End}\,G$. As $(g)(\alpha + \alpha') = (g)\alpha + (g)\alpha' = (g)\alpha' + (g)\alpha = (g)(\alpha' + \alpha)$ for all $g \in G$ we see that the endomorphisms $\alpha + \alpha'$ and $\alpha' + \alpha$ are equal, that is, $\alpha + \alpha' = \alpha' + \alpha$. In fact $(\text{End}\,G, +)$, the set of all endomorphisms of G together with the binary operation of addition, is an abelian group, the additive group of $\text{End}\,G$ (Exercises 3.3, Question 1(a)). Let α'' belong to $\text{End}\,G$. The distributive law $(\alpha + \alpha')\alpha'' = \alpha\alpha'' + \alpha'\alpha''$ holds as

$$(g)((\alpha + \alpha')\alpha'') = ((g)(\alpha + \alpha'))\alpha'' = ((g)\alpha + (g)\alpha')\alpha''$$

$$= ((g)\alpha)\alpha'' + ((g)\alpha')\alpha'' = (g)\alpha\alpha'' + (g)\alpha'\alpha'' = (g)(\alpha\alpha'' + \alpha'\alpha'')$$

for all $g \in G$. The remaining laws of a ring may be verified in the same way and we leave this to the reader. Note that the *zero endomorphism* 0, defined by $(g)0 = 0$ for all $g \in G$, is the 0-element of $\text{End}\,G$. The identity mapping $\iota : G \to G$, defined by $(g)\iota = g$ for all $g \in G$, is the 1-element of $\text{End}\,G$.

The elements of the group $U(\text{End}\,G)$ are the invertible endomorphisms α of G, that is, those $\alpha \in \text{End}\,G$ such that there is $\beta \in \text{End}\,G$ satisfying $\alpha\beta = \iota = \beta\alpha$. So each $\alpha \in U(\text{End}\,G)$ is invertible and its inverse $\alpha^{-1} = \beta$ is also an endomorphism of G. From Definition 2.4 we see $U(\text{End}\,G) \subseteq \text{Aut}\,G$ as each $\alpha \in U(\text{End}\,G)$ is bijective. Conversely consider $\alpha \in \text{Aut}\,G$. From Exercises 2.1, Question 4(d) we deduce $\alpha^{-1} \in \text{Aut}\,G$. As $\text{Aut}\,G \subseteq \text{End}\,G$ we conclude $\alpha^{-1} \in \text{End}\,G$. So $\alpha \in U(\text{End}\,G)$ showing $\text{Aut}\,G \subseteq U(\text{End}\,G)$. Therefore $U(\text{End}\,G) = \text{Aut}\,G$. □

Just as $\text{Aut}\,G$ is usually a non-abelian group, so $\text{End}\,G$ is in general a non-commutative ring. We now look at three particular types of f.g. abelian groups G and show that in each case $\text{End}\,G$ is a ring already familiar to the reader.

First suppose G is free of rank t. Let $v_1, v_2, \ldots, v_t$ be a ℤ-basis of G. The reader is reminded that $\mathfrak{M}_t(\mathbb{Z})$ denotes the ring of all $t \times t$ matrices over ℤ. Each endomorphism α of G gives rise to a matrix $A = (a_{ij})$ in $\mathfrak{M}_t(\mathbb{Z})$ where

$$(v_i)\alpha = \sum_{j=1}^{t} a_{ij} v_j \quad \text{for } 1 \leq i \leq t. \tag{♣♣♣}$$

It is usual to call A *the matrix of* α *relative to* $v_1, v_2, \ldots, v_t$. You are certain to have met this concept in the context of linear mappings of finite-dimensional vector spaces (this topic is revised in Definition 5.1). The mapping $\theta : \text{End } G \rightarrow \mathfrak{M}_t(\mathbb{Z})$ is defined by $(\alpha)\theta = A$ for all $\alpha \in \text{End } G$, that is, each endomorphism α of G is mapped by θ to its matrix A relative to the $\mathbb{Z}$-basis $v_1, v_2, \ldots, v_t$ of G. It's reasonable to expect – even take for granted – that if the endomorphisms α and α' of G have matrices A and A' respectively relative to $v_1, v_2, \ldots, v_t$ then $\alpha\alpha'$ has matrix AA' relative to $v_1, v_2, \ldots, v_t$. Indeed this property is included in our next lemma. However the reader should be aware that the innocent-looking equations (♣♣♣) only 'work' because we have adopted the notation $(g)\alpha$ (rather than the more usual $\alpha(g)$) for the image of g under the mapping α; so throughout $\alpha\alpha'$ means: first apply α and secondly apply α'.

Theorem 3.15

Let G be a free abelian group of rank t and let $v_1, v_2, \ldots, v_t$ be a $\mathbb{Z}$-basis of G. Let $\theta : \text{End } G \rightarrow \mathfrak{M}_t(\mathbb{Z})$ be defined by $(\alpha)\theta = A$ for all $\alpha \in \text{End } G$ where $A = (a_{ij})$ in $\mathfrak{M}_t(\mathbb{Z})$ is given by (♣♣♣) above. Then $\theta : \text{End } G \cong \mathfrak{M}_t(\mathbb{Z})$ is a ring isomorphism.

Proof

Consider α and α' in $\text{End } G$ having matrices $A = (a_{ij})$ and $A' = (a'_{ij})$ respectively relative to the $\mathbb{Z}$-basis $v_1, v_2, \ldots, v_t$ of G. So $(\alpha)\theta = A$ and $(\alpha')\theta = A'$. Then $(v_i)\alpha = \sum_{j=1}^{t} a_{ij} v_j$ and $(v_i)\alpha' = \sum_{j=1}^{t} a'_{ij} v_j$ for $1 \leq i \leq t$ by (♣♣♣). Adding these equations together gives

$$(v_i)(\alpha + \alpha') = (v_i)\alpha + (v_i)\alpha' = \sum_{j=1}^{t} a_{ij} v_j + \sum_{j=1}^{t} a'_{ij} v_j = \sum_{j=1}^{t} (a_{ij} + a'_{ij}) v_j$$

for $1 \leq i \leq t$ showing that $\alpha + \alpha'$ has matrix $(a_{ij} + a'_{ij}) = A + A'$ relative to $v_1, v_2, \ldots, v_t$. So $(\alpha + \alpha')\theta = A + A' = (\alpha)\theta + (\alpha')\theta$, that is, θ is additive.

In order to find $(\alpha\alpha')\theta$ we first replace the dummy suffixes i, j in $(v_i)\alpha' = \sum_{j=1}^{t} a'_{ij} v_j$ by j, k giving $(v_j)\alpha' = \sum_{k=1}^{t} a'_{jk} v_k$ for $1 \leq j \leq t$. Then

$$(v_i)\alpha\alpha' = ((v_i)\alpha)\alpha' = \left(\sum_{j=1}^{t} a_{ij} v_j\right)\alpha' = \sum_{j=1}^{t} a_{ij}(v_j)\alpha'$$

$$= \sum_{j=1}^{t} a_{ij}\left(\sum_{k=1}^{t} a'_{jk} v_k\right) = \sum_{k=1}^{t}\left(\sum_{j=1}^{t} a_{ij} a'_{jk}\right) v_k$$

for $1 \leq i \leq t$ which shows, on replacing α and the dummy suffix j in (♣♣♣) by $\alpha\alpha'$ and k, that the matrix of $\alpha\alpha'$ relative to $v_1, v_2, \ldots, v_t$ is AA' as its (i, k)-entry is $\sum_{j=1}^{t} a_{ij} a'_{jk}$. We've shown $(\alpha\alpha')\theta = AA' = (\alpha)\theta(\alpha')\theta$ for all α and α' in $\text{End } G$.

As $(\iota)\theta = I$, that is, θ maps the 1-element of $\operatorname{End} G$ to the 1-element of $\mathfrak{M}_t(\mathbb{Z})$, we conclude that θ is a ring homomorphism.

Finally we show that θ is bijective. Consider $A = (a_{ij}) \in \mathfrak{M}_t(\mathbb{Z})$. Suppose there is an α in $\operatorname{End} G$ having matrix A relative to $v_1, v_2, \ldots, v_t$, that is, $(\alpha)\theta = A$. Let $g \in G$. There are unique integers $\lambda_1, \lambda_2, \ldots, \lambda_t$ such that $g = \lambda_1 v_1 + \lambda_2 v_2 + \cdots + \lambda_t v_t$. Write $(\lambda_1, \lambda_2, \ldots, \lambda_t)A = (\mu_1, \mu_2, \ldots, \mu_t)$, that is, $\sum_{i=1}^t \lambda_i a_{ij} = \mu_j$ for $1 \leq j \leq t$. From (♣♣♣) we know $(v_i)\alpha = \sum_{j=1}^t a_{ij} v_j$ and so

$$(g)\alpha = \left(\sum_{i=1}^t \lambda_i v_i\right)\alpha = \sum_{i=1}^t \lambda_i (v_i)\alpha = \sum_{i=1}^t \lambda_i \left(\sum_{j=1}^t a_{ij} v_j\right)$$

$$= \sum_{j=1}^t \left(\sum_{i=1}^t \lambda_i a_{ij}\right) v_j = \sum_{j=1}^t \mu_j v_j.$$

This tells us that α maps $g = \sum_{i=1}^t \lambda_i v_i$ to the element $\sum_{j=1}^t \mu_j v_j$ of G where $(\lambda_1, \lambda_2, \ldots, \lambda_t)A = (\mu_1, \mu_2, \ldots, \mu_t)$. So there is *at most* one endomorphism α of G with $(\alpha)\theta = A$, showing θ to be injective.

We have still to prove that there is *at least* one endomorphism α of G with $(\alpha)\theta = A$. To do this let $\beta : G \to G$ be defined by $(g)\beta = \sum_{j=1}^t \mu_j v_j$ where $g = \sum_{i=1}^t \lambda_i v_i$ and $\sum_{i=1}^t \lambda_i a_{ij} = \mu_j$ for $1 \leq j \leq t$. Is β an endomorphism of G? Consider $g' \in G$. There are integers $\lambda_i'(1 \leq i \leq t)$ with $g' = \sum_{i=1}^t \lambda_i' v_i$. Then $(g')\beta = \sum_{j=1}^t \mu_j' v_j$ where $\sum_{i=1}^t \lambda_i' a_{ij} = \mu_j'$ for $1 \leq j \leq t$. Then $g + g' = \sum_{i=1}^t (\lambda_i + \lambda_i')v_i$ and $\sum_{i=1}^t (\lambda_i + \lambda_i')a_{ij} = \mu_j + \mu_j'$ for $1 \leq j \leq t$. So

$$(g + g')\beta = \sum_{j=1}^t (\mu_j + \mu_j')v_j = \sum_{j=1}^t \mu_j v_j + \sum_{j=1}^t \mu_j' v_j = (g)\beta + (g')\beta$$

showing that β is indeed an endomorphism of G. Is β the endomorphism we are looking for? Taking $g = v_i$ gives $(\lambda_1, \lambda_2, \ldots, \lambda_t) = e_i$ which is row i of the $t \times t$ identity matrix I over $\mathbb{Z}$ for $1 \leq i \leq t$. Hence

$$(\lambda_1, \lambda_2, \ldots, \lambda_t)A = e_i A = (a_{i1}, a_{i2}, \ldots, a_{it}) = (\mu_1, \mu_2, \ldots, \mu_t)$$

and so $(v_i)\beta = \sum_{j=1}^t a_{ij} v_j$ for $1 \leq i \leq t$. We have obtained (♣♣♣) with β in place of α. So β is the endomorphism of G we want, that is, $(\beta)\theta = A$ showing θ to be surjective. Therefore θ is a ring isomorphism.   □

Let G be a free abelian group with $\mathbb{Z}$-basis $v_1, v_2, \ldots, v_t$ as in Theorem 3.15. Restricting θ to the group $\operatorname{Aut} G = U(\operatorname{End} G)$ gives the isomorphism $\theta| : \operatorname{Aut} G \cong GL_t(\mathbb{Z})$ in which each automorphism α of G corresponds to its invertible $t \times t$ matrix $A = (\alpha)\theta$ relative to $v_1, v_2, \ldots, v_t$. So

> The automorphism group of every free abelian group of rank t is isomorphic to the multiplicative group of $t \times t$ invertible matrices over $\mathbb{Z}$

Taking $t = 1$ we obtain $\operatorname{End}\langle v_1 \rangle \cong \mathbb{Z}$ and $\operatorname{Aut}\langle v_1 \rangle \cong \{1, -1\}$ for the infinite cyclic group $\langle v_1 \rangle$. Endomorphism rings of finite cyclic groups will be dealt with shortly. But first we discuss the structure of finite elementary abelian p-groups (their definition follows Corollary 3.12) which is, in effect, no more than the modulo p version of Theorem 3.15.

Theorem 3.16

Let p be prime and let G be an elementary abelian group of order p^t. Then G is the additive group of a t-dimensional vector space over $\mathbb{Z}_p$. Let $v_1, v_2, \ldots, v_t$ be a basis of this vector space. The subgroups of G are precisely the subspaces of $\langle v_1, v_2, \ldots, v_t \rangle$. Also $\theta : \operatorname{End} G \cong \mathfrak{M}_t(\mathbb{Z}_p)$ is a ring isomorphism where $(\alpha)\theta = A$ is the matrix of $\alpha \in \operatorname{End} G$ relative to $v_1, v_2, \ldots, v_t$. Restricting θ to $\operatorname{Aut} G$ gives a group isomorphism $\theta| : \operatorname{Aut} G \cong GL_t(\mathbb{Z}_p)$.

Proof

As $pg = 0$ for all $g \in G$ we obtain $mg = m'g$ for all integers m, m' with $m \equiv m' \pmod{p}$. So it makes sense (and is very pertinent) to introduce the product of scalar $\overline{m} \in \mathbb{Z}_p$ and vector $g \in G$ by

$$\overline{m}g = mg. \qquad (\blacklozenge\blacklozenge)$$

The breath-taking simplicity of $(\blacklozenge\blacklozenge)$ is matched by its significance: the seven laws of a $\mathbb{Z}$-module, listed at the start of Section 2.1, immediately become the laws of a vector space over $\mathbb{Z}_p$, showing that G, together with scalar multiplication as in $(\blacklozenge\blacklozenge)$, is a vector space over $\mathbb{Z}_p$. As $|G| = p^t$ we see that this vector space, which we continue to denote by G, has dimension t and so has basis $v_1, v_2, \ldots, v_t$ (the reader will know that every finite-dimensional vector space has a basis). Also the reader should have no qualms about the equation $G = \langle v_1, v_2, \ldots, v_t \rangle$ because $v_1, v_2, \ldots, v_t$ generate the $\mathbb{Z}$-module G and $v_1, v_2, \ldots, v_t$ span the $\mathbb{Z}_p$-module G. It is easy to check, using $(\blacklozenge\blacklozenge)$, that subgroups (=submodules) of the $\mathbb{Z}$-module G coincide with subspaces of the $\mathbb{Z}_p$-module G.

Consider $\alpha \in \operatorname{End} G$. As α is $\mathbb{Z}$-linear, we deduce from $(\blacklozenge\blacklozenge)$ that $(\overline{m}g)\alpha = (mg)\alpha = m((g)\alpha) = \overline{m}((g)\alpha)$ for all $m \in \mathbb{Z}$, $g \in G$ showing that α is a linear mapping of the vector space G. Conversely every linear mapping of the vector space G belongs to $\operatorname{End} G$. The last part of Theorem 3.16 concerning θ is standard linear algebra and is left as an exercise for the reader. $\qquad \square$

From the theory at the start of Section 3.2, we deduce that the endomorphism ring of each finite abelian group is isomorphic to the external direct sum of the endomorphism rings of its primary components (Exercises 3.3, Question 1(f)).

We next determine the structure of the multiplicative group F^* of non-zero elements of a finite field F. It turns out that F^* is cyclic as mentioned in Example 2.8. However the proof is non-constructive: it shows that F^* must be generated by a single element without explicitly specifying a generator.

Corollary 3.17

Let F be a field. Every finite subgroup G of F^* is cyclic. In particular the multiplicative group of every finite field is cyclic.

Proof

We need one fact, proved independently in Corollary 4.2(ii), which should be familiar to the reader, concerning polynomials $f(x)$ with coefficients in F. Suppose $f(x) = a_0 + a_1 x + \cdots + a_n x^n$ where each a_i belongs to F for $0 \leq i \leq n$ and $a_n \neq 0$, that is, $f(x)$ is a polynomial in x of degree n over F. Then $f(x)$ has at most n zeros in F; equivalently the equation $f(c) = 0$ has at most n roots: there are at most n elements c in F satisfying $f(c) = 0$, that is, $a_0 + a_1 c + \cdots + a_n c^n = 0$. A quadratic equation with real coefficients has at most two real roots, a cubic equation with real coefficients has at most three real roots and so on. But note that, working in the ring $\mathbb{Z}_6$, the quadratic equation $x^2 = x$ has four roots, namely $\overline{0}, \overline{1}, \overline{3}, \overline{4}$. Of course $\mathbb{Z}_6$ is not a field.

We apply Theorem 3.4 to the finite abelian group G expressing the result in multiplicative notation: there are t non-trivial cyclic subgroups H_i of G such that H_i has order d_i ($1 \leq i \leq t$) where $d_i | d_{i+1}$ ($1 \leq i < t$) and each element of G is uniquely expressible as a product $h_1 h_2 \cdots h_t$ ($h_i \in H_i$). So $G = H_1 \times H_2 \times \cdots \times H_t$ meaning that G is the internal direct product of its subgroups $H_1, H_2, \ldots, H_t$ (it is the multiplicative version of Lemma 2.15). We ask the question: is it possible for t to be larger than 1? If so, by the multiplicative version of the $|G|$-lemma of Section 2.2, we see that the d_1 elements h_1 of H_1 satisfy $h_1^{d_1} = 1$, the d_2 elements h_2 of H_2 satisfy $h_2^{d_2} = 1$, and hence the $d_1 d_2$ elements $h_1 h_2$ of G satisfy $(h_1 h_2)^{d_2} = 1$ as $d_1 | d_2$. So the polynomial $x^{d_2} - 1$ with coefficients in F has $d_1 d_2$ zeros $h_1 h_2$ in F. Now $d_1 > 1$ as H_1 is non-trivial and so $d_1 d_2 > d_2$, showing that the polynomial $x^{d_2} - 1$ over F has more zeros in F than its degree! This contradiction shows $t \leq 1$. Therefore G is either trivial ($t = 0$) or $G = H_1$ is cyclic and non-trivial ($t = 1$). In any case G is cyclic. $\square$

You are certain to have met the group G of fourth roots of 1, that is, $G = \{z \in \mathbb{C} : z^4 = 1\}$. As $G = \{1, -1, i, -i\}$ where $i^2 = -1$ we see that G is cyclic with generator i. Similarly $(1+i)/\sqrt{2}$ generates the group of eighth roots of 1. It follows from Corollary 3.17 that the only finite subgroups of the group $\mathbb{C}*$, of non-zero complex numbers, are of isomorphism type C_n where n is a positive integer. In fact the subgroup generated by $\cos(2\pi/n) + i\sin(2\pi/n)$ is the only subgroup of $\mathbb{C}*$ having order n.

Let p be prime. As $\mathbb{Z}_p$ is a field Corollary 3.17 tells us that $\mathbb{Z}_p^*$ is cyclic. However there is no clue as to how a generator of $\mathbb{Z}_p^*$ might be found, and one may have to resort to 'trial and error' in the hunt for a generator of this group as in Example 2.8. For a group G of prime order p we obtain $\operatorname{End} G \cong \mathbb{Z}_p$ and $\operatorname{Aut} G \cong \mathbb{Z}_p^*$ on taking $t = 1$ in Theorem 3.16. Hence from Corollary 3.17 we deduce

Aut G has isomorphism type C_{p-1} where G is a group of prime order p.

For example let G be the additive group of $\mathbb{Z}_7$. Then $G = \langle \bar{1} \rangle$ is a 1-dimensional vector space over $\mathbb{Z}_7$. The element $a \in \mathbb{Z}_7$ corresponds to the endomorphism μ_a of G defined by $(g)\mu_a = ag$ for all $g \in G$. In fact the ring isomorphism $\theta : \operatorname{End} G \cong \mathbb{Z}_p$ of Theorem 3.16 is given by $(\mu_a)\theta = a$ for all $a \in \mathbb{Z}_7$, the 1×1 matrix of μ_a relative to the basis $\langle \bar{1} \rangle$ of G being simply a (matrix brackets being omitted). We leave the reader to verify that $\bar{3}$ generates $\mathbb{Z}_7^*$ and so $\alpha = \mu_{\bar{3}}$ generates $\operatorname{Aut} G = \{\mu_{\bar{1}}, \mu_{\bar{2}}, \mu_{\bar{3}}, \mu_{\bar{4}}, \mu_{\bar{5}}, \mu_{\bar{6}}, \} = \{\alpha^6, \alpha^2, \alpha, \alpha^4, \alpha^5, \alpha^3\}$.

Next we discuss $\operatorname{End} G$ and $\operatorname{Aut} G$ in the case of G being finite cyclic. We assume $|G| = p_1^{n_1} p_2^{n_2} \cdots p_k^{n_k}$ and show how $\operatorname{Aut} G$ decomposes into the direct sum of at most $k + 1$ cyclic subgroups.

Lemma 3.18

Let G be an additive abelian group and let m be an integer. The mapping $\mu_m : G \to G$, given by $(g)\mu_m = mg$ for all $g \in G$, is an endomorphism of G. The mapping $\chi : \mathbb{Z} \to \operatorname{End} G$, where $(m)\chi = \mu_m$ for all $m \in \mathbb{Z}$, is a ring homomorphism.

Suppose G is finite and cyclic. Then $\operatorname{im} \chi = \operatorname{End} G$ and $\ker \chi = \langle |G| \rangle$. Hence $\tilde{\chi} : \mathbb{Z}_{|G|} \cong \operatorname{End} G$ is a ring isomorphism where $(\bar{m})\tilde{\chi} = \mu_m$ for all $m \in \mathbb{Z}$.

Proof

As $(g_1 + g_2)\mu_m = m(g_1 + g_2) = mg_1 + mg_2 = (g_1)\mu_m + (g_2)\mu_m$ for all $g_1, g_2 \in G$ we see $\mu_m \in \operatorname{End} G$. From $(g)\mu_{m+m'} = (m+m')g = mg + m'g = (g)\mu_m + (g)\mu_{m'} = (g)(\mu_m + \mu_{m'})$ for $m, m' \in \mathbb{Z}$ we deduce that χ is additive as $(m + m')\chi = \mu_{m+m'} = \mu_m + \mu_{m'} = (m)\chi + (m')\chi$. Also $(g)\mu_{mm'} = mm'g = m'mg = (g)\mu_m\mu_{m'}$ for $m, m' \in \mathbb{Z}$ and so χ respects multiplication as $(mm')\chi = \mu_{mm'} = \mu_m\mu_{m'} = (m)\chi(m')\chi$. As $(1)\chi = \mu_1 = \iota$ we conclude that χ is a ring homomorphism.

Suppose G is cyclic with generator g_0 and consider $\alpha \in \operatorname{End} G$. As $(g_0)\alpha \in G = \langle g_0 \rangle$ there is an integer m with $(g_0)\alpha = mg_0$. Each $g \in G$ can be expressed as $g = m'g_0$

where $m' \in \mathbb{Z}$ and so $(g)\alpha = (m'g_0)\alpha = m'(g_0)\alpha = m'mg_0 = mm'g_0 = m(m'g_0) = mg = (g)\mu_m$ showing that $\alpha = \mu_m$. We've shown that each endomorphism α of the cyclic group G is of the form μ_m and so belongs to the image of χ, that is, im $\chi =$ End G.

Now suppose that G is finite and cyclic. By Exercises 2.3, Question 3(b) the kernel of χ is an ideal of $\mathbb{Z}$. By Theorem 1.15 there is a non-negative integer d with ker $\chi = \langle d \rangle$. The $|G|$-lemma of Section 2.2 gives $(|G|)\chi = \mu_{|G|} = 0$ and so $|G| \in \ker \chi$. Hence $d \,|\, |G|$. On the other hand $\mu_d = 0$ and so $dg_0 = 0$. As g_0 has order $|G|$ from Theorem 2.5 we deduce $|G| \,|\, d$ and so $|G| = d$. We have shown ker $\chi = \langle |G| \rangle$, that is, the kernel of χ is the principal ideal of $\mathbb{Z}$ generated by $|G|$. Now the quotient ring $\mathbb{Z}/\langle |G| \rangle$ is the same as the ring $\mathbb{Z}_{|G|}$ of integers modulo $|G|$. By the first isomorphism theorem for rings (Exercises 2.3, Question 3(b) again), we see

$$\tilde{\chi} : \mathbb{Z}_{|G|} \cong \text{End } G \quad \text{where } (\overline{m})\tilde{\chi} = \mu_m \text{ for all } m \in \mathbb{Z},$$

is a ring isomorphism. $\qquad\qquad\qquad\qquad\qquad\qquad\qquad\qquad\qquad\qquad\qquad\qquad\qquad\square$

Restricting $\tilde{\chi}$ to the group of $\phi(|G|)$ units of $\mathbb{Z}_{|G|}$ gives a group isomorphism $U(\mathbb{Z}_{|G|}) \cong \text{Aut } G$. Restricting the ring isomorphism α of the generalised Chinese remainder theorem to $U(\mathbb{Z}_{|G|})$ gives a group isomorphism

$$U(\mathbb{Z}_{|G|}) \cong U(\mathbb{Z}_{q_1}) \times U(\mathbb{Z}_{q_2}) \times \cdots \times U(\mathbb{Z}_{q_k})$$

where every two of the prime powers $q_j = p_j^{n_j}$ have gcd 1 $(1 \leq j \leq k)$.

Our next theorem reveals the structure of the abelian groups $U(\mathbb{Z}_{q_j})$.

Theorem 3.19

Let $n \geq 3$. Then $U(\mathbb{Z}_{2^n}) = \langle \overline{-1} \rangle \times \langle \overline{3} \rangle$, that is, $U(\mathbb{Z}_{2^n})$ is the internal direct product of its cyclic subgroups $\langle \overline{-1} \rangle$ of order 2 and $\langle \overline{3} \rangle$ of order 2^{n-2}.

Let p be an odd prime and $n \geq 2$. Then $U(\mathbb{Z}_{p^n})$ is cyclic of order $p^n - p^{n-1}$.

Proof

Our first job is to determine the order of $\overline{3}$ in the multiplicative group $U(\mathbb{Z}_{2^n})$. To do this we show by induction on n, where $n \geq 3$, that

$$3^{2^{n-2}} - 1 = 2^n \times m \quad \text{for odd } m. \qquad\qquad (\heartsuit\heartsuit)$$

As $3^2 - 1 = 2^3 \times 1$ we see that ($\heartsuit\heartsuit$) holds for $n = 3$. Suppose $n > 3$ and ($\heartsuit\heartsuit$) holds with n replaced by $n - 1$, that is, $3^{2^{n-3}} - 1 = 2^{n-1} \times m$ where m is odd.

Now $3 \equiv -1 \pmod 4$ and so $3^{2^{n-3}} + 1 \equiv (-1)^{2^{n-3}} + 1 \pmod 4$, that is, $3^{2^{n-3}} + 1 \equiv 2 \pmod 4$ which means $3^{2^{n-3}} + 1 = 2 \times m'$ where m' is odd. Factorising a difference of two squares gives $3^{2^{n-2}} - 1 = (3^{2^{n-3}} - 1)(3^{2^{n-3}} + 1) = 2^{n-1} \times m \times 2 \times m' = 2^n \times mm'$ which shows that (❤❤) holds with m replaced by the odd integer mm'. The induction is now complete. Therefore $3^{2^{n-2}} \equiv 1 \pmod{2^n}$ and so the order of $\overline{3}$ in $U(\mathbb{Z}_{2^n})$ is a divisor of 2^{n-2}. However for $n \geq 4$ the above induction shows $3^{2^{n-3}} - 1 = 2^{n-1}m = 2^{n-1}(2l+1)$ for some integer l and so $3^{2^{n-3}} \equiv 2^{n-1} + 1 \pmod{2^n}$. Hence $3^{2^{n-3}} \not\equiv 1 \pmod{2^n}$ for $n \geq 4$. As $3 \not\equiv 1 \pmod 8$ we conclude that the order of $\overline{3}$ in $U(\mathbb{Z}_{2^n})$ is 2^{n-2} as this order is not a divisor of 2^{n-3} for $n \geq 3$.

The cyclic subgroup $\langle \overline{3} \rangle$ of $U(\mathbb{Z}_{2^n})$ contains a unique element of order 2, namely $(\overline{3})^{2^{n-3}}$. Also $\overline{-1}$ has order 2 in $U(\mathbb{Z}_{2^n})$. Is it possible for these two elements of order 2 to coincide? We suppose for a moment that $(\overline{3})^{2^{n-3}} = \overline{-1}$ in $\mathbb{Z}_{2^n}$. As $\overline{3} \neq \overline{-1}$ in $\mathbb{Z}_8$ we see $n \geq 4$. Hence $2^{n-1} + 1 \equiv -1 \pmod{2^n}$ from the above paragraph; so 2^n is a divisor of $2^{n-1} + 2$ which happens only for $n = 2$. The conclusion is: $\langle \overline{-1} \rangle \cap \langle \overline{3} \rangle = \langle \overline{1} \rangle$, that is, the cyclic subgroups $\langle \overline{-1} \rangle$ and $\langle \overline{3} \rangle$, of orders 2 and 2^{n-2}, have trivial intersection. Hence the two cosets $\langle \overline{3} \rangle$ and $\overline{-1}\langle \overline{3} \rangle$ are different and so disjoint. Together these cosets account for $2^{n-2} + 2^{n-2} = 2^{n-1}$ elements of $U(\mathbb{Z}_{2^n})$. But $|U(\mathbb{Z}_{2^n})| = \phi(2^n) = 2^n - 2^{n-1} = 2^{n-1}$ and so $\langle \overline{-1} \rangle \langle \overline{3} \rangle = U(\mathbb{Z}_{2^n})$, that is, the product of the subgroups $\langle \overline{-1} \rangle$ and $\langle \overline{3} \rangle$ is the whole group $U(\mathbb{Z}_{2^n})$. Therefore $U(\mathbb{Z}_{2^n}) = \langle \overline{-1} \rangle \times \langle \overline{3} \rangle$, that is, $U(\mathbb{Z}_{2^n})$ is the direct product of $\langle \overline{-1} \rangle$ and $\langle \overline{3} \rangle$.

Let p be an odd prime. By Corollary 3.17 there is a generator $\overline{a}$ of $\mathbb{Z}_p^* = U(\mathbb{Z}_p)$. As $|\mathbb{Z}_p^*| = p - 1$ we obtain $(\overline{a})^{p-1} = \overline{1}$ and so $a^{p-1} \equiv 1 \pmod p$. It turns out, as we prove in a moment, that the integer a can always be chosen with $a^{p-1} \not\equiv 1 \pmod{p^2}$. In fact suppose $a^{p-1} \equiv 1 \pmod{p^2}$. Consider $b = a + p$. Then $\overline{a} = \overline{b}$ in $\mathbb{Z}_p$ and so $\overline{b}$ generates $\mathbb{Z}_p^*$. Using the binomial theorem

$$b^{p-1} = (a+p)^{p-1} = \sum_{i=0}^{p-1} \frac{(p-1)!}{i!(p-1-i)!} a^{p-1-i} p^i.$$

All terms in the summation with $i \geq 2$ have factor p^2 and so

$$b^{p-1} \equiv 1 + (p-1)a^{p-2}p \pmod{p^2}$$

on using $a^{p-1} \equiv 1 \pmod{p^2}$. Hence $b^{p-1} \not\equiv 1 \pmod{p^2}$ as $(p-1)a^{p-2}p \not\equiv 0 \pmod{p^2}$ since $(p-1)a^{p-2}$ is not divisible by p. We have shown that if the original integer a fails to satisfy $a^{p-1} \not\equiv 1 \pmod{p^2}$, then $b = a + p$ satisfies $b^{p-1} \not\equiv 1 \pmod{p^2}$.

So there is an integer a such that $\overline{a}$ generates $\mathbb{Z}_p^*$ and $a^{p-1} \not\equiv 1 \pmod{p^2}$. Write x_n for the congruence class of a modulo p^n and write e_n for the congruence class of 1 modulo p^n. So x_n and e_n are elements of $\mathbb{Z}_{p^n}$. As $\gcd\{a, p\} = 1$ we see that x_n belongs to the multiplicative group $U(\mathbb{Z}_{p^n})$ which has identity element e_n. The proof is finished by showing that x_n generates $U(\mathbb{Z}_{p^n})$ for all $n \geq 2$. To do this we show by induction on n

$$a^{(p-1)p^{n-2}} \not\equiv 1 \pmod{p^n} \quad \text{for all } n \geq 2. \tag{$\spadesuit$}$$

Now ($\spadesuit$) holds for $n = 2$ by our choice of the integer a. So suppose $n \geq 3$ and $a^{(p-1)p^{n-3}} \not\equiv 1 \pmod{p^{n-1}}$. Now x_{n-2} belongs to $U(\mathbb{Z}_{p^{n-2}})$ which is a multiplicative group of order $\phi(p^{n-2}) = p^{n-2} - p^{n-3} = (p-1)p^{n-3}$ with identity element e_{n-2}. Using the multiplicative version of the $|G|$-lemma we obtain $(x_{n-2})^{(p-1)p^{n-3}} = e_{n-2}$ which gives $a^{(p-1)p^{n-3}} \equiv 1 \pmod{p^{n-2}}$. So $a^{(p-1)p^{n-3}} = 1 + rp^{n-2}$ where, by inductive hypothesis, r is not divisible by p. We raise this last equation to the power p and use the binomial expansion

$$a^{(p-1)p^{n-2}} = (a^{(p-1)p^{n-3}})^p = (1 + rp^{n-2})^p = \sum_{i=0}^{p} \frac{p!}{i!(p-i)!} r^i p^{i(n-2)}.$$

The term given by $i = 2$ in the sum is divisible by $p^{1+2(n-2)}$ and $1 + 2(n-2) \geq n$ as $n \geq 3$. Each of the following terms (with $3 \leq i \leq p$) is divisible by $p^{i(n-2)}$ and $i(n-2) \geq 3(n-2) \geq n$ as $n \geq 3$. So all terms, except the first two, in the above sum are divisible by p^n and hence $a^{(p-1)p^{n-2}} \equiv 1 + rp^{n-1} \pmod{p^n}$. As $rp^{n-1} \not\equiv 0 \pmod{p^n}$ we conclude that $a^{(p-1)p^{n-2}} \not\equiv 1 \pmod{p^n}$ completing the inductive step. So ($\spadesuit$) is established.

The end of the proof is now in sight! Let r be the order of x_n. Then x_n generates a cyclic subgroup of order r. By Lagrange's theorem r is a divisor of $\phi(p^n) = (p-1)p^{n-1} = |U(\mathbb{Z}_{p^n})|$ and so $r = sp^t$ where $s|p-1$ and $0 \leq t < n$. From ($\spadesuit$) we deduce $x_n(p-1)p^{n-2} \neq e_n$. Hence r is not a divisor of $(p-1)p^{n-2}$ and so $t \leq n-2$ is impossible, that is, $t = n-1$. Now $x_n^r = e_n$ which means $a^r \equiv 1 \pmod{p^n}$, and so $a^r \equiv 1 \pmod{p}$, that is, $(\bar{a})^r = \bar{1}$ in $\mathbb{Z}_p^*$. Therefore $(p-1)|r$ as $p-1$ is the order of $\bar{a}$ in $\mathbb{Z}_p^*$. As $\gcd\{p-1, p^{n-1}\} = 1$ we deduce $(p-1)|s$ and so $s = p-1$. We have shown that x_n has order $r = (p-1)p^{n-1} = |U(\mathbb{Z}_{p^n})|$ and so x_n generates $U(\mathbb{Z}_{p^n})$. $\square$

The reader should check that $U(\mathbb{Z}_4)$ has order 2 and so is cyclic; in fact $U(\mathbb{Z}_4) = \langle \bar{3} \rangle$. Let k be the number of different prime divisors of n. Combining Lemma 3.14, Corollary 3.17, Lemma 3.18, Theorem 3.19 and the Chinese remainder theorem we see that $\operatorname{Aut} \mathbb{Z}_n$ is the direct product of k non-trivial cyclic subgroups if either n is odd or $n \equiv 4 \pmod 8$. Further $\operatorname{Aut} \mathbb{Z}_n$ is the direct product of $k-1$ or $k+1$ non-trivial cyclic subgroups according as $n \equiv 2 \pmod 4$ or $n \equiv 0 \pmod 8$.

We look at a few particular cases. The automorphisms η and μ of $G = \mathbb{Z}_{16}$ defined by $(g)\eta = -g$, $(g)\mu = 3g$ for all $g \in G$ correspond to the elements $\overline{-1}$ and $\bar{3}$ of $U(\mathbb{Z}_{16})$ under the isomorphism $\tilde{\chi}$ of Lemma 3.18, that is, $(\overline{-1})\tilde{\chi} = \eta$, $(\bar{3})\tilde{\chi} = \mu$. The orders of η and μ are 2 and 4 respectively and $\operatorname{Aut} G = \langle \eta \rangle \times \langle \mu \rangle$ consists of the eight automorphisms

$$\iota, \quad \eta, \quad \mu, \quad \eta\mu, \quad \mu^2, \quad \eta\mu^2, \quad \mu^3, \quad \eta\mu^3, \quad \mu^4, \quad \eta\mu^4$$

by Theorem 3.19. The isomorphism class of the abelian group $\text{Aut}\,\mathbb{Z}_{16}$ is $C_2 \oplus C_4$.

The element $\bar{2}$ of $\mathbb{Z}_3$ generates $\mathbb{Z}_3^*$. Here $a = 2$, $p = 3$ as in Theorem 3.19. As $a^{p-1} = (2)^2 = 4$ and $4 \not\equiv 1 \pmod 9$ we see $U(\mathbb{Z}_3^n)$ is cyclic with generator $\bar{2} \in \mathbb{Z}_{3^n}$ for all positive integers n. In particular

$$U(\mathbb{Z}_9) = \{\bar{2}, (\bar{2})^2, (\bar{2})^3, (\bar{2})^4, (\bar{2})^5, (\bar{2})^6\} = \{\bar{2}, \bar{4}, \bar{8}, \bar{7}, \bar{5}, \bar{1}\}.$$

The reader may verify that $U(\mathbb{Z}_{27})$ consists of the 18 powers of $\bar{2} \in \mathbb{Z}_{27}$.

As $800 = 2^5 \times 5^2$ we see from Theorem 3.19 and the discussion preceding it that $\text{Aut}\,\mathbb{Z}_{800}$ is the direct product of cyclic groups of order 2, 8 and 20. By Theorem 2.11

$$C_2 \oplus C_8 \oplus C_{20} = C_2 \oplus C_8 \oplus C_4 \oplus C_5 = C_2 \oplus C_4 \oplus (C_8 \oplus C_5)$$

$$= C_2 \oplus C_4 \oplus C_{40}$$

showing that $\text{Aut}\,\mathbb{Z}_{800}$ has invariant factor sequence $(2, 4, 40)$.

Let G be an f.g. abelian group and let H be a subgroup of G. What is the connection between the invariant factors of G, the invariant factors of H and those of G/H? We begin by looking at the torsion-free rank of these groups. Remember that the torsion-free rank of G, which we denote by $\text{tf rank}\,G$, is the number of zero invariant factors of G.

Lemma 3.20

Let H be a subgroup of a finitely generated abelian group G. Then H and G/H are finitely generated and $\text{tf rank}\,G/H + \text{tf rank}\,H = \text{tf rank}\,G$.

Proof

Let $g_1, g_2, \ldots, g_t$ generate G and let $\theta : \mathbb{Z}^t \to G$ be the surjective $\mathbb{Z}$-linear mapping defined, as usual, by $(m_1, m_2, \ldots, m_t)\theta = m_1 g_1 + m_2 g_2 + \cdots + m_t g_t$ for all integers $m_1, m_2, \ldots, m_t$. Write $z = (m_1, m_2, \ldots, m_t)$ and let $K' = \{z \in \mathbb{Z}^t : (z)\theta \in H\}$. Then K' is a submodule of $\mathbb{Z}^t$ and so K' is free of rank s say where $s \leq t$ by Theorem 3.1. Also $K \subseteq K'$ where $K = \ker\theta$ since H contains the zero element of G. So $\text{rank}\,K \leq s$ by Theorem 3.1. From the proof of Theorem 3.4 we see $\text{tf rank}\,G = t - \text{rank}\,K$.

Let $z_1, z_2, \ldots, z_s$ be a $\mathbb{Z}$-basis of K' and let θ' be the restriction of θ to K'. Then $\theta' : K' \to H$ is the $\mathbb{Z}$-linear mapping defined by $(z)\theta' = (z)\theta$ for all $z \in K'$ and $\langle (z_1)\theta', (z_2)\theta', \ldots, (z_s)\theta' \rangle = \text{im}\,\theta' = H$ is finitely generated. As $\ker\theta' = K$ we see, as above, $\text{tf rank}\,H = s - \text{rank}\,K$.

Consider the composite $\mathbb{Z}$-linear mapping $\theta\eta : \mathbb{Z}^t \to G/H$ where $\eta : G \to G/H$ is the natural mapping. Now $\langle(e_1)\theta\eta, (e_2)\theta\eta, \ldots, (e_t)\theta\eta\rangle = \operatorname{im}\theta\eta = G/H$ is finitely generated. As $\ker\theta\eta = K'$ we see, once again as above, $\operatorname{tf\,rank} G/H = t - s$. So $\operatorname{tf\,rank} G = t - \operatorname{rank} K = (t - s) + (s - \operatorname{rank} K) = \operatorname{tf\,rank} G/H + \operatorname{tf\,rank} H.$ $\qquad\square$

From Lemma 3.20 we deduce

$$\operatorname{tf\,rank} H \leq \operatorname{tf\,rank} G \quad \text{and} \quad \operatorname{tf\,rank} G/H \leq \operatorname{tf\,rank} G.$$

Definition 3.21

A sequence $(d_1, d_2, \ldots, d_s)$ of non-negative integers with $d_i | d_{i+1}$ for $1 \leq i < s$ is called a *divisor sequence of length s*. The divisor sequence $(d'_1, d'_2, \ldots, d'_s)$ is called a *subsequence* of the divisor sequence $(d_1, d_2, \ldots, d_s)$ if $d'_i | d_i$ for $1 \leq i \leq s$.

For instance the divisor sequence $(2, 2, 6)$ of length 3 has subsequences $(1, 1, 1)$, $(1, 1, 2)$, $(1, 1, 3)$, $(1, 1, 6)$, $(1, 2, 2)$, $(1, 2, 6)$, $(2, 2, 2)$, $(2, 2, 6)$. It follows from our next theorem that each abelian group of isomorphism type $C_2 \oplus C_2 \oplus C_6$ has subgroups of the eight isomorphism types C_1, C_2, C_3, C_6, $C_2 \oplus C_2$, $C_2 \oplus C_6$, $C_2 \oplus C_2 \oplus C_2$, $C_2 \oplus C_2 \oplus C_6$ – no more and no less – corresponding to the eight subsequences of $(2, 2, 6)$.

In effect we have already met divisor sequences in Chapter 1. Let A be an $s \times t$ matrix over $\mathbb{Z}$ where $s \leq t$. Then the invariant factors d_i of A form a divisor sequence $(d_1, d_2, \ldots, d_s)$ of length s. By Corollary 1.20 the equivalence classes of such matrices A correspond bijectively in this way to divisor sequences of length s.

Theorem 3.22

Let H be a subgroup of the finite abelian group G. Let $(d_1, d_2, \ldots, d_t)$ be the invariant factor sequence of G. Then H has invariant factor sequence $(d'_1, d'_2, \ldots, d'_s)$ where $s \leq t$ and $d'_k | d_{t-s+k}$ for $1 \leq k \leq s$.

Proof

By Theorem 3.4 there are cyclic subgroups H_j of G such that H_j has isomorphism type C_{d_j} for $1 \leq j \leq t$ and $G = H_1 \oplus H_2 \oplus \cdots \oplus H_t$. Let $\theta : \mathbb{Z}^t \to G$ be defined by $(m_1, m_2, \ldots, m_t)\theta = m_1 h_1 + m_2 h_2 + \cdots + m_t h_t$ for all $z = (m_1, m_2, \ldots, m_t) \in \mathbb{Z}^t$ where h_j generates H_j for $1 \leq j \leq t$. Then $K = \ker\theta$ has $\mathbb{Z}$-basis consisting of the rows of the $t \times t$ matrix $D = \operatorname{diag}(d_1, d_2, \ldots, d_t)$. As in the proof of Lemma 3.20 write $K' = \{z \in \mathbb{Z}^t : (z)\theta \in H\}$. Then K' is a free submodule of $\mathbb{Z}^t$ of rank t by Theorems 3.1 and 3.4. Let the rows of the $t \times t$ matrix B form a $\mathbb{Z}$-basis of K'. As $0 \in H$ we see $K \subseteq K'$ and so each row of D is an integer linear combination of the rows of B. In other words there is a $t \times t$ matrix A over $\mathbb{Z}$ with $D = AB$. As $\det D = d_1 d_2 \cdots d_t > 0$ we see $\det A \neq 0$ and so the hypothesis of Theorem 1.21 is satisfied: the matrices A, B, AB have positive invariant factors. By Theorem 1.21 the invariant factors of A form a subsequence $(1, 1, \ldots, 1, d_1', d_2', \ldots, d_s')$ of the divisor sequence $(d_1, d_2, \ldots, d_t)$ of invariant factors of AB where $d_1' > 1$; so s is the number of invariant factors of A which are different from 1 and $d_k' | d_{t-s+k}$ for $1 \leq k \leq s$.

Consider the restriction of θ to K', that is, the $\mathbb{Z}$-linear mapping $\theta' : K' \to H$ defined by $(k')\theta' = (k')\theta$ for all $k' \in K'$. From the proof of Lemma 3.20, we have $\operatorname{im}\theta' = H$ and $\ker\theta' = K$. Write $z_i = e_i B$ for row i of B and $h_i = (z_i)\theta'$ $(1 \leq i \leq t)$. Now $\varphi : \mathbb{Z}^t \cong K'$, defined by $(z)\varphi = zB$ for all $z = (m_1, m_2, \ldots, m_t) \in \mathbb{Z}^t$, is an isomorphism since $z_1, z_2, \ldots, z_t$ is a $\mathbb{Z}$-basis of K'. As $h_1, h_2, \ldots, h_t$ generate H and $(e_i)\varphi\theta' = (z_i)\theta' = h_i$ for $1 \leq i \leq t$, the composite $\mathbb{Z}$-linear mapping

$$\varphi\theta' : \mathbb{Z}^t \to H \text{ is given by } (m_1, m_2, \ldots, m_t)\varphi\theta' = m_1 h_1 + m_2 h_2 + \cdots + m_t h_t$$

and so is the analogue of $\theta : \mathbb{Z}^t \to G$. Therefore the invariant factors of H can be found using Theorem 3.4 with $\varphi\theta'$ in place of θ. In fact there is just one more piece of the jigsaw to be put in place! As φ is an isomorphism we see $\operatorname{im}\varphi\theta' = H$ and $(\ker\varphi\theta')\varphi = \ker\theta' = K$. The equation $AB = D$ gives the t row equations $(e_i A)\varphi = e_i AB = e_i D = d_i e_i$ for $1 \leq i \leq t$. From $K = \langle d_1 e_1, d_2 e_2, \ldots, d_t e_t \rangle$ we deduce that the rows $e_1 A, e_2 A, \ldots, e_t A$ of A form a $\mathbb{Z}$-basis of $\ker\varphi\theta'$. By Theorem 3.4 the invariant factors of H are the invariant factors $\neq 1$ of A, that is, $d_1', d_2', \ldots, d_s'$. $\qquad\square$

Let H be a subgroup of the f.g. abelian group G. From Lemma 3.20 and Theorem 3.22 it is possible to give a complete description of the invariant factors of H in terms of those of G. Suppose that G has torsion-free rank r and the torsion subgroup $T(G)$ of G has invariant factor sequence $(d_1, d_2, \ldots, d_t)$. Then H has torsion-free rank at most r by Lemma 3.20 and its torsion subgroup $T(H)$ has invariant factor sequence $(d_1', d_2', \ldots, d_s')$ where $s \leq t$ and $d_i' | d_{t-s+i}$ for $1 \leq i \leq s$ by Theorem 3.22, as $T(H)$ is a subgroup of $T(G)$.

For example suppose G has invariant factor sequence $(2, 2, 6, 0, 0)$ and let H be a subgroup of G. Comparing torsion subgroups, there are 8 possible isomorphism types for $T(H)$, namely those listed after Definition 3.21, as $T(H)$ is a subgroup of $T(G)$ and $T(G)$ has isomorphism type $C_2 \oplus C_2 \oplus C_6$. There are 3 possible values for tfrank H, namely 0, 1, 2 as $0 \leq \text{tfrank } H \leq \text{tfrank } G = 3$ by Lemma 3.20. So G has exactly $8 \times 3 = 24$ isomorphism types of subgroups H.

Finally we discuss the connection between the invariant factors of an f.g. abelian group G and those of its homomorphic images G/H. As one might expect G/H cannot have more invariant factors than G and the invariant factors of G/H are divisors of the last so many corresponding invariant factors of G.

Theorem 3.23

Let $(d_1, d_2, \ldots, d_t)$ be the invariant factor sequence of the finitely generated abelian group G. Let H be a subgroup of G. Then G/H has invariant factor sequence $(d'_1, d'_2, \ldots, d'_{t'})$ where $t' \leq t$ and $d'_k | d_{t-t'+k}$ for $1 \leq k \leq t'$.

Proof

By Theorem 3.4 there are cyclic subgroups H_j of G such that H_j has isomorphism type C_{d_j} for $1 \leq j \leq t$ and $G = H_1 \oplus H_2 \oplus \cdots \oplus H_t$. Let $\theta : \mathbb{Z}^t \to G$ be defined by $(m_1, m_2, \ldots, m_t)\theta = m_1 h_1 + m_2 h_2 + \cdots + m_t h_t$ for all $z = (m_1, m_2, \ldots, m_t) \in \mathbb{Z}^t$ where h_j generates H_j for $1 \leq j \leq t$. Then $K = \ker \theta$ has $\mathbb{Z}$-basis consisting of the rows of the $r \times t$ matrix $D = \mathrm{diag}(d_1, d_2, \ldots, d_r)$ where $d_1, d_2, \ldots, d_r$ are the non-zero invariant factors of G. As before write $K' = \{z \in \mathbb{Z}^t : (z)\theta \in H\}$ and consider the composite $\mathbb{Z}$-linear mapping $\theta\eta : \mathbb{Z}^t \to G/H$ where $\eta : G \to G/H$ is the natural mapping. Then $\mathrm{im}\,\theta\eta = G/H$ and $\ker \theta\eta = K'$. Write $s = \mathrm{rank}\, K'$. Then $r \leq s \leq t$ and $d'_k = 0$ for $s - t + t' < k \leq t'$ as the torsion-free rank of G/H is $t - s$. There is an $s \times t$ matrix B over $\mathbb{Z}$ such that its rows form a $\mathbb{Z}$-basis of K'. Using the method of Theorem 3.4 with G, θ, A replaced by G/H, $\theta\eta$, B respectively, we see that the first $t - t'$ invariant factors of B are equal 1 and the remainder are $d'_1, d'_2, \ldots d'_{s-t+t'}$.

As $K \subseteq K'$ each row of D is an integer linear combination of the rows of B, that is, there is an $r \times s$ matrix A over $\mathbb{Z}$ such that $AB = D$. As the rows of D are $\mathbb{Z}$-independent so also are the rows of A. Hence the conditions of Theorem 1.21 are fulfilled: the invariant factors of A, B and D are positive. By Theorem 1.21 we see $d'_k | d_{t-t'+k}$ for $1 \leq k \leq r + t' - t$. But $d'_k | d_{t-t'+k}$ as $d_{t-t'+k} = 0$ for $r + t' - t < k \leq t'$, completing the proof.   □

As an illustration suppose G has invariant factor sequence $(2, 6, 0, 0)$. Then G has homomorphic images G/H with invariant factor sequences $(3, 6, 0)$ and $(2, 2, 2, 2)$ for instance. But neither $(4, 4, 8)$ nor $(3, 3, 6, 0)$ can arise as the invariant factor sequence of a homomorphic image of G.

EXERCISES 3.3

1. (a) Let α and α' be endomorphisms of the abelian group G. Show that $\alpha + \alpha'$ is an endomorphism of G where $(g)(\alpha + \alpha') = (g)\alpha + (g)\alpha'$ for all $g \in G$.

   Show that the set $\mathrm{End}\, G$ of all endomorphisms of the abelian group G, with the binary operation of addition as above, is itself an abelian group.

Verify the distributive law $\alpha(\alpha' + \alpha'') = \alpha\alpha' + \alpha\alpha''$ where $\alpha, \alpha', \alpha'' \in$ End G.

(b) Find the smallest abelian group G such that End G is a non-commutative ring. How many automorphisms does G have?

(c) Let G be the additive group $\mathbb{Z}^2$ and let $\alpha_0 \in$ End G be given by $(m_1, m_2)\alpha_0 = (0, m_1)$ for all $m_1, m_2 \in \mathbb{Z}$. Write down the matrix of α_0 relative to the standard $\mathbb{Z}$-basis $e_1 = (1, 0)$, $e_2 = (0, 1)$ of G. Write $Z(\alpha_0) = \{\alpha \in$ End $G : \alpha\alpha_0 = \alpha_0\alpha\}$. Determine the matrices relative to e_1, e_2 of the endomorphisms α in $Z(\alpha_0)$. Use Theorem 3.15 to decide whether or not $Z(\alpha_0)$ is a commutative subring of End G. If so, find the isomorphism type of the additive groups of End G, $Z(\alpha_0)$ and that of the multiplicative group $U(Z(\alpha_0))$.

(d) Let G be the additive group of the ring $\mathbb{Z}_3 \oplus \mathbb{Z}_3$ and let $\alpha_0 \in$ End G be given by $(x_1, x_2)\alpha_0 = (-x_2, x_1)$ for all $x_1, x_2 \in \mathbb{Z}_3$. Determine the matrices relative to e_1, e_2 of the endomorphisms α in $Z(\alpha_0) = \{\alpha \in$ End $G : \alpha\alpha_0 = \alpha_0\alpha\}$. Use Theorem 3.16 to find the isomorphism types of the additive groups of End G and $Z(\alpha_0)$. Is $Z(\alpha_0)$ a field? What is the isomorphism type of the multiplicative group $U(Z(\alpha_0))$. *Hint*: Show $a^2 + b^2 = 0$ where $a, b \in \mathbb{Z}_3$ has only the one solution $a = b = 0$.

(e) Let G be the additive group of the ring $\mathbb{Z}_5 \oplus \mathbb{Z}_5$ and let $\alpha_0 \in$ End G be given by $(x_1, x_2)\alpha_0 = (-x_2, x_1)$ for all $x_1, x_2 \in \mathbb{Z}_5$. Find the isomorphism types of the additive groups of End G and $Z(\alpha_0)$. Show that there are 9 ordered pairs $(a, b) \in \mathbb{Z}_5 \times \mathbb{Z}_5$ satisfying $a^2 + b^2 = 0$. Hence show $|U(Z(\alpha_0))| = 16$. By considering the multiplicative cyclic subgroups generated by $\left(\begin{smallmatrix} \bar{2} & \bar{0} \\ \bar{0} & \bar{2} \end{smallmatrix}\right)$ and by $\left(\begin{smallmatrix} \bar{1} & \bar{1} \\ -\bar{1} & \bar{1} \end{smallmatrix}\right)$ determine the isomorphism type of $U(Z(\alpha_0))$.

(f) Let G be a finite additive abelian group with $|G| = p_1^{n_1} p_2^{n_2} \cdots p_k^{n_k}$ where $n_1, n_2, \ldots, n_k$ are positive integers and $p_1, p_2, \ldots, p_k$ are distinct primes. Let $\alpha_j \in$ End G_{p_j} for $1 \le j \le k$ where G_{p_j} is the p_j-component of G. Use Theorem 3.10 to show that $\alpha = \alpha_1 \oplus \alpha_2 \oplus \cdots \oplus \alpha_k$ is an endomorphism of G where $(g)\alpha = \sum_{j=1}^{k} (g_j)\alpha_j$ with $g = \sum_{j=1}^{k} g_j$ and $g_j \in G_{p_j}$ for $1 \le j \le k$. Show also that each endomorphism $\beta \in$ End G can be expressed as $\beta = \beta_1 \oplus \beta_2 \oplus \cdots \oplus \beta_k$ for unique $\beta_j \in$ End G_{p_j}. Deduce that End $G \cong$ End $G_{p_1} \oplus$ End $G_{p_2} \oplus \cdots \oplus$ End G_{p_k}, that is, the ring End G is isomorphic to the external direct sum of the endomorphism rings of the primary components of G. *Hint*: Mimic the method of Exercises 3.2, Question 5(b) substituting endomorphisms for automorphisms.

2. (a) Show that the 16 elements of the form $(-\bar{1})^r \times (\bar{3})^s$ where $0 \le r < 2$, $0 \le s < 8$ and $-\bar{1}, \bar{3} \in \mathbb{Z}_{32}$ are distinct. Is each element of $U(\mathbb{Z}_{32})$ of

this type? (Yes/No). State the isomorphism type of the multiplicative group $U(\mathbb{Z}_{32})$.

(b) Verify that $\bar{2} \in \mathbb{Z}_3$ has multiplicative order 2 and $2^2 \not\equiv 1 \pmod 9$. Verify that $\bar{2} \in U(\mathbb{Z}_{81})$ has order 54.

Hint: It is enough to show $2^{18} \not\equiv 1 \pmod{81}$, $2^{27} \equiv -1 \pmod{81}$. Use Theorem 3.19 to find the order of $\bar{2}$ in $\mathbb{Z}_{243}$.

(c) Find an integer a with $1 \leq a < 7$ such that $\bar{a}$ in $\mathbb{Z}_7$ has multiplicative order 6 and $a^6 \not\equiv 1 \pmod{49}$. Deduce from Theorem 3.19 that $\bar{a}$ in $U(\mathbb{Z}_{49})$ has order 42. What is the order of $\bar{a}$ in $U(\mathbb{Z}_{(49)^2})$?

(d) Let p be an odd prime and a an integer with $1 \leq a < p$ such that $\bar{a}$ in $\mathbb{Z}_p$ has multiplicative order $p - 1$. Is it necessarily the case that $a^{p-1} \not\equiv 1 \pmod{p^2}$?

Hint: Consider $p = 29$, $a = 14$.

(e) Let p be an odd prime and let a, n be integers with $n \geq 2$. Denote the congruence class of a modulo p^n by x_n. Use the proof of Theorem 3.19 and the ring homomorphism $\theta_2 : \mathbb{Z}_{p^n} \rightarrow \mathbb{Z}_{p^2}$ given by $(x_n)\theta_2 = x_2$ to show that x_n generates $U(\mathbb{Z}_{p^n})$ if and only if x_2 generates $U(\mathbb{Z}_{p^2})$.

(f) Let p be an odd prime. Show by induction on $i \geq 0$ that there is an integer k_i with $\gcd\{k_i, p\} = 1$ satisfying $(1 + p)^{p^i} = 1 + k_i p^{i+1}$. Let g denote the congruence class of $1 + p$ modulo p^n, $n \geq 2$. Taking $i = n - 2$ and $i = n - 1$ in turn, deduce that g has order p^{n-1} in $U(\mathbb{Z}_{p^n})$. Show that $U(\mathbb{Z}_{p^n})$ contains an element h of order $p - 1$ and $\langle gh \rangle = U(\mathbb{Z}_{p^n})$.

Hint: Use Corollary 3.17 and the first parts of Exercises 2.1, Question 4(b) and Exercises 2.2, Question 4(e).

3. (a) In each of the following cases express $U(\mathbb{Z}_{|G|})$ as a direct product of cyclic groups and hence find the invariant factors of $\text{Aut}\, G$ where G is a finite cyclic group.

    (i) $|G| = 105$;   (ii) $|G| = 100$;   (iii) $|G| = 98$;

    (iv) $|G| = 96$.

(b) Let G be cyclic of order n and let $H = \{\theta \in \text{Aut}\, G : \theta = \theta^{-1}\}$. Is H an elementary abelian 2-group? (Yes/No). Show that the l invariant factors of $\text{Aut}\, G$ are even and $|H| = 2^l$.

(c) Let G be a cyclic group of order p^n where p is an odd prime and $n \geq 1$. Show that $\text{Aut}\, G$ has a unique element of order 2. Are there any other finite cyclic groups G with this property?

(d) Specify, in terms of their prime power factorisation, the orders of all finite cyclic groups G such that $\text{Aut}\, G$ has exactly two invariant factors.

(e) Find the six integers $|G|$ such that G is cyclic and $\operatorname{Aut} G$ is a non-trivial elementary abelian 2-group.

(f) Verify that the cyclic groups of orders 21, 28, 36, 42 have isomorphic automorphism groups. Find the eight integers $|G|$ such that $\operatorname{Aut} \mathbb{Z}_{|G|}$ has isomorphism type $C_2 \oplus C_{12}$.

4. (a) List the subsequences of the divisor sequence $(2, 4, 4)$ and those of the divisor sequence $(1, 3, 9)$. How many subsequences does the divisor sequence $(2, 12, 36)$ have? (You needn't list them all! Use the conclusion of the following paragraph.) How many isomorphism types of abelian groups occur among the subgroups of an abelian group of isomorphism type $C_2 \oplus C_{12} \oplus C_{36}$?

Let $(d_1, d_2, \ldots, d_s)$ and $(\delta_1, \delta_2, \ldots, \delta_s)$ be divisor sequences of length s where d_s and δ_s are positive integers with $\gcd\{d_s, \delta_s\} = 1$. Explain why every subsequence of the divisor sequence $(d_1\delta_1, d_2\delta_2, \ldots, d_s\delta_s)$ is uniquely expressible in the form $(d_1'\delta_1', d_2'\delta_2', \ldots, d_s'\delta_s')$ where $(d_1', d_2', \ldots, d_s')$ is a subsequence of $(d_1, d_2, \ldots, d_s)$ and $(\delta_1', \delta_2', \ldots, \delta_s')$ is a subsequence of $(\delta_1, \delta_2, \ldots, \delta_s)$. What is the connection between the numbers of these subsequences?

Find the number of subsequences of $(6, 60, 600)$.

(b) Can an abelian group G of isomorphism type $C_2 \oplus C_{12} \oplus C_{84}$ have a subgroup H of isomorphism type

   (i) C_{14};   (ii) $C_3 \oplus C_3 \oplus C_{14}$;   (iii) C_8?

(c) Let G denote the additive group of $\mathbb{Z}_2 \oplus \mathbb{Z}_8$. Find a subgroup H of G such that H and G/H are isomorphic.

(d) Let G be an additive abelian group $\mathbb{Z}_2 \oplus \mathbb{Z}_8 \oplus \mathbb{Z}$. List the 14 isomorphism types of subgroups H of G. Find three subgroups H of G, no two being isomorphic, such that G/H is of isomorphism type $C_2 \oplus C_{12}$. Specify the isomorphism types of the non-cyclic quotient groups G/H where $|G/H| \leq 100$ and state their total number.

5. (Counting endomorphisms and automorphisms of a finite abelian group.)

(a) Let $G = H_1 \oplus H_2 \oplus \cdots \oplus H_s$ be a finite abelian group where $H_i = \langle h_i \rangle$, d_i is the order of h_i for $1 \leq i \leq s$, $d_1 > 1$ and $d_i | d_j$ for $1 \leq i \leq j \leq s$. So G has invariant factor sequence $(d_1, d_2, \ldots, d_s)$. Let $\alpha \in \operatorname{End} G$. There are integers a_{ij} with $(h_i)\alpha = \sum_{j=1}^{s} a_{ij}h_j$ for $1 \leq i \leq s$. Show that the $s \times s$ matrix $A = (a_{ij})$ over $\mathbb{Z}$ satisfies the *endomorphism condition*

$$d_i a_{ij} \equiv 0 \pmod{d_j} \quad \text{for } 1 \leq i, j \leq s.$$

For $i \geq j$ show $d_i a_{ij} \equiv 0 \pmod{d_j}$ holds for arbitrary integers a_{ij}.
For $i < j$ show $d_i a_{ij} \equiv 0 \pmod{d_j} \Leftrightarrow a_{ij} \equiv 0 \pmod{d_j/d_i}$.
Show that the matrices in the ring $\mathfrak{M}_s(\mathbb{Z})$ which satisfy the endomorphism condition form a subring R_G.
Conversely for each $A = (a_{ij}) \in R_G$ let $\alpha : G \to G$ be defined by $(g)\alpha = \sum_{j=1}^{s} y_j h_j$ where $g = x_1 h_1 + x_2 h_2 + \cdots + x_s h_s$ and $y_j = \sum_{i=1}^{s} x_i a_{ij}$. Show that α is unambiguously defined and $\alpha \in \mathrm{End}\, G$. Write $(A)\varphi = \alpha$. Show that $\varphi : R_G \to \mathrm{End}\, G$ is a ring homomorphism with $\mathrm{im}\,\varphi = \mathrm{End}\, G$ and $\ker \varphi = \{C = (c_{ij}) : c_{ij} \equiv 0 \pmod{d_j}$ for $1 \leq i, j \leq s\}$. Deduce $R_G/\ker \varphi \cong \mathrm{End}\, G$. Show that each element of $R_G/\ker \varphi$ contains a unique matrix $A = (a_{ij})$ with $0 \leq a_{ij} < d_j$. Hence prove the analogue of *Frobenius' theorem* Corollary 6.34, namely

$$|\mathrm{End}\, G| = \prod_{i=1}^{s} d_i^{2s-2i+1}.$$

Hint: Each α corresponds to its 'reduced' matrix $A = (a_{ij})$ as above, there being d_i or d_j choices for each a_{ij} according as $i < j$ or $i \geq j$.

(b) Let π denote a permutation of the set $S = \{1, 2, \ldots, s\}$ and suppose $i_0, j_0 \in S$ satisfy $(j_0)\pi = i_0$. Show that there is a positive integer k with $(i_0)\pi^k = j_0$. Show that the least such integer k is such that $i_0, (i_0)\pi, (i_0)\pi^2, \ldots, (i_0)\pi^{k-1}$ are distinct.
Hint: $K = \{l \in \mathbb{Z} : (i_0)\pi^l = i_0\}$ is an ideal of $\mathbb{Z}$.
Let $A = (a_{ij})$ belong to the ring R_G of (a) above. The (i_0, j_0)-entry in the adjugate matrix $\mathrm{adj}\, A$ is the cofactor $A_{j_0 i_0}$ and a typical term (apart from sign) in $A_{j_0 i_0}$ is $t_\pi = \prod_{j \neq j_0} a_{j(j)\pi}$. Show $d_{i_0} t_\pi \equiv 0 \pmod{d_{j_0}}$ and deduce $d_{i_0} A_{j_0 i_0} \equiv 0 \pmod{d_{j_0}}$, that is, $\mathrm{adj}\, A \in R_G$.
Hint: t_π has factor $a_{i_0(i_0)\pi} a_{(i_0)\pi(i_0)\pi^2} \cdots a_{(i_0)\pi^{k-1} j_0}$.

(c) Let G be a finite abelian p-group, p prime. Suppose G has m_1 invariant factors p^{t_1}, m_2 invariant factors $p^{t_2}, \ldots, m_l$ invariant factors p^{t_l} where $t_1 < t_2 < \cdots < t_l$. Let the $s \times s$ matrix $A = (a_{ij})$ over $\mathbb{Z}$ be reduced and belong to the ring R_G of (a) above where $s = m_1 + m_2 + \cdots + m_l$. Show

$$\alpha = (A)\varphi \in \mathrm{Aut}\, G \quad \Leftrightarrow \quad |A| \not\equiv 0 \pmod{p}.$$

Hint: For $\Leftarrow$ consider $b \in \mathbb{Z}$ with $b|A| \equiv 1 \pmod{p^{t_l}}$. Show that $B = b\,\mathrm{adj}\, A$ over $\mathbb{Z}$ satisfies $AB \equiv I \pmod{p^{t_l}}$, i.e. corresponding entries in AB and I are congruent modulo p^{t_l}. Deduce $AB \equiv I \pmod{\ker \varphi}$ and using (b) above conclude $\alpha^{-1} = (B)\varphi$.

Partition

$$A = \begin{pmatrix} M_{11} & M_{12} & \cdots & M_{1l} \\ M_{21} & M_{22} & \cdots & M_{2l} \\ \vdots & \vdots & \ddots & \vdots \\ M_{l1} & M_{l2} & \cdots & M_{ll} \end{pmatrix}$$

where M_{ij} is the indicated $m_i \times m_j$ submatrix of A. Show that all entries in M_{ij} ($i < j$) are divisible by p and deduce $|A| \equiv |M_{11}| \times |M_{22}| \times \cdots \times |M_{ll}|$ (mod p). Write $r_m = |GL_m(\mathbb{Z}_p)|/p^{(m^2)}$. Show $|\operatorname{Aut} G| = r_{m_1} r_{m_2} \cdots r_{m_l} |\operatorname{End} G|$ by counting the number of choices for each M_{ij}.

(d) Find a formula involving p, t_i, m_i ($1 \le i \le l$) for the number $|\operatorname{End} G|$ of endomorphisms of the finite abelian p-group G of (c) above.

(e) Let G have isomorphism type $C_3 \oplus C_9 \oplus C_9$. Using the terminology of (a) above, verify that

$$A = \begin{pmatrix} 2 & 3 & 6 \\ 1 & 7 & 4 \\ 0 & 1 & 5 \end{pmatrix}$$

satisfies the endomorphism condition and is reduced. Verify $|A| \not\equiv 0$ (mod 3) and find the 3×3 reduced matrix B satisfying $AB \equiv I$ (mod $\ker\varphi$). Find the multiplicative order of the automorphism $\alpha = (A)\varphi$ of G. Is $\alpha^{-1} = (B)\varphi$? (Yes/No).

(f) Calculate, in factorised form, $|\operatorname{End} G|$ and $|\operatorname{Aut} G|$ the isomorphism class of G being:

    (i) $C_4 \oplus C_8 \oplus C_8$;   (ii) $C_3 \oplus C_3 \oplus C_9$;

    (iii) $C_{12} \oplus C_{24} \oplus C_{72}$.

(g) The abelian group G has invariant factor sequence (d_1, d_2). Suppose G finite. By considering

$$A = \begin{pmatrix} 1 & 0 \\ 1 & 1 \end{pmatrix} \quad \text{and} \quad B = \begin{pmatrix} 1 & d_2/d_1 \\ 0 & 1 \end{pmatrix}$$

show that $\operatorname{Aut} G$ is non-abelian.

Suppose G infinite and so $d_2 = 0$. Find d_1 with $\operatorname{Aut} G$ abelian. Show that d_1 is unique.

6. (a) Let G and G' be additive abelian groups and let $\alpha : G \to G'$ and $\beta : G \to G'$ be group homomorphisms. Show that $\alpha + \beta : G \to G'$, defined by $(g)(\alpha + \beta) = (g)\alpha + (g)\beta$ for all $g \in G$, is also a group

homomorphism. Hence show that the set $\text{Hom}(G, G')$ of all group homomorphisms $\alpha : G \to G'$, with the above binary operation of addition, is an additive abelian group.

(b) Let G, G_1, G_2, G', G'_1, G'_2 be additive abelian groups. Establish group isomorphisms:

$$\text{Hom}(G_1 \oplus G_2, G') \cong \text{Hom}(G_1, G') \oplus \text{Hom}(G_2, G'),$$
$$\text{Hom}(G, G'_1 \oplus G'_2) \cong \text{Hom}(G, G'_1) \oplus \text{Hom}(G, G'_2).$$

(c) Let m and n be positive integers. Use Exercises 2.1, Question 4(b) to show $\text{Hom}(\mathbb{Z}_m, \mathbb{Z}_n)$ is cyclic of order $\gcd\{m, n\}$. Show also that $\text{Hom}(\mathbb{Z}, \mathbb{Z}_n)$, $\text{Hom}(\mathbb{Z}_m, \mathbb{Z})$ and $\text{Hom}(\mathbb{Z}, \mathbb{Z})$ are cyclic and state their isomorphism types. Suppose the abelian groups G and G' are finitely generated. Using Theorem 3.4 deduce that $\text{Hom}(G, G')$ is finitely generated.

(d) The finite abelian groups G and G' have invariant factor sequences $(d_1, d_2, \ldots, d_s)$ and $(d'_1, d'_2, \ldots, d'_{s'})$ respectively. Express $\text{Hom}(G, G')$ as a direct sum of cyclic groups. Are the groups $\text{Hom}(G, G')$ and $\text{Hom}(G', G)$ isomorphic? Specify the invariant factors of $\text{End}\, G = \text{Hom}(G, G)$.
List the invariant factors of $\text{Hom}(G, G')$, $\text{Hom}(G, G \oplus G')$, $\text{Hom}(G \oplus G', G')$ and $\text{End}\, G$ in the case $(d_1, d_2, \ldots, d_s) = (2, 6, 12)$, $(d'_1, d'_2, \ldots, d'_{s'}) = (3, 3, 6, 24)$.

(e) The f.g. abelian groups G and G' decompose $G = T(G) \oplus M(G)$ and $G' = T(G') \oplus M(G')$ where $T(G)$, $T(G')$ are the torsion subgroups of G, G' and $M(G)$, $M(G')$ are free of rank r, r' respectively as in Corollary 3.5. Express the torsion subgroup and torsion-free rank of $\text{Hom}(G, G')$ and $\text{End}\, G$ using $T(G)$, $T(G')$, r, r' and Hom.
List the invariant factors of $\text{Hom}(G, G')$, $\text{Hom}(G', G)$, $\text{End}\, G$, and $\text{End}\, G'$ in the case of $T(G)$, $T(G')$ having invariant factor sequences $(2, 4)$, $(2, 2, 4, 4)$ respectively and $r = 1$, $r' = 2$.

Part II

Similarity of Square Matrices over a Field

A Bird's-Eye View of Similarity of $t \times t$ Matrices over a Field

The term 'similarity' in this context has a specific meaning: the $t \times t$ matrices A and B over a field F are called *similar* if there is an invertible $t \times t$ matrix X over F such that $XA = BX$ and so $XAX^{-1} = B$. Similarity is an equivalence relation on the set $\mathfrak{M}_t(F)$ of all $t \times t$ matrices A over F. The multiplicative group of all invertible $t \times t$ matrices X over F is denoted by $GL_t(F)$. So *the similarity class of the $t \times t$ matrix A* over F is the set $\{XAX^{-1} : X \in GL_t(F)\}$ of all $t \times t$ matrices over F which are similar to A. How does similarity in this sense arise? Let V be a t-dimensional vector space over the field F and let $\alpha : V \rightarrow V$ be a linear mapping of V. Also let $v_1, v_2, \ldots, v_t$ be a basis $\mathcal{B}$ of V. The reader will know that many problems in linear algebra are solved by the choice of a suitable basis. The $t \times t$ matrix $A = (a_{ij})$ over F defined by

$$(v_i)\alpha = a_{i1}v_1 + a_{i2}v_2 + \cdots + a_{it}v_t \quad \text{for } 1 \leq i \leq t$$

is called *the matrix of α relative to $\mathcal{B}$* as in Theorem 3.15. The question is: how does the matrix of α change when the basis $\mathcal{B}$ of V is changed? The answer is: A changes to a similar matrix XAX^{-1}. In fact the matrices XAX^{-1} of a given linear mapping α of V relative to bases of V are precisely the elements of the similarity class of A. The reason will be known to the reader: each ordered pair of bases of V is connected by an invertible matrix X. The details are revised in Chapter 5.

How can one tell whether two $t \times t$ matrices over F are similar or not? It turns out that this question is analogous to a problem we have already solved, namely: how to determine whether two finite abelian groups are isomorphic or not. As we saw in Section 3.1, this problem is solved by expressing each finite $\mathbb{Z}$-module G as a direct sum of cyclic submodules from which the invariant factors and the isomorphism type of G can be read off. What is the analogous procedure starting with a given $t \times t$ matrix A over a field F? The first step is to construct

the $F[x]$-module $M(A)$ determined by A.

The elements of $M(A)$ are t-tuples $v = (a_1, a_2, \ldots, a_t)$ where $a_i \in F$ for $1 \leq i \leq t$. So $M(A) = F^t$ (set equality) and we'll refer to the elements of $M(A)$ as vectors; indeed $M(A)$ is a t-dimensional vector space over F when the extra structure we are about to introduce is ignored! Here $F[x]$ denotes the ring of polynomials in x over the field F. We review the properties of $F[x]$ in Chapter 4, the point being that $F[x]$ behaves in the same way as the ring $\mathbb{Z}$ of integers: there is a division law for polynomials over F, the Euclidean algorithm can be used to calculate gcds of pairs of polynomials over F, and $F[x]$ is a principal ideal domain. So it is reasonable to expect $F[x]$-modules to behave in the same way as $\mathbb{Z}$-modules and this is indeed the case. The extra structure referred to above is *the module product* $f(x)v$ which belongs to $M(A)$ for all $f(x) \in F[x]$ and all $v \in M(A)$. We write

$$xv = vA \quad \text{for all } v \in M(A)$$

and so multiplication of v by x (on the left) is defined to be the matrix product vA; as A is a $t \times t$ matrix over F and $v \in F^t$ we see $vA \in F^t$, that is, $vA \in M(A)$. In the same way

$$x^2v = vA^2 \quad \text{for all } v \in M(A)$$

the general rule being

$$(a_nx^n + \cdots + a_1x + a_0)v = v(a_nA^n + \cdots + a_1A + a_0I)$$
$$\text{for all } v \in M(A) \text{ and } a_n, \ldots, a_1, a_0 \in F$$

where I denotes the $t \times t$ identity matrix over F. The above equation can be expressed

$$f(x)v = vf(A)$$

where $f(x) = a_nx^n + \cdots + a_1x + a_0$ is a typical element of $F[x]$ and $f(A) = a_nA^n + \cdots + a_1A + a_0I$ is the corresponding $t \times t$ matrix over F obtained from $f(x)$ by substituting A for x. Let us leave to one side the routine proof that $M(A)$ is an $F[x]$-module. Notice that $M(A) = \langle e_1, e_2, \ldots, e_t \rangle$ showing that $M(A)$ is finitely generated (Definition 2.19). Suppose there is $v_0 \in M(A)$ such that every element of the module $M(A)$ is of the form $f(x)v_0$ for some $f(x) \in F[x]$. Then $M(A)$ is called *cyclic* with *generator* v_0 and using bold pointed brackets we write $M(A) = \langle v_0 \rangle$. We prove in Chapter 5 that the elements

$$v_0, xv_0, x^2v_0, \ldots, x^{t-1}v_0 \quad \text{form a basis of } F^t.$$

Now $-x^t v_0$ belongs to $M(A) = F^t$ and so there are unique scalars (elements of F) $b_0, b_1, \ldots, b_{t-1}$ such that $-x^t v_0 = b_0 v_0 + b_1 x v_0 + \cdots + b_{t-1} x^{t-1} v_0$ which rearranges to give

$$d_0(x) v_0 = 0 \quad \text{where } d_0(x) = x^t + b_{t-1} x^{t-1} + \cdots + b_1 x + b_0.$$

The monic (leading coefficient 1) polynomial $d_0(x)$ is called *the order of v_0 in the $F[x]$-module* $M(A)$. In fact $d_0(x)$ is the monic polynomial of least degree over F satisfying $d_0(x) v_0 = 0$. Whether or not $M(A)$ is cyclic each of its elements v has an *order* $d(x)$, namely the monic generator of the

$$\text{order ideal } \{ f(x) \in F[x] : f(x) v = 0 \} \text{ of } v.$$

Using the notation introduced above we write

$$\langle v \rangle = \{ f(x) v : f(x) \in F[x] \} \quad \text{for the cyclic submodule of } M(A) \text{ generated by } v.$$

The reader will be familiar with the analogous concept in group theory: each element of a group generates a cyclic subgroup. Here the vectors $x^i v$ for $0 \le i < \deg d(x)$ form an F-basis of $\langle v \rangle$. So $\dim \langle v \rangle = \deg d(x)$ where v has order ideal $\langle d(x) \rangle$.

   It is high time to look at some examples of cyclic modules $M(A)$. Suppose that the above basis $v_0, x v_0, x^2 v_0, \ldots, x^{t-1} v_0$ of F^t is the standard basis $\mathcal{B}_0$ of F^t consisting of the rows $e_1, e_2, \ldots, e_t$ of the $t \times t$ identity matrix I over F. Then $v_0 = e_1, x v_0 = e_2$, $x^2 v_0 = e_3, \ldots, x^{t-1} v_0 = e_t$. Using consecutive pairs of these equations, on substituting $x^{i-1} v_0 = e_i$ in $x(x^{i-1} v_0) = x^i v_0 = e_{i+1}$, we obtain $x e_i = e_{i+1}$ for $1 \le i < t$. So $e_i A = e_{i+1}$ which shows that row i of A is e_{i+1} for $1 \le i < t$. The above equations combine with $d_0(x) v_0 = 0$ to give

$$e_t A = x e_t = x(x^{t-1} v_0) = x^t v_0$$
$$= -(b_0 v_0 + b_1 x v_0 + b_2 x^2 v_0 + \cdots + b_{t-1} x^{t-1} v_0)$$
$$= -b_0 e_1 - b_1 e_2 - b_2 e_3 - \cdots - b_{t-1} e_{t-1}$$

showing that the last row of A is $e_t A = -(b_0, b_1, b_2, \ldots, b_{t-1})$. In this case

$$A \text{ is the companion matrix of the monic polynomial}$$
$$d_0(x) = x^t + b_{t-1} x^{t-1} + \cdots + b_1 x + b_0$$

and denoted by $C(d_0(x))$. So

$$C(d_0(x)) = \begin{pmatrix} 0 & 1 & 0 & 0 & \cdots & 0 \\ 0 & 0 & 1 & 0 & \cdots & 0 \\ 0 & 0 & 0 & 1 & \cdots & 0 \\ \vdots & \vdots & \vdots & \vdots & \ddots & \vdots \\ 0 & 0 & 0 & 0 & \cdots & 1 \\ -b_0 & -b_1 & -b_2 & -b_3 & \cdots & -b_{t-1} \end{pmatrix}.$$

For example

$$C(x^3 + 4x^2 + 7x + 6) = \begin{pmatrix} 0 & 1 & 0 \\ 0 & 0 & 1 \\ -6 & -7 & -4 \end{pmatrix},$$

$$C(x^2 - 1) = \begin{pmatrix} 0 & 1 \\ 1 & 0 \end{pmatrix}, \qquad C(x - 6) = (6).$$

We will see that companion matrices have many remarkable properties and we now mention one. The reader will know that *the characteristic polynomial* $|xI - A|$ of the $t \times t$ matrix A is a monic polynomial of degree t and typically some calculation is required to determine its coefficients. However in the case of companion matrices things could not be easier! The characteristic polynomial of $C(d_0(x))$ is simply $d_0(x)$ (see Theorem 5.26). Retracing the steps in the above discussion we see

$M(C(d_0(x)))$ is a cyclic $F[x]$-module with generator e_1 of order $d_0(x)$.

Indeed $F[x]/\langle d_0(x) \rangle \cong M(C(d_0(x))) = \langle e_1 \rangle$ is the standard example of such a cyclic module just as the additive group $\mathbb{Z}_n = \langle \bar{1} \rangle$ is the standard example of a finite cyclic group of order n.

We now take a closer look at the module $M = M(C(d_0(x)))$ where $d_0(x) = x^3 + 4x^2 + 7x + 6$ and $F = \mathbb{Q}$ is the rational field. As $e_2 = xe_1$ and $e_3 = x^2e_1$ we see $v = (a_0, a_1, a_2) \in \mathbb{Q}^3$ can be expressed as

$$v = a_0e_1 + a_1e_2 + a_2e_3 = a_0e_1 + a_1xe_1 + a_2x^2e_1 = (a_2x^2 + a_1x + a_0)e_1$$

showing that M is cyclic with generator e_1. As $d_0(x) = (x + 2)(x^2 + 2x + 3)$ is the characteristic polynomial of $C(d_0(x))$, we see $d_0(-2) = 0$ showing that -2 is a zero of this polynomial, that is, -2 is an eigenvalue of $C(d_0(x))$. Let $w_1 = (x^2 + 2x + 3)e_1 = e_3 + 2e_2 + 3e_1 = (3, 2, 1)$. As e_1 has order $d_0(x)$ in M we see that w_1 is non-zero and satisfies $(x + 2)w_1 = 0$, that is, $C(d_0(x))w_1 = -2w_1$ showing that w_1 is a row eigenvector of $C(d_0(x))$ with associated eigenvalue -2. More generally, the row eigenvectors of any $t \times t$ matrix A are precisely the elements w having order $x - \lambda$ in $M(A)$, the associated eigenvalue being λ. Let $w_2 = (x + 2)e_1$ and so $w_2 = e_2 + 2e_1 = (2, 1, 0)$. Then w_2 has order $x^2 + 2x + 3$ in M (the reasoning is analogous to: from an element g of order 6 in an additive abelian group we obtain elements $2g$ of order 3 and $3g$ of order 2) and working in M we find $xw_2 = x(x+2)e_1 = x^2e_1 + 2xe_1 = e_3 + 2e_2 = (0, 2, 1)$. We write $N_1 = \{f(x)w_1 : f(x) \in \mathbb{Q}[x]\} = \langle w_1 \rangle$ and $N_2 = \{f(x)w_2 : f(x) \in \mathbb{Q}[x]\} = \langle w_2 \rangle$ for the cyclic submodules of M generated by w_1 and w_2 respectively. As $x^2 + 2x + 3$ is irreducible over $\mathbb{Q}$ (it does not factorise into a product of polynomials of smaller degree over $\mathbb{Q}$) we see that $d_0(x)$ has exactly four monic divisors over $\mathbb{Q}$ namely 1, $x + 2$, $x^2 + 2x + 3$, $x^3 + 4x^2 + 7x + 6$. The analogue of Lemma 2.2 now tells us that M has exactly four submodules – no more and no less! These submodules are $\{(0, 0, 0)\}$, N_1, N_2, M each being cyclic and the order of a generator being one of the above monic divisors of $d_0(x)$. Further

$$M = N_1 \oplus N_2$$

showing that M decomposes as the internal direct sum of its cyclic submodules N_1 and N_2. What is the significance of this decomposition in terms of matrices?

Let A be an $t \times t$ matrix over F. We call $\alpha : F^t \to F^t$, defined by $(v)\alpha = vA$ for all $v \in F^t$, the linear mapping determined by A. It turns out that the submodules N of $M(A)$ are precisely those subspaces N of F^t which are α-invariant, that is,

$$(N)\alpha \subseteq N.$$

Returning to our example, $N_1 = \langle w_1 \rangle$ is a 1-dimensional α-invariant subspace of $\mathbb{Q}^3$ and $N_2 = \langle w_2, xw_2 \rangle = \langle w_2 \rangle$ is a 2-dimensional α-invariant subspace of $\mathbb{Q}^3$ where α is the linear mapping of $\mathbb{Q}^3$ determined by $C(d_0(x))$. Further N_1 and N_2 are complementary subspaces of $\mathbb{Q}^3$ and so w_1, w_2, xw_2 is a basis $\mathcal{B}$ of $\mathbb{Q}^3$. Let

$$X = \begin{pmatrix} w_1 \\ w_2 \\ xw_2 \end{pmatrix} = \begin{pmatrix} 3 & 2 & 1 \\ 2 & 1 & 0 \\ 0 & 2 & 1 \end{pmatrix}$$

be the invertible matrix over $\mathbb{Q}$ having the vectors in $\mathcal{B}$ as its rows. Let B denote the matrix of α relative to $\mathcal{B}$. As $(w_1)\alpha = -2w_1$, $(w_2)\alpha = xw_2$ and $(xw_2)\alpha = x^2w_2 = -3w_2 - 2xw_2$ we obtain

$$B = \left(\begin{array}{c|cc} -2 & 0 & 0 \\ \hline 0 & 0 & 1 \\ 0 & -3 & -2 \end{array} \right) = \left(\begin{array}{c|c} C(x+2) & 0 \\ \hline 0 & C(x^2+2x+3) \end{array} \right).$$

The right-hand matrix above is *the direct sum* of $C(x+2)$ and $C(x^2+2x+3)$ and we write $B = C(x+2) \oplus C(x^2+2x+3)$; we explain this concept below. Since $C(d_0(x))$ is the matrix of α relative to the standard basis $\mathcal{B}_0$ of $\mathbb{Q}^3$, as we mentioned earlier, the matrices $C(d_0(x))$ and B are similar. In fact $XC(d_0(x))X^{-1} = B$ (the reader should check this matrix equality by verifying $\det X \neq 0$ and $XC(d_0(x)) = BX$; there is no need to find the entries in X^{-1}). The matrix $C(x+2) \oplus C(x^2+2x+3)$, being the direct sum of companion matrices of powers of irreducible polynomials over $\mathbb{Q}$, is in *primary canonical form (pcf)*.

The determination of a matrix in pcf, similar to a given square matrix A over F, requires the factorisation of the characteristic polynomial of A into irreducible polynomials over F; there is no known algorithm for obtaining this factorisation except in a handful of special cases and so matrices in pcf are of more theoretical than practical use. Luckily this difficulty can be avoided when tackling similarity problems: we use the method of Section 3.1, where isomorphisms between finite abelian groups were thoroughly analysed without using prime factorisation.

Let A_1 be an $t_1 \times t_1$ matrix over F and let A_2 be an $t_2 \times t_2$ matrix over F. We construct the $(t_1 + t_2) \times (t_1 + t_2)$ partitioned matrix over F

$$A_1 \oplus A_2 = \left(\begin{array}{c|c} A_1 & 0 \\ \hline 0 & A_2 \end{array} \right)$$

which is called *the direct sum* of A_1 and A_2. So $A_1 \oplus A_2$ is a diagonal block matrix with A_1 and A_2 in the diagonal positions and with rectangular blocks of zeros elsewhere. More generally given s square matrices $A_1, A_2, \ldots, A_s$ over F, their direct sum is

$$A_1 \oplus A_2 \oplus \cdots \oplus A_s = \left(\begin{array}{c|c|c|c} A_1 & 0 & \cdots & 0 \\ \hline 0 & A_2 & & \vdots \\ \hline \vdots & & \ddots & 0 \\ \hline 0 & \cdots & 0 & A_s \end{array} \right)$$

that is, a partitioned matrix with $A_1, A_2, \ldots, A_s$ on the diagonal and zeros elsewhere. We are now ready to meet the similarity theorem.

A matrix of the type $C(d_1(x)) \oplus C(d_2(x)) \oplus \cdots \oplus C(d_s(x))$, where $d_1(x), d_2(x), \ldots, d_s(x)$ are monic polynomials of positive degree over F with $d_j(x) | d_{j+1}(x)$ for $1 \le j < s$, is said to be in *rational canonical form* (*rcf*). So a matrix in rcf is a direct sum of companion matrices of monic polynomials which are successive divisors of each other. The similarity theorem is refreshingly brief:

Let A be a $t \times t$ matrix over a field F. Then A is similar to
a unique matrix in rational canonical form.

So starting from a given $t \times t$ matrix A over F there is a $t \times t$ invertible matrix X over F such that $XAX^{-1} = C(d_1(x)) \oplus C(d_2(x)) \oplus \cdots \oplus C(d_s(x))$ is in rcf. The polynomials $d_1(x), d_2(x), \ldots, d_s(x)$ are called *the invariant factors* of A and the $t \times t$ matrix $C(d_1(x)) \oplus C(d_2(x)) \oplus \cdots \oplus C(d_s(x))$ is called *the rational canonical form of A*.

How can a suitable matrix X, and hence the invariant factors of A, be found? It turns out, as we now outline, that the analogue of Theorem 3.4 will do the job! Instead of decomposing the f.g. $\mathbb{Z}$-module G into a direct sum of cyclic submodules, the $F[x]$-module $M(A)$ is decomposed in the same way. First we need the $F[x]$-module $F[x]^t$, the elements of which are t-tuples $(f_1(x), f_2(x), \ldots, f_t(x))$ of polynomials over F. Denote row i of the $t \times t$ identity matrix over $F[x]$ by $e_i(x)$ $(1 \le i \le t)$. Then $e_1(x), e_2(x), \ldots, e_t(x)$ form the standard $F[x]$-basis of $F[x]^t$, and so $F[x]^t$ is a free $F[x]$-module of rank t. Notice

$$\big(f_1(x), f_2(x), \ldots, f_t(x)\big) = \sum_{i=1}^{t} f_i(x) e_i(x).$$

The evaluation homomorphism

$$\theta_A : F[x]^t \to M(A) \quad \text{defined by} \quad \left(\sum_{i=1}^{t} f_i(x)e_i(x)\right)\theta_A = \sum_{i=1}^{t} e_i f_i(A)$$

provides the connection between $F[x]^t$ and $M(A)$. So θ_A maps each t-tuple $(f_1(x),$ $f_2(x), \ldots, f_t(x))$ of polynomials over F to the t-tuple of scalars in $M(A) = F^t$ obtained by adding together row 1 of $f_1(A)$, row 2 of $f_2(A), \ldots,$ row t of $f_t(A)$. We study θ_A thoroughly in Chapter 6. For the moment notice $(e_i(x))\,\theta_A = e_i$ $(1 \le i \le t)$ and so θ_A maps the standard $F[x]$-basis of $F[x]^t$ to the standard F-basis $\mathcal{B}_0$ of F^t. Hence θ_A is surjective, that is, $\operatorname{im}\theta_A = M(A)$. The reader's antennae will know what to expect next, namely a scrutiny of the kernel of θ_A. Now $\ker\theta_A$ is a submodule of the free $F[x]$-module $F[x]^t$ and so is free using the polynomial analogue of Theorem 3.1. However it should come as something of a surprise that

the t rows of the characteristic matrix $xI - A$ form an $F[x]$-basis of $\ker\theta_A$.

We are given an $F[x]$-basis of $\ker\theta_A$ on a plate! The entries in $xI - A$ are polynomials (admittedly of degree at most1) over F and so $xI - A$ is a matrix over $F[x]$. Just as each matrix over $\mathbb{Z}$ can be reduced to its Smith normal form, the same is true, by analogy, of each matrix over $F[x]$. In particular there are invertible $t \times t$ matrices $P(x)$ and $Q(x)$ over $F[x]$ such that

$$P(x)(xI - A)Q(x)^{-1} = \operatorname{diag}\bigl(1, 1, \ldots, 1, d_1(x), d_2(x), \ldots, d_s(x)\bigr) = S(xI - A)$$

where the $t - s$ entries 1 on the diagonal of the Smith normal form $S(xI - A)$ of $xI - A$ are followed by the s (monic and non-constant) invariant factors $d_j(x)$ of A. The determinants $|P(x)|$ and $|Q(x)|$ are non-zero scalars (polynomials of zero degree over F) and so taking determinants of the above matrix equation gives

$$|xI - A| = d_1(x)d_2(x)\cdots d_s(x)$$

as $|P(x)||Q(x)|^{-1} = 1$ on comparing coefficients of x^t, both sides of the above equation being monic polynomials of degree t over F. We have shown that the characteristic polynomial of A is the product of the invariant factors of A. This interesting connection is, however, eclipsed by a more important fact: the invertible $t \times t$ matrices $P(x)$ and $Q(x)$ over $F[x]$ can be found algorithmically by modifying the method of Chapter 1. In other words, $P(x)$ and $Q(x)$ arise from the row and column operations used in reducing $xI - A$ to $S(xI - A)$. The equation

$$P(x)(xI - A) = S(xI - A)Q(x)$$

'says it all' although it will take us some time to appreciate what it is saying: the rows of the *lhs* form an $F[x]$-basis of $\ker\theta_A$ which, being equal to the *rhs*, consist of monic polynomial multiples (the diagonal entries in $S(xI - A)$) of the elements of an $F[x]$-basis of $F[x]^t$ (the rows of $Q(x)$). We have met the analogous situation in

Theorem 3.4 where $\mathbb{Z}$-bases of $\ker\theta$ and $\mathbb{Z}^t$, related in the same way, led directly to the invariant factor decomposition of $\mathbb{Z}^t / \ker\theta$.

It is a relatively small step to derive an invertible matrix X over F with XAX^{-1} in rational canonical form. Let $\rho_i(x)$ denote row i of $Q(x)$ for $1 \le i \le t$. Write $v_j = (\rho_{t-s+j}(x))\theta_A$ and let $N_j = \langle v_j \rangle$ be the cyclic submodule of $M(A)$ generated by v_j for $1 \le j \le s$. Then

$$M(A) = N_1 \oplus N_2 \oplus \cdots \oplus N_s$$

which expresses $M(A)$ as a direct sum of non-zero cyclic submodules N_j. We will see the full proof in Chapter 6. Also $(\rho_i(x))\theta_A = 0$ for $1 \le i \le t-s$ and v_j has order $d_j(x)$ in $M(A)$ for $1 \le j \le s$. The $\deg d_j(x)$ vectors $x^i v_j$ for $0 \le i < \deg d_j(x)$ form a basis $\mathcal{B}_{v_j}$ of the α-invariant subspace N_j of $F^t (1 \le j \le s)$. As $t = \deg d_1(x) + \deg d_2(x) + \cdots + \deg d_s(x)$ we see that the vectors in the ordered set $\mathcal{B} = \mathcal{B}_{v_1} \cup \mathcal{B}_{v_2} \cup \cdots \cup \mathcal{B}_{v_s}$ form a basis of F^t. The matrix X, having the vectors of $\mathcal{B}$ as its rows, is invertible over F and satisfies

$$XAX^{-1} = C(d_1(x)) \oplus C(d_2(x)) \oplus \cdots \oplus C(d_s(x))$$

which is *the rational canonical form of A*.

We work through the particular case

$$A = \begin{pmatrix} 3 & -1 & 1 \\ 4 & -2 & 4 \\ 3 & -3 & 5 \end{pmatrix}$$

over $\mathbb{Q}$. The first job is to reduce $xI - A$ to its Smith normal form $S(xI - A)$, noting the *ecos* and *eros* used in the reduction. We mimic the reduction method in Theorem 1.11:

$$xI - A = \begin{pmatrix} x-3 & 1 & -1 \\ -4 & x+2 & -4 \\ -3 & 3 & x-5 \end{pmatrix}$$

$$\underset{\substack{c_1 - (x-3)c_2 \\ c_1 \leftrightarrow c_2}}{\equiv} \begin{pmatrix} 1 & 0 & -1 \\ x+2 & -x^2+x+2 & -4 \\ 3 & -3x+6 & x-5 \end{pmatrix}$$

$$\underset{c_3 + c_1}{\equiv} \begin{pmatrix} 1 & 0 & 0 \\ x+2 & -x^2+x+2 & x-2 \\ 3 & -3x+6 & x-2 \end{pmatrix}$$

$$\underset{\substack{r_2 - (x+2)r_1 \\ r_3 - 3r_1}}{\equiv} \begin{pmatrix} 1 & 0 & 0 \\ 0 & -x^2+x+2 & x-2 \\ 0 & -3x+6 & x-2 \end{pmatrix}$$

$$\underset{\substack{c_2+(x+1)c_3 \\ c_2 \leftrightarrow c_3}}{\equiv} \begin{pmatrix} 1 & 0 & 0 \\ 0 & x-2 & 0 \\ 0 & x-2 & (x-2)^2 \end{pmatrix}$$

$$\underset{r_3-r_2}{\equiv} \begin{pmatrix} 1 & 0 & 0 \\ 0 & x-2 & 0 \\ 0 & 0 & (x-2)^2 \end{pmatrix} = S(xI - A).$$

From $S(xI - A)$ we immediately see $d_1(x) = x - 2$, $d_2(x) = (x-2)^2 = x^2 - 4x + 4$ are the invariant factors of A and

$$C(x-2) \oplus C((x-2)^2) = \begin{pmatrix} 2 & 0 & 0 \\ \hline 0 & 0 & 1 \\ 0 & -4 & 4 \end{pmatrix}$$

is the rcf of A. The sequence of *ecos* used in the above reduction is: $c_1 - (x-3)c_2$, $c_1 \leftrightarrow c_2$, $c_3 + c_1$, $c_2 + (x+1)c_3$, $c_2 \leftrightarrow c_3$. Mimicking the theory of Chapter 1, the invertible matrix $Q(x)$ over $\mathbb{Q}[x]$ is found by applying the conjugate sequence, namely $r_2 + (x-3)r_1, r_1 \leftrightarrow r_2, r_1 - r_3, r_3 - (x+1)r_2, r_2 \leftrightarrow r_3$ to the 3×3 identity matrix I over $\mathbb{Q}[x]$:

$$I = \begin{pmatrix} 1 & 0 & 0 \\ 0 & 1 & 0 \\ 0 & 0 & 1 \end{pmatrix} \underset{\substack{r_2+(x-3)r_1 \\ r_1 \leftrightarrow r_2}}{\equiv} \begin{pmatrix} x-3 & 1 & 0 \\ 1 & 0 & 0 \\ 0 & 0 & 1 \end{pmatrix}$$

$$\underset{\substack{r_1-r_3 \\ r_3-(x+1)r_2}}{\equiv} \begin{pmatrix} x-3 & 1 & -1 \\ 1 & 0 & 0 \\ -(x+1) & 0 & 1 \end{pmatrix}$$

$$\underset{r_2 \leftrightarrow r_3}{\equiv} \begin{pmatrix} x-3 & 1 & -1 \\ -(x+1) & 0 & 1 \\ 1 & 0 & 0 \end{pmatrix} = Q(x).$$

The invertible matrix $P(x)$ over $\mathbb{Q}[x]$ is found (although we don't really need it) by applying the sequence of *eros* used in the above reduction, namely $r_2 - (x+2)r_1$, $r_3 - 3r_1, r_3 - r_2$ to I as above:

$$I = \begin{pmatrix} 1 & 0 & 0 \\ 0 & 1 & 0 \\ 0 & 0 & 1 \end{pmatrix} \underset{\substack{r_2-(x+2)r_1 \\ r_3-3r_1}}{\equiv} \begin{pmatrix} 1 & 0 & 0 \\ -(x+2) & 1 & 0 \\ -3 & 0 & 1 \end{pmatrix}$$

$$\underset{r_3-r_2}{\equiv} \begin{pmatrix} 1 & 0 & 0 \\ -(x+2) & 1 & 0 \\ x-1 & -1 & 1 \end{pmatrix} = P(x).$$

Note that a square matrix over $F[x]$ is invertible over $F[x]$ if and only if its determinant is a non-zero constant polynomial, that is, an invertible element of $F[x]$. The reader can now verify $P(x)(xI - A)Q(x)^{-1} = S(xI - A)$ by checking $\det Q(x) = 1$ (so $Q(x)$ is invertible over $\mathbb{Q}[x]$) and $P(x)(xI - A) = S(xI - A)Q(x)$.

Finally we use the rows of $Q(x)$ and $\theta_A : \mathbb{Q}[x]^3 \to \mathbb{Q}^3$ to construct a matrix X with the property we are looking for. The theory says that the (i, i)-entry in $S(xI - A)$ is the order of $(\rho_i(x))\theta_A$ in $M(A)$ for $1 \le i \le 3$ where $\rho_i(x)$ is row i of $Q(x)$, but we'll reassure ourselves of this fact as we go along. First

$$(\rho_1(x))\theta_A = (x - 3, 1, -1)\theta_A = ((x - 3)e_1(x) + e_2(x) - e_3(x))\theta_A$$

$$= e_1(A - 3I) + e_2 - e_3 = e_1 A - 3e_1 + e_2 - e_3$$

$$= (3, -1, 1) - (3, 0, 0) + (0, 1, 0) - (0, 0, 1)$$

$$= (0, 0, 0)$$

showing $(\rho_1(x))\theta_A = 0$. So the $(1, 1)$-entry in $S(xI - A)$ is the order of the zero vector $(\rho_1(x))\theta_A$ in $M(A)$ since both are 1. Secondly

$$(\rho_2(x))\theta_A = (-(x + 1), 0, 1)\theta_A = (-(x + 1)e_1(x) + e_3(x))\theta_A$$

$$= -e_1(A + I) + e_3 = -e_1 A - e_1 + e_3$$

$$= -(3, -1, 1) - (1, 0, 0) + (0, 0, 1) = (-4, 1, 0) = v_1 \quad \text{say.}$$

As $v_1 \ne 0$ and

$$d_1(x)v_1 = (x - 2)v_1 = v_1(A - 2I) = (-4, 1, 0) \begin{pmatrix} 1 & -1 & 1 \\ 4 & -4 & 4 \\ 3 & -3 & 3 \end{pmatrix} = (0, 0, 0)$$

we see that v_1 has order $d_1(x) = x - 2$ in $M(A)$. So v_1 is a row eigenvector of A with associated eigenvalue 2 and $N_1 = \langle v_1 \rangle = \langle v_1 \rangle$.

Lastly $(\rho_3(x))\theta_A = ((e_1(x))\theta_A = e_1 = (1, 0, 0) = v_2$ say. Then $xv_2 = v_2 A = (3, -1, 1)$ and so v_2, xv_2 are linearly independent vectors of $\mathbb{Q}^3$. However

$$x^2 v_2 = x(xv_2) = x(3, -1, 1) = (3, -1, 1)A = (8, -4, 4)$$

$$= 4(3, -1, 1) - 4(1, 0, 0) = 4xv_2 - 4v_2$$

and so $(x^2 - 4x + 4)v_2 = 0$ showing that $d_2(x) = (x - 2)^2$ is the order of v_2 in $M(A)$. So $N_2 = \langle v_2 \rangle = \langle v_2, xv_2 \rangle$. In this case $\mathcal{B} = \mathcal{B}_{v_1} \cup \mathcal{B}_{v_2}$ is the ordered set v_1; v_2, xv_2 of vectors. As $M(A) = N_1 \oplus N_2$, it follows that $\mathcal{B}$ is a basis of $\mathbb{Q}^3$ and

$$X = \begin{pmatrix} v_1 \\ \hline v_2 \\ xv_2 \end{pmatrix} = \begin{pmatrix} -4 & 1 & 0 \\ 1 & 0 & 0 \\ 3 & -1 & 1 \end{pmatrix}$$

is invertible over $\mathbb{Q}$. Then $\det X \ne 0$ and $XA = (C(x - 2) \oplus C((x - 2)^2))X$, which on postmultiplying by X^{-1} gives $XAX^{-1} = C(x - 2) \oplus C((x - 2)^2)$, the rcf of A. We have achieved our aim! Starting from A we have found an invertible matrix X over $\mathbb{Q}$ such that XAX^{-1} is in rational canonical form.

The matrix X can be modified to give other canonical forms of A. In this example

$$X_1 = \left(\begin{array}{c} v_1 \\ \hline v_2 \\ \hline (x-2)v_2 \end{array} \right) = \begin{pmatrix} -4 & 1 & 0 \\ 1 & 0 & 0 \\ 1 & -1 & 1 \end{pmatrix}$$

is invertible over $\mathbb{Q}$ and

$$X_1 A X_1^{-1} = \left(\begin{array}{c|cc} 2 & 0 & 0 \\ \hline 0 & 2 & 1 \\ 0 & 0 & 2 \end{array} \right)$$

which is *the Jordan normal form (Jnf) of* A being first described in 1870 by the French mathematician Camille Jordan. Notice that v_1, $(x-2)v_2$, the first and third rows of X_1, are eigenvectors of A and that the Jnf is nearly diagonal. Incidentally it's worth noting that $M(A)$ has a large number of submodules and among these $N' = \langle (x-2)v_2 \rangle$ and $N'' = \langle v_1, (x-2)v_2 \rangle$ are significant: N'' is the eigenspace of A with associated eigenvector 2 and all subspaces of N'' are submodules of $M(A)$. It is also true that all subspaces of $\mathbb{Q}^3$ which contain N' are submodules of $M(A)$. The particular reduction process of $xI - A$ to its Smith normal form $S(xI - A)$ selects a suitable pair of submodules N_1, N_2 from these. The conclusion is that there are, in this case, an infinity of matrices X and X_1 as above.

The last invariant factor $d_s(x)$ of a $t \times t$ matrix A over F has special significance being the analogue of the exponent (Definition 3.11) of a finite abelian group. In fact $d_s(x)$ is the minimum polynomial of A, being the monic polynomial of least degree over F satisfying $d_s(A) = 0$ the zero $t \times t$ matrix over F. The reader should verify $(A - 2I)^2 = 0$ and $A - 2I \neq 0$ in the above example, showing that A has minimum polynomial $(x-2)^2$.

We will see that, of all the canonical forms under similarity, the rational canonical form is in many ways the best: it exists without any conditions on either A or F, it can be found by a systematic application of the polynomial division algorithm, the structure of the linear mapping α determined by A is laid bare, and two $t \times t$ matrices over F are similar if and only if their rcfs are identical.

EXERCISES

1. Use the sequence

$$c_1 + (x-3)c_2, \quad c_1 \leftrightarrow c_2, \quad -c_1, \quad c_3 + c_1, \quad r_2 + (x+3)r_1,$$

$$r_3 - 2r_1, \quad c_2 + (x+1)c_3, \quad c_2 - c_3, \quad -c_2, \quad r_3 + r_2$$

of row and column operations over $\mathbb{Q}[x]$ to reduce $xI - A$ to its Smith normal form $D(x)$, where

$$A = \begin{pmatrix} 3 & 1 & 1 \\ -8 & -3 & -4 \\ 4 & 2 & 3 \end{pmatrix}$$

over $\mathbb{Q}$. Write down the rational canonical form C of A. Find invertible 3×3 matrices $P(x)$ and $Q(x)$ over $\mathbb{Q}[x]$ satisfying $P(x)(xI - A) = D(x)Q(x)$. Verify that the (i,i)-entry in $D(x)$ is the order of $(\rho_i(x))\theta_A$ in $M(A)$, where $\rho_i(x)$ is row i of $Q(x)$ for $i = 1, 2, 3$. Hence construct an invertible 3×3 matrix X over $\mathbb{Q}$ satisfying $XAX^{-1} = C$. By modifying row 3 of X, find an invertible 3×3 matrix X_1 over $\mathbb{Q}$ satisfying $X_1 A X_1^{-1} = J$ in Jordan normal form. Which rows of X_1 are eigenvectors of A?

2. Use row and column operations over $\mathbb{Q}[x]$ to reduce $xI - A$ to its Smith normal form $D(x)$, where

$$A = \begin{pmatrix} 2 & -2 & 1 \\ 2 & -3 & 2 \\ 1 & -2 & 2 \end{pmatrix}$$

over $\mathbb{Q}$.

Hint: Start: $r_1 \leftrightarrow r_3$, $-c_1$, $c_2 - 2c_1$, $c_3 - (x - 2)c_1$.

Write down the rational canonical form C of A. Find invertible 3×3 matrices $P(x)$ and $Q(x)$ over $\mathbb{Q}[x]$ satisfying $P(x)(xI - A) = D(x)Q(x)$. Verify that the (i,i)-entry in $D(x)$ is the order of $(\rho_i(x))\theta_A$ in $M(A)$, where $\rho_i(x)$ is row i of $Q(x)$ for $i = 1, 2, 3$. Hence construct an invertible 3×3 matrix X over $\mathbb{Q}$ satisfying $XAX^{-1} = C$. Modify rows 2 and 3 of X to obtain an invertible 3×3 matrix X_1 over $\mathbb{Q}$ such that $X_1 A X_1^{-1}$ is diagonal.

3. Use row and column operations over $\mathbb{Q}[x]$ to reduce $xI - A$ to its Smith normal form $D(x) = \text{diag}(1, 1, (x - 1)^3)$, where

$$A = \begin{pmatrix} 3 & 2 & -2 \\ -2 & 0 & 3 \\ 1 & 1 & 0 \end{pmatrix}$$

over $\mathbb{Q}$. Write down the rational canonical form C of A. Find invertible 3×3 matrices $P(x)$ and $Q(x)$ over $\mathbb{Q}[x]$ satisfying $P(x)(xI - A) = D(x)Q(x)$. Verify that the (i,i)-entry in $D(x)$ is the order of $(\rho_i(x))\theta_A$ in $M(A)$, where $\rho_i(x)$ is row i of $Q(x)$ for $i = 1, 2, 3$. Hence construct an invertible 3×3 matrix X over $\mathbb{Q}$ satisfying $XAX^{-1} = C$. Is the module $M(A)$ cyclic? If so specify a generator of each of the submodules of $M(A)$. Which (if any) of these generators are eigenvectors of A?

4

The Polynomial Ring $F[x]$ and Matrices over $F[x]$

The $s \times t$ matrices $A(x)$ and $B(x)$ over the polynomial ring $F[x]$ are called *equivalent*
Definition 4.11 if there is an invertible $s \times s$ matrix $P(x)$ over $F[x]$ and an invertible
$t \times t$ matrix $Q(x)$ over $F[x]$ satisfying

$$P(x)A(x)Q(x)^{-1} = B(x).$$

In Chapter 1 the analogous concept of equivalence of matrices over $\mathbb{Z}$ was discussed
and led to the classification (by isomorphism) of f.g. abelian groups in Section 3.1. We
will see that the theory of equivalence of matrices over $F[x]$, developed in Section 4.2,
leads to the classification (by similarity) of $t \times t$ matrices over the field F which is
carried to its conclusion in Section 6.1. The reader is assured that there is a pay-off
to come! In Section 4.1 the ring $F[x]$ of polynomials in x with coefficients in F is
studied in detail, this being a necessary preliminary.

The properties of the ring $\mathbb{Z}$ of integers used in the theory of the Smith normal form
are shared by many other rings and in particular by $F[x]$. This boils down to one fact
(which the reader already knows): it is always possible to divide one polynomial over
F by a non-zero polynomial over F obtaining quotient and remainder polynomials
over F. The *polynomial division property* Theorem 4.1 leads to a constructive way
of determining the Smith normal form of every matrix over $F[x]$ which amounts to a
generalisation of the Euclidean algorithm. As a consequence it is possible to determine
in a finite number of steps, each involving the division of one polynomial by another,
whether any two matrices over $F[x]$ are equivalent or not: either their Smith normal
forms are equal or different!

C. Norman, *Finitely Generated Abelian Groups and Similarity of Matrices over a Field*,   165
Springer Undergraduate Mathematics Series,
DOI 10.1007/978-1-4471-2730-7_4, © Springer-Verlag London Limited 2012

4.1 The Polynomial Ring $F[x]$ where F is a Field

We begin with a resumé of polynomial rings. Let F be a given field. For instance F might be $\mathbb{Z}_2$, $\mathbb{Z}_3$, $\mathbb{Z}_5$, ...or one of $\mathbb{Q}$, $\mathbb{R}$, $\mathbb{C}$ (the rational, real, complex fields) and Theorem 4.9 will give us more examples of fields. *A polynomial over F in the indeterminate x is a sum* $f(x) = \sum_{i \geq 0} a_i x^i$ where all the *coefficients* a_i belong to F and only a finite number of the a_i are non-zero. Two such polynomials $f(x) = \sum_{i \geq 0} a_i x^i$ and $g(x) = \sum_{i \geq 0} b_i x^i$ are decreed to be equal if and only if corresponding coefficients are equal, that is, $\sum_{i \geq 0} a_i x^i = \sum_{i \geq 0} b_i x^i \Leftrightarrow a_i = b_i$ for all $i \geq 0$. The set of all such polynomials is denoted by $F[x]$. The symbol x satisfies $ax = xa$ for all $a \in F$ and so elements of $F[x]$ can be added and multiplied in the usual way producing further elements of $F[x]$, that is,

$$f(x) + g(x) = \sum_{i \geq 0}(a_i + b_i)x^i \quad \text{and} \quad f(x)g(x) = \sum_{i \geq 0}(a_0 b_i + a_1 b_{i-1} + \cdots + a_i b_0)x^i.$$

So for $f(x) = a_0 + a_1 x$ and $g(x) = b_0 + b_1 x + b_2 x^2$ we obtain

$$f(x) + g(x) = a_0 + b_0 + (a_1 + b_1)x + b_2 x^2,$$

$$f(x)g(x) = a_0 b_0 + (a_0 b_1 + a_1 b_0)x + (a_0 b_2 + a_1 b_1)x^2 + a_1 b_2 x^3.$$

In fact $F[x]$ is an *integral domain* (a non-trivial commutative ring with an identity element and without divisors of zero) (see Exercises 4.1, Question 2(c)). The 0-element of $F[x]$ is the *zero polynomial* $0(x)$ over F, that is, all the coefficients in $0(x)$ are zero. Therefore a *non-zero* polynomial has at least one non-zero coefficient.

The *degree* $\deg f(x)$ of the non-zero polynomial $f(x) = \sum_{i \geq 0} a_i x^i$ is the largest non-negative integer i with $a_i \neq 0$. For example $f(x) = 2x^4 + 3x - 1/5$ over $\mathbb{Q}$ has degree 4. So $\deg f(x)$ is the highest power of x appearing in the non-zero polynomial $f(x)$. Let $f(x)$ and $g(x)$ be non-zero polynomials over F and write $m = \deg f(x)$, $n = \deg g(x)$. Then

$$f(x)g(x) = (a_m x^m + a_{m-1}x^{m-1} + \cdots + a_0)(b_n x^n + b_{n-1}x^{n-1} + \cdots + b_0)$$

$$= a_m b_n x^{m+n} + (a_{m-1}b_n + a_m b_{n-1})x^{m+n-1} + \cdots + a_0 b_0$$

showing $\deg f(x)g(x) = m + n$ and establishing the *degree formula*

$$\deg f(x)g(x) = \deg f(x) + \deg g(x)$$

as $a_m b_n \neq 0$. Can the zero polynomial $0(x)$ be assigned a degree so that the above formula holds for all polynomials (non-zero and zero) over F? The degrees of the polynomials $x^2 + x + 1$, $x + 1$, 1 are 2, 1, 0 respectively and so it is reasonable to expect $\deg 0(x)$ to be negative. For the above degree formula to hold with $g(x) = 0(x)$ we need $\deg 0(x) = \deg f(x) + \deg 0(x)$ as $f(x)0(x) = 0(x)$. It is customary to define $\deg 0(x) = -\infty$; then every polynomial has a degree and the above degree formula holds for all polynomials over F since $m + (-\infty) = -\infty$ for $m \in \mathbb{Z}$ and $(-\infty) + (-\infty) = -\infty$. The reader shouldn't be put off by this strange convention – it is adopted solely to make the theory a bit easier to express.

Let $f(x) = \sum_{i \geq 0} a_i x^i$ be a polynomial over the field F and let a be an element of F. The scalar $f(a) = \sum_{i \geq 0} a_i a^i$ is called *the evaluation of* $f(x)$ *at* a. Should $f(a) = 0$ we say that $f(x)$ has *zero* a. Notice that $f(0) = a_0$ is the *constant term* in $f(x)$. The reader can check that the *evaluation mapping*

$$\varepsilon_a : F[x] \to F, \quad \text{given by } (f(x))\varepsilon_a = f(a) \text{ for all } f(x) \in F[x],$$

is a surjective ring homomorphism (see Exercises 4.1, Question 2(a)).

The polynomial $f(x) = \sum_{i \geq 0} a_i x^i$ is called *constant* if $a_i = 0$ for all $i > 0$. So $f(x)$ is a constant polynomial if and only if $\deg f(x) \leq 0$. We identify each constant polynomial $f(x)$ with its constant term $f(0) = f(a)$ obtaining $F \subseteq F[x]$, that is, the integral domain $F[x]$ contains F as the subring of constant polynomials. The constant polynomial $1(x) = 1$, the 1-element of F, is also the 1-element of $F[x]$. Using the degree formula we see that the invertible elements of $F[x]$ are precisely the polynomials of degree 0, that is, $U(F[x]) = F^*$.

The polynomials $f(x)$ and $g(x)$ over F are called *associate* if there is $a \in F^*$ with $f(x) = ag(x)$ in which case we write $f(x) \equiv g(x)$. It is straightforward to verify that $\equiv$ is an equivalence relation on $F[x]$ (Exercises 3.1, Question 7(a)(i)).

The non-zero polynomial $f(x) = \sum_{i=0}^{m} a_i x^i$ of degree m has *leading term* $a_m x^m$ and *leading coefficient* a_m; should $a_m = 1$ then $f(x)$ is said to be *monic* and we write $f(x) = x^m + a_{m-1} x^{m-1} + \cdots + a_1 x + a_0$. The product of monic polynomials over F is itself monic. The reader has already been alerted to the analogy between $F[x]$ and $\mathbb{Z}$; monic polynomials are analogous to positive integers. Just as each non-zero integer m is associate to a unique positive integer namely $|m| = \pm m$, so each non-zero polynomial $f(x)$ over F has a unique monic associate namely $\overline{f}(x) = a_m^{-1} f(x)$ where $f(x)$ has leading coefficient a_m.

Next the familiar process of 'long' division of one polynomial $f(x)$ by another $g(x)$, obtaining quotient $q(x)$ and remainder $r(x)$, is detailed.

Theorem 4.1 (The division property for polynomials over a field)

Let $f(x)$ and $g(x)$ be polynomials over a field F with $g(x) \neq 0(x)$. There are unique polynomials $q(x)$ and $r(x)$ over F satisfying

$$f(x) = q(x)g(x) + r(x) \quad \text{where } \deg r(x) < \deg g(x).$$

Proof

Write $m = \deg f(x), n = \deg g(x)$ and suppose $f(x) = \sum_{i \geq 0} a_i x^i$, $g(x) = \sum_{i \geq 0} b_i x^i$. So $b_n \neq 0$. We first show that there are polynomials $q(x)$ and $r(x)$ as stated. For $m < n$ the polynomials $q(x) = 0(x)$ and $r(x) = f(x)$ satisfy the above conditions. We use induction on $m - n$. The initial case $m - n < 0$ is covered above and so we assume $m \geq n$. The polynomial $f_1(x) = f(x) - (a_m/b_n)x^{m-n}g(x)$ has no x^m term and so $\deg f_1(x) < m$. By the inductive hypothesis, applied to $f_1(x)$ and $g(x)$, there are $q_1(x), r_1(x) \in F[x]$ with $f_1(x) = q_1(x)g(x) + r_1(x)$ where $\deg r_1(x) < n$. Substituting for $f_1(x)$ and rearranging gives $f(x) = ((a_m/b_n)x^{m-n} + q_1(x))g(x) + r_1(x)$ showing that $q(x) = (a_m/b_n)x^{m-n} + q_1(x)$ and $r(x) = r_1(x)$ satisfy Theorem 4.1. The induction is now complete.

To show that $q(x)$ and $r(x)$ are unique, suppose that $q'(x), r'(x) \in F[x]$ also satisfy $f(x) = q'(x)g(x) + r'(x)$ where $\deg r'(x) < n$. As $\deg(r(x) - r'(x)) < n$ the degree formula gives $\deg(q(x) - q'(x)) + n < n$ and so $\deg(q(x) - q'(x)) < 0$. The zero polynomial is the only polynomial of negative degree and so $q(x) - q'(x) = 0(x)$. Therefore $q(x) = q'(x)$ and hence $r(x) = r'(x)$ also.   □

As an illustration of Theorem 4.1 take $F = \mathbb{Q}$, $f(x) = 3x^4 - 2x^3 + 5$, $g(x) = x^3 + 2x^2$. As in Theorem 4.1 we construct first $f(x) - 3xg(x) = -8x^3 + 5$ of degree $3 < \deg f(x) = 4$ on comparing leading terms in $f(x)$ and $g(x)$. Then construct $f(x) - 3xg(x) + 8g(x) = 16x^2 + 5$ of degree $2 < \deg(f(x) - 3xg(x)) = 3$ on comparing leading terms in $f(x) - 3xg(x)$ and $g(x)$. As $\deg(f(x) - 3xg(x) + 8g(x)) = 2 < 3 = \deg g(x)$ the process terminates with $f(x) = (3x - 8)g(x) + 16x^2 + 5$, that is, $q(x) = 3x - 8$, $r(x) = 16x^2 + 5$.

For $r(x) = 0(x)$ in Theorem 4.1 we obtain $f(x) = q(x)g(x)$, that is, $g(x)$ is a *factor* or *divisor* of $f(x)$ and we write $g(x)|f(x)$.

Corollary 4.2

Let $f(x)$ be a polynomial over the field F and let $c \in F$.
 (i) The remainder on dividing $f(x)$ by $x - c$ is $f(c)$. Also $x - c$ is a divisor of $f(x)$ if and only if $f(c) = 0$.
(ii) Suppose $f(x) \neq 0(x)$. There are at most $\deg f(x)$ elements c in F with $f(c) = 0$, that is, the non-zero polynomial $f(x)$ has at most $\deg f(x)$ zeros in F.

Proof

(i) Taking $g(x) = x - c$ in Theorem 4.1, there are polynomials $q(x)$, $r(x)$ over F satisfying $f(x) = q(x)(x - c) + r(x)$ where $\deg r(x) < \deg(x - c) = 1$. So $r(x)$ is a constant polynomial, that is, $r(x) = r(c)$. Using the ring homomorphism ε_c we obtain $f(c) = (f(x))\varepsilon_c = (q(x)(x-c)+r(x))\varepsilon_c = q(c)(c-c)+r(c) = q(c) \times 0 + r(c) = r(c)$ and so $f(x) = q(x)(x-c) + f(c)$.

Suppose that $x - c$ is a divisor of $f(x)$. There is $q'(x) \in F[x]$ with $f(x) = q'(x)(x-c)$. On comparing this equation with $f(x) = q(x)(x-c) + f(c)$ we obtain $q(x) = q'(x)$ and $f(c) = 0$ using the uniqueness in Theorem 4.1. Conversely suppose $f(c) = 0$. Then $f(x) = q(x)(x-c) + f(c) = q(x)(x-c)$ using the first part of the proof. So $x - c$ is a divisor of $f(x)$.

(ii) We use induction on deg $f(x) = n \geq 1$. Suppose $n = 1$. Then $f(x) = a_1 x + a_0$ where $a_1 \neq 0$. Then $f(c_1) = 0 \Leftrightarrow c_1 = -a_0/a_1$ showing that $f(x)$ has a unique zero c_1 in F. Now suppose $n > 1$ and every polynomial of degree $n-1$ over F has at most $n-1$ zeros in F. Consider $f(x)$ of degree n over F. Could there be $n+1$ distinct elements c_i in F where $1 \leq i \leq n+1$ satisfying $f(c_i) = 0$? By Corollary 4.2(i) there is $q(x) \in F[x]$ with $f(x) = q(x)(x-c_1)$ and the degree formula gives $\deg q(x) = n-1$. Using the evaluation homomorphisms ε_{c_i} for $2 \leq i \leq n+1$ gives $0 = f(c_i) = (f(x))\varepsilon_{c_i} = (q(x)(x-c_1))\varepsilon_{c_i} = q(c_i)(c_i - c_1)$. So $q(c_i) = 0$ as $c_i - c_1 \neq 0$ and F has no divisors of zero. We have shown that $q(x)$ of degree $n-1$ over F has n zeros in F namely $c_2, c_3, \ldots, c_{n+1}$, contrary to the inductive hypothesis. The conclusion is that $f(x)$ has at most n zeros in F, completing the induction. $\square$

The reader will remember that Corollary 4.2(ii) was used in the proof of Corollary 3.17. It is worth convincing oneself that the proof of Corollary 4.2(ii) does not rely on Corollary 3.17.

Definition 4.3

Let K be a subgroup of the additive group of $F[x]$, that is, K is closed under addition, K contains the zero polynomial $0(x)$ and K is closed under negation. Suppose also

$$q(x)k(x) \in K \quad \text{for all } q(x) \in F[x], \; k(x) \in K,$$

that is, K is closed under polynomial multiplication. Then K is called *an ideal of the ring $F[x]$*.

The reader should compare Definition 4.3 with the concept of an ideal of $\mathbb{Z}$ introduced at the beginning of Section 1.3. Ideals occur as kernels of ring homomorphisms (Exercises 2.3, Question 3(b)). In particular the kernel of the evaluation homomorphism $\varepsilon_a : F[x] \to F \, (a \in F)$ is, by Corollary 4.2(i), the ideal $\langle x - a \rangle$ consisting of all polynomials over F having divisor $x - a$. More generally an ideal K of $F[x]$ is called *principal* if there is a polynomial $d(x) \in K$ such that $K = \{q(x)d(x) : q(x) \in F[x]\}$, that is, the elements of K are precisely the polynomials over F having divisor $d(x)$.

In this case we write $K = \langle d(x) \rangle$ and call $d(x)$ a *generator of* K. Conversely let $d(x)$ be any polynomial over F; it is straightforward to verify that the subset $K = \{q(x)d(x) : q(x) \in F[x]\}$ of $F[x]$ satisfies Definition 4.3 and so is an ideal of $F[x]$, that is, $d(x)$ generates the principal ideal $K = \langle d(x) \rangle$ of $F[x]$. Notice

$$\langle d_1(x) \rangle \subseteq \langle d_2(x) \rangle \quad \Leftrightarrow \quad d_2(x)|d_1(x) \quad \text{for } d_1(x), d_2(x) \in F[x].$$

For example $\langle x^2 \rangle \subseteq \langle x \rangle \subseteq \langle 1 \rangle = F[x]$ as $1|x$ and $x|x^2$.

We next establish the polynomial analogue of Theorem 1.15.

Theorem 4.4

Let F be a field. Every ideal K of $F[x]$ is principal. Every non-zero ideal of $F[x]$ has a unique monic generator.

Proof

Let K be an ideal of $F[x]$. The zero ideal $\{0(x)\} = \langle 0(x) \rangle$ of $F[x]$ is principal with generator the zero polynomial $0(x)$. Suppose K to be non-zero, that is, K contains a non-zero polynomial. Consider a non-zero polynomial $d(x)$ in K with $m = \deg d(x)$ as small as possible. So $0(x)$ is the only polynomial in K of degree less than m. Let a be the leading coefficient of $d(x)$. Then $\bar{d}(x) = (1/a)d(x)$ is in K by Definition 4.3 with $q(x) = 1/a, k(x) = d(x)$. We show that $\bar{d}(x)$ generates K. As $\bar{d}(x) \in K$, by Definition 4.3 we see $q(x)\bar{d}(x) \in K$ for all $q(x) \in F[x]$, that is, $\langle \bar{d}(x) \rangle \subseteq K$. To show $K \subseteq \langle \bar{d}(x) \rangle$ start with $f(x) \in K$ and use Theorem 4.1 with $g(x) = \bar{d}(x)$: there are $q(x), r(x) \in F[x]$ with $f(x) = q(x)\bar{d}(x) + r(x)$ where $\deg r(x) < \deg \bar{d}(x) = m$. As $f(x)$ and $\bar{d}(x)$ both belong to K, from Definition 4.3 we deduce that $f(x) - q(x)\bar{d}(x) = r(x)$ also belongs to K. So $r(x) = 0(x)$ as $r(x) \in K$ and $\deg r(x) < m$. Therefore $f(x) = q(x)\bar{d}(x) \in \langle \bar{d}(x) \rangle$ showing $K \subseteq \langle \bar{d}(x) \rangle$. So in fact $\langle \bar{d}(x) \rangle = K$, that is, K is the principal ideal with monic generator $\bar{d}(x)$.

Let K be a non-zero ideal of $F[x]$. Suppose $K = \langle d_1(x) \rangle = \langle d_2(x) \rangle$ where $d_1(x)$ and $d_2(x)$ are monic polynomials over F. Then $d_2(x) \in \langle d_1(x) \rangle$ and so there is $q_1(x) \in F[x]$ with $d_2(x) = q_1(x)d_1(x)$. In the same way we see that there is $q_2(x) \in F[x]$ with $d_1(x) = q_2(x)d_2(x)$. Hence $d_2(x) = q_1(x)q_2(x)d_2(x)$. The reader will know that cancellation is legitimate in an integral domain; cancelling $d_2(x)$ gives $1(x) = q_1(x)q_2(x)$ and so $q_1(x)$ and $q_2(x)$ are non-zero constant polynomials. Let $q_1(x) = a_1 \in F^*$. Comparing leading coefficients in $d_2(x) = q_1(x)d_1(x)$ gives $1 = a_1 \times 1$. We conclude $a_1 = 1$ and $q_1(x) = 1(x)$ which gives $d_2(x) = 1(x)d_1(x) = d_1(x)$. So K has a unique monic generator. $\square$

We now know from Theorem 4.4 that, for all fields F, the polynomial ring $F[x]$ is a principal ideal domain (PID). Our next task is to introduce the concept of the greatest common divisor (gcd) of a finite set of polynomials over F. This should not give the reader a headache: the theory is almost the same for polynomials as for integers. Also polynomial gcds can be found using the Euclidean algorithm.

Definition 4.5

Let $X = \{f_1(x), f_2(x), \ldots, f_t(x)\}$ be a set of polynomials over the field F where t is a positive integer. A polynomial $d(x)$ over F is called a *greatest common divisor* (*gcd*) *of* X if
(i) $d(x)|f_i(x)$ for $1 \le i \le t$,
(ii) each $d'(x) \in F[x]$ with $d'(x)|f_i(x)$ for $1 \le i \le t$ satisfies $d'(x)|d(x)$.
Could a given set X as in Definition 4.5 have two monic gcds $d_1(x)$ and $d_2(x)$? If so then $d_1(x)|d_2(x)$ and $d_2(x)|d_1(x)$ by Definition 4.5. Arguing as in the final paragraph of the proof of Theorem 4.4 we see $d_2(x) = d_1(x)$. The conclusion is: the set X has at most one monic gcd. The only gcd of $X = \{0(x)\}$ is $0(x)$. But should X contain a non-zero polynomial we show next, by mimicking Corollary 1.16, that X does have a monic gcd.

Corollary 4.6

Let F be a field and let $X = \{f_1(x), f_2(x), \ldots, f_t(x)\} \subseteq F[x]$ where $f_i(x) \ne 0(x)$ for at least one i with $1 \le i \le t$. Then X has a unique monic gcd $d(x)$. Also there are polynomials $a_1(x), a_2(x), \ldots, a_t(x)$ over F satisfying

$$d(x) = a_1(x)f_1(x) + a_2(x)f_2(x) + \cdots + a_t(x)f_t(x).$$

Proof

We outline the main steps and leave the reader to fill in the gaps. Note first

$$K(X) = \{b_1(x)f_1(x) + b_2(x)f_2(x) + \cdots + b_t(x)f_t(x) : b_i(x) \in F[x], 1 \le i \le t\} \ (\clubsuit)$$

is an ideal of $F[x]$; it is usual to write $K(X) = \langle f_1(x), f_2(x), \ldots, f_t(x) \rangle$ and call $K(X)$ the ideal generated by $f_1(x), f_2(x), \ldots, f_t(x)$. Secondly for each i with $1 \le i \le t$ take $b_i(x) = 1(x)$, $b_j(x) = 0(x)$ where $j \ne i$, $1 \le j \le t$ to show $f_i(x) \in K(X)$. By hypothesis $K(X)$ is a non-zero ideal of $F[x]$ and so $K(X)$ has a unique monic generator $d(x)$ by Theorem 4.4. As $K(X) = \langle d(x) \rangle$ we see $d(x)|f_i(x)$ for $1 \le i \le t$. As $d(x) \in K(X)$ there are $a_i(x) \in F[x]$ for $1 \le i \le t$ with $d(x) = a_1(x)f_1(x) + a_2(x)f_2(x) + \cdots + a_t(x)f_t(x)$. Lastly suppose $d'(x) \in F[x]$ satisfies $d'(x)|f_i(x)$ for $1 \le i \le t$. There are $q_i(x) \in F[x]$ with $q_i(x)d'(x) = f_i(x)$ for $1 \le i \le t$. Hence $d(x) = q(x)d'(x)$ where $q(x) = a_1(x)q_1(x) + a_2(x)q_2(x) + \cdots + a_t(x)q_t(x) \in F[x]$

showing $d'(x)|d(x)$. Therefore $d(x)$ satisfies Definition 4.5 and so is *the* monic gcd of X.   □

We use the notation $d(x) = \gcd X = \gcd\{f_1(x), f_2(x), \ldots, f_t(x)\}$ for X as in Definition 4.5. Also write $\gcd\{0(x)\} = 0(x)$ which is consistent with Definition 4.5. Using (♣) we obtain

$$K(X) \subseteq K(Y) \quad \Leftrightarrow \quad \gcd Y \,|\, \gcd X$$

where X and Y are finite subsets of $F[x]$ and $K(\emptyset) = \langle 0(x) \rangle$ is the zero ideal of $F[x]$.

As $\gcd\{f_1(x), f_2(x), \ldots, f_t(x)\} = \gcd\{f_1(x), \gcd\{f_2(x), \ldots, f_t(x)\}\}$ for $t \geq 2$ (Exercises 4.1, Question 2(b)), the calculation of $\gcd X$ reduces to the case $t = 2$. Suppose therefore $X = \{f_1(x), f_2(x)\}$. The *Euclidean algorithm* is an efficient method of finding a gcd of X in the case of $f_1(x)$ *not* being a gcd of X, $f_1(x) \neq 0(x)$. We remind the reader of the technique: write $r_1(x) = f_1(x)$, $r_2(x) = f_2(x)$. Then $r_2(x) \neq 0(x)$. Start by dividing $r_1(x)$ by $r_2(x)$ obtaining $r_1(x) = q_2(x)r_2(x) + r_3(x)$ with $\deg r_2(x) > \deg r_3(x)$ as in Theorem 4.1. Should $r_3(x) = 0(x)$ then $r_2(x)|r_1(x)$ and $r_2(x)$ is a gcd of $X = \{r_1(x), r_2(x)\}$; as above we see that $r_2(x)$ and $\gcd X$ are associate, that is, $\bar{r}_2(x) = \gcd X$ where $\bar{r}_2(x)$ is $r_2(x)$ *made monic* (by dividing $r_2(x)$ by its leading coefficient). For $r_3(x) \neq 0(x)$ we see $\gcd\{r_2(x), r_3(x)\} = \gcd X$ by Exercises 4.1, Question 2(b), and so it makes sense to repeat the process with $r_2(x)$, $r_3(x)$ in place of $r_1(x)$, $r_2(x)$. We suppose that i non-zero polynomials $r_1(x), r_2(x), \ldots, r_i(x)$ have been found with

$$\gcd\{r_1(x), r_2(x)\} = \gcd\{r_2(x), r_3(x)\} = \cdots = \gcd\{r_{i-1}(x), r_i(x)\}$$

where $\deg r_2(x) > \deg r_3(x) > \cdots > \deg r_i(x)$ and $i \geq 3$. These degrees decrease by at least 1 at each step, that is, $0 \leq \deg r_i(x) \leq \deg r_2(x) - (i - 2)$. So $r_i(x) \neq 0(x)$ gives $2 \leq i \leq \deg r_2(x) + 2$. Dividing $r_{i-1}(x)$ by $r_i(x)$ as in Theorem 4.1 produces polynomials $q_i(x)$ and $r_{i+1}(x)$ over F with $r_{i-1}(x) = q_i(x)r_i(x) + r_{i+1}(x)$ where $\deg r_i(x) > \deg r_{i+1}(x)$. Using Exercises 4.1, Question 2(c) again we obtain $\gcd\{r_{i-1}(x), r_i(x)\} = \gcd\{r_i(x), r_{i+1}(x)\}$. As i is bounded above the sequence of divisions must terminate in a zero remainder: there is an integer k with $2 \leq k \leq \deg r_2(x) + 2$ such that $r_{k+1}(x) = 0(x)$, $r_k(x) \neq 0(x)$. We conclude that $r_k(x)$ is a gcd of $\{r_{k-1}(x), r_k(x)\}$ and

$$\bar{r}_k(x) = \gcd\{r_1(x), r_2(x)\}$$

where $\bar{r}_k(x)$ is the monic associate of $r_k(x)$.

The algorithm can be used to find a particular pair of polynomials $a_1(x)$, $a_2(x)$ over F satisfying $d(x) = a_1(x)f_1(x) + a_2(x)f_2(x)$ as in Corollary 4.6 with $t = 2$. The connection between consecutive pairs of remainder polynomials is expressed by the matrix equation

$$\begin{pmatrix} r_{i-1}(x) \\ r_i(x) \end{pmatrix} = T_i \begin{pmatrix} r_i(x) \\ r_{i+1}(x) \end{pmatrix} \quad \text{for } 2 \le i \le k$$

where $T_i = \begin{pmatrix} q_i(x) & 1 \\ 1 & 0 \end{pmatrix}$ is invertible over $F[x]$ as $\det T_i = -1$. The above $k-1$ matrix equations combine to give

$$\begin{pmatrix} r_1(x) \\ r_2(x) \end{pmatrix} = T \begin{pmatrix} r_k(x) \\ 0 \end{pmatrix}$$

where $T = T_2 T_3 \cdots T_k$. Writing $T = \begin{pmatrix} t_{11} & t_{12} \\ t_{21} & t_{22} \end{pmatrix}$ we obtain $r_1(x) = t_{11} r_k(x)$, $r_2(x) = t_{21} r_k(x)$ showing that the entries in col 1 of T are the quotients $t_{11} = r_1(x)/r_k(x)$ and $t_{21} = r_2(x)/r_k(x)$ obtained on dividing the original polynomials $r_1(x)$ and $r_2(x)$ by their (possibly non-monic) gcd $r_k(x)$. We now demonstrate that the entries in col 2 of T are associates (non-zero scalar multiples) of the polynomials $a_1(x)$, $a_2(x)$ we are looking for. By the multiplicative property of determinants $\det T = (-1)^{k-1}$, as T is the product of $k-1$ matrices T_i $(2 \le i \le k)$ each of which has determinant -1. Therefore $t_{11}t_{22} - t_{12}t_{21} = (-1)^{k-1}$. On multiplying this equation by $(-1)^{k-1}(1/c)r_k(x)$, where c is the leading coefficient of $r_k(x)$, we get $(-1)^{k-1}(1/c)t_{22}r_1(x) + (-1)^k(1/c)t_{12}r_2(x) = (1/c)r_k(x) = \bar{r}_k(x)$. So the polynomials

$$a_1(x) = (-1)^{k-1}(1/c)t_{22} \quad \text{and} \quad a_2(x) = (-1)^k(1/c)t_{12}$$

satisfy $a_1(x)r_1(x) + a_2(x)r_2(x) = \bar{r}_k(x)$.

As an example consider $r_1(x) = x^4 + 3x^2 - 2x + 2$, $r_2(x) = x^3 + x^2 + 3x$ over the rational field $\mathbb{Q}$. Using Theorem 4.1 we carry out the following sequence of divisions.
Divide $r_1(x)$ by $r_2(x)$: $x^4 + 3x^2 - 2x + 2 = (x-1)(x^3 + x^2 + 3x) + x^2 + x + 2$
   giving $q_2(x) = x - 1$, $r_3(x) = x^2 + x + 2$.
Divide $r_2(x)$ by $r_3(x)$: $x^3 + x^2 + 3x = x(x^2 + x + 2) + x$ giving $q_3(x) = x$, $r_4(x) = x$.
Divide $r_3(x)$ by $r_4(x)$: $x^2 + x + 2 = (x+1)x + 2$ giving $q_4(x) = x + 1$, $r_5(x) = 2$.
Divide $r_4(x)$ by $r_5(x)$: $x = (x/2) \times 2 + 0$ giving $q_5(x) = (1/2)x$, $r_6(x) = 0(x)$.

In this case $k = 5$ and $r_5(x) = 2$ is a gcd of $\{r_1(x), r_2(x)\}$ by Definition 4.3. So $c = 2$ and $\gcd\{r_1(x), r_2(x)\} = (1/2)r_5(x) = 1$. The reader can check

$$T = T_2 T_3 T_4 T_5 = \begin{pmatrix} x-1 & 1 \\ 1 & 0 \end{pmatrix} \begin{pmatrix} x & 1 \\ 1 & 0 \end{pmatrix} \begin{pmatrix} x+1 & 1 \\ 1 & 0 \end{pmatrix} \begin{pmatrix} x/2 & 1 \\ 1 & 0 \end{pmatrix}$$

$$= \begin{pmatrix} r_1(x)/2 & x^3 + x \\ r_2(x)/2 & x^2 + x + 1 \end{pmatrix}$$

and hence $a_1(x) = (x^2 + x + 1)/2$, $a_2(x) = -(x^3 + x)/2$ satisfy

$$a_1(x)r_1(x) + a_2(x)r_2(x) = 1.$$

As a second example let $r_1(x) = x^3 + \bar{3}x^2 + \bar{4}x + \bar{2}$, $r_2(x) = x^3 + \bar{2}x^2 + \bar{3}x + \bar{3}$ over the field $\mathbb{Z}_5$. So the coefficients of the powers of x in $r_1(x)$ and $r_2(x)$ are elements of $\mathbb{Z}_5$, the powers themselves being non-negative integers.

Divide $r_1(x)$ by $r_2(x)$: $r_1(x) = r_2(x) + x^2 + x + \bar{4}$ giving $q_2(x) = \bar{1}$, $r_3(x) = x^2 + x + \bar{4}$.

Divide $r_2(x)$ by $r_3(x)$: $r_2(x) = (x + \bar{1})r_3(x) + \bar{3}x + \bar{4}$ giving $q_3(x) = x + \bar{1}$, $r_4(x) = \bar{3}x + \bar{4}$.

Divide $r_3(x)$ by $r_4(x)$: $r_3(x) = (\bar{2}x + \bar{1})r_4(x)$ giving $q_4(x) = \bar{2}x + \bar{1}$, $r_5(x) = 0(x)$. In this case $k = 4$ and $r_4(x) = \bar{3}x + \bar{4}$ is a gcd of $\{r_1(x), r_2(x)\}$ by Definition 4.3. So $c = \bar{3}$ and as $1/c = \bar{2}$ we see that $\gcd\{r_1(x), r_2(x)\} = \bar{2}r_4(x) = \bar{2}(\bar{3}x + \bar{4}) = x + \bar{3}$ is the monic gcd of $r_1(x)$ and $r_2(x)$. Also

$$T = T_2 T_3 T_4 = \begin{pmatrix} \bar{1} & \bar{1} \\ \bar{1} & 0 \end{pmatrix} \begin{pmatrix} x + \bar{1} & \bar{1} \\ \bar{1} & 0 \end{pmatrix} \begin{pmatrix} \bar{2}x + \bar{1} & \bar{1} \\ \bar{1} & 0 \end{pmatrix}$$

$$= \begin{pmatrix} \bar{2}x^2 + \bar{3} & x + \bar{2} \\ \bar{2}x^2 + \bar{3}x + \bar{2} & x + \bar{1} \end{pmatrix}.$$

The reader can check $\bar{2}x^2 + \bar{3} = r_1(x)/r_4(x)$, $\bar{2}x^2 + \bar{3}x + \bar{2} = r_2(x)/r_4(x)$ and $(x + \bar{1})r_1(x) - (x + \bar{2})r_2(x) = -r_4(x)$. Multiplying the last equation through by $\bar{3}$ produces the monic polynomial $\bar{3}(-r_4(x)) = x + \bar{3}$ and so $a_1(x) = \bar{3}(x + \bar{1}) = \bar{3}x + \bar{3}$ and $a_2(x) = -\bar{3}(x + \bar{2}) = \bar{2}x + \bar{4}$ satisfy $a_1(x)r_1(x) + a_2(x)r_2(x) = x + \bar{3} = \gcd\{r_1(x), r_2(x)\}$.

Definition 4.7

Let $p(x)$ be a polynomial of positive degree n over the field F. Suppose there do *not* exist $f(x), g(x) \in F[x]$ with $\deg f(x) < n$, $\deg g(x) < n$ satisfying $p(x) = f(x)g(x)$. Then $p(x)$ is said to be *irreducible over F*.

So a polynomial of positive degree over the field F, which is not a product of two polynomials of lower degree over F, is irreducible over F. Every polynomial $p(x)$ of degree 1 over any field F is irreducible over F, since $p(x)$ is not a product of constant polynomials.

The *fundamental theorem of algebra* states that each non-constant polynomial $f(x)$ over the complex field $\mathbb{C}$ has a zero z_0 in $\mathbb{C}$, that is, $f(z_0) = 0$. The reader may have met this deep theorem, first proved by the German mathematician Gauss in 1799, as an application of complex analysis. Suppose $p(x)$ is irreducible over $\mathbb{C}$. Then $p(x)$ has a complex zero z_0 and so $x - z_0$ is a divisor of $p(x)$ by Corollary 4.2(i). From Definition 4.7 we deduce $p(x) = a(x - z_0)$ where $a \in \mathbb{C}$. Therefore the *only* irreducible polynomials over $\mathbb{C}$ are those of degree 1.

Our next lemma deals with the connection between irreducibility and the non-existence of zeros.

Lemma 4.8

Let $p(x)$ be a polynomial of degree n over a field F.
(i) Suppose $p(x)$ is irreducible over F and $n \geq 2$. Then $p(x)$ has no zeros in F.
(ii) Suppose $p(x)$ has no zeros in F and $n = 2$ or $n = 3$. Then $p(x)$ is irreducible over F.

Proof

(i) Suppose to the contrary that $p(x)$ has zero c in F. So $p(c) = 0$ and $x - c$ is a factor of $p(x)$ by Corollary 4.2(i), that is, $p(x) = (x - c)q(x)$ where $q(x) \in F[x]$. Comparing degrees gives $\deg q(x) = n - 1$ and so $p(x)$ is a product of polynomials of degree at most $n - 1$, showing that $p(x)$ is not irreducible. This contradiction shows that $p(x)$ has no zeros in F.

(ii) Suppose $p(x)$ is not irreducible, that is, $p(x) = f(x)g(x)$ where $f(x), g(x) \in F[x]$ and $\deg f(x) = l < n$, $\deg g(x) = m < n$. We may assume $l \leq m$. Comparing degrees gives $l + m = n$. For $n = 2$ the only possibility is $l = m = 1$: for $n = 3$ the only possibility is $l = 1, m = 2$. So in any case $f(x) = ax + b, a \neq 0$ as $\deg f(x) = 1$. Hence $f(c) = 0$ where $c = -b/a \in F$. Therefore $p(c) = f(c)g(c) = 0 \times g(c) = 0$ showing that $p(x)$ has a zero in F, contrary to hypothesis. The conclusion is: $p(x)$ is irreducible over F. $\qquad\square$

Which polynomials $p(x)$ are irreducible over the real field $\mathbb{R}$? We know that all polynomials $ax + b$ of degree 1 are irreducible over $\mathbb{R}$. Consider $p(x) = ax^2 + bx + c$ of degree 2 over $\mathbb{R}$ with $b^2 < 4ac$. On completing the square we see

$$ap(r) = a^2 r^2 + abr + ac = (ar + b/2)^2 + (4ac - b^2)/4 > 0 \quad \text{for } r \in \mathbb{R},$$

as $(ar + b/2)^2 \geq 0$ and $(4ac - b^2)/4 > 0$. So $p(r) \neq 0$ showing that $p(x)$ has no real zeros. Therefore $p(x)$ is irreducible over $\mathbb{R}$ by Lemma 4.8(ii). We now use the fundamental theorem of algebra to show that there are no further polynomials $p(x)$ which are irreducible over $\mathbb{R}$. Such a polynomial can be regarded as being over $\mathbb{C}$ simply because $\mathbb{R}$ is a subfield of $\mathbb{C}$. So there is $z_0 \in \mathbb{C}$ with $p(z_0) = 0$. We may assume $\deg p(x) \geq 2$ and so $z_0 \notin \mathbb{R}$ by Lemma 4.8(i). The complex conjugation mapping $z \to \bar{z}$ is an automorphism of $\mathbb{C}$ having $\mathbb{R}$ as its fixed field, that is, $z = \bar{z} \Leftrightarrow z \in \mathbb{R}$. Applying this automorphism to $p(z_0) = 0$ produces $p(\bar{z}_0) = 0$. As $z_0 \neq \bar{z}_0$ the polynomials $x - z_0$ and $x - \bar{z}_0$ are coprime (their gcd is 1) and both are divisors of $p(x)$ by Corollary 4.2(i). So $(x - z_0)(x - \bar{z}_0)$ is a divisor of $p(x)$. What is more $(x - z_0)(x - \bar{z}_0) = x^2 - (z_0 + \bar{z}_0)x + z_0 \bar{z}_0$ is a polynomial with real coefficients: writing $z_0 = x_0 + iy_0$ ($x_0, y_0 \in \mathbb{R}$) we have $\bar{z}_0 = x_0 - iy_0$ and so $-(z_0 + \bar{z}_0) = -2x_0 \in \mathbb{R}$, $z_0 \bar{z}_0 = |z_0|^2 = x_0^2 + y_0^2 \in \mathbb{R}$. From Definition 4.7 we deduce $p(x) = a(x - z_0)(x - \bar{z}_0)$

for some $a \in \mathbb{R}$. Let $b = -a(z_0 + \bar{z}_0) = -2ax_0$ and $c = az_0\bar{z}_0 = a(x_0^2 + y_0^2)$. Then $p(x) = ax^2 + bx + c$ and $b^2 - 4ac = 4a^2x_0^2 - 4a^2(x_0^2 + y_0^2) = -4a^2y_0^2 < 0$ as $a \neq 0$, $y_0 \neq 0$. Therefore $b^2 < 4ac$ showing that the irreducible polynomials over $\mathbb{R}$ are $ax + b$ of degree 1 and $ax^2 + bx + c$ of degree 2 with $b^2 < 4ac$.

A polynomial of degree at least 2 over F is called *reducible over F* if it is not irreducible over F. So the reducible polynomials over F are of the type $f(x)g(x)$ where $f(x)$ and $g(x)$ are non-constant polynomials over F.

By Lemma 4.8(ii) $x^2 - 2$ is irreducible over $\mathbb{Q}$ as $\sqrt{2} \notin \mathbb{Q}$, but $x^2 - 2 = (x - \sqrt{2})(x + \sqrt{2})$ is reducible over $\mathbb{R}$. Similarly $x^2 + 1$ is irreducible over $\mathbb{Q}$ and over $\mathbb{R}$, but $x^2 + 1 = (x - i)(x + i)$ is reducible over $\mathbb{C}$. There are four quadratic (degree 2) polynomials over $\mathbb{Z}_2$ namely $x^2, x^2 + \bar{1}, x^2 + x, x^2 + x + \bar{1}$; as $x^2 + \bar{1} = (x + \bar{1})^2$ we see that the first three are reducible over $\mathbb{Z}_2$, but $x^2 + x + \bar{1}$ is irreducible over $\mathbb{Z}_2$ by Lemma 4.8(ii). There are eight cubic (degree 3) polynomials over $\mathbb{Z}_2$ and, as the reader can check, just two of these are irreducible over $\mathbb{Z}_2$, namely $x^3 + x + \bar{1}$ and $x^3 + x^2 + \bar{1}$. Notice that $x^4 + x^2 + \bar{1} = (x^2 + x + \bar{1})^2$ is the only reducible quartic (degree 4) polynomial over $\mathbb{Z}_2$ having no zeros in $\mathbb{Z}_2$; hence

$$x^4 + x + \bar{1}, \ x^4 + x^3 + \bar{1}, \ x^4 + x^3 + x^2 + x + \bar{1} \text{ are irreducible over } \mathbb{Z}_2.$$

The monic irreducible polynomials $p(x)$ over a given field F are analogous to the (positive) prime integers p. The polynomial analogue of *the fundamental theorem of arithmetic* (see the end of Section 1.2) states that each monic polynomial $f(x)$ over F can be expressed

$$f(x) = p_1(x)^{n_1} p_2(x)^{n_2} \cdots p_k(x)^{n_k}$$

where $p_1(x), p_2(x), \ldots, p_k(x)$ are distinct (no two are equal) monic irreducible polynomials over F, $n_1, n_2, \ldots, n_k$ are positive integers, $k \geq 0$, and also the above factorisation of $f(x)$ is unique apart from the order in which the factors $p_i(x)$ occur. This theorem is of more theoretic than practical use, as for most $f(x)$ it is difficult if not impossible to find its irreducible factors $p_i(x)$. An exception is $x^p - x$ over $\mathbb{Z}_p$: by the $|G|$-lemma of Section 2.2 applied to the multiplicative abelian group $\mathbb{Z}_p^*$ we obtain $c^{p-1} = \bar{1}$ for all $c \in \mathbb{Z}_p^*$, that is, each element c of $\mathbb{Z}_p^*$ is a zero of $x^{p-1} - \bar{1}$ over $\mathbb{Z}_p$. So $x - c$ is an irreducible factor of $x^{p-1} - \bar{1}$ over $\mathbb{Z}_p$ by Corollary 4.2(i). Hence

$$x^p - x = x(x^{p-1} - \bar{1}) = x(x - \bar{1})(x - \bar{2}) \cdots (x - \overline{(p-1)})$$

is the factorisation of $x^p - x$ into irreducible polynomials over $\mathbb{Z}_p$. In the same way

$$x^{|F|} - x = \prod_{c \in F}(x - c) \quad \text{over each finite field } F.$$

Lastly we discuss the rings which are homomorphic images of the ring $F[x]$. The rings $\mathbb{Z}_n$ for $n \geq 0$ are, up to isomorphism, the homomorphic images of the ring $\mathbb{Z}$

(Exercises 2.3, Question 3(c)). We now discuss the analogous theory, replacing $\mathbb{Z}$ by $F[x]$.

Theorem 4.9

Let R be a ring and let $\theta : F[x] \to R$ be a ring homomorphism, where F is a field. Then there is $p(x) \in F[x]$ such that

$$\tilde{\theta} : F[x]/\langle p(x) \rangle \cong \operatorname{im} \theta$$

is a ring isomorphism, where $(\langle p(x) \rangle + f(x))\tilde{\theta} = (f(x))\theta$ for all $f(x) \in F[x]$. Further $\operatorname{im} \theta$ is a field if and only if $p(x)$ is irreducible over F.

Proof

Write $K = \ker \theta = \{k(x) \in F[x] : (k(x))\theta = 0\}$. Then K is an ideal of $F[x]$ (Exercises 2.3, Question 3(b)) and so by Theorem 4.4 there is a polynomial $p(x)$ over F with $K = \langle p(x) \rangle$. As in Theorem 2.16 the ring homomorphism θ gives rise to the ring isomorphism $\tilde{\theta}$ as above (Exercises 2.3, Question 3(b) again). Therefore every homomorphic image $\operatorname{im} \theta$ of $F[x]$ is isomorphic to a quotient ring of the type

$$F[x]/\langle p(x) \rangle \quad \text{for some } p(x) \in F[x].$$

Suppose $p(x) = 0(x)$. Then $F[x]/\langle p(x) \rangle \cong F[x]$ which is not a field and so $\operatorname{im} \theta$ is not a field in this case.

Suppose that $p(x)$ is a non-zero constant polynomial. Then $p(x)$ is an invertible element of $F[x]$ giving $K = \langle p(x) \rangle = F[x]$. Hence $F[x]/\langle p(x) \rangle = F[x]/F[x]$ is trivial which means that $\operatorname{im} \theta$ is also trivial and so not a field.

Suppose $p(x)$ is reducible of degree n over F. Then $p(x) = g(x)h(x)$ where $g(x), h(x) \in F[x]$ and $\deg g(x) < n, \deg h(x) < n$. The cosets $K + g(x)$ and $K + h(x)$ are non-zero elements of $F[x]/K = F[x]/\langle p(x) \rangle$ but $(K + h(x))(K + g(x)) = K + g(x)h(x) = K + p(x) = K$ showing that their product is the zero element of $F[x]/\langle p(x) \rangle$, that is, $F[x]/\langle p(x) \rangle$ has divisors of zero. Hence the ring $\operatorname{im} \theta$ also has zero divisors and so is not a field.

Suppose now that $\operatorname{im} \theta$ is a field. By the preceding three paragraphs $p(x)$ has positive degree n and cannot be a product of two polynomials of degree less than n over F, that is, $p(x)$ is irreducible over F.

Conversely suppose that $p(x)$ is irreducible over F. By Theorem 4.4 we may assume that $p(x)$ is monic. Then $F[x]/K$ is a non-trivial commutative ring with 1-element $K + 1$ where $K = \langle p(x) \rangle$. Consider a typical non-zero element $K + f(x)$ of $F[x]/K$. Then $f(x) \notin K$ which means that $p(x)$ is not a divisor of $f(x)$. What could $\gcd\{p(x), f(x)\}$ be? As this gcd is a monic divisor of $p(x)$ the only possibilities are

1 and $p(x)$. As $\gcd\{p(x), f(x)\}$ is a divisor of $f(x)$ we see $\gcd\{p(x), f(x)\} \neq p(x)$. So $\gcd\{p(x), f(x)\} = 1$. By Corollary 4.6 there are $a_1(x), a_2(x) \in F[x]$ satisfying $a_1(x)p(x) + a_2(x)f(x) = 1$. Hence $K + 1 = K + a_2(x)f(x)$ as $1 - a_2(x)f(x) = a_1(x)p(x) \in K$. Therefore $K + 1 = (K + a_2(x))(K + f(x))$ showing that $K + f(x)$ has inverse $K + a_2(x)$ in $F[x]/K$. So $F[x]/K$ is a non-trivial commutative ring in which each non-zero element has an inverse, that is, $F[x]/K$ is a field.   □

Let $p(x)$ be an irreducible polynomial of degree $n \geq 2$ over F and as above let $K = \langle p(x) \rangle$. The field $F[x]/K$ is an important concept (it is analogous to the field $\mathbb{Z}_p$) and it is customary to simplify the notation for $F[x]/K$ and its elements as we now explain.

First $F' = \{K + a : a \in F\}$, that is, the set of cosets with constant polynomial representatives, is closed under addition and multiplication and contains the 0-element $K + 0$ and the 1-element $K + 1$ of $F[x]/K$. In fact the correspondence $F \to F'$ in which $a \to K + a$ (for all $a \in F$) is a ring isomorphism. So F' is a *subfield* of $F[x]/K$ with $F \cong F'$. It is usual to replace each coset $K + a$ by its representative a for $a \in F$ and so F' is replaced by F. Therefore F is a subfield of $F[x]/K$, or in other words, $F[x]/K$ is an *extension field* of F.

Next write $K + x = c$. As $\deg p(x) \geq 2$, a typical element $g(x)p(x) + x$ of $K + x$ has degree at least 1 and so $K + x = c \notin F$. Suppose

$$p(x) = b_n x^n + b_{n-1} x^{n-1} + \cdots + b_1 x + b_0.$$

Then using sums and products of cosets, that is, working in $F[x]/K$ we obtain

$$p(c) = (K + b_n)(K + x)^n + (K + b_{n-1})(K + x)^{n-1} + \cdots$$

$$+ (K + b_1)(K + x) + (K + b_0)$$

$$= (K + b_n x^n) + (K + b_{n-1} x^{n-1}) + \cdots + (K + b_1 x) + (K + b_0)$$

$$= K + (b_n x^n + b_{n-1} x^{n-1} + \cdots + b_1 x + b_0) = K + p(x) = K = K + 0$$

showing that *c is a zero of $p(x)$*. This technique (of extending F to a field which contains a zero of an irreducible polynomial over F) is important in Galois Theory.

Consider a typical element $K + f(x)$ of $F[x]/K$. By Theorem 4.1 there are unique $q(x), r(x) \in F[x]$ with $f(x) = q(x)p(x) + r(x)$ and $\deg r(x) < n$. Then $f(x) \equiv r(x) \pmod{K}$ as in Section 2.2 and so $K + f(x) = K + r(x)$. Write $r(x) = a_{n-1} x^{n-1} + a_{n-2} x^{n-2} + \cdots + a_1 x + a_0$. Manipulating cosets we see

$$K + f(x) = K + (a_{n-1} x^{n-1} + a_{n-2} x^{n-2} + \cdots + a_1 x + a_0)$$

$$= (K + a_{n-1} x^{n-1}) + (K + a_{n-2} x^{n-2}) + \cdots + (K + a_1 x) + (K + a_0)$$

$$= (K + a_{n-1})(K + x^{n-1}) + (K + a_{n-2})(K + x^{n-2}) + \cdots$$

$$+ (K + a_1)(K + x) + (K + a_0)$$

$$= (K + a_{n-1})(K + x)^{n-1} + (K + a_{n-2})(K + x)^{n-2} + \cdots$$

$$+ (K + a_1)(K + x) + (K + a_0)$$

$$= a_{n-1}c^{n-1} + a_{n-2}c^{n-2} + \cdots + a_1 c + a_0 = r(c).$$

So each element of $F[x]/K$ is uniquely expressible in the form

$$r(c) = a_{n-1}c^{n-1} + a_{n-2}c^{n-2} + \cdots + a_1 c + a_0$$

where $p(c) = 0$, $\deg p(x) = n$ and $a_i \in F$ $(0 \le i < n)$. Notice that $F[x]/K$ is an n-dimensional vector space over F with basis $1, c, c^2, \ldots, c^{n-1}$. Finally write $F[x]/K = F(c)$ and so

$F(c)$ is the extension field obtained by adjoining the zero c of $p(x)$ to F.

The polynomial $p(x) = x^2 + 1$ is irreducible over the real field $\mathbb{R} = F$; with $c = i$, and so $i^2 = -1$, we obtain $\mathbb{C} = \mathbb{R}(i) = \{a + ib : a, b \in \mathbb{R}\}$, the familiar construction of the complex field. Similarly $x^2 - 2$ is irreducible over the rational field $\mathbb{Q}$ and so $\mathbb{Q}(\sqrt{2}) = \{a + b\sqrt{2} : a, b \in \mathbb{Q}\}$ is the field obtained by adjoining $\sqrt{2}$ to $\mathbb{Q}$.

As we saw earlier, $p(x) = x^2 + x + \bar{1}$ is irreducible over $\mathbb{Z}_2$. The element $c = \langle p(x) \rangle + x$ of the field $\mathbb{Z}_2(c) = \mathbb{Z}_2[x]/\langle p(x) \rangle$ satisfies $p(c) = \bar{0}$, that is, $c^2 = c + \bar{1}$. This equation (the analogue of $i^2 = -1$ above) is all one needs to manipulate the four elements of $\mathbb{Z}_2(c) = \{\bar{0}, \bar{1}, c, c + \bar{1}\}$. The reader can check that the addition and multiplication tables of $\mathbb{Z}_2(c)$ are:

$+$	$\bar{0}$	$\bar{1}$	c	$c + \bar{1}$
$\bar{0}$	$\bar{0}$	$\bar{1}$	c	$c + \bar{1}$
$\bar{1}$	$\bar{1}$	$\bar{0}$	$c + \bar{1}$	c
c	c	$c + \bar{1}$	$\bar{0}$	$\bar{1}$
$c + \bar{1}$	$c + \bar{1}$	c	$\bar{1}$	$\bar{0}$

and

$\times$	$\bar{0}$	$\bar{1}$	c	$c + \bar{1}$
$\bar{0}$	$\bar{0}$	$\bar{0}$	$\bar{0}$	$\bar{0}$
$\bar{1}$	$\bar{0}$	$\bar{1}$	c	$c + \bar{1}$
c	$\bar{0}$	c	$c + \bar{1}$	$\bar{1}$
$c + \bar{1}$	$\bar{0}$	$c + \bar{1}$	$\bar{1}$	c

It is usual to denote this field by $\mathbb{F}_4$ as it can be shown that every field with exactly four elements is isomorphic to $\mathbb{Z}_2(c)$. Another common notation is $GF(4)$, the Galois field of order 4. The additive group of $\mathbb{F}_4$ is the Klein 4-group of isomorphism type $C_2 \oplus C_2$ and the multiplicative group $\mathbb{F}_4^*$ is cyclic of order 3. From the tables $(c + \bar{1})^2 = c = (c + \bar{1}) + \bar{1} = -(c + \bar{1}) - \bar{1}$ showing $p(c + \bar{1}) = \bar{0}$, that is, the irreducible polynomial $p(x)$ over $\mathbb{Z}_2$ has zeros $c, c + \bar{1}$ in $\mathbb{F}_4$ and so factorises $p(x) = (x - c)(x - c - \bar{1})$ over $\mathbb{F}_4$.

We know (Exercises 2.3, Question 4(a)) that the number of elements in every finite field F is a prime power $q = p^n$. Conversely for every prime power $q = p^n$ it can be shown that there is, up to isomorphism, a unique field $\mathbb{F}_q$ (also denoted $GF(q)$, the *Galois field of order q*) having exactly q elements. By the $|G|$-lemma, the q elements of $\mathbb{F}_q$ are the zeros of $x^q - x$ over $\mathbb{Z}_p$. What is more, all the monic irreducible polynomials of degree n over $\mathbb{Z}_p$ occur once and once only in the factorisation of $x^{p^n} - x$ over $\mathbb{Z}_p$ (Exercises 4.1, Question 3(c)).

In Chapter 6 the field $F[x]/\langle p(x) \rangle$ is used to describe the $F[x]$-module $M(A)$ where A is a $t \times t$ matrix over F satisfying $p(A) = 0$.

EXERCISES 4.1

1. (a) For each of the following pairs $r_1(x)$, $r_2(x)$ of polynomials over $\mathbb{Q}$, use the Euclidean algorithm to determine $d(x) = \gcd\{r_1(x), r_2(x)\}$ and $r_1'(x) = r_1(x)/d(x)$, $r_2'(x) = r_2(x)/d(x)$. Also find $a_1(x), a_2(x) \in \mathbb{Q}[x]$ satisfying $a_1(x)r_1(x) + a_2(x)r_2(x) = d(x)$.
   (i) $r_1(x) = x^4 + x^2 + 1$, $r_2(x) = x^3 + 1$;
   (ii) $r_1(x) = x^4 + x^3 - 2x^2 - x + 1$, $r_2(x) = x^3 - 1$;
   (iii) $r_1(x) = x^3 + x^2 + 1$, $r_2(x) = x^2 - 1$;
   (iv) $r_1(x) = x^{46} - 1$, $r_2(x) = x^{32} - 1$; first use the Euclidean algorithm to find $\gcd\{46, 32\}$.
   (b) In each of the following cases, working over the indicated field, calculate $d(x) = \gcd\{r_1(x), r_2(x)\}$ and the polynomials $r_1(x)/d(x)$ and $r_2(x)/d(x)$.
   (i) $r_1(x) = x^5 + x^4 + x^3 + \bar{1}$, $r_2(x) = x^5 + x^2 + x + \bar{1}$ over $\mathbb{Z}_2$;
   (ii) $r_1(x) = x^4 + x^3 + x^2 - \bar{1}$, $r_2(x) = x^4 + \bar{1}$ over $\mathbb{Z}_3$;
   (iii) $r_1(x) = \bar{2}x^3 + x^2 + \bar{3}x + \bar{4}$, $r_2(x) = \bar{3}x^3 + \bar{4}x^2 + \bar{2}x + \bar{1}$ over $\mathbb{Z}_5$.
   (c) Let m, n, q and r be integers with $m = qn + r$ where $m \geq n > r \geq 0$. Let F be a field with 1-element e. Find the quotient $q(x)$ and remainder $r(x)$ on dividing $x^m - e$ by $x^n - e$ over F. Hence show $(x^n - e)|(x^m - e)$ if and only if $n|m$. Use the Euclidean algorithm to conclude $\gcd\{x^m - e, x^n - e\} = x^d - e$ where $d = \gcd\{m, n\}$.
   (d) Let $f_1(x)$ and $f_2(x)$ be non-zero polynomials over a field F and let $l(x)$ be the monic generator of the ideal $\langle f_1(x) \rangle \cap \langle f_2(x) \rangle$ of $F[x]$.

Show that $l(x)$ is the *least common multiple* (*lcm*) of $f_1(x)$ and $f_2(x)$, that is, $l(x)$ is the unique monic polynomial over F satisfying
   (i) $f_1(x)|l(x)$, $f_2(x)|l(x)$ and
   (ii) $f_1(x)|l'(x)$, $f_2(x)|l'(x) \Rightarrow l(x)|l'(x)$ for $l'(x) \in F[x]$. Use Corollary 4.6 to show $l(x) = f_1(x)f_2(x)/\gcd\{f_1(x), f_2(x)\}$ in the case $f_1(x)$, $f_2(x)$ both monic.

(e) Let $f(x)$ be a monic polynomial of degree t over a field F. Use the polynomial analogue of the fundamental theorem of arithmetic to show that $f(x)$ has at most 2^t monic divisors over F.

Hint: Consider first the case $f(x) = p(x)^n$ where $p(x)$ is monic and irreducible over F.

2. (a) Let F be a field with an element a. Show that $\varepsilon_a : F[x] \to F$, defined by $(f(x))\varepsilon_a = f(a)$ for all $f(x) \in F[x]$, is a surjective ring homomorphism.

(b) Let $f(x), g(x), q(x), r(x)$ be polynomials over a field F satisfying $f(x) = q(x)g(x) + r(x)$. Show that the ideals $\langle f(x), g(x)\rangle$ and $\langle g(x), r(x)\rangle$ are equal (as sets). Deduce $\gcd\{f(x), g(x)\} = \gcd\{g(x), r(x)\}$.

Let $f_1(x), f_2(x), \ldots, f_t(x)$ be polynomials over a field F where $t \geq 3$. Show

$$\gcd\{f_1(x), f_2(x), \ldots, f_t(x)\} = \gcd\{f_1(x), \gcd\{f_2(x), \ldots, f_t(x)\}\}.$$

(c) (Polynomials in one indeterminate over a ring.) Let R be a ring and let $P(R)$ denote the set of infinite sequences $(a_i) = (a_0, a_1, \ldots, a_i, \ldots)$ where the entries a_j all belong to R and only a finite number of the a_j are non-zero. The sum and product of elements (a_i) and (b_i) of $P(R)$ are defined by

$$(a_0, a_1, \ldots, a_i, \ldots) + (b_0, b_1, \ldots, b_i, \ldots)$$
$$= (a_0 + b_0, a_1 + b_1, \ldots, a_i + b_i, \ldots),$$
$$(a_0, a_1, \ldots, a_i, \ldots)(b_0, b_1, \ldots, b_i, \ldots)$$
$$= (a_0 b_0, a_0 b_1 + a_1 b_0, \ldots, a_0 b_i + a_1 b_{i-1} + \cdots + a_i b_0, \ldots).$$

Show that $P(R)$ is closed under sum and product. Show that $P(R)$ is a ring.

Write $(a_0)\iota' = (a_0, 0, 0, \ldots, 0, \ldots)$ for $a_0 \in R$ show that $\iota' : R \to P(R)$ is an injective ring homomorphism and deduce that $R' = \operatorname{im} \iota'$ is a subring of $P(R)$ with $R \cong R'$. Show R is an integral domain $\Leftrightarrow P(R)$ is an integral domain.

For $i \geq 0$ let $e_i = (0, 0, \ldots, 0, 1, 0, \ldots) \in P(R)$, that is, $e_i = (a_0, a_1, \ldots, a_i, \ldots)$ where $a_j = 0$ for $j \neq i$ and $a_i = 1$. Write $x = e_1$. Verify $(a_0)\iota' x = x(a_0)\iota'$ for all $a_0 \in R$. Show $x^i = e_i$ by induction on i. Deduce

$$(a_0, a_1, \ldots, a_i, \ldots)$$
$$= (a_0)\iota' x^0 + (a_1)\iota' x + (a_2)\iota' x^2 + \cdots + (a_i)\iota' x^i + \cdots,$$

that is, elements of $P(R)$ are polynomials in x over R'. (It is customary to identify $(a_0)\iota'$ with a_0 and write $P(R) = R[x]$.)

3. (a) Let F be a finite field.

(i) For $|F|$ odd, use Lagrange's theorem, Corollary 3.17 and Lemma 4.8 to show that the polynomial $x^2 + \bar{1}$ over F is reducible if and only if $|F| \equiv 1 \pmod 4$.

(ii) Over which of the fields $\mathbb{Z}_2, \mathbb{Z}_3, \mathbb{Z}_5, \mathbb{Z}_7$ is $x^2 + x + \bar{1}$ irreducible? Determine the integers $|F|$ such that $x^2 + x + \bar{1}$ irreducible over F.
Hint: Consider the three congruence classes of $|F| \pmod 3$ separately.

(b) Show that $x^3 - \bar{2}$ over $\mathbb{Z}_7$ is irreducible. List the irreducible cubics over $\mathbb{Z}_7$ of the type $x^3 - a$.
Hint: First find the elements of $\mathbb{Z}_7^*$ of the form b^3 where $b \in \mathbb{Z}_7^*$.
Let F be a finite field. By considering the endomorphism θ of F^* defined by $(b)\theta = b^3$ for all $b \in F^*$, show that all cubics over F of the form $x^3 - a$ are reducible provided $|F| \not\equiv 1 \pmod 3$. Determine, in terms of $|F|$, the number of reducible cubics of the form $x^3 - a$ over F in the case $|F| \equiv 1 \pmod 3$.

(c) Let $p(x)$ be a monic irreducible polynomial of degree n over a finite field F. Write $|F| = q$. Use the $|G|$-lemma of Section 2.2 to show that each element b of the field $E = F[x]/\langle p(x) \rangle$ is a zero of $x^{q^n} - x$. Hence show that $x^{q^n} - x$ *splits* (is the product of factors of degree 1) over E. By considering $\gcd\{p(x), x^{q^n} - x\}$, show that $p(x)$ is a divisor of $x^{q^n} - x$ and deduce that $p(x)$ splits over E. Let $p'(x)$ be also a monic irreducible polynomial of degree n over F. Show that $E \cong F[x]/\langle p'(x) \rangle$. Is it possible for an irreducible polynomial over a finite field F to have a *repeated* zero (a factor $(x - b)^2$) in an extension field E?

(d) The polynomial $p(x)$ of degree 5 over $\mathbb{Z}_2$ satisfies $p(\bar{0}) = \bar{1}$ and $\gcd\{p(x), x^3 + \bar{1}\} = \bar{1}$. Show that $p(x)$ is irreducible over $\mathbb{Z}_2$. Show that exactly one of $x^5 + x^3 + x + \bar{1}, x^5 + x^4 + \bar{1}, x^5 + x^2 + \bar{1}, x^5 + x + \bar{1}$ is an irreducible polynomial $p(x)$ over $\mathbb{Z}_2$. What is $\gcd\{p(x), x^{31} - \bar{1}\}$ over $\mathbb{Z}_2$?
Hint: Either use the Euclidean algorithm or the theory in (c) above.

4. (a) Let F be a field of characteristic 0 (see Exercises 2.3, Question 5(a))
       with multiplicative identity e. Let $F_0 = \{me/ne : m, n \in \mathbb{Z}, n \neq 0\}$
       denote its prime subfield. Let θ be an automorphism of F (so θ is a
       bijective ring homomorphism of F to itself). Show that $(a_0)\theta = a_0$ for
       all $a_0 \in F_0$. Show that $L = \{a \in F : (a)\theta = a\}$ is a subfield of F (L is
       called the *fixed field* of θ).
       Let $f(x)$ be a polynomial over F_0. Show that $(f(c))\theta = f((c)\theta)$ for
       all $c \in F$. Deduce that c is a zero of $f(x)$ if and only if $(c)\theta$ is a zero
       of $f(x)$.
   (b) The mapping $\theta : \mathbb{Q}(\sqrt{2}) \to \mathbb{Q}(\sqrt{2})$ is defined by $(a + b\sqrt{2})\theta = a - b\sqrt{2}$ for all $a, b \in \mathbb{Q}$. Show that θ is an automorphism of $\mathbb{Q}(\sqrt{2})$.
       Use (a) above with $f(x) = x^2 - 2$ to show that θ is the only non-identity automorphism of $\mathbb{Q}(\sqrt{2})$.
   (c) Let $p(x) = x^2 + a_1 x + a_0$ be irreducible over an arbitrary field F.
       Let $F(c)$ be the extension field obtained by adjoining a zero c of $p(x)$
       to F. Show $-a_1 - c$ is also a zero of $p(x)$. Generalise (b) showing that
       $\theta : F(c) \to F(c)$, given by $(a + bc)\theta = a - a_1 b - bc$ for all $a, b \in F$,
       is a self-inverse ($\theta = \theta^{-1}$) automorphism of $F(c)$.
       Hint: Show θ respects multiplication – this surprising fact is important
       in Galois Theory.
       Let L denote the fixed field of θ. Show $-a_1 - c \neq c$ if and only if
       $L = F$.

5. (a) Let F be a finite field of characteristic p, let F_0 be its prime subfield,
       and let θ be an automorphism of F. Show that $(a_0)\theta = a_0$ for all
       $a_0 \in F_0$. Show that $L = \{a \in F : (a)\theta = a\}$ is a subfield of F (L is the
       fixed field of θ) and $F_0 \subseteq L \subseteq F$.
       Suppose $|F| = p^n$. Use Corollary 4.2(ii) to show that the *Frobenius
       automorphism* θ of F, defined by $(a)\theta = a^p$ for all $a \in F$ (see Exercises 2.3, Question 5(a)), has multiplicative order n.
   (b) Let $p(x) = x^3 + x + \bar{1}$ over $\mathbb{Z}_2$. Verify that $p(x)$ is irreducible
       over $\mathbb{Z}_2$. Write out the addition and multiplication tables of the field
       $\mathbb{Z}_2(c) = \{a_0 + a_1 c + a_2 c^2 : a_0, a_1, a_2 \in \mathbb{Z}_2\}$ where $p(c) = \bar{0}$. Express
       each element of $\mathbb{Z}_2(c)^*$ as a power of c. Factorise $p(x)$ into a product
       of polynomials of degree 1 over $\mathbb{Z}_2(c)$. Show that $p'(x) = x^3 + x^2 + 1$
       is irreducible over $\mathbb{Z}_2$ and splits over $\mathbb{Z}_2(c)$. Hence factorise $x^8 - x$
       into irreducible polynomials over $\mathbb{Z}_2(c)$ and into irreducible polynomials over $\mathbb{Z}_2$. Are $p(x)$ and $p'(x)$ the only irreducible polynomials
       of degree 3 over $\mathbb{Z}_2$?
   (c) Use Question 3(d) above to show that $p(x) = x^5 + x^3 + \bar{1}$ is irreducible over $\mathbb{Z}_2$. Use the Frobenius automorphism θ of the field
       $\mathbb{Z}_2(c)$ to express the zeros of $p(x)$ as linear combinations over

$\mathbb{Z}_2$ of $\bar{1}, c, c^2, c^3, c^4$ where $p(c) = \bar{0}$ (remember $(a)\theta = a^2$ for all $a \in \mathbb{Z}_2(c)$). Find the irreducible polynomial $p'(x)$ over $\mathbb{Z}_2$ such that $p'(c + \bar{1}) = \bar{0}$ in $\mathbb{Z}_2(c)$ and factorise $p'(x)$ into irreducible polynomials over $\mathbb{Z}_2(c)$.

Hint: Find the connection between the zeros of $f(x)$ and those of $f(x - a)$?

6. (a) Verify that $x^2 + \bar{1}, x^2 + x - \bar{1}, x^2 - x - \bar{1}$ are irreducible over $\mathbb{Z}_3$. Calculate their product and hence resolve $x^9 - x$ into monic irreducible polynomials over $\mathbb{Z}_3$. Are there any further monic irreducible polynomials of degree 2 over $\mathbb{Z}_3$? Factorise each of the above quadratic polynomials into monic irreducible polynomials over the field $\mathbb{Z}_3(i)$ where $i^2 = -\bar{1}$. Simplify $(a + bi)^3$ where $a, b \in \mathbb{Z}_3$ and hence find the connection between the Frobenius automorphism θ of $\mathbb{Z}_3(i)$ and 'conjugation' $a + bi \rightarrow a - bi$. Find a generator of the cyclic group $\mathbb{Z}_3(i)^*$. Write out the addition and multiplication tables of $\mathbb{Z}_3(i)$.

   (b) Verify that $p(x) = x^3 - x - \bar{1}$ is irreducible over $\mathbb{Z}_3$. Let c be a zero of $p(x)$ and so $c^3 = c + \bar{1}$. Does c generate the multiplicative group of the field $\mathbb{Z}_3(c)$? Is $p(x)$ a divisor of $x^{13} - \bar{1}$ over $\mathbb{Z}_3$? Does $-c$ generate the cyclic group $\mathbb{Z}_3(c)^*$? Is $-p(-x) = x^3 - x + \bar{1}$ a divisor of $x^{13} + \bar{1}$ over $\mathbb{Z}_3$?

   Use $x^3 - x$ over $\mathbb{Z}_3$ to show that $(a)\theta = a \Leftrightarrow a \in \mathbb{Z}_3$ where θ is the Frobenius automorphism of $\mathbb{Z}_3(c)$. Hence show that $p_a(x) = (x - a)(x - a^3)(x - a^9)$ is a polynomial over $\mathbb{Z}_3$ for $a \in \mathbb{Z}_3(c)$. For $a \notin \mathbb{Z}_3$ is $p_a(x)$ irreducible over $\mathbb{Z}_3$? How many monic irreducible polynomials of degree 3 over $\mathbb{Z}_3$ are there? Factorise $x^{13} - \bar{1}$ and $x^{13} + \bar{1}$ into monic irreducible polynomials over $\mathbb{Z}_3$.

   (c) Verify that $p(x) = x^4 + x^2 - \bar{1}$ over $\mathbb{Z}_3$ is not divisible by any monic irreducible quadratic over $\mathbb{Z}_3$ (see (a) above). Hence show that $p(x)$ is irreducible over $\mathbb{Z}_3$. Let c satisfy $p(c) = \bar{0}$ and write $i = c^2 - \bar{1}$. Verify $i^2 = -\bar{1}$ and deduce that $\mathbb{Z}_3(i)$ is a subfield of $\mathbb{Z}_3(c)$. Find the factorisations of $p(x)$ into monic irreducible polynomials over $\mathbb{Z}_3(c)$ and into monic irreducible quadratics over $\mathbb{Z}_3(i)$. How many monic irreducible polynomials of degree 4 over $\mathbb{Z}_3$ are there?

7. (a) Let E be a field with subfield F. Suppose that E, which is a vector space over F, has finite dimension $[E : F]$. Then E is called a *finite extension* of F. Let c be an element of E. By considering the 'vectors' c^i in E for $0 \leq i \leq [E : F]$, show that the evaluation homomorphism $\varepsilon_c : F[x] \rightarrow E$ has a non-zero kernel K. The monic generator $m_c(x)$ of K (see Theorem 4.4) is called *the minimum polynomial* of c. Show that $m_c(x)$ is irreducible over F and deduce that $F(c) = \text{im}\,\varepsilon_c$ is a subfield of E with $F(c) \cong F[x]/K$ and $[F(c) : F] = \deg m_c(x)$.

(b) Let E be a field with subfield F and let L be an *intermediate subfield*, that is, L is a subfield of E and $F \subseteq L \subseteq E$. Suppose that L is a finite extension of F with basis $u_1, u_2, \ldots, u_m$ where $m = [L : F]$ and suppose also that E is a finite extension of L with basis $v_1, v_2, \ldots, v_n$ where $n = [E : L]$. Prove that the mn elements $u_i v_j$ of E ($1 \leq i \leq m$, $1 \leq j \leq n$) form a basis of E over F. Deduce that E is a finite extension of F and $[E : F] = [E : L][L : F]$.

(c) Let E be a finite field with subfield F of order q and let $[E : F] = n$.
(i) Show that every subfield L of E satisfies $|L| = q^d$ where $d | n$. Conversely show $(q^d - 1) | (q^n - 1)$ and $(x^{q^d} - x) | (x^{q^n} - x)$ for each positive divisor d of n, and hence show that $L = \{c \in E : c^{q^d} = c\}$ is the unique subfield of E with $|L| = q^d$.
Hint: Consider the fixed field of θ^d where θ is the Frobenius automorphism of E.
(ii) Let L and M be subfields of E with $F \subseteq L \cap M$ and so $|L| = q^d$ and $|L| = q^e$. Deduce from (i) above that $|L \cap M| = q^{\gcd\{d,e\}}$.
(iii) Let $n = p_1^{n_1} p_2^{n_2} \cdots p_k^{n_k}$ be the factorisation of $n > 1$ into positive powers of distinct primes $p_1, p_2, \ldots, p_k$. For each subset X of $\{1, 2, \ldots, k\}$ write $\pi_X = \prod_{j \in X} p_j$, i.e. π_X is the product of the primes p_j for $j \in X$ and $\pi_\emptyset = 1$. Let L_j be the subfield of E with $|L_j| = q^n / p_j$ ($1 \leq j \leq k$). Use induction on $s = |X|$ and (ii) above to show $|\bigcap_{j \in X} L_j| = q^n / \pi_X$. The *sieve formula* now asserts that the number of elements of E which are not in any subfield L of E with $F \subseteq L \neq E$ is $r = \sum_X (-1)^{|X|} q^n / \pi_X$, the summation being over the 2^k subsets X of $\{1, 2, \ldots, k\}$. Using Questions 3(c) and 7(a) above, explain why r/n (*Dedekind's formula*) is the number of monic irreducible polynomials of degree n over $F \cong \mathbb{F}_q$.
Verify that there are 335 monic irreducible polynomials of degree 12 over $\mathbb{Z}_2$. Verify that there are 670 monic irreducible polynomials of degree 6 over $\mathbb{F}_4$. Calculate the numbers of monic irreducible polynomials of degree 12 over $\mathbb{Z}_3$ and of degree 6 over $\mathbb{F}_9$.

8. (a) Let $p(x)$ be a monic irreducible polynomial of degree n over a field F and let m be a positive integer. Write $K = \langle p(x)^m \rangle$ for the principal ideal of $F[x]$ generated by $p(x)^m$ and let $R_m = F[x]/K$. Also let $G_m = U(R_m)$ denote the multiplicative group of invertible elements of the ring R_m. Show $K + f(x) \in G_m \Leftrightarrow \gcd\{f(x), p(x)\} = 1$. Show also that $H_m = \{K + f(x) : f(x) \equiv 1 \pmod{p(x)}\}$ is a subgroup of G_m.
Suppose F is a finite field of order q. Show $|R_m| = q^{mn}$, $|G_m| = q^{(m-1)n}(q^n - 1)$ and $|H_m| = q^{(m-1)n}$. By applying Corollary 3.17 to $R_m/\langle K + p(x) \rangle$ show that $G_m \cong H_m \times H_0$ (external direct product, Exercises 2.3, Question 4(d)) where H_0 is cyclic of order

$q^n - 1$. Express the invariant factors of G_m in terms of the invariant factors of H_m.

(b) Let $R_m = \mathbb{Z}_2[x]/\langle x^m \rangle$ and let $G_m = U(R_m)$ where m is a positive integer. Show $|G_m| = 2^{m-1}$.

Find the invariant factors of G_2, G_3, G_4, G_5 and G_6. Adapting the notation used in the proof of Theorem 3.7 to multiplicative abelian groups, write $^n G_m = \{g \in G_m : g^n = \langle x^m \rangle + 1\}$. Find the invariant factors of the subgroups $^2 G_6$, $^4 G_6$, $^8 G_6$ of G_6.

(c) Let j and m be positive integers and let $\lfloor y \rfloor$ denote the integer part of the real number y. Show that there are $\lfloor (j-1)m/j \rfloor$ integers i in the range $m/j \leq i < m$.

Suppose $2^{r-1} < m \leq 2^r$. Show

$$|^{2^j} G_m| = 2^{t_j} \quad \text{where } t_j = \lfloor (2^j - 1)m/2^j \rfloor$$

and G_m is defined in (b) above. Show also $t_j = m - 1$ for $j \geq r$. Find a formula for the number s_j of invariant factors 2^j of G_m for $j \geq 1$.
Hint: You should find $s_j = 0$ for $j > r$.

List the invariant factors of G_{25} and G_{32}.

(d) Let $g(x)$ and $h(x)$ be polynomials over a field F. Show that the mapping $\alpha : F[x]/\langle g(x)h(x) \rangle \to (F[x]/\langle g(x) \rangle) \oplus (F[x]/\langle h(x) \rangle)$, given by

$$(\langle g(x)h(x) \rangle + f(x))\alpha = (\langle g(x) \rangle + f(x), \langle h(x) \rangle + f(x))$$

for all $f(x) \in F[x]$,

is unambiguously defined and is a ring homomorphism. Establish

the polynomial version of the Chinese remainder theorem 2.11,

namely $\gcd\{g(x), h(x)\} = 1$ implies that α is a ring isomorphism. Show that the rings $R_m = \mathbb{Z}_2[x]/\langle x^m \rangle$ and $R'_m = \mathbb{Z}_2[x]/\langle (x-1)^m \rangle$ are isomorphic and deduce $G_m = U(R_m) \cong U(R'_m)$. Let l and m be positive integers. Show that the rings $\mathbb{Z}_2[x]/\langle x^l(x-1)^m \rangle$ and $R_l \oplus R_m$ are isomorphic. Show that the multiplicative abelian groups $U(\mathbb{Z}_2[x]/\langle x^l(x-1)^m \rangle)$ and $G_l \times G_m$ are isomorphic.
Hint: Use the automorphism $: f(x) \to f(x-1)$, for all $f(x) \in \mathbb{Z}_2[x]$, of the ring $\mathbb{Z}_2[x]$.

List the invariant factors of $U(\mathbb{Z}_2[x]/\langle x^{25}(x-1)^{32} \rangle)$.

(e) Let $p(x)$ be a polynomial of positive degree n over a field F and let m be a positive integer. Let $f(x)$ be a polynomial of degree less than mn over F. Show that there are unique polynomials $r_i(x)$ over F with

$\deg r_i(x) < n$ for $0 \le i < m$ such that

$$f(x) = \sum_{i=0}^{m-1} r_i(x) p(x)^i.$$

Let $p(x)$ be monic irreducible polynomial of degree n over the finite field $\mathbb{F}_q$ where $q = p_0^l$ (p_0 prime) and let m be a positive integer. Using the notation of (a) above write $R_m = \mathbb{F}_q[x]/\langle p(x)^m \rangle$ and $G_m = U(R_m) = H_m \oplus H_0$. Using the method of (c) above show $|p_0^j H_m| = q^{nt_j}$ where $t_j = \lfloor (p_0^j - 1)m/p_0^j \rfloor$ for $j \ge 0$.

Find a formula for the number s_j of invariant factors p_0^j of H_m.

Specify the invariant factors of H_{11} and G_{11} in the case $n = 2$, $q = 9$.

4.2 Equivalence of Matrices over $F[x]$

The equivalence of matrices over $\mathbb{Z}$ is covered in Chapter 1: by Theorem 1.11 and Corollary 1.20: every $s \times t$ matrix A over $\mathbb{Z}$ is equivalent to a unique $s \times t$ matrix $S(A)$ in Smith normal form Definition 1.6. Also the reduction of A to $S(A)$ can be carried out by a finite sequence of elementary operations over $\mathbb{Z}$, this process being a generalisation of the Euclidean algorithm. Here we show that this method applies, almost unchanged, to matrices with entries from the ring $F[x]$ of polynomials in x over a given field F. The analogy rests on the fact that polynomials over F have the division property Theorem 4.1 just as integers do.

Let F be a field and let $\mathfrak{M}_{s \times t}(F[x])$ denote the set of all $s \times t$ matrices $A(x) = (a_{ij}(x))$ over $F[x]$. The (i, j)-entry in $A(x)$ is the polynomial $a_{ij}(x)$ over F for $1 \le i \le s$, $1 \le j \le t$. We consider the effect of applying *elementary row operations* (*eros*) *over* $F[x]$ and *elementary column operations* (*ecos*) *over* $F[x]$ to the matrix $A(x)$, that is, operations of the following types:

(i) interchange of two rows or two columns

(ii) multiplication of a row or column by a non-zero element of F

(iii) addition of a multiple $f(x)$ of a row/column to a different row/column where $f(x) \in F[x]$.

The reader should compare the above with the familiar *eros* and *ecos* over $\mathbb{Z}$ of Section 1.1. Notice that (ii) says multiplication of a row or column by any non-zero scalar (invertible element of $F[x]$) is a permitted elementary operation over $F[x]$.

As before $r_i \leftrightarrow r_{i'}$ denotes the interchange of row i and row i' ($i \ne i'$) and $c_j \leftrightarrow c_{j'}$ denotes the interchange of col j and col j' ($j \ne j'$). For $a \in F^*$ denote by ar_i and ac_j respectively the elementary operations over $F[x]$ of multiplying row i and col j by a. Also $r_i + f(x)r_{i'}$ denotes the addition to row i of $f(x)$ times row i' ($i \ne i'$), and $c_j + f(x)c_{j'}$ denotes adding $f(x)$ times col j' to col j ($j \ne j'$).

It should come as no surprise that all elementary operations over $F[x]$, regarded as mappings of the set $\mathfrak{M}_{s \times t}(F[x])$ of $s \times t$ matrices over $F[x]$, are invertible and their inverses are elementary operations over $F[x]$ of the same type. For instance the inverse of ar_i is $(1/a)r_i$, the inverse of $c_j + f(x)c_{j'}$ is $c_j - f(x)c_{j'}$, and the inverse of $r_i \leftrightarrow r_{i'}$ is $r_i \leftrightarrow r_{i'}$ again.

A $t \times t$ matrix which results on applying a single *ero* over $F[x]$ to the $t \times t$ identity matrix I is called *an elementary matrix over* $F[x]$. For example

$$\begin{pmatrix} 0 & 1 \\ 1 & 0 \end{pmatrix}, \quad \begin{pmatrix} 1 & 0 \\ 0 & 2 \end{pmatrix}, \quad \begin{pmatrix} 1 & 0 \\ 0 & 1/2 \end{pmatrix}, \quad \begin{pmatrix} 1 & -x \\ 0 & 1 \end{pmatrix}, \quad \begin{pmatrix} 1 & 0 \\ 2x + 1/3 & 1 \end{pmatrix}$$

are the elementary matrices over $\mathbb{Q}[x]$ which result on applying respectively the *eros* $r_1 \leftrightarrow r_2, 2r_2, (1/2)r_2, r_1 - xr_2, r_2 + (2x + 1/3)r_1$ to the 2×2 identity matrix I. Each elementary matrix over $F[x]$ can be obtained equally well by applying a single *eco* over $F[x]$ to I. The preceding five elementary matrices over $\mathbb{Q}[x]$ arise from I by applying the *ecos* $c_1 \leftrightarrow c_2, 2c_2, (1/2)c_2, c_2 - xc_1, c_1 + (2x + 1/3)c_2$ respectively. Elementary matrices over $F[x]$ are invertible over $F[x]$ and their inverses are again elementary matrices over $F[x]$.

An *ero* and an *eco* over $F[x]$ which produce equal matrices on being applied to I are said to be *paired*. Therefore $r_i \leftrightarrow r_j, c_i \leftrightarrow c_j$ are paired, ar_i, ac_i are paired, and (watch those suffices!) $r_i + f(x)r_j, c_j + f(x)c_i$ are paired where $1 \le i \le t, 1 \le j \le t$, $i \ne j$, and $a \in F^*$. Conversely each elementary matrix over $F[x]$ corresponds to (arises from) a paired *ero* and *eco* over $F[x]$.

As in Chapter 1 an *ero* and an *eco* over $F[x]$ which produce inverse matrices on being applied to I are said to be *conjugate*. Thus $r_i \leftrightarrow r_j, c_i \leftrightarrow c_j$ are conjugate, ar_i, $a^{-1}c_i$ are conjugate, and (watch the suffices and signs) $r_i + f(x)r_j, c_j - f(x)c_i$ are conjugate where $1 \le i \le t, 1 \le j \le t, i \ne j$, and $a \in F^*$.

Having spelt out the analogous details in Chapter 1, the reader will soon become aware of a marked 'do-it-yourself' attitude to the proofs here. The analogy between the rings $\mathbb{Z}$ and $F[x]$ is close (they are examples of *Euclidean rings*, that is, PIDs in which gcds can be calculated by a generalised Euclidean algorithm). Further, the theory of equivalence of matrices over $F[x]$ is almost identical to the theory covered in Chapter 1. Thus the following principle holds:

pre/postmultiplication of a matrix $A(x)$ over $F[x]$ by an elementary matrix over $F[x]$ carries out the corresponding *ero/eco* over $F[x]$.

Our next lemma is the polynomial analogue of Lemma 1.4.

Lemma 4.10

Let $A(x)$ be an $s \times t$ matrix over $F[x]$ where F is a field. Let $P_1(x)$ be an elementary $s \times s$ matrix over $F[x]$ and let $Q_1(x)$ be an elementary $t \times t$ matrix over $F[x]$. The result of applying to $A(x)$ the *ero* corresponding to $P_1(x)$ is $P_1(x)A(x)$. Applying to $A(x)$ the *eco* corresponding to $Q_1(x)$ produces $A(x)Q_1(x)$.

We leave the reader to construct a proof of Lemma 4.10 by referring back to Lemma 1.4 and Exercises 1.2, Questions 1(c) and (d) (see Exercises 4.2, Question 6(a)). The determinants of elementary matrices corresponding to type (i) and type (iii) elementary operations over $F[x]$ are -1 and $+1$ respectively. Also elementary matrices over $F[x]$ corresponding to the type (ii) operations ar_i and ac_j have determinant $a \in F^*$.

The polynomial analogue of Definition 1.5 is:

Definition 4.11

Let F be a field. The $s \times t$ matrices $A(x)$ and $B(x)$ over $F[x]$ are called *equivalent* and we write $A(x) \equiv B(x)$ if there is an $s \times s$ matrix $P(x)$ and a $t \times t$ matrix $Q(x)$, both invertible over $F[x]$, satisfying

$$P(x)A(x)Q(x)^{-1} = B(x).$$

As the name and notation suggest $\equiv$ is an equivalence relation on the set $\mathfrak{M}_{s \times t}(F[x])$ of all $s \times t$ matrices over $F[x]$. Notice that two 1×1 matrices over $F[x]$ are equivalent if and only if their entries are associate (Exercises 3.1, Question 7a(i)). Is there a method of determining whether two given $s \times t$ matrices $A(x)$ and $B(x)$ over $F[x]$ are equivalent or not? The answer is: Yes! We discuss the details next. It boils down to finding the 'simplest' matrix $S(A(x))$ in the equivalence class of $A(x)$ and the simplest matrix $S(B(x))$ in the equivalence class of $B(x)$. Then

$$A(x) \equiv B(x) \quad \Leftrightarrow \quad S(A(x)) = S(B(x)).$$

The reader should not be surprised to learn that

$$S(A(x)) \text{ is called the Smith normal form of } A(x).$$

The polynomial analogue of Definition 1.6 is:

Definition 4.12

Let $D(x)$ be an $s \times t$ matrix over $F[x]$ such that
(i) all (i, j)-entries in $D(x)$ are zero for $i \neq j$, that is, $D(x)$ is a diagonal matrix,
(ii) each (i, i)-entry $d_i(x)$ in $D(x)$ is either monic or zero,
(iii) $d_i(x) | d_{i+1}(x)$ for $1 \leq i < \min\{s, t\}$.
Then $D(x) = \text{diag}(d_1(x), d_2(x), \ldots, d_{\min\{s, t\}}(x))$ is said to be in *Smith normal form*.

It turns out (see Theorem 4.16 and Corollary 4.19) that the equivalence class of $A(x)$ contains a *unique* matrix $D(x)$ as in Definition 4.12 and so it makes sense to write $D(x) = S(A(x))$ and refer to $d_1(x), d_2(x), \ldots, d_{\min\{s,t\}}(x)$ as *the invariant factors of* $A(x)$.

It follows from Theorem 4.16 that $A(x)$ can be changed into (reduced to) $S(A(x))$ by means of a finite sequence of elementary operations over $F[x]$. Write

$$P(x) = P_u(x) P_{u-1}(x) \cdots P_2(x) P_1(x)$$

where $P_k(x)$ is the elementary matrix over $F[x]$ corresponding to the kth *ero* used in the reduction. Write

$$Q(x) = Q_v(x)^{-1} Q_{v-1}(x)^{-1} \cdots Q_2(x)^{-1} Q_1(x)^{-1}$$

where $Q_l(x)$ is the elementary matrix corresponding to the lth *eco* used in the reduction and so $Q(x)^{-1} = Q_1(x) Q_2(x) \cdots Q_{v-1}(x) Q_v(x)$. Both $P(x)$ and $Q(x)$ are invertible over $F[x]$ being products of elementary matrices over $F[x]$. Applying Lemma 4.10 $u + v$ times we obtain the polynomial analogue of Corollary 1.13 namely

$$P_u(x) \cdots P_2(x) P_1(x) A(x) Q_1(x) Q_2(x) \cdots Q_v(x) = S(A(x)),$$

that is,

$$P(x) A(x) Q(x)^{-1} = S(A(x))$$

as in Definition 4.11. So the above equation shows $A(x) \equiv S(A(x))$. The equation $P(x) = P_u(x) P_{u-1}(x) \cdots P_2(x) P_1(x) I$ together with u applications of Lemma 4.10 tell us that $P(x)$ can be calculated by applying in sequence the u *eros* used in the reduction to the $s \times s$ identity matrix I over $F[x]$. Also the equation $Q(x) = Q_v(x)^{-1} Q_{v-1}(x)^{-1} \cdots Q_2(x)^{-1} Q_1(x)^{-1} I$ and v applications of Lemma 4.10 tell us that $Q(x)$ can be calculated by applying in sequence the *conjugates* of the v *ecos* used in the reduction to the $t \times t$ identity matrix I over $F[x]$. So it is 'business as usual' with polynomials over F being the matrix entries in place of integers.

Following closely the theory in Section 1.2 we detail the reduction of 1×2 matrices over $F[x]$.

Lemma 4.13

Let $A(x) = (r_1(x), r_2(x))$ be a 1×2 matrix over $F[x]$ where F is a field.
  (i) There is a sequence of *ecos* over $F[x]$ which reduces $A(x)$ to $S(A(x)) = (d(x), 0)$ where $d(x) = \gcd\{r_1(x), r_2(x)\}$.
 (ii) There is a corresponding invertible matrix

$$Q(x) = \begin{pmatrix} q_{11}(x) & q_{12}(x) \\ q_{21}(x) & q_{22}(x) \end{pmatrix}$$

over $F[x]$ satisfying $A(x) = S(A(x))Q(x)$.
Also $q_{22}(x)r_1(x) - q_{21}(x)r_2(x) = |Q(x)|d(x)$.

Proof

(i) The 1×2 matrix $(d(x), 0)$ is in Smith normal form Definition 4.12 as $d(x)$ is either monic or zero. Suppose $r_1(x) = r_2(x) = 0$; in this case no $ecos$ over $F[x]$ are needed as $A(x) = (0, 0) = S(A(x))$. Otherwise we may assume $r_1(x) \neq 0$ applying $c_1 \leftrightarrow c_2$ if necessary. Suppose $r_1(x)$ and $d(x)$ are associate; in this case the $ecos$ $c_2 - (r_2(x)/r_1(x))c_1$, $(d(x)/r_1(x))c_1$ change $A(x) = (r_1(x), r_2(x))$ into $(d(x), 0)$.

Suppose now that $r_1(x) \neq 0$ and $d(x)$ are *not* associate. Then $r_2(x) \neq 0$ also. We apply the Euclidean algorithm to $(r_1(x), r_2(x))$ as set out following Corollary 4.6: there are polynomials $r_3(x), r_4(x), \ldots, r_k(x), r_{k+1}(x) = 0$ over F such that $r_{i-1}(x) = q_i(x)r_i(x) + r_{i+1}(x)$ where $\deg r_{i+1}(x) < \deg r_i(x)$ and $q_i(x) \in F[x]$ for $1 \leq i \leq k$. Then $r_k(x) \equiv d(x)$, that is, $d(x) = \bar{r}_k(x)$. Write $a = r_k(x)/d(x)$ for the leading coefficient of $r_k(x)$.

Applying to $A(x)$ the sequence of $k - 1$ $ecos$:

$$c_1 - q_2(x)c_2, \quad\quad c_2 - q_3(x)c_1, \quad\quad c_1 - q_4(x)c_2, \quad\quad c_2 - q_5(x)c_1, \quad \cdots$$

terminating in either $c_2 - q_k(x)c_1$ (k odd) or $c_1 - q_k(x)c_2$ (k even) produces

$$A(x) = (r_1(x), r_2(x)) \equiv (r_3(x), r_2(x)) \equiv (r_3(x), r_4(x))$$

$$\equiv (r_5(x), r_4(x)) \equiv (r_5(x), r_6(x)) \equiv \cdots$$

terminating in either $(r_k(x), 0)$ or $(0, r_k(x))$. In the first case (k odd) the eco $a^{-1}c_1$ finishes the reduction by changing $(r_k(x), 0)$ into $(d(x), 0)$. In the second case (k even) the $ecos$ $c_1 \leftrightarrow c_2$ and $a^{-1}c_1$ change $(0, r_k(x))$ into $(d(x), 0)$.

So $A(x)$ can be reduced to $S(A(x))$ by a finite sequence of v $ecos$ over $F[x]$.

(ii) In the case $A(x) = S(A(x))$ we take $Q(x) = I$ the 2×2 identity matrix over $F[x]$. Otherwise let $Q_l(x)$ be the elementary matrix corresponding to the lth eco used in the reduction of $A(x)$ to $S(A(x))$ for $1 \leq l \leq v$. By the foregoing theory $Q(x) = Q_v(x)^{-1}Q_{v-1}(x)^{-1} \cdots Q_2(x)^{-1}Q_1(x)^{-1}$ and also $\det Q(x) = \pm a$. Comparing entries in $A(x) = S(A(x))Q(x)$ gives $r_1(x) = d(x)q_{11}(x)$, $r_2(x) = d(x)q_{12}(x)$. Multiplying $q_{11}(x)q_{22}(x) - q_{12}(x)q_{21}(x) = |Q(x)| = \pm a$ by $d(x)$ gives $q_{22}(x)r_1(x) - q_{21}(x)r_2(x) = |Q(x)|d(x)$. $\square$

Note that the matrices $T_i = \begin{pmatrix} q_i(x) & 1 \\ 1 & 0 \end{pmatrix}$ used in the description of the Euclidean algorithm following Corollary 4.6 are not elementary. In fact $T_i = \begin{pmatrix} 0 & 1 \\ 1 & 0 \end{pmatrix}\begin{pmatrix} 1 & 0 \\ q_i(x) & 1 \end{pmatrix}$ shows that T_i is the product of two elementary matrices over $F[x]$. The reader can check that at most $k + 1$ $ecos$ are needed in the reduction of the non-zero matrix

$A(x) = (r_1(x), r_2(x))$ to $S(A(x))$ where $k \leq \max\{\deg r_1(x), \deg r_2(x)\} + 2$. From Lemmas 4.10 and 4.13 a method of constructing $Q(x)$ step by step is obtained, one step for each *eco* used in the reduction. We will see in Section 6.1 that $Q(x)$ is key to the theory of similarity.

As an example of Lemma 4.13 consider $A(x) = (r_1(x), r_2(x))$ where $r_1(x) = x^4 + 3x^2 - 2x + 2$, $r_2(x) = x^3 + x^2 + 3x$ over $\mathbb{Q}[x]$. We have already applied the Euclidean algorithm to these polynomials (before Definition 4.7): in this case $k = 5$ and $q_2(x) = x - 1$, $q_3(x) = x$, $q_4(x) = x + 1$, $q_5(x) = (1/2)x$ and $r_5(x) = 2$. By Lemma 4.13 the sequence of *ecos*: $c_1 - (x - 1)c_2$, $c_2 - xc_1$, $c_1 - (x + 1)c_2$, $c_2 - (x/2)c_1$, $(1/2)c_1$ reduces $A(x)$ to $S(A(x)) = (1, 0)$. Applying in sequence the *conjugates* of these *ecos* to I produces

$$
\begin{pmatrix} 1 & 0 \\ 0 & 1 \end{pmatrix} \underset{r_2 + (x-1)r_1}{\equiv} \begin{pmatrix} 1 & 0 \\ x - 1 & 1 \end{pmatrix} \underset{r_1 + xr_2}{\equiv} \begin{pmatrix} x^2 - x + 1 & x \\ x - 1 & 1 \end{pmatrix}
$$

$$
\underset{r_2 + (x+1)r_1}{\equiv} \begin{pmatrix} x^2 - x + 1 & x \\ x^3 + x & x^2 + x + 1 \end{pmatrix}
$$

$$
\underset{r_1 + (x/2)r_2}{\equiv} \begin{pmatrix} (x^4 + 3x^2 - 2x + 2)/2 & (x^3 + x^2 + 3x)/2 \\ x^3 + x & x^2 + x + 1 \end{pmatrix}
$$

$$
\underset{2r_1}{\equiv} \begin{pmatrix} x^4 + 3x^2 - 2x + 2 & x^3 + x^2 + 3x \\ x^3 + x & x^2 + x + 1 \end{pmatrix} = Q(x)
$$

which is invertible over $\mathbb{Q}[x]$ and satisfies $A(x)Q(x)^{-1} = S(A(x))$. The reader should check $\det Q(x) = 2$ which gives $((x^2 + x + 1)/2)r_1(x) - ((x^3 + x)/2)r_2(x) = 1 = d(x)$, that is, the reduction produces polynomials $a_1(x) = (x^2 + x + 1)/2$ and $a_2(x) = -(x^3 + x)/2$ satisfying $a_1(x)r_1(x) + a_2(x)r_2(x) = \gcd\{r_1(x), r_2(x)\}$ as in Corollary 4.6.

In fact the reduction of every 1×2 matrix $A(x)$ over $F[x]$ to its Smith normal form $S(A(x))$ can be carried out using only ecos of type (iii) over $F[x]$. So the matrix $Q(x)$ of Lemma 4.13(ii) can be chosen to have determinant 1, establishing the polynomial analogue of Lemma 1.8 (Exercises 4.2, Question 3(e)).

Suppose now that $A(x)$ is a 2×1 matrix over $F[x]$. The matrix transpose of Lemma 4.13 tells us that there is a finite sequence of *eros* over $F[x]$ which reduces $A(x)$ to its Smith normal form $S(A(x))$; also there is a corresponding matrix $P(x)$, invertible over $F[x]$, satisfying $P(x)A(x) = S(A(x))$. Comparing entries gives $a_1(x)r_1(x) + a_2(x)r_2(x) = \gcd\{r_1(x), r_2(x)\}$ where $(a_1(x), a_2(x))$ is row 1 of $P(x)$.

As an example consider $A = \begin{pmatrix} r_1(x) \\ r_2(x) \end{pmatrix}$ where $r_1(x) = x^3 + \bar{3}x^2 + \bar{4}x + \bar{2}$ and $r_2(x) = x^3 + \bar{2}x^2 + \bar{3}x + \bar{3}$ over the field $\mathbb{Z}_5$. Applying the Euclidean algorithm to $r_1(x)$ and $r_2(x)$ (see before Definition 4.7) gives $k = 4$, $q_2(x) = \bar{1}$, $q_3(x) = x + \bar{1}$, $q_4(x) = \bar{2}x + \bar{1}$ and $r_4(x) = \bar{3}x + \bar{4}$. So $a = \bar{3}$ and $a^{-1} = \bar{2}$. The sequence of *eros*:

$$
r_1 - r_2, \qquad r_2 - (x + \bar{1})r_1, \qquad r_1 - (\bar{2}x + \bar{1})r_2, \qquad r_1 \leftrightarrow r_2, \qquad \bar{2}r_1
$$

changes $A(x)$ into $S(A(x)) = \left(\begin{smallmatrix} x+\bar{3} \\ 0 \end{smallmatrix}\right)$. Applying this (unchanged) sequence of *eros* to the 2×2 identity matrix I over $\mathbb{Z}_5[x]$, produces:

$$\begin{pmatrix} \bar{1} & \bar{0} \\ \bar{0} & \bar{1} \end{pmatrix} \underset{r_1 - r_2}{\equiv} \begin{pmatrix} \bar{1} & \bar{4} \\ \bar{0} & \bar{1} \end{pmatrix} \underset{r_2 - (x+\bar{1})r_1}{\equiv} \begin{pmatrix} \bar{1} & \bar{4} \\ \bar{4}x + \bar{4} & x + \bar{2} \end{pmatrix}$$

$$\underset{r_1 - (\bar{2}x+\bar{1})r_2}{\equiv} \begin{pmatrix} \bar{2}x^2 + \bar{3}x + \bar{2} & \bar{3}x^2 + \bar{2} \\ \bar{4}x + \bar{4} & x + \bar{2} \end{pmatrix}$$

$$\underset{r_1 \leftrightarrow r_2}{\equiv} \begin{pmatrix} \bar{4}x + \bar{4} & x + \bar{2} \\ \bar{2}x^2 + \bar{3}x + \bar{2} & \bar{3}x^2 + \bar{2} \end{pmatrix}$$

$$\underset{\bar{2}r_1}{\equiv} \begin{pmatrix} \bar{3}x + \bar{3} & \bar{2}x + \bar{4} \\ \bar{2}x^2 + \bar{3}x + \bar{2} & \bar{3}x^2 + \bar{2} \end{pmatrix} = P(x)$$

satisfying $P(x)A(x) = \left(\begin{smallmatrix} x+\bar{3} \\ 0 \end{smallmatrix}\right) = S(A(x))$.

We now tackle the reduction to Smith normal form of a general matrix over $F[x]$.

Lemma 4.14

Let $A(x)$ be an $s \times t$ matrix over $F[x]$ where F is a field. Then $A(x)$ can be changed into $B(x) = (b_{ij}(x))$ using elementary operations over $F[x]$ where $b_{1j}(x) = b_{i1}(x) = 0$ for $1 < i \leq s$, $1 < j \leq t$ and $b_{11}(x)$ is either monic or zero.

As with Lemma 4.10 we leave the diligent reader to construct a proof of Lemma 4.14 based on its integer analogue Lemma 1.9 using Lemma 4.13(i) in place of Lemma 1.7 (see Exercises 4.2, Question 6(b)). Using induction on $\min\{s, t\}$ and Lemma 4.14, a general $s \times t$ matrix $A(x)$ over $F[x]$ can be reduced to a diagonal matrix over $F[x]$. How then can a diagonal matrix over $F[x]$ be reduced to Smith normal form? The answer is provided by the polynomial analogue of Lemma 1.10. For once we give the proof in detail if only to show how close it is to the integer version.

Lemma 4.15

Let F be a field. The diagonal matrix

$$A(x) = \begin{pmatrix} f_1(x) & 0 \\ 0 & f_2(x) \end{pmatrix}$$

over $F[x]$, where $f_i(x)$ is either monic or zero $(i = 1, 2)$, can be reduced to Smith normal form by at most five elementary operations of type (iii) over $F[x]$.

Proof

We mimic the proof of Lemma 1.10 step by step. There is nothing to do in the case $f_1(x)|f_2(x)$ as $A(x)$ is already in Smith normal form. Otherwise let $d(x) = \gcd\{f_1(x), f_2(x)\}$. By Corollary 4.6 there are polynomials $a_1(x)$ and $a_2(x)$ over F with $a_1(x)f_1(x) + a_2(x)f_2(x) = d(x)$. Then the sequence of *eros* and *ecos* over $F[x]$

$$c_2 + a_1(x)c_1, \qquad r_1 + a_2(x)r_2, \qquad c_1 - (f_1(x)/d(x) - 1)c_2,$$

$$c_2 - c_1, \qquad r_2 + (f_2(x)/d(x))(f_1(x)/d(x) - 1)r_1$$

changes $A(x)$ into

$$\begin{pmatrix} d(x) & 0 \\ 0 & f_1(x)f_2(x)/d(x) \end{pmatrix}$$

which is in Smith normal form.   □

The polynomial $f_1(x)f_2(x)/d(x)$ is *the least common multiple* of the monic polynomials $f_1(x)$ and $f_2(x)$ over F where $d(x) = \gcd\{f_1(x), f_2(x)\}$ and we write

$$\mathrm{lcm}\{f_1(x), f_2(x)\} = f_1(x)f_2(x)/d(x)$$

(see Exercises 4.1, Question 1(d)). Note that $\mathrm{lcm}\{f_1(x), 0\} = 0$ for all $f_1(x) \in F[x]$. It follows from Lemma 4.15 that

$$S(\mathrm{diag}(f_1(x), f_2(x))) = \mathrm{diag}(\gcd\{f_1(x), f_2(x)\}, \mathrm{lcm}\{f_1(x), f_2(x)\})$$

for all polynomials $f_1(x)$ and $f_2(x)$ over F.

From Theorem 4.4 all ideals of $F[x]$ are principal. The set equalities

$$\langle f_1(x)\rangle + \langle f_2(x)\rangle = \langle \gcd\{f_1(x), f_2(x)\}\rangle,$$

$$\langle f_1(x)\rangle \cap \langle f_2(x)\rangle = \langle \mathrm{lcm}\{f_1(x), f_2(x)\}\rangle$$

encapsulate the divisor properties of gcd and lcm.

Lemmas 4.10, 4.14 and 4.15 can be used to establish the polynomial analogues of Theorem 1.11 and Corollary 1.13 (see Exercises 4.2, Question 6(c)):

Theorem 4.16 (The existence of the Smith normal form over $F[x]$)

Let $A(x)$ be an $s \times t$ matrix over $F[x]$ where F is a field. There is a sequence of elementary operations over $F[x]$ which changes $A(x)$ into $S(A(x))$ in Smith normal form. There are invertible matrices $P(x)$ and $Q(x)$ over $F[x]$ satisfying

$$P(x)A(x)Q(x)^{-1} = S(A(x)).$$

As before we leave the conscientious reader to construct a proof Theorem 4.16 based on that of Theorem 1.11 (see Exercises 4.2, Question 6(c)). As explained following Definition 4.12, suitable matrices $P(x)$ and $Q(x)$ can be calculated from the *eros* and *ecos* used in the reduction of $A(x)$ to Smith normal form.

We illustrate the process by reducing the following 3×3 matrix $A(x)$ over $\mathbb{Q}[x]$:

$$A(x) = \begin{pmatrix} x^2 & x & x \\ x^3 + 2x^2 & x^2 + x & x^2 + 2x \\ 2x^3 + x + 1 & x^2 & 2x^2 \end{pmatrix}$$

$$\underset{\substack{c_1 - xc_2 \\ c_1 \leftrightarrow c_2}}{\equiv} \begin{pmatrix} x & 0 & x \\ x^2 + x & x^2 & x^2 + 2x \\ x^2 & x^3 + x + 1 & 2x^2 \end{pmatrix}$$

$$\underset{c_3 - c_1}{\equiv} \begin{pmatrix} x & 0 & 0 \\ x^2 + x & x^2 & x \\ x^2 & x^3 + x + 1 & x^2 \end{pmatrix} \underset{\substack{r_2 - (x+1)r_1 \\ r_3 - xr_1}}{\equiv} \begin{pmatrix} x & 0 & 0 \\ 0 & x^2 & x \\ 0 & x^3 + x + 1 & x^2 \end{pmatrix}$$

$$\underset{\substack{c_2 - xc_3 \\ c_2 \leftrightarrow c_3}}{\equiv} \begin{pmatrix} x & 0 & 0 \\ 0 & x & 0 \\ 0 & x^2 & x + 1 \end{pmatrix} \underset{r_3 - xr_2}{\equiv} \begin{pmatrix} x & 0 & 0 \\ 0 & x & 0 \\ 0 & 0 & x + 1 \end{pmatrix} = \operatorname{diag}(x, x, x + 1).$$

Applying the sequence: $r_2 + r_3$, $c_3 - c_2$, $c_2 \leftrightarrow c_3$, $c_3 - xc_2$, $r_3 - (x+1)r_2$, $-r_3$ to $\operatorname{diag}(x, x, x+1)$ produces $\operatorname{diag}(x, 1, x(x+1))$. Finally applying $r_1 \leftrightarrow r_2$, $c_1 \leftrightarrow c_2$ to $\operatorname{diag}(x, 1, x(x+1))$ produces $S(A(x)) = \operatorname{diag}(1, x, x(x+1))$ in Smith normal form.

A matrix $P(x)$ as in Theorem 4.16 is now found by applying, in sequence, the *eros* used above to the 3×3 identity matrix I over $\mathbb{Q}[x]$:

$$\begin{pmatrix} 1 & 0 & 0 \\ 0 & 1 & 0 \\ 0 & 0 & 1 \end{pmatrix} \underset{\substack{r_2 - (x+1)r_1 \\ r_3 - xr_1}}{\equiv} \begin{pmatrix} 1 & 0 & 0 \\ -x - 1 & 1 & 0 \\ -x & 0 & 1 \end{pmatrix} \underset{\substack{r_3 - xr_2 \\ r_2 + r_3}}{\equiv} \begin{pmatrix} 1 & 0 & 0 \\ x^2 - x - 1 & -x + 1 & 1 \\ x^2 & -x & 1 \end{pmatrix}$$

$$\underset{r_3 - (x+1)r_2}{\equiv} \begin{pmatrix} 1 & 0 & 0 \\ x^2 - x - 1 & -x + 1 & 1 \\ -x^3 + x^2 + 2x + 1 & x^2 - x - 1 & -x \end{pmatrix}$$

$$\underset{\substack{-r_3 \\ r_1 \leftrightarrow r_2}}{\equiv} \begin{pmatrix} x^2 - x - 1 & -x + 1 & 1 \\ 1 & 0 & 0 \\ x^3 - x^2 - 2x - 1 & -x^2 + x + 1 & x \end{pmatrix} = P(x).$$

A matrix $Q(x)$ as in Theorem 4.16 is now found by applying, in sequence, the conjugates of the *ecos* used above to the 3×3 identity matrix I over $\mathbb{Q}[x]$:

$$I = \begin{pmatrix} 1 & 0 & 0 \\ 0 & 1 & 0 \\ 0 & 0 & 1 \end{pmatrix} \underset{\substack{r_2+xr_1 \\ r_1 \leftrightarrow r_2}}{\equiv} \begin{pmatrix} x & 1 & 0 \\ 1 & 0 & 0 \\ 0 & 0 & 1 \end{pmatrix} \underset{\substack{r_1+r_3 \\ r_3+xr_2}}{\equiv} \begin{pmatrix} x & 1 & 1 \\ 1 & 0 & 0 \\ x & 0 & 1 \end{pmatrix} \underset{\substack{r_2 \leftrightarrow r_3 \\ r_2+r_3}}{\equiv} \begin{pmatrix} x & 1 & 1 \\ x+1 & 0 & 1 \\ 1 & 0 & 0 \end{pmatrix}$$

$$\underset{\substack{r_2 \leftrightarrow r_3 \\ r_2+xr_3}}{\equiv} \begin{pmatrix} x & 1 & 1 \\ x^2+x+1 & 0 & x \\ x+1 & 0 & 1 \end{pmatrix} \underset{r_1 \leftrightarrow r_2}{\equiv} \begin{pmatrix} x^2+x+1 & 0 & x \\ x & 1 & 1 \\ x+1 & 0 & 1 \end{pmatrix} = Q(x).$$

It is routine to check $P(x)A(x) = S(A(x))Q(x)$. In this case $\det P(x) = \det Q(x) = 1$ and so $P(x)$ and $Q(x)$ are invertible over $\mathbb{Q}[x]$ and satisfy $P(x)A(x)Q(x)^{-1} = S(A(x))$ as in Theorem 4.16.

The polynomial analogue of Corollary 1.14 is:

Lemma 4.17

Let $P(x)$ be an invertible $s \times s$ matrix over $F[x]$ where F is a field. Then $P(x)$ is expressible as a product of elementary matrices over $F[x]$. The invertible matrix $P(x)$ can be reduced to the $s \times s$ identity matrix $I = S(P(x))$ using only *eros* over $F[x]$. Also $P(x)$ reduces to the $s \times s$ identity matrix I using only *ecos* over $F[x]$.

The proof of Lemma 4.17 is left as an exercise for the reader (Exercises 4.2, Question 6(d)). As an example consider the following 3×3 matrix $P(x)$ over $\mathbb{Q}[x]$. It is not 'obvious' at the outset that $P(x)$ is invertible over $\mathbb{Q}[x]$, but this will become clear on reducing $P(x)$ to its Smith normal form (which turns out to be I).

$$P(x) = \begin{pmatrix} 1 & 1 & 2x \\ 0 & 1 & x \\ x+1 & -x-1 & 1 \end{pmatrix}$$

$$\underset{\substack{c_2-c_1 \\ c_3-2xc_1}}{\equiv} \begin{pmatrix} 1 & 0 & 0 \\ 0 & 1 & x \\ x+1 & -2x-2 & -2x^2-2x+1 \end{pmatrix}$$

$$\underset{r_3-(x+1)r_1}{\equiv} \begin{pmatrix} 1 & 0 & 0 \\ 0 & 1 & x \\ 0 & -2x-2 & -2x^2-2x+1 \end{pmatrix} \underset{\substack{c_3-xc_2 \\ r_3+(2x+2)r_2}}{\equiv} \begin{pmatrix} 1 & 0 & 0 \\ 0 & 1 & 0 \\ 0 & 0 & 1 \end{pmatrix} = I.$$

Then $P_2(x)P_1(x)P(x)Q_1(x)Q_2(x)Q_3(x) = I$ where $P_i(x)$ and $Q_j(x)$ are the elementary matrices corresponding to the ith *ero* and jth *eco* used in the above reduction. So

$$P(x) = (P_2(x)P_1(x))^{-1} I (Q_1(x)Q_2(x)Q_3(x))^{-1}$$

$$= P_1(x)^{-1} P_2(x)^{-1} Q_3(x)^{-1} Q_2(x)^{-1} Q_1(x)^{-1}$$

showing that $P(x)$ is the product of six elementary matrices over $\mathbb{Q}[x]$, since $P_i(x)^{-1}$ and $Q_j(x)^{-1}$ are elementary matrices over $\mathbb{Q}[x]$. The equation $(P_2(x)P_1(x))(P(x)Q_1(x)Q_2(x)Q_3(x)) = I$, bracketed as indicated, shows that $P_2(x)P_1(x)$ and $P(x)Q_1(x)Q_2(x)Q_3(x)$ are inverse matrices. By Lemma 2.18 we obtain $P(x)Q_1(x)Q_2(x)Q_3(x)P_2(x)P_1(x) = I$ and hence the sequence: $c_2 - c_1$, $c_3 - 2xc_1$, $c_3 - xc_2$, $c_2 + (2x+2)c_3$, $c_1 - (x+1)c_3$, consisting of *ecos* only, reduces $P(x)$ to I by Lemma 4.10; here the two *eros* in the above reduction are replaced by their paired *ecos*. In the same way $P_2(x)P_1(x)P(x)$ and $Q_1(x)Q_2(x)Q_3(x)$ are inverses of each other. So $Q_1(x)Q_2(x)Q_3(x)P_2(x)P_1(x)P(x) = I$ showing that the sequence of *eros*: $r_3 - (x+1)r_1$, $r_3 + (2x+2)r_2$, $r_2 - xr_3$, $r_1 - 2xr_3$, $r_1 - r_2$ reduces $P(x)$ to I; here the three *ecos* in the original reduction are replaced by their paired *eros*.

Let $A(x)$ and $B(x)$ be matrices over $F[x]$ where F is a field. Combining Lemma 4.10, Definition 4.11 and Lemma 4.17 we see

$A(x) \equiv B(x) \quad \Leftrightarrow \quad A(x)$ can be obtained from $B(x)$ by a sequence of elementary operations over $F[x]$.

We next deal with the uniqueness of the Smith normal form in this context. The hard work has been done in Section 1.3, and so it is a matter of convincing the reader that no extra problems arise on replacing matrices over $\mathbb{Z}$ by matrices over $F[x]$.

Let $A(x)$ be an $s \times t$ matrix over $F[x]$ where F is a field and let l be an integer with $1 \leq l \leq \min\{s, t\}$. Suppose l rows and l columns of $A(x)$ are selected. Recall that the determinant of the $l \times l$ matrix which remains on deleting the unselected $s - l$ rows and $t - l$ columns is called an l-*minor of* $A(x)$. The 1-minors of $A(x)$ are simply the st entries in $A(x)$. The number of l-minors of $A(x)$ is $\binom{s}{l}\binom{t}{l}$ and each is a polynomial over F. As in Section 1.3 we introduce

$$g_l(A(x)) = \gcd\{\text{all } l\text{-minors of } A(x)\},$$

that is, $g_l(A(x))$ is the greatest common divisor of the set of l-minors of $A(x)$. Notice that, being a gcd, the polynomial $g_l(A(x))$ is either zero or monic ($g_l(A(x)) = 0$ $\Leftrightarrow$ all the l-minors of $A(x)$ are zero). It follows from Corollary 4.19 that $g_l(A(x))$ remains *unchanged* on applying *eros* and *ecos* to $A(x)$; so these polynomials are of great importance.

As an illustration consider

$$A(x) = \begin{pmatrix} 1 & 1 & 1 \\ 1 & x & x \\ 1 & x & x^2 \end{pmatrix}$$

over $\mathbb{Q}[x]$. Here $g_1(A(x)) = 1$ and the 2-minors are: $x - 1$, $x - 1$, 0, $x - 1$, $x^2 - 1$, $x^2 - x$, 0, $x^2 - x$, $x^3 - x^2$. Hence $g_2(A(x)) = x - 1$. The reader can verify that $g_3(A(x)) = \det A(x) = (x-1)^2 x$. It's straightforward to find the matrix $D(x) =$

$\mathrm{diag}(d_1(x), d_2(x), d_3(x))$ over $\mathbb{Q}[x]$ which is in Smith normal form and satisfies $g_l(D(x)) = g_l(A(x))$ for $l = 1, 2, 3$. In fact from the l-minors of $D(x)$ we deduce $g_1(D(x)) = d_1(x)$, $g_2(D(x)) = d_1(x)d_2(x)$, $g_3(D(x)) = d_1(x)d_2(x)d_3(x)$ and so $D(x) = \mathrm{diag}(1, x - 1, x(x - 1))$. Anticipating Corollary 4.19 we conclude $D(x) = S(A(x))$ as $D(x)$ is the only matrix which is in Smith normal form and equivalent to $A(x)$.

The polynomial analogue of Corollary 1.19 is:

Corollary 4.18

Let $B(x)$ be an $s \times r$ matrix over $F[x]$ and let $C(x)$ be an $r \times t$ matrix over $F[x]$ where F is a field. Then $g_l(B(x))$ and $g_l(C(x))$ are divisors of $g_l(B(x)C(x))$ for $1 \leq l \leq \min\{r, s, t\}$.

Notice that Corollary 4.18 is a direct consequence of the Cauchy–Binet theorem over $F[x]$ (see the discussion after Theorem 1.18). The next corollary is the polynomial analogue of Corollary 1.20 and it is just what we're looking for!

Corollary 4.19

Let $A(x)$ and $B(x)$ be $s \times t$ matrices over $F[x]$ where F is a field. Then

$$A(x) \equiv B(x) \quad \Leftrightarrow \quad g_l(A(x)) = g_l(B(x)) \quad \text{for } 1 \leq l \leq \min\{s, t\}.$$

Further $A(x)$ is equivalent to a unique matrix $S(A(x))$ over $F[x]$ in Smith normal form.

The proof of Corollary 4.19 follows closely that of its integer analogue Corollary 1.20 (see Exercises 4.2, Question 6(e)).

Definition 4.20

Let $A(x)$ be an $s \times t$ matrix over $F[x]$ where F is a field. The matrix $S(A(x)) = \mathrm{diag}(d_1(x), d_2(x), \ldots, d_{\min\{s,t\}}(x))$ is called *the Smith normal form of* $A(x)$. The polynomials $d_l(x)$ for $1 \leq l \leq \min\{s, t\}$ are called *the invariant factors of* $A(x)$.

We will see in Chapter 6 that all the invariant factors of matrices of the type $xI - A$ over $F[x]$, where A is a $t \times t$ matrix over the field F, are non-zero. Our last theorem in this section is the polynomial analogue of Theorem 1.21.

Theorem 4.21

Let $A(x)$ be an $r \times s$ matrix over $F[x]$ and let $B(x)$ be an $s \times t$ matrix over $F[x]$ where F is a field and $r \le s \le t$. Suppose all the invariant factors of $A(x)$, $B(x)$ and $A(x)B(x)$ are non-zero. Then the kth invariant factors of $A(x)$ and $B(x)$ are divisors of the kth invariant factor of $A(x)B(x)$ for $1 \le k \le r$.

As usual the reader should verify that 'nothing goes wrong' with the proof of Theorem 1.21 on replacing $\mathbb{Z}$ by $F[x]$ (see Exercises 4.2, Question 6(f)).

The theory of similarity of $t \times t$ matrices A over a field F, which is the main topic of Chapters 5 and 6, depends on the reduction of the $t \times t$ matrix $xI - A$ over $F[x]$ to its Smith normal form $S(xI - A)$. For scalar matrices, that is, matrices $A = \lambda I$ where $\lambda \in F$, no reduction is needed as

$$xI - A = (x - \lambda)I = \operatorname{diag}(x - \lambda, x - \lambda, \ldots, x - \lambda) = S(xI - A).$$

For non-scalar matrices A, the entries in $xI - A$ have gcd 1 which is the $(1, 1)$-entry in $S(xI - A)$. The worked example in 'A bird's-eye view of similarity' illustrates this point.

EXERCISES 4.2

1. (a) List the eight elementary 2×2 matrices over $\mathbb{Z}_2[x]$ having entries of degree at most 1 (this list should include the identity matrix) and the corresponding *eros*. Find a formula in terms of the prime p for the number of elementary 2×2 matrices over $\mathbb{Z}_p[x]$ having entries of degree at most 1.

   (b) Write down the *eros* over $\mathbb{Q}[x]$ which produce the elementary matrices

   $$\begin{pmatrix} 0 & 1 \\ 1 & 0 \end{pmatrix}, \quad \begin{pmatrix} 1 & 0 \\ 1 & 1 \end{pmatrix}, \quad \begin{pmatrix} 1 & x^2 \\ 0 & 1 \end{pmatrix}, \quad \begin{pmatrix} 3 & 0 \\ 0 & 1 \end{pmatrix},$$

   $$\begin{pmatrix} 1 & 0 \\ 0 & 1/3 \end{pmatrix}, \quad \begin{pmatrix} 1 & -x^2 \\ 0 & 1 \end{pmatrix}$$

   over $\mathbb{Q}[x]$. Write down the six *ecos* over $\mathbb{Q}[x]$ which are paired to these *eros*. Write down the six *ecos* over $\mathbb{Q}[x]$ which are conjugate to these *eros*.

   (c) Let F be a field. The *ero* $r_i \leftrightarrow r_j$ and the *eco* $c_i \leftrightarrow c_j$ over $F[x]$ are paired and conjugate. Which other paired (non-identity) *eros* and *ecos* over $F[x]$ are also conjugate?

Hint: Treat the cases $\chi(F) = 2$ and $\chi(F) \neq 2$ separately (see Exercises 2.3, Question 5(a)).

2. (a) Let F be a field. Use Definition 4.11 to show that $\equiv$ is an equivalence relation on the set $\mathfrak{M}_{s \times t}(F(x))$ of all $s \times t$ matrices over $F[x]$. Which elements of $\mathfrak{M}_{s \times t}(F(x))$ belong to the equivalence class of the $s \times t$ zero matrix 0 over $F[x]$? Which elements of $\mathfrak{M}_t(F(x))$ belong to the equivalence class of the $t \times t$ identity matrix I over $F[x]$?

   (b) Let F be a field and let $t \geq 3$. List the sequences of invariant factors $(d_1(x), d_2(x), d_3(x))$ of $3 \times t$ matrices over $F[x]$ where $d_3(x)|x^3$. How many equivalence classes of such matrices are there?

   (c) Let $A(x)$ be an $s \times t$ matrix over $F[x]$ where F is a field with $\chi(F) \neq 2$. Let $A(-x)$ be the matrix obtained from $A(x)$ by replacing x by $-x$ throughout. Write $S(A(x)) = \text{diag}(d_1(x), d_2(x), \ldots, d_{\min\{s,t\}}(x))$. Show that $P(x)$ is an elementary matrix over $F[x]$ if and only if $P(-x)$ is an elementary matrix over $F[x]$ and deduce from Lemma 4.17 that $A(-x) \equiv \text{diag}(d_1(-x), d_2(-x), \ldots, d_{\min\{s,t\}}(-x))$. Hence find the Smith normal form $S(A(-x))$ of $A(-x)$. State a necessary and sufficient condition on each $d_i(x)$ for $A(x)$ and $A(-x)$ to be equivalent.

   Use Corollary 4.19 to find the Smith normal form of

   $$A(x) = \begin{pmatrix} x^2 + 1 & x^2 + x + 1 \\ x^2 & x^2 + x \end{pmatrix}.$$

   Is $A(x) \equiv A(-x)$? Find $S(B(x))$ where

   $$B(x) = \begin{pmatrix} x^2 - x - 1 & x^2 \\ x^2 - x & x^2 + 1 \end{pmatrix}.$$

   Is $B(x) \equiv B(-x)$?

3. (a) Using Lemma 4.15 reduce

   $$A(x) = \begin{pmatrix} x^2(x + 1) & 0 \\ 0 & (x + 1)^2 x \end{pmatrix}$$

   over $F[x]$ to Smith normal form where F is an arbitrary field. Find 2×2 matrices $P(x)$ and $Q(x)$, invertible over $F[x]$, satisfying $P(x)A(x) = S(A(x))Q(x)$. What are the invariant factors of $A(x)^2$?

   (b) Reduce

   $$\begin{pmatrix} 0 & x^2 & 0 \\ 0 & 0 & 1 \\ x & 0 & 0 \end{pmatrix}$$

   to its Smith normal form using four elementary operations.

(c) Use Lemmas 4.14 and 4.15 to find invertible 3×3 matrices $P(x)$ and $Q(x)$ over $\mathbb{Q}[x]$ satisfying $P(x)A(x)Q(x)^{-1} = S(A(x))$ where

$$
A(x) = \begin{pmatrix} x^2 & x & x \\ x^3 - 2x^2 & x^2 - x & x^2 - 2x \\ 2x^3 + x - 1 & x^2 & 2x^2 \end{pmatrix}.
$$

(d) Let $A = (a_{ij})$ be an invertible $t \times t$ matrix over the field F and let $A(x) = (a_{ij}x^j)$. Find $S(A(x))$.

Hint: Express $A(x)$ as a product.

(e) Let F be a field and let $a \in F^*$. Reduce $\begin{pmatrix} a & 0 \\ 0 & a^{-1} \end{pmatrix}$ to $\begin{pmatrix} 1 & 0 \\ 0 & 1 \end{pmatrix}$ using *ecos* over F of type (iii).

Hint: Start with $c_2 + (a^{-1} - 1)c_1, c_1 + c_2$.

Hence show that every 1×2 matrix $A(x)$ over $F[x]$ can be reduced to its Smith normal form $S(A(x))$ by a sequence of *ecos* over $F[x]$ of type (iii). Deduce $A(x) = S(A(x))Q(x)$ where $Q(x)$ is over $F[x]$ and $\det Q(x) = 1$.

Hint: Use Exercises 1.1, Question 5.

4. (Polynomial version of Exercises 1.2, Question 6(b).) Let F be a field. A sequence $a_0(x), a_1(x), a_2(x), \ldots$ of monic polynomials over F is defined by:

$$
a_0(x) = 1, \qquad a_1(x) = x \quad \text{and}
$$

$$
a_n(x) = a_{n-1}(x)(x^2 a_{n-2}(x) + 1) \quad \text{for } n \geq 2.
$$

Write $K_n = \langle x^2 a_n(x) \rangle$. Show $\gcd\{a_n(x), xa_{n-1}(x)\} = a_{n-1}(x)$ for $n \geq 2$. Show $a_n(x)/a_{n-r+1}(x) \equiv 1 \pmod{K_{n-r}}$ where $1 \leq r \leq n$.

Using the polynomial analogue of the proof of Lemma 1.9, that is, applying Lemma 4.13 alternately to row 1 and the transpose of col 1, find a sequence of eight elementary operations over $F[x]$ which reduces $\begin{pmatrix} a_4(x) & xa_3(x) \\ 0 & -1 \end{pmatrix}$ to $\mathrm{diag}(x, -a_4(x)/x)$.

Hint: Start with $c_1 - xa_2(x)c_2$.

Let

$$
A_n(x) = \begin{pmatrix} a_n(x) & xa_{n-1}(x) \\ 0 & -1 \end{pmatrix}
$$

where $n \geq 3$. Using the polynomial analogue of the proof of Lemma 1.9, show that this method requires $2 + 2 + 3(n-3) + 1$ elementary operations over $F[x]$ to reduce $A_n(x)$ to $\mathrm{diag}(x, -a_n(x)/x)$.

Hint: Use the polynomial analogue of the matrices B_r in Exercises 1.2, Question 6(b). You'll need the *eco* $c_2 - xa_{n-r}(x)q_{n-r+1}(x)c_1$ for odd integers $r \geq 3$ and the *ero* $r_2 - xa_{n-r}(x)q_{n-r+1}(x)r_1$ for even integers $r \geq 4$ where $a_n(x)/a_{n-r+2}(x) = 1 + q_{n-r+1}(x)a_{n-r+1}(x)$, $3 \leq r < n$.

List a sequence of five elementary operations over $F[x]$ which reduce
$\mathrm{diag}(x, -a_n(x)/x)$ to $S(A_n(x)) = \mathrm{diag}(1, a_n(x))$.

Find a sequence of four elementary operations over $F[x]$ which reduces
$A_n(x)$ to $S(A_n(x))$.

5. (Polynomial version of Exercises 1.2, Question 7.) Let F be a field and let
s, t be positive integers. Write $G = GL_s(F[x]) \times GL_t(F[x])$, the external
direct product group (Exercises 2.3, Question 4(d)).

   (a) Let $D(x)$ be an $s \times t$ matrix over $F[x]$. Verify that $Z(D(x)) =$
   $\{(P(x), Q(x)) \in G : P(x)D(x) = D(x)Q(x)\}$ is a subgroup of G.

   (b) Suppose that the $s \times t$ matrix $D(x) = \mathrm{diag}(d_1(x), d_2(x), \ldots, d_s(x))$
   over $F[x]$ is in Smith normal form where $s \le t$ and $d_s(x) \ne 0$. Let
   $(P(x), Q(x)) \in G$ and write $P(x) = (p_{ij}(x))$, $Q(x) = (q_{kl}(x))$ for
   $1 \le i, j \le s, 1 \le k, l \le t$. Show

   $$(P(x), Q(x)) \in Z(D(x)) \quad \Leftrightarrow \quad p_{ij}(x)d_j(x) = d_i(x)q_{ij}(x)$$
   $$\text{for } 1 \le i, j \le s, \text{ where } q_{il}(x) = 0 \text{ for } 1 \le i \le s < l \le t.$$

   (c) You and your classmate independently reduce the $s \times t$ matrix $A(x)$
   to $D(x)$. You get $P'(x)A(x) = D(x)Q'(x)$ and your classmate gets
   $P''(x)A(x) = D(x)Q''(x)$. Verify the coset equality
   $Z(D(x))(P'(x), Q'(x))) = Z(D(x))(P''(x), Q''(x))$.

   (d) Use the two reductions of

   $$A_2(x) = \begin{pmatrix} x(x^2 + 1) & x^2 \\ 0 & -1 \end{pmatrix}$$

   suggested in Question 4 above to find an element of $Z(\mathrm{diag}(1, a_2(x))$
   of the form $(P'(x), Q'(x))(P''(x), Q''(x))^{-1}$.

   (e) Modify part (b) to cover the case $d_s(x) = 0$.
   Hint: Consider the largest non-negative integer r with $d_r(x) \ne 0$.

6. (a) Considering each type of *ero* and *eco* over $F[x]$ in turn, write out
   a proof of Lemma 4.10 based on Theorem 1.11 and Exercises 1.2,
   Questions 1(c) and 1(d).

   (b) Write out a proof of Lemma 4.14 analogous to that of Lemma 1.9
   using Lemma 4.13(i) in place of Lemma 1.7.

   (c) Write out a proof of Theorem 4.16 based on the proof of Theorem 1.11 using Lemmas 4.10, 4.14 and 4.15 in place of Lemmas 1.4,
   1.9 and 1.10.

   (d) Write out a proof of Lemma 4.17 analogous to that of Corollary 1.14.

   (e) Write out a proof of Corollary 4.19 analogous to the proof of Corollary 1.20 using Corollary 4.18 in place of Corollary 1.19.

   (f) Write out a proof of Theorem 4.21 based on the proof of Theorem 1.21.

$F[x]$-Modules: Similarity of $t \times t$ Matrices over a Field F

We begin with a review of the concept of *similarity* in the context of matrix theory. For the reader familiar with diagonalisation there is little new here. However, as we'll see in Chapter 6, diagonalisation is something of a 'red herring' as far as the general theory of similarity is concerned. The matrix $xI - A$ plays a major role but there's no need to find the zeros of the polynomial $|xI - A|$, that is, the eigenvalues of the $t \times t$ matrix A, to get started – instead the calculation of the Smith normal form of $xI - A$ is the initial objective, but this is deferred to the next chapter as there is preliminary work to be done. This preliminary work is analogous to the discussion of $\mathbb{Z}$-modules carried out in Chapter 2. Each $t \times t$ matrix A over a field F gives rise to an $F[x]$-module $M(A)$ and such modules behave in an analogous way to finite abelian groups; in particular they decompose into a direct sum of cyclic submodules. This theory comes to a climax in Section 6.1 where problems concerning similarity are miraculously solved using $M(A)$. Meanwhile here the necessary building blocks are introduced: the *order* of a module element, the *direct sum* of matrices and the important concept of the *companion matrix* of a monic polynomial with positive degree.

5.1 The $F[x]$-Module $M(A)$

Let V denote a finite-dimensional vector space over a field F and suppose given a linear mapping $\alpha : V \to V$. Write $t = \dim V$ and let $\mathcal{B}$ denote a basis $v_1, v_2, \ldots, v_t$ of V. The reader will know that the action of α on V can be expressed by a matrix.

C. Norman, *Finitely Generated Abelian Groups and Similarity of Matrices over a Field*, 203
Springer Undergraduate Mathematics Series,
DOI 10.1007/978-1-4471-2730-7_5, © Springer-Verlag London Limited 2012

Definition 5.1

The $t \times t$ matrix $A = (a_{ij})$ over F, where $(v_i)\alpha = a_{i1}v_1 + a_{i2}v_2 + \cdots + a_{it}v_t$ for $1 \leq i \leq t$, is called *the matrix of α relative to the basis $\mathcal{B}$*.

Our first task is to discover how the matrix of α changes on changing the basis $\mathcal{B}$.

Lemma 5.2

Let $\alpha : V \to V$ be a linear mapping of the t-dimensional vector space V over the field F. Let $\mathcal{B}$ and $\mathcal{B}'$ be bases of V. Denote by A and A' the $t \times t$ matrices of α relative to $\mathcal{B}$ and $\mathcal{B}'$ respectively. There is then an invertible $t \times t$ matrix X over F such that $XAX^{-1} = A'$. Suppose $V = F^t$ and $\mathcal{B} = \mathcal{B}_0$, the standard basis of F^t. In this case the rows of X are, in order, the vectors of $\mathcal{B}'$.

Proof

Let $v_1, v_2, \ldots, v_t$ denote the vectors in $\mathcal{B}$ and let $v_1', v_2', \ldots, v_t'$ denote the vectors in $\mathcal{B}'$. Each v_i' is expressible as a linear combination of the vectors in $\mathcal{B}$ and so there is a $t \times t$ matrix $X = (x_{ij})$ over F such that $v_i' = x_{i1}v_1 + x_{i2}v_2 + \cdots + x_{it}v_t$ for $1 \leq i \leq t$. In the same way each v_j is expressible as a linear combination of the vectors in $\mathcal{B}'$: so there is a $t \times t$ matrix $Y = (y_{jk})$ over F such that $v_j = y_{j1}v_1' + y_{j2}v_2' + \cdots + y_{jt}v_t'$ for $1 \leq j \leq t$. As in the proof of Theorem 2.20 the matrices X and Y are related by the equations $XY = I$ and $YX = I$ showing that X is invertible over F and $X^{-1} = Y$.

It is possible to express $(v_i')\alpha$ in terms of $\mathcal{B}$ in two ways: first using the linearity of α followed by the matrix $A = (a_{jk})$ gives

$$(v_i')\alpha = \left(\sum_{j=1}^{t} x_{ij}v_j\right)\alpha = \sum_{j=1}^{t} x_{ij}(v_j)\alpha = \sum_{j=1}^{t} x_{ij}\left(\sum_{k=1}^{t} a_{jk}v_k\right) = \sum_{k=1}^{t}\left(\sum_{j=1}^{t} x_{ij}a_{jk}\right)v_k.$$

Secondly using the matrices $A' = (a_{ij}')$ and X (in that order) we obtain

$$(v_i')\alpha = \sum_{i=1}^{t} a_{ij}'v_j' = \sum_{i=1}^{t} a_{ij}'\left(\sum_{j=1}^{t} x_{jk}v_k\right) = \sum_{i=1}^{t}\left(\sum_{j=1}^{t} a_{ij}'x_{jk}\right)v_k.$$

Equating the coefficients of v_k in the above equations gives

$$\sum_{j=1}^{t} x_{ij}a_{jk} = \sum_{j=1}^{t} a_{ij}'x_{jk}$$

and so $XA = A'X$ as the (i, k)-entries agree for $1 \leq i, k \leq t$. Postmultiplying by X^{-1} produces $XAX^{-1} = A'$.

Suppose $V = F^t$ and $v_j = e_j$ for $1 \le j \le t$, that is, the basis $\mathcal{B}$ is the standard basis $\mathcal{B}_0$ of F^t. The above equation expressing v_i' in $\mathcal{B}'$ as a linear combination of the vectors in $\mathcal{B} = \mathcal{B}_0$ is $v_i' = x_{i1}e_1 + x_{i2}e_2 + \cdots + x_{it}e_t = e_i X$ which is row i of X for $1 \le i \le t$. So the basis $\mathcal{B}'$ consists of the rows of X in their given order. $\square$

Definition 5.3

The $t \times t$ matrices A and A' over the field F are called *similar* if there is an invertible $t \times t$ matrix X over F such that $XAX^{-1} = A'$ in which case we write $A \sim A'$.

From Lemma 5.2 we see that similar matrices arise by relating a given linear mapping $\alpha : V \to V$ to different bases of V. We will see in Theorem 5.13 that the concept of similarity Definition 5.3 is analogous to that of isomorphism (in the context of finite abelian groups). For the moment we note that similarity $\sim$ is an equivalence relation on the set $\mathfrak{M}_t(F)$ of all $t \times t$ matrices over F (Exercises 5.1, Question 1(d)).

Remember (see the discussion after Lemma 2.18) that the multiplicative group of all invertible $t \times t$ matrices X over F is denoted by $GL_t(F)$.

Definition 5.4

Let $A \in \mathfrak{M}_t(F)$. The subset $\{XAX^{-1} : X \in GL_t(F)\}$ of $\mathfrak{M}_t(F)$ is called *the similarity class of* A.

So $\{A' \in \mathfrak{M}_t(F) : A' \sim A\}$ is the similarity class of A and consists of all $t \times t$ matrices A' over F which are similar to the given $t \times t$ matrix A over F. As $X0X^{-1} = 0$ for all $X \in GL_t(F)$, we see that the similarity class of the zero $t \times t$ matrix 0 over F consists of 0 alone. In the same way, as $XIX^{-1} = I$ for all $X \in GL_t(F)$, we have $\{XIX^{-1} : X \in GL_t(F)\} = \{I\}$, that is, the similarity class of the $t \times t$ identity matrix I over F has only one element, namely I itself. In Chapter 6 we develop a method of determining whether or not two matrices are similar.

The reader will know that multiplication by invertible matrices leaves the rank of a matrix unchanged and so

similar matrices have equal rank.

For instance $\left(\begin{smallmatrix} 1 & 0 \\ 0 & 0 \end{smallmatrix}\right)$ and $\left(\begin{smallmatrix} 1 & 1 \\ 0 & 1 \end{smallmatrix}\right)$ are not similar as their ranks are 1 and 2 respectively. Also $\left(\begin{smallmatrix} 1 & 1 \\ 0 & 1 \end{smallmatrix}\right)$ and $\left(\begin{smallmatrix} 1 & 0 \\ 0 & 1 \end{smallmatrix}\right) = I$ are not similar although they both have rank 2 since the first matrix does not belong to the similarity class of I.

The reader will be familiar with *the characteristic polynomial*

$$\chi_A(x) = \det(xI - A) \quad \text{of the } t \times t \text{ matrix } A \text{ over } F.$$

The zeros of $\chi_A(x)$, which is a monic polynomial of degree t over F, are the eigen-values of A. The coefficient of x^{t-1} in $\chi_A(x)$ is trace A where trace $A = a_{11} + a_{22} + \cdots + a_{tt}$ is the sum of the diagonal entries in A. Also the constant term in $\chi_A(x)$ is $(-1)^t|A|$ (Exercises 5.1, Question 2(a)).

Lemma 5.5

Similar matrices have equal characteristic polynomials.

Proof

Let A and A' be similar $t \times t$ matrices over F. By Definition 5.3 there is an invertible $t \times t$ matrix X with $XAX^{-1} = A'$. This factorisation of A' (as the product XAX^{-1}) gives rise to the factorisation of the *characteristic matrix* $xI - A'$ of A'

$$xI - A' = xXIX^{-1} - XAX^{-1} = X(xI)X^{-1} - XAX^{-1} = X(xI - A)X^{-1}.$$

Using the multiplicative property Theorem 1.18 of determinants we obtain

$$\chi_{A'}(x) = |xI - A'| = |X(xI - A)X^{-1}| = |X||xI - A||X^{-1}|$$

$$= |xI - A||X||X^{-1}| = |xI - A| = \chi_A(x)$$

as $|X||X^{-1}| = |XX^{-1}| = |I| = 1$ and the scalar $|X|$ commutes with the polynomial $|xI - A|$. So $\chi_{A'}(x) = \chi_A(x)$.   □

It follows from Lemma 5.5 that similar matrices have equal traces and determi-nants. Note that $\left(\begin{smallmatrix} 1 & 1 \\ 0 & 1 \end{smallmatrix}\right)$ and $\left(\begin{smallmatrix} 1 & 0 \\ 0 & 1 \end{smallmatrix}\right)$ are not similar but both have rank 2 and characteristic polynomial

$$(x - 1)^2 = \begin{vmatrix} x - 1 & -1 \\ 0 & x - 1 \end{vmatrix} = \begin{vmatrix} x - 1 & 0 \\ 0 & x - 1 \end{vmatrix}.$$

On the other hand

$$\begin{pmatrix} a & 0 \\ 0 & b \end{pmatrix} \sim \begin{pmatrix} b & 0 \\ 0 & a \end{pmatrix}$$

using $X = \left(\begin{smallmatrix} 0 & 1 \\ 1 & 0 \end{smallmatrix}\right)$ for all $a, b \in F$.

So what exactly must two matrices have in common in order to be similar? The answer will have to wait until Definition 6.8. Suffice it to say we already know the answer to the analogous question: what exactly must two finite abelian groups G and

G' have in common in order to be isomorphic? By Theorem 3.7 we know $G \cong G'$ if and only if their invariant factor sequences are equal.

It is time to introduce the main concept in this section. Let V be a vector space over the field F and let $\alpha : V \to V$ be a linear mapping. So V is an F-module and α is an F-linear mapping in the terminology of Section 2.3. We prepare to show Lemma 5.7 that V can be given the extra structure of an $F[x]$-module by using α.

Definition 5.6

Let V be a vector space over F and let $\alpha : V \to V$ be a linear mapping. For each polynomial $f(x) = a_n x^n + a_{n-1}x^{n-1} + \cdots + a_1 x + a_0$ over F the linear mapping $f(\alpha) : V \to V$ given by

$$(v)f(\alpha) = a_n((v)\alpha^n) + a_{n-1}((v)\alpha^{n-1}) + \cdots + a_1((v)\alpha) + a_0 v \quad \text{for all } v \in V$$

is called *the evaluation of $f(x)$ at α*.

The linear mapping α can be composed with itself a finite number of times to give $\alpha\alpha = \alpha^2$, $\alpha\alpha\alpha = \alpha^3$, ... and these positive integer powers of α are also linear mappings $\alpha^n : V \to V$. The reader can verify that $f(\alpha)$ is linear. Also

$$\text{the evaluation at } \alpha \text{ mapping } \varepsilon_\alpha : F[x] \to \text{End } V$$
$$\text{where } (f(x))\varepsilon_\alpha = f(\alpha) \text{ for all } f(x) \in F[x]$$

is a ring homomorphism (Exercises 5.1, Question 2(e)).

Lemma 5.7

Let V be a vector space over F and let $\alpha : V \to V$ be a linear mapping. Write $f(x)v = (v)f(\alpha)$ for all $f(x) \in F[x]$ and $v \in V$. Then V, with vector addition and the above product (of polynomial and vector), is an $F[x]$-module $M(\alpha)$ called the $F[x]$-*module determined by α*.

Proof

We verify the seven laws of an R-module M listed before Definition 2.19 with $R = F[x]$ and $M = M(\alpha)$. The elements of $M(\alpha)$ are no more than the vectors of V. Also addition in $M(\alpha)$ is no more than addition of vectors in V and so laws 1, 2, 3 and 4 of an abelian group are obeyed. Consider polynomials $f(x)$, $f_1(x)$, $f_2(x)$ over F and vectors v, v_1, v_2 in V. Notice $f(x)v \in V$, that is, each polynomial multiple of a vector in V is again a vector in V. As $f(\alpha)$ is additive: $f(x)(v_1 + v_2) = (v_1 + v_2)f(\alpha) =$

$(v_1)f(\alpha) + (v_2)f(\alpha) = f(x)v_1 + f(x)v_2$. Now the linear mapping $(f_1(x) + f_2(x))\varepsilon_\alpha$ is the evaluation of $f_1(x) + f_2(x)$ at α. As ε_α is additive we obtain

$$(f_1(x) + f_2(x))v = (v)((f_1(x) + f_2(x))\varepsilon_\alpha) = (v)((f_1(x))\varepsilon_\alpha + (f_2(x))\varepsilon_\alpha$$

$$= (v)(f_1(\alpha) + f_2(\alpha)) = (v)f_1(\alpha) + (v)f_2(\alpha) = f_1(x)v + f_2(x)v.$$

So module law 5 (the distributive law) is obeyed in $M(\alpha)$.

As ε_α is multiplicative and $F[x]$ is a commutative ring we see

$$(f_1(x)f_2(x))v = (f_2(x)f_1(x))v = (v)(f_2(x)f_1(x))\varepsilon_\alpha = (v)((f_2(x))\varepsilon_\alpha(f_1(x))\varepsilon_\alpha)$$

$$= (v)(f_2(\alpha)f_1(\alpha)) = ((v)f_2(\alpha))f_1(\alpha)$$

$$= (f_2(x)v)f_1(\alpha) = f_1(x)(f_2(x)v)$$

showing that module law 6 is obeyed in $M(\alpha)$.

The 1-element of $F[x]$ is the constant polynomial $1(x)$ and $(1(x))\varepsilon_\alpha = 1(\alpha)$ is the identity mapping of V as $(v)1(\alpha) = v$ for all $v \in V$ by Definition 5.6. Therefore $1(x)v = (v)1(\alpha) = v$ for all $v \in V$ showing that module law 7 is obeyed in $M(\alpha)$. Therefore $M(\alpha)$ is an $F[x]$-module.   □

The module $M(\alpha)$ is an abstract version of the module $M(A)$ which we define next.

Definition 5.8

Let A be a $t \times t$ matrix over the field F and let $\alpha : F^t \to F^t$ denote *the linear mapping determined by* A, that is, $(v)\alpha = vA$ for all $v \in F^t$. We write $M(\alpha) = M(A)$ which is called *the $F[x]$-module determined by* A. Also write

$$f(A) = a_n A^n + a_{n-1} A^{n-1} + \cdots + a_1 A + a_0 I \in \mathfrak{M}_t(F)$$

$$\text{where } f(x) = a_n x^n + a_{n-1} x^{n-1} + \cdots + a_1 x + a_0 \in F[x].$$

Denote by $\varepsilon_A : F[x] \to \mathfrak{M}_t(F)$ the ring homomorphism given by $(f(x))\varepsilon_A = f(A)$ for all $f(x) \in F[x]$. We call ε_A *the evaluation at A homomorphism*.

So $M(A)$ is a concrete version of $M(\alpha)$. Notice that $(v)f(\alpha) = (v)f(A)$ for all $v \in F^t$ where α is the linear mapping determined by A, and the matrix of α relative to the standard basis $\mathcal{B}_0$ of F^t is again A. In $M(A)$ the product of the polynomial $f(x)$ and vector (module element) v is

$$f(x)v = vf(A).$$

In effect the left of the above equation is defined to be the t-tuple (element of F^t) on the right. In particular $xv = vA$ showing that in $M(A)$ multiplication by x on the left means multiplication by A on the right. It is high time for some examples!

Example 5.9a

Suppose $F = \mathbb{Q}$, $A = \left(\begin{smallmatrix} 0 & 1 \\ 1 & 0 \end{smallmatrix}\right)$ and so $t = 2$. The elements of $M(A)$ are ordered pairs (q_1, q_2) of rational numbers, that is, $M(A) = \mathbb{Q}^2$ as sets, and we shall continue to refer to these elements as vectors. There's nothing mysterious about the sum of vectors nor about the scalar multiple of a vector – both are exactly what you expect. The novelty is x (q_1, q_2) which is also a vector. Which vector is it? Using the above rule and the given A we obtain

$$x(q_1, q_2) = (q_1, q_2)A = (q_1, q_2) \begin{pmatrix} 0 & 1 \\ 1 & 0 \end{pmatrix} = (q_2, q_1)$$

showing that multiplication by x in this $\mathbb{Q}[x]$-module $M(A)$ amounts to interchanging q_1 and q_2. Thus $xe_1 = x(1, 0) = (0, 1) = e_2$, $xe_2 = x(0, 1) = (1, 0) = e_1$ and $x(5, 1/2) = (1/2, 5)$ etc. Also $(x + 1)e_1 = xe_1 + e_1 = (0, 1) + (1, 0) = (1, 1)$ and $(x - 1)e_1 = xe_1 - e_1 = (0, 1) - (1, 0) = (-1, 1)$. Notice that every element (q_1, q_2) of $M(A)$ can be expressed as a polynomial multiple of e_1: specifically $(q_1, q_2) = q_1e_1 + q_2e_2 = q_1e_1 + q_2xe_1 = (q_1 + q_2x)e_1$. So $M(A)$ is described as being *cyclic* with *generator* e_1 (see Definition 5.21 for the general case). In other words, e_1 generates $M(A) = \{f(x)e_1 : f(x) \in \mathbb{Q}[x]\}$. It is significant that there is no need to use quadratic or polynomials of higher degree: in fact

$$x^2(q_1, q_2) = x(x(q_1, q_2)) = x(q_2, q_1) = (q_1, q_2)$$

showing that multiplying elements of $M(A)$ by x^2 has no effect. Put in a different way, which is more dramatic and significant, multiplication by $x^2 - 1$ *annihilates* all elements of $M(A)$, that is,

$$(x^2 - 1)(q_1, q_2) = x^2(q_1, q_2) - (q_1, q_2) = (q_1, q_2) - (q_1, q_2) = 0.$$

We will see that A is *the companion matrix* Definition 5.25 *of* $x^2 - 1$. Although $\mathbb{Q}^2$ has an infinite number of subspaces there are by Theorem 5.28 just four of them which are *submodules* Definition 5.14 of $M(A)$, that is, subspaces which are closed under polynomial multiplication. From Lemma 2.2 and Theorem 2.5 all subgroups of a finite cyclic group G are themselves cyclic and each such subgroup corresponds to a positive divisor of the order $|G|$ of a generator of G. Now *the order of e_1 in $M(A)$ is $x^2 - 1$*, that is, the monic polynomial $d(x)$ of smallest degree over $\mathbb{Q}$ satisfying $d(x)e_1 = 0$ is $d(x) = x^2 - 1$ (see Definition 5.11 for the details). As e_1 generates $M(A)$ we see that $x^2 - 1 = \chi_A(x)$ is the module analogue of $|G|$. So it is reasonable to expect the four submodules of $M(A)$ to be themselves cyclic and to correspond to the four monic divisors $1, x + 1, x - 1, x^2 - 1$ of $x^2 - 1$. In fact this is true (see Theorem 5.28 for the general case and its proof) and so

$$M(A) = \langle e_1, e_2 \rangle, \qquad N_1 = \langle e_1 + e_2 \rangle, \qquad N_2 = \langle e_2 - e_1 \rangle, \qquad \langle 0 \rangle$$

are the four submodules of $M(A)$ having generators e_1, $(x + 1)e_1$, $(x - 1)e_1$, $(x^2 - 1)e_1$; these submodules are subspaces of $\mathbb{Q}^2$ and have dimensions 2, 1, 1, 0 respectively. The submodules N_1 and N_2 are the row eigenspaces of A as $e_1 + e_2$ and $e_2 - e_1$ are row eigenvectors of A with eigenvalues 1 and -1. Further $M(A) = N_1 \oplus N_2$ (internal direct sum) and

$$XAX^{-1} = \operatorname{diag}(1, -1) \quad \text{where } X = \begin{pmatrix} e_1 + e_2 \\ e_2 - e_1 \end{pmatrix} = \begin{pmatrix} 1 & 1 \\ -1 & 1 \end{pmatrix}$$

showing that A is similar to a diagonal matrix, the rows of the invertible matrix X forming a basis of $\mathbb{Q}^2$ consisting of eigenvectors of A. Finally note $e_1 + e_2$ and $e_2 - e_1$ have orders $x - 1$ and $x + 1$ respectively.

More generally v is a *row eigenvector* of the $t \times t$ matrix A over F associated with the eigenvalue $\lambda \in F$ if and only if the order Definition 5.11 of v in $M(A)$ is $x - \lambda$.

Example 5.9b

Take $F = \mathbb{Q}$, $A = \begin{pmatrix} 0 & 1 \\ -1 & 0 \end{pmatrix}$ and so $t = 2$. In the $\mathbb{Q}[x]$-module $M(A)$ multiplication by x is $x(q_1, q_2) = (q_1, q_2)A = (q_1, q_2)\begin{pmatrix} 0 & 1 \\ -1 & 0 \end{pmatrix} = (-q_2, q_1)$ for all $(q_1, q_2) \in \mathbb{Q}^2$. In particular $xe_1 = e_2$, $xe_2 = -e_1$. As in Example 5.9a above, $M(A)$ is cyclic with generator e_1. Also

$$x^2(q_1, q_2) = x(x(q_1, q_2)) = x(-q_2, q_1) = (-q_2, q_1)A$$

$$= (-q_2, q_1)\begin{pmatrix} 0 & 1 \\ -1 & 0 \end{pmatrix} = (-q_1, -q_2)$$

that is, $x^2 v = -v$ for all $v = (q_1, q_2) \in \mathbb{Q}^2$. Therefore $(x^2 + 1)v = 0$ in this module, showing that $x^2 + 1 = \chi_A(x)$ annihilates every vector v of $M(A)$. Now $x^2 + 1$ is irreducible over $\mathbb{Q}$ and so its only monic factors over $\mathbb{Q}$ are 1 and $x^2 + 1$. Anticipating Theorem 5.28 we see that $M(A)$ has only two submodules, namely $M(A)$ and $\{(0, 0)\}$ corresponding to these two divisors. In fact every non-zero vector in $M(A)$ is a generator and $x^2 + 1$ is its order Definition 5.11. Also A is the companion matrix Definition 5.25 of $x^2 + 1$ over $\mathbb{Q}$. Write $\mathbb{Q}' = \mathbb{Q}[x]/\langle x^2 + 1 \rangle$ which is an extension field of $\mathbb{Q}$ (see Theorem 4.9); the elements of $\mathbb{Q}'$ are $q + q'\iota$ where the coset $\iota = \langle x^2 + 1 \rangle + x$ satisfies $\iota^2 = -1$ and $q, q' \in \mathbb{Q}$. So $\mathbb{Q}'$ is the field of complex numbers having rational real parts and rational imaginary parts. As $(x^2 + 1)v = 0$ for all $v \in M(A)$ it is possible to turn $\mathbb{Q}^2$ into a $\mathbb{Q}'$-module M' by writing $(q + q'\iota)v = (q + q'x)v$ (Exercises 5.1, Question 3(e)). But a $\mathbb{Q}'$-module is no more than a vector space over $\mathbb{Q}'$. What is more M' is a 1-dimensional vector space over $\mathbb{Q}'$ with basis e_1 as

$$(q_1, q_2) = q_1 e_1 + q_2 e_2 = q_1 e_1 + q_2 x e_1 = (q_1 + x q_2)e_1 = (q_1 + \iota q_2)e_1.$$

Extending the ground field F (from $\mathbb{Q}$ to $\mathbb{Q}'$ in this case) is, where appropriate, very helpful in the analysis of A. Notice that doubling the field ($\mathbb{Q}'$ has degree 2 over $\mathbb{Q}$) causes the dimension to halve: $M(A)$ has dimension 2 as a vector space over $\mathbb{Q}$ and M' has dimension 1 as a vector space over $\mathbb{Q}'$.

Example 5.9c

Suppose $F = \mathbb{Q}$, $A = \begin{pmatrix} 3 & 0 \\ 0 & 3 \end{pmatrix}$ and so $t = 2$. Here $xv = vA = 3v$ for all $v \in M(A)$, that is, multiplication by x in the $\mathbb{Q}[x]$-module $M(A)$ is multiplication by the scalar 3. More generally $f(x)v = f(3)v \in \langle v \rangle$ for all $v \in M(A)$ and so $M(A)$ is not cyclic: there is no vector $v_0 \in M(A)$ with $M(A) = \{f(x)v_0 : f(x) \in F[x]\}$ as $M(A) \neq \langle v_0 \rangle$. Each subspace N of $\mathbb{Q}^2$ is closed under polynomial multiplication, that is, $f(x)v = f(3)v \in N$ for all $f(x) \in F[x]$ and all $v \in N$. Therefore N is a submodule of $M(A)$ by Definition 5.14, showing that, in this case, subspaces and submodules coincide.

More generally in the case of A being a $t \times t$ scalar matrix, that is, $A = cI$ for $c \in F$, subspaces of F^t and submodules of $M(A)$ coincide. In this case there is 'nothing to do' since the matrix A is already in canonical form and so it is not surprising that the $F[x]$-module $M(A)$ is virtually identical to the F-module F^t. Notice that $\chi_A(x) = (x - c)^t$ and every non-zero vector in $M(A)$ has order $x - c$.

We now describe in detail the new concepts arising in the above examples.

Lemma 5.10

Let A be a $t \times t$ matrix over the field F and let $v \in M(A)$. There is a unique monic polynomial $d(x)$ of degree at most t over F such that $d(x)v = 0$ and $d(x)|f(x)$ for all $f(x) \in F[x]$ with $f(x)v = 0$ in $M(A)$.

Proof

The $t + 1$ vectors $v, vA, vA^2, \ldots, vA^t$ belong to the t-dimensional vector space F^t over F and so are linearly dependent: there are scalars $b_0, b_1, b_2, \ldots, b_t \in F$, not all zero, satisfying $b_0 v + b_1 vA + b_2 vA^2 + \cdots + b_t vA^t = 0$. In the module $M(A)$ this equation becomes $b_0 v + b_1 xv + b_2 x^2 v + \cdots + b_t x^t v = 0$, that is, $g(x)v = 0$ where $g(x) = b_0 + b_1 x + b_2 x^2 + \cdots + b_t x^t \in F[x]$. Also $g(x)$ is non-zero with $\deg g(x) \le t$.

Write $K = \{f(x) \in F[x] : f(x)v = 0\}$. The notation suggests that K is an ideal of $F[x]$ and this is true (Exercises 5.1, Question 3(d)). The above paragraph shows that K is non-zero. So K has a monic generator $d(x)$ by Theorem 4.4. Therefore $d(x)v = 0$ and $d(x)|f(x)$ for all $f(x) \in K$. In particular $d(x)|g(x)$ and so $\deg d(x) \le t$. So $d(x)$ satisfies the conditions of Lemma 5.10. Conversely suppose $d(x)$ satisfies the conditions of Lemma 5.10. Then $d(x)$ is a monic generator of the ideal K of $F[x]$. Therefore $d(x)$ is unique by Theorem 4.4.   □

Definition 5.11

Let v be an element of an $F[x]$-module M. The ideal

$$K = \{f(x) \in F[x] : f(x)v = 0\} \quad \text{of } F[x]$$

is called *the order ideal of v in M*. So $K = \langle d(x) \rangle$ where $d(x)$ is monic or zero by Theorem 4.4. The polynomial $d(x)$ is called *the order of v in M*.

From Lemma 5.10 we see that K is non-zero for all $v \in M(A)$ and so each element of $M(A)$ has a monic order. In other words, the order of $v \in M(A)$ is the unique monic polynomial $d(x)$ of *least* degree over F satisfying $d(x)v = 0$. This concept is analogous to the order n of $g \in G$ where G is an additive finite abelian group since n is the *smallest* positive integer satisfying $ng = 0$.

Notice that $K = F[x]$ if and only if $v = 0$, that is, the zero vector is the only vector of M having the constant polynomial $1(x)$ as its order since $F[x] = \langle 1(x) \rangle$. On the other hand, taking $M = F[x]$, all non-zero elements of the free $F[x]$-module $F[x]$ have order $0(x)$ since $F[x]$ is an integral domain. More generally each non-zero element of a free $F[x]$-module has order $0(x)$.

From the $|G|$-lemma we know $|G|g = 0$ for all elements g of the finite additive abelian group G. It'll take us some time to prove the analogous result for $t \times t$ matrices A over a field F, namely *the Cayley–Hamilton theorem* (Corollary 6.11)

$$\chi_A(A) = 0$$

that is, A satisfies its own characteristic polynomial. This beautiful equation was in fact first established for general A by the German mathematician Frobenius. As a consequence $\chi_A(x)v = 0$ for all $v \in M(A)$, which tells us

the order of each v in $M(A)$ is a monic divisor of $\chi_A(x)$.

As $\chi_A(x)$ has at most 2^t monic divisors (Exercises 4.1, Question 1(e)) we see that the number of possible orders of elements in $M(A)$ cannot exceed 2^t.

As preparation for our next theorem consider two $t \times t$ matrices A and B over F with $A \sim B$. Is it true that $A^2 \sim B^2$? Is it true that $A^3 \sim B^3$ etc.? What about

$A^2 + I \sim B^2 + I$ and $A^{23} - A + I \sim B^{23} - B + I$ etc.? We show next that all these similarities *are* true.

Lemma 5.12

Let A and B be $t \times t$ matrices over F and suppose that X is an invertible $t \times t$ matrix over F with $XAX^{-1} = B$. Then $Xf(A)X^{-1} = f(B)$ for all $f(x) \in F[x]$.

Proof

Let's start with the case $f(x) = x^2$. Then

$$Xf(A)X^{-1} = XA^2X = XAAX^{-1} = XAIAX^{-1} = XAX^{-1}XAX^{-1}$$

$$= BB = B^2 = f(B)$$

where the factor $I = X^{-1}X$ has been inserted at a strategic point. On the one hand this factor makes no difference and on the other hand it 'does the trick'. More generally suppose $f(x) = x^i$ where i is a positive integer. Inserting $i - 1$ factors $I = X^{-1}X$ gives

$$Xf(A)X^{-1} = XA^iX^{-1} = XAA \cdots AX^{-1} = XAIAI \cdots IAX^{-1}$$

$$= XAX^{-1}XAX^{-1} \cdots XAX^{-1} = (XAX^{-1})^i = B^i = f(B).$$

From the above equations we see $XA^i = B^i X$ which will help us in the general case. Suppose now $f(x) = a_n x^n + a_{n-1}x^{n-1} + \cdots + a_1 x + a_0$. Then

$$f(A) = a_n A^n + a_{n-1}A^{n-1} + \cdots + a_1 A + a_0 I \quad \text{and}$$

$$f(B) = a_n B^n + a_{n-1}B^{n-1} + \cdots + a_1 B + a_0 I$$

by Definition 5.8. Now $X(a_i A^i) = X(a_i I)A^i = a_i I(XA^i) = a_i I(B^i X) = (a_i B^i)X$ for $1 \leq i \leq n$ and $X(a_0 I) = (a_0 I)X$ as the scalar matrix $a_i I$ commutes with all matrices in $\mathfrak{M}_t(F)$. Adding up these $n + 1$ equations and using the distributive laws in the matrix ring $\mathfrak{M}_t(F)$ leads to

$$Xf(A) = X(a_n A^n) + X(a_{n-1}A^{n-1}) + \cdots + X(a_1 A) + X(a_0 I)$$

$$= (a_n B^n)X + (a_{n-1}B^{n-1})X + \cdots + (a_1 B)X + (a_0 I)X = f(B)X$$

and so $Xf(A)X^{-1} = f(B)$. $\square$

From Lemma 5.12 we see: $A \sim B$ implies $f(A) \sim f(B)$ for all $f(x) \in F[x]$.

Our next theorem establishes the close connection between similarity of matrices and isomorphism of the corresponding modules.

Theorem 5.13

Let A and B be $t \times t$ matrices over the field F. Then A and B are similar if and only if $M(A)$ and $M(B)$ are isomorphic $F[x]$-modules.

Proof

Suppose $A \sim B$. There is an invertible $t \times t$ matrix X with $X^{-1}AX = B$; to facilitate the notation we have replaced A, A', X in Definition 5.3 by A, B, X^{-1} respectively. Let $\theta : F^t \to F^t$ be the F-linear mapping determined by X, that is, $(v)\theta = vX$ for all $v \in F^t$. We show $\theta : M(A) \to M(B)$ to be $F[x]$-linear. As v is an element of $M(A)$ and $(v)\theta$ is an element of $M(B)$ we have $f(x)v = vf(A)$ and $f(x)((v)\theta) = ((v)\theta)f(B)$. Using Lemma 5.12 and inserting a factor $I = XX^{-1}$ in the 'right' place gives

$$(f(x)v)\theta = (vf(A))\theta = vf(A)X = vIf(A)X = vXX^{-1}f(A)X$$

$$= vX(X^{-1}f(A)X) = vXf(X^{-1}AX) = vXf(B)$$

$$= ((v)\theta)f(B) = f(x)((v)\theta).$$

So $\theta : M(A) \to M(B)$ is $F[x]$-linear. In the same way $\varphi : M(B) \to M(A)$ is $F[x]$-linear, where $\varphi : F^t \to F^t$ is the F-linear mapping Definition 5.8 determined by X^{-1}. As the matrices X and X^{-1} are inverses of each other the same is true of the $F[x]$-linear mappings θ and φ, that is, $\varphi = \theta^{-1}$. Therefore $\theta : M(A) \cong M(B)$, that is, the $F[x]$-modules $M(A)$ and $M(B)$ are isomorphic.

Conversely suppose that the $F[x]$-modules $M(A)$ and $M(B)$ are isomorphic. Let $\theta : M(A) \cong M(B)$ be an isomorphism. As θ is $F[x]$-linear, then $\theta : F^t \to F^t$ is certainly F-linear. Let X be the matrix of θ relative to the standard basis $\mathcal{B}_0$ of F^t. Then θ is the F-linear mapping determined by X. As θ is invertible so also is X. Consider $e_i \in M(A)$ and so $xe_i = e_i A$ for $1 \leq i \leq t$. Then $(e_i)\theta \in M(B)$ and so $x((e_i)\theta) = x(e_i X) = e_i XB$. As θ is $F[x]$-linear we have $(xe_i)\theta = x((e_i)\theta)$, that is, $e_i AX = e_i XB$ for $1 \leq i \leq t$ showing that corresponding rows of AX and XB are equal. Therefore $AX = XB$, that is, A and B are similar. $\qquad\square$

The three matrices A in Examples 5.9a–5.9c have different determinants. No two of them are similar by Lemma 5.5 and so no two of the $\mathbb{Q}[x]$-modules $M(A)$ in Examples 5.9a–5.9c are isomorphic by Theorem 5.13.

The concepts of *submodule* (Definition 2.26) and *quotient module* (Lemma 2.27) were introduced in the context of R-modules where R is a non-trivial commutative ring. We now discuss submodules N of the 'parent' $F[x]$-modules $M(\alpha)$ and $M(A)$. Submodules play a crucial role in understanding how their parent modules are built up. Just as finite abelian groups are direct sums of cyclic subgroups (Theorem 3.4) we show in Theorem 6.5 that $M(A)$ decomposes analogously into a direct sum of cyclic submodules.

Definition 5.14

Let N be a subset of an $F[x]$-module M such that N is a subgroup of the additive group of M and N is closed under polynomial multiplication: $f(x)v \in N$ for all $v \in N$ and all $f(x) \in F[x]$. Then N is called a *submodule* of M.

The reader should check that Definition 2.26 becomes Definition 5.14 in the case $R = F[x]$. We have already met examples of submodules of $\mathbb{Q}[x]$-modules $M(A)$ in Examples 5.9a–5.9c. The next lemma tells us that submodules in this context are a special type of subspace.

Lemma 5.15

Let F be a field. Let N be a subset of an $F[x]$-module M. Then N is a submodule of M if and only if N is a subspace of M (regarded as vector space over F) such that $xv \in N$ for all $v \in N$.

Let M and M' be $F[x]$-modules and let $\theta : M \to M'$ be an F-linear mapping satisfying $(xv)\theta = x((v)\theta)$ for all $v \in M$. Then θ is $F[x]$-linear.

Proof

Note that M has the structure of a vector space over F on ignoring the products $f(x)v$ for all polynomials $f(x)$ of degree at least 1 over F but retaining the products av for all constant polynomials a over F and all $v \in M$. In other words, on suppressing all occurrences of the indeterminate x, the $F[x]$-module M drops its status to that of a mere F-module.

Let N be a submodule of the $F[x]$-module M. Then N is a subgroup of the additive group of M by Definition 5.14. Taking $f(x) = a$ in Definition 5.14 gives $av \in N$ for all $a \in F$ and all $v \in N$, showing that N is a subspace of the vector space M. Taking $f(x) = x$ in Definition 5.14 gives $xv \in N$ for all $v \in N$.

Conversely suppose that N is a subspace of the vector space M satisfying $xv \in N$ for all $v \in N$. Then N is a subgroup of the additive group of M by Definition 5.14 and $av \in N$ for all $a \in F$ and all $v \in N$. By induction $x^n v = x(x^{n-1}v) \in N$ for all positive integers n and all $v \in N$. Consider a general polynomial $f(x) = a_n x^n + a_{n-1} x^{n-1} + \cdots + a_1 x + a_0$ over F. As $v \in N$ implies $x^n v, x^{n-1} v, \ldots, xv, v$ all belong to N, it follows that all linear combinations of these $n + 1$ elements of N also belong to N, that is, $f(x)v = a_n x^n v + a_{n-1} x^{n-1} v + \cdots + a_1 xv + a_0 v \in N$. We conclude that N is a submodule of the $F[x]$-module M by Definition 5.14.

Both M and M' are vector spaces over F, as explained above. As θ satisfies $(xv)\theta = x((v)\theta)$ for all $v \in M$ we deduce $(x^2v)\theta = (x(xv))\theta = x((xv)\theta) = x(x((v)\theta)) = x^2((v)\theta)$. More generally $(x^iv)\theta = x^i((v)\theta)$ by induction on i. With $f(x)$ as above we obtain

$$(f(x)v)\theta = ((a_nx^n + \cdots + a_1x + a_0)v)\theta = (a_nx^nv + \cdots + a_1xv + a_0v)\theta$$

$$= a_n((x^nv)\theta) + \cdots + a_1((xv)\theta) + a_0((v)\theta)$$

$$= a_nx^n((v)\theta) + \cdots + a_1x((v)\theta) + a_0((v)\theta)$$

$$= (a_nx^n + \cdots + a_1x + a_0)((v)\theta) = f(x)((v)\theta)$$

using the F-linearity of θ. So θ is $F[x]$-linear. $\square$

Let N be a subset of a vector space V over the field F and let $\alpha : V \to V$ be F-linear. Then N is a submodule of the $F[x]$-module $M(\alpha)$ if and only if N is a subspace of V with $xv \in N$ for all $v \in N$ by Lemma 5.15. As $xv = (v)\alpha$ in $M(\alpha)$, the set inclusion $(N)\alpha \subseteq N$ shows that N is closed under multiplication by x. A subspace N of V is called α-*invariant* if $(N)\alpha \subseteq N$. Therefore

the submodules of $M(\alpha)$ are precisely the α-invariant subspaces of V.

Let A be a $t \times t$ matrix over the field F. A subspace N of F^t is called A-*invariant* if $vA \in N$ for all $v \in N$. As above we obtain:

the submodules of $M(A)$ are precisely the A-invariant subspaces of F^t.

The 1-dimensional α-invariant subspaces of V are $\langle v \rangle$ where v is an eigenvector of the linear mapping $\alpha : V \to V$ and the 1-dimensional A-invariant subspaces of F^t are $\langle v \rangle$ where v is a row eigenvector of the $t \times t$ matrix A over F.

Definition 5.16

Let $\alpha : V \to V$ be a linear mapping of a vector space V over a field F. Let N be a submodule of the $F[x]$-module $M(\alpha)$. The mapping $\alpha|_N : N \to N$, defined by $(w)\alpha|_N = (w)\alpha$ for all $w \in N$, is called *the restriction of α to N*.

Note $\alpha|_N$, as defined above, is a linear mapping of N regarded as a vector space over F; in fact $\alpha|_N$ is an $F[x]$-linear mapping of the module N. Also $(w)\alpha|_N = xw$ for all $w \in N$ and $N = M(\alpha|_N)$ by Lemma 5.7. We need this set-theoretic concept and Definition 5.17 below to describe decompositions of $M(\alpha)$ into a direct sum of submodules.

Definition 5.17

Let A_1 be a $t_1 \times t_1$ matrix over R and let A_2 be a $t_2 \times t_2$ matrix over R where R is a commutative ring. The partitioned $(t_1 + t_2) \times (t_1 + t_2)$ matrix

$$A_1 \oplus A_2 = \left(\begin{array}{c|c} A_1 & 0 \\ \hline 0 & A_2 \end{array} \right) \quad \text{over } R \text{ is called } \textit{the direct sum of } A_1 \textit{ and } A_2,$$

that is, the (i, j)-entries in $A_1 \oplus A_2$ and A_1 are equal for $1 \le i, j \le t_1$, the $(t_1 + i, t_1 + j)$-entry in $A_1 \oplus A_2$ equals the (i, j)-entry in A_2 for $1 \le i, j \le t_2$, all other entries in $A_1 \oplus A_2$ being zero.

For example with $R = \mathbb{Z}$

$$\begin{pmatrix} 1 & 2 \\ 3 & 4 \end{pmatrix} \oplus \begin{pmatrix} 5 & 6 & 7 \\ 8 & 9 & 10 \\ 11 & 12 & 13 \end{pmatrix} = \left(\begin{array}{cc|ccc} 1 & 2 & 0 & 0 & 0 \\ 3 & 4 & 0 & 0 & 0 \\ \hline 0 & 0 & 5 & 6 & 7 \\ 0 & 0 & 8 & 9 & 10 \\ 0 & 0 & 11 & 12 & 13 \end{array} \right).$$

Suppose that A_j is a $t_j \times t_j$ matrix over a commutative ring R for $1 \le j \le s$ where $s \ge 3$. The direct sum of matrices is associative, that is, $(A_1 \oplus A_2) \oplus A_3 = A_1 \oplus (A_2 \oplus A_3)$ and so this matrix is unambiguously denoted by $A_1 \oplus A_2 \oplus A_3$. More generally

$$A_1 \oplus A_2 \oplus \cdots \oplus A_s = \left(\begin{array}{c|c|c|c} A_1 & 0 & \cdots & 0 \\ \hline 0 & A_2 & \cdots & 0 \\ \hline \vdots & \vdots & \ddots & \vdots \\ \hline 0 & 0 & \cdots & A_s \end{array} \right)$$

is the partitioned $t \times t$ matrix over R having the given s matrices A_j on the diagonal and rectangular zero matrices elsewhere where $t = t_1 + t_2 + \cdots + t_s$. Notice that $A_1 \oplus A_2 \sim A_2 \oplus A_1$ (see Exercises 5.1, Question 2(c)). The reader should know $\det(A_1 \oplus A_2) = (\det A_1)(\det A_2)$ (see Exercises 5.1, Question 2(b)). As

$$xI - (A_1 \oplus A_2 \oplus \cdots \oplus A_s) = (xI - A_1) \oplus (xI - A_2) \oplus \cdots \oplus (xI - A_s)$$

on taking determinants we obtain

$$|xI - (A_1 \oplus A_2 \oplus \cdots \oplus A_s)| = |xI - A_1||xI - A_2| \cdots |xI - A_s|$$

showing that the characteristic polynomial of a direct sum of matrices is the product of the characteristic polynomials of the individual matrices.

The direct sum of matrices has a further useful property: it respects matrix addition and matrix multiplication. Specifically let A_j and B_j be $t_j \times t_j$ matrices over a commutative ring R for $1 \leq j \leq s$. Then

$$(A_1 \oplus A_2 \oplus \cdots \oplus A_s) + (B_1 \oplus B_2 \oplus \cdots \oplus B_s)$$

$$= \begin{pmatrix} A_1 + B_1 & 0 & \cdots & 0 \\ 0 & A_2 + B_2 & \cdots & 0 \\ \vdots & \vdots & \ddots & \vdots \\ 0 & 0 & \cdots & A_s + B_s \end{pmatrix},$$

$$(A_1 \oplus A_2 \oplus \cdots \oplus A_s)(B_1 \oplus B_2 \oplus \cdots \oplus B_s)$$

$$= \begin{pmatrix} A_1 B_1 & 0 & \cdots & 0 \\ 0 & A_2 B_2 & \cdots & 0 \\ \vdots & \vdots & \ddots & \vdots \\ 0 & 0 & \cdots & A_s B_s \end{pmatrix},$$

that is,

$$(A_1 \oplus A_2 \oplus \cdots \oplus A_s) + (B_1 \oplus B_2 \oplus \cdots \oplus B_s)$$
$$= (A_1 + B_1) \oplus (A_2 + B_2) \oplus \cdots \oplus (A_s + B_s),$$

$$(A_1 \oplus A_2 \oplus \cdots \oplus A_s)(B_1 \oplus B_2 \oplus \cdots \oplus B_s)$$
$$= (A_1 B_1) \oplus (A_2 B_2) \oplus \cdots \oplus (A_s B_s).$$

Taking $A_j = B_j$ and using induction the preceding equations combine to give

$$f(A_1 \oplus A_2 \oplus \cdots \oplus A_s)$$
$$= f(A_1) \oplus f(A_2) \oplus \cdots \oplus f(A_s) \quad \text{for all } f(x) \in R[x]$$

(see Exercises 5.1, Question 2(d)).

For instance taking $f(x) = x^3 + 1$ and $s = 2$ we obtain $(A_1 \oplus A_2)^3 + I = (A_1^3 + I) \oplus (A_2^3 + I)$ and so if $f(A_1) = 0$ and $f(A_2) = 0$, then also $f(A_1 \oplus A_2) = 0$. The reader can verify that

$$A_1 = \begin{pmatrix} 0 & 1 \\ -1 & 1 \end{pmatrix} \quad \text{and} \quad A_2 = \begin{pmatrix} 0 & 1 & 0 \\ 0 & 0 & 1 \\ -1 & 0 & 0 \end{pmatrix}$$

are two such matrices. Taking $R = \mathbb{R}$ (the real field) we obtain $(x^3 + 1)v = 0$ for all v in $M(A_1 \oplus A_2)$ and so the only possibilities for the order of v in the $\mathbb{R}[x]$-module $M(A_1 \oplus A_2)$ are the monic divisors of $x^3 + 1$, that is, $1, x + 1, x^2 - x + 1, x^3 + 1$.

How do direct sums of matrices arise? The answer is: from decompositions of $M(A)$ into a direct sum of submodules, as we explain in Lemma 5.19. However we must first deal with an important method of constructing bases. Let V be a vector space and suppose that $\mathcal{B}_1$ and $\mathcal{B}_2$ are bases of the finite-dimensional subspaces U_1 and U_2 of V. The *ordered set* $\mathcal{B}_1 \cup \mathcal{B}_2$ consists of the vectors in $\mathcal{B}_1$ followed by the vectors in $\mathcal{B}_2$. The reader will be aware that, although the vectors in $\mathcal{B}_1 \cup \mathcal{B}_2$ span $U_1 + U_2$, the totality of these vectors may *not* be linearly independent. In fact $\mathcal{B}_1 \cup \mathcal{B}_2$ is a basis of $U_1 + U_2$ if and only if $U_1 \cap U_2 = 0$, that is, if and only if U_1 and U_2 are independent (Definition 2.14) as additive subgroups of the additive group of V. Our next lemma deals with the general case.

Lemma 5.18

Let V be a vector space and let $\mathcal{B}_j$ be a basis of the finite-dimensional subspace U_j of V for $1 \leq j \leq s$. Then $\mathcal{B} = \mathcal{B}_1 \cup \mathcal{B}_2 \cup \cdots \cup \mathcal{B}_s$ is a basis of $U = U_1 + U_2 + \cdots + U_s$ if and only if $U_1, U_2, \ldots, U_s$ are independent.

Proof

Suppose that $\mathcal{B}$ is a basis of U. Let $v_1, v_2, \ldots, v_t$ be the vectors of $\mathcal{B}$. Write $t_0' = 0$ and $t_j' = t_1 + t_2 + \cdots + t_j$ where $t_j = \dim U_j$ for $1 \leq j \leq s$. The basis $\mathcal{B}$ of U is built up as follows: first come the t_1 vectors of $\mathcal{B}_1$ in order, next come the t_2 vectors of $\mathcal{B}_2$ in order, $\ldots$, and lastly come the t_s vectors of $\mathcal{B}_s$ in order. So $t_s' = t_1 + t_2 + \cdots + t_s = \dim U$ and $\mathcal{B}_j$ consists of the vectors v_i for $t_{j-1}' < i \leq t_j'$ where $1 \leq j \leq s$. To test $U_1, U_2, \ldots, U_s$ for independence suppose $u_1 + u_2 + \cdots + u_s = 0$ where $u_j \in U_j$ for $1 \leq j \leq s$. As u_j is a linear combination of the vectors in $\mathcal{B}_j$ there are scalars a_i with $u_j = \sum_{v_i \in \mathcal{B}_j} a_i v_i$ for $1 \leq j \leq s$. Adding these s equations gives

$$\sum_{v_i \in \mathcal{B}} a_i v_i = \sum_{j=1}^{s} \left(\sum_{v_i \in \mathcal{B}_j} a_i v_i \right) = \sum_{j=1}^{s} u_j = 0.$$

By the linear independence of $v_1, v_2, \ldots, v_t$ we deduce $a_1 = a_2 = \cdots = a_t = 0$. So $u_j = 0$ for $1 \leq j \leq s$ proving the independence Definition 2.14 of $U_1, U_2, \ldots, U_s$.

Conversely suppose the subspaces $U_1, U_2, \ldots, U_s$ of V to be independent. Consider $u \in U$. There are $u_j \in U_j$ with $u = u_1 + u_2 + \cdots + u_s$ as $U = U_1 + U_2 + \cdots + U_s$. Now $\mathcal{B}_j$ is a basis of U_j and so there are scalars b_i with $u_j = \sum_{v_i \in \mathcal{B}_j} b_i v_i$ for $1 \leq j \leq s$. Adding these s equations gives

$$u = \sum_{j=1}^{s} u_j = \sum_{j=1}^{s} \left(\sum_{v_i \in \mathcal{B}_j} b_i v_i \right) = \sum_{i=1}^{t} b_i v_i$$

showing that $v_1, v_2, \ldots, v_t$ span U. To show that $v_1, v_2, \ldots, v_t$ are linearly independent, suppose there are scalars $a_1, a_2, \ldots, a_t$ with $\sum_{i=1}^{t} a_i v_i = 0$. Write $u_j = \sum_{v_i \in \mathcal{B}_j} a_i v_i$ and then the preceding equation becomes $\sum_{j=1}^{s} u_j = 0$. As $u_j \in U_j$ we deduce $u_1 = u_2 = \cdots = u_s = 0$ from the independence of $U_1, U_2, \ldots, U_s$. The vectors in $\mathcal{B}_j$ are linearly independent and so $u_j = 0$ implies $a_i = 0$ for $t'_{j-1} < i \leq t'_j$ and $1 \leq j \leq s$. Therefore $a_1 = a_2 = \cdots = a_t = 0$ proving that $v_1, v_2, \ldots, v_t$ are linearly independent. So $\mathcal{B}$ is a basis of U. $\square$

Let M be an $F[x]$-module which, as a vector space over the field F, has finite dimension t. Suppose M to be the internal direct sum of its submodules $N_1, N_2, \ldots, N_s$, that is, $M = N_1 \oplus N_2 \oplus \cdots \oplus N_s$ meaning that each $v \in M$ can be *uniquely* expressed as $v = w_1 + w_2 + \cdots + w_s$ where $w_j \in N_j$ for $1 \leq j \leq s$. The submodule N_j, being a subspace of the vector space M, is also finite-dimensional. Write $t_j = \dim N_j$ and let $\mathcal{B}_j$ denote a basis of N_j for $1 \leq j \leq s$. Applying Lemma 5.18 with $U = V = M$ and $U_j = N_j$ for $1 \leq j \leq s$ we see that $\mathcal{B} = \mathcal{B}_1 \cup \mathcal{B}_2 \cup \cdots \cup \mathcal{B}_s$ is a basis of M. What is the matrix of $\alpha : M \to M$, given by $(v)\alpha = xv$ for all $v \in M$, relative to $\mathcal{B}$? The answer, which involves the direct sum Definition 5.17 of matrices over F, is in our next lemma.

Lemma 5.19

Using the above notation

$$A_1 \oplus A_2 \oplus \cdots \oplus A_s$$

is the matrix of $\alpha : M \to M$ relative to the basis $\mathcal{B} = \mathcal{B}_1 \cup \mathcal{B}_2 \cup \cdots \cup \mathcal{B}_s$ of $M = N_1 \oplus N_2 \oplus \cdots \oplus N_s$ where A_j is the matrix of $\alpha|_{N_j} : N_j \to N_j$ relative to $\mathcal{B}_j$ for $1 \leq j \leq s$.

Proof

Let $v_1, v_2, \ldots, v_t$ be the vectors of $\mathcal{B}$. Denote the matrix of α relative to $\mathcal{B}$ by $B = (b_{ik})$ where $1 \leq i, k \leq t$. As in the proof of Lemma 5.18 write $t'_0 = 0$ and $t'_j = t_1 + \cdots + t_j$ for $1 \leq j \leq s$. Then $\mathcal{B}_j$ consists of the vectors v_i for $t'_{j-1} < i \leq t'_j$. For $t'_{j-1} < i \leq t'_j$ we know $(v_i)\alpha = b_{i1}v_1 + b_{i2}v_2 + \cdots + b_{it}v_t \in N_j$ as N_j is α-invariant. So the second suffix k of any *non-zero* b_{ik} satisfies $t'_{j-1} < k \leq t'_j$. The above equation can be expressed: $(v_i)\alpha|_{N_j} = \sum_{t'_{j-1} < k \leq t'_j} b_{ik}v_k$ showing that b_{ik} is the $(i - t'_{j-1}, k - t'_{j-1})$-entry in A_j for such i and k by Definitions 5.1 and 5.16. We've now identified b_{ik} for all $1 \leq i, k \leq t$ and so $B = A_1 \oplus A_2 \oplus \cdots \oplus A_s$ by Definition 5.17. $\square$

Finally we apply the preceding theory to the case $M = M(A)$.

Corollary 5.20

Let A be a $t \times t$ matrix over a field F. Let $\alpha : F^t \to F^t$ be the linear mapping determined by A. Suppose the $F[x]$-module $M(A)$ has submodules $N_1, N_2, \ldots, N_s$ such that $M(A) = N_1 \oplus N_2 \oplus \cdots \oplus N_s$. Let A_j be the $t_j \times t_j$ matrix of $\alpha|_{N_j} : N_j \to N_j$ relative to a basis $\mathcal{B}_j$ of N_j where $t_j = \dim N_j$ for $1 \leq j \leq s$. Let X be the invertible matrix over F having the vectors of $\mathcal{B}_1 \cup \mathcal{B}_2 \cup \cdots \cup \mathcal{B}_s$ as its rows. Then $XAX^{-1} = A_1 \oplus A_2 \oplus \cdots \oplus A_s$.

Proof

First note that α has matrix A relative to the standard basis $\mathcal{B}_0$ of F^t. By Lemma 5.18 we see $\mathcal{B}' = \mathcal{B}_1 \cup \mathcal{B}_2 \cup \cdots \cup \mathcal{B}_s$ is a basis of F^t. By Lemma 5.19 the matrix of α relative to $\mathcal{B}'$ is $A' = A_1 \oplus A_2 \oplus \cdots \oplus A_s$. From Lemma 5.2 we conclude that X satisfies $XAX^{-1} = A'$. $\square$

Using the above notation $N_j = M(\alpha|_{N_j})$ it is true that $M(\alpha|_{N_j}) \cong M(A_j)$, that is, the $F[x]$-modules N_j and $M(A_j)$ are isomorphic for $1 \leq j \leq s$ (Exercises 5.1, Question 5). The decomposition of $M(A)$ in Corollary 5.20 can therefore be expressed as the external direct sum

$$M(A) \cong M(A_1) \oplus M(A_2) \oplus \cdots \oplus M(A_s).$$

As an illustration consider

$$A = \begin{pmatrix} 0 & -1 & 1 & 0 \\ 1 & -1 & 1 & 1 \\ 1 & 0 & 0 & 1 \\ 0 & 1 & -1 & 0 \end{pmatrix}$$

over $\mathbb{Q}$. We will see in Corollary 5.29 that each vector v_0 of $\mathbb{Q}^4$ generates a submodule N of $M(A)$ and N is an A-invariant space of dimension equal to the degree of the order of v_0 in $M(A)$. Let $N_1 = \{f(x)e_1 : f(x) \in F[x]\}$, that is, the elements of N_1 are polynomial multiples of $e_1 = (1, 0, 0, 0)$. Now $xe_1 = e_1 A = (0, -1, 1, 0)$ and $x^2 e_1 = x(xe_1) = (0, -1, 1, 0)A = (0, 1, -1, 0)$ and so $(x^2 + x)e_1 = 0$. As e_1, xe_1 are linearly independent we see that e_1 has order $x^2 + x$ in $M(A)$. On dividing $f(x)$ by $x^2 + x$ we obtain $f(x) = q(x)(x^2 + x) + r(x)$ where $q(x), r(x) \in \mathbb{Q}[x]$ and $r(x) = a_0 + a_1 x$. So

$$f(x)e_1 = (q(x)(x^2 + x) + r(x))e_1 = q(x)(x^2 + x)e_1 + r(x)e_1$$

$$= q(x)0 + r(x)e_1 = 0 + r(x)e_1 = r(x)e_1 = (a_0 + a_1 x)e_1 = a_0 e_1 + a_1 x e_1$$

showing that e_1, xe_1 span N_1. So the ordered set e_1, xe_1 is a basis $\mathcal{B}_1$ of the A-invariant subspace N_1 of $\mathbb{Q}^4$. In the same way let $N_2 = \{f(x)e_3 : f(x) \in F[x]\}$, that is, the elements of N_2 are polynomial multiples of $e_3 = (0, 0, 1, 0)$. Now $xe_3 = e_3A = (1, 0, 0, 1)$ and $x^2 e_3 = x(xe_3) = (1, 0, 0, 1)A = (0, 0, 0, 0) = 0$. As e_3, xe_3 are linearly independent we conclude that e_3 has order x^2 in $M(A)$. As before e_3, xe_3 span N_2 and so e_3, xe_3 is a basis $\mathcal{B}_2$ of N_2. The reader may wonder why we have chosen e_1 and e_3 as generators (why not e_1 and e_2 for example?). The answer is: e_1 and e_3 are the simplest vectors such that $M(A) = N_1 \oplus N_2$ as we now demonstrate. Construct the matrix X having the vectors of $\mathcal{B}_1 \cup \mathcal{B}_2$ as its rows, that is

$$
X = \begin{pmatrix} e_1 \\ xe_1 \\ e_3 \\ xe_3 \end{pmatrix} = \begin{pmatrix} 1 & 0 & 0 & 0 \\ 0 & -1 & 1 & 0 \\ 0 & 0 & 1 & 0 \\ 1 & 0 & 0 & 1 \end{pmatrix}.
$$

As $\det X = -1 \neq 0$ the rows of X are the vectors of the basis $\mathcal{B} = \mathcal{B}_1 \cup \mathcal{B}_2$ of $\mathbb{Q}^4$. So the union of $\mathcal{B}_1$ and $\mathcal{B}_2$ is a basis of $\mathbb{Q}^4$ and this fact guarantees that N_1 and N_2 are independent and $M(A) = N_1 + N_2$, that is, $M(A) = N_1 \oplus N_2$. Then

$$
XAX^{-1} = \left(\begin{array}{cc|cc} 0 & 1 & 0 & 0 \\ 0 & -1 & 0 & 0 \\ \hline 0 & 0 & 0 & 1 \\ 0 & 0 & 0 & 0 \end{array} \right) = C(x(x+1)) \oplus C(x^2)
$$

where $C((x+1)x) = \begin{pmatrix} 0 & 1 \\ 0 & -1 \end{pmatrix}$ and $C(x^2) = \begin{pmatrix} 0 & 1 \\ 0 & 0 \end{pmatrix}$ are companion matrices Definition 5.25. In fact $N_1 \cong M(C(x^2+x))$ and $N_2 \cong M(C(x^2))$ using Exercises 5.1, Question 5. Finally note that e_1, xe_1, e_2, xe_2 are linearly dependent and so e_2 couldn't have been used instead of e_3.

Here is a tip: similarity problems can generally be solved *without* explicitly finding the entries in X^{-1}; specifically it is good enough to verify $\det X \neq 0$ and $XA = BX$ in order to show $XAX^{-1} = B$ where A, B and X are $t \times t$ matrices over a field.

In Section 6.1 a systematic method of determining generators of cyclic submodules N_j as in Corollary 5.20 is developed.

EXERCISES 5.1

1. (a) Let

$$
A = \begin{pmatrix} 1 & 3 \\ 2 & 4 \end{pmatrix} \quad \text{and} \quad X = \begin{pmatrix} 2 & 1 \\ 5 & 3 \end{pmatrix}
$$

over $\mathbb{Q}$. Calculate the entries in $B = XAX^{-1}$ and verify that $\det A = \det B$ and trace $A = $ trace B. What is the characteristic polynomial of X^2AX^{-2}?

(b) The matrices

$$A = \begin{pmatrix} 3-a & 7 \\ -1 & a \end{pmatrix} \quad \text{and} \quad B = \begin{pmatrix} b & 1 \\ -1 & 2 \end{pmatrix}$$

over $\mathbb{Q}$ are similar. Find the possible values of a and b.

(c) Verify that

$$XAX^{-1} = \begin{pmatrix} 1 & a \\ 0 & -1 \end{pmatrix} \quad \text{where}$$

$$A = \begin{pmatrix} 1 & 0 \\ 0 & -1 \end{pmatrix} \quad \text{and} \quad X = \begin{pmatrix} 1 & -a/2 \\ 0 & 1 \end{pmatrix}$$

over the field $\mathbb{R}$ of real numbers. Does the similarity class of A contain an infinite number of elements? (Yes/No). Show $A \sim B$ where

$$B = (1/\sqrt{2}) \begin{pmatrix} 1 & 1 \\ 1 & -1 \end{pmatrix}.$$

(d) Show that similarity $\sim$ is an equivalence relation on the set $\mathfrak{M}_t(F)$ of all $t \times t$ matrices over the field F.

(e) The similarity class of $A \in \mathfrak{M}_2(F)$ consists of A alone where F is a field. Use the invertible matrices $\begin{pmatrix} 1 & 1 \\ 0 & 1 \end{pmatrix}$ and $\begin{pmatrix} 1 & 0 \\ 1 & 1 \end{pmatrix}$ to show

$$A = \begin{pmatrix} a & 0 \\ 0 & a \end{pmatrix}$$

for some $a \in F$.

2. (a) Let $A = (a_{ij})$ be a $t \times t$ matrix over a non-trivial commutative ring R. For each permutation π of $\{1, 2, \ldots, t\}$ write sign $\pi = \pm 1$ according as π is even/odd. Then the determinant of A is defined by $\det A = \sum_\pi (\text{sign } \pi) a_{1(1)\pi} a_{2(2)\pi} \cdots a_{t(t)\pi}$ where the summation is over all $t!$ permutations π of $\{1, 2, \ldots, t\}$. Show that the coefficient of x^{t-1} in the monic polynomial $\chi_A(x) = \det(xI - A)$ of degree t over R is $-$ trace A where trace $A = a_{11} + a_{22} + \cdots + a_{tt}$. Show that the constant term in $\chi_A(x)$ is $\chi_A(0) = (-1)^t \det A$.

(b) Let

$$A = (a_{ij}) = \left(\begin{array}{c|c} A_1 & B \\ \hline 0 & A_2 \end{array} \right)$$

be a $t \times t$ matrix over the commutative ring R, partitioned as indicated, where A_1 and A_2 are respectively $t_1 \times t_1$ and $t_2 \times t_2$ matrices, 0 is the $t_2 \times t_1$ zero submatrix and $t = t_1 + t_2$. Use the above definition of determinant to show

$$\det A = (\det A_1)(\det A_2).$$

Deduce $\det(A_1 \oplus A_2) = (\det A_1)(\det A_2)$.

Hint: Suppose first $(i)\pi \leq t_1 < i$ for some $i \in \{1, 2, \ldots, t\}$. What is the value of $a_{i(i)\pi}$? Next consider π such that $i > t_1 \Rightarrow (i)\pi > t_1$ and hence $i \leq t_1 \Rightarrow (i)\pi \leq t_1$.

(c) Let A_1 and A_2 be respectively $t_1 \times t_1$ and $t_2 \times t_2$ matrices over a field F. Specify an invertible $t \times t$ matrix X over F such that $X(A_1 \oplus A_2) = (A_2 \oplus A_1)X$ where $t = t_1 + t_2$. Deduce $A_1 \oplus A_2 \sim A_2 \oplus A_1$.

Let A and B be respectively 2×2 and 3×3 matrices over F satisfying $A \oplus B = B \oplus A$. Show that $A \oplus B$ is a scalar matrix. More generally suppose $A_1 \oplus A_2 = A_2 \oplus A_1$ and let $t = \gcd\{t_1, t_2\}$. Show that there is a $t \times t$ matrix A over F with $A_1 = A \oplus A \oplus \cdots \oplus A$ (t_1/t terms) and $A_2 = A \oplus A \oplus \cdots \oplus A$ (t_2/t terms).

Hint: Suppose not and consider the least integer $t_1 + t_2$ for which A does not exist.

(d) For $1 \leq j \leq s$ let A_j denote a $t_j \times t_j$ matrix over a non-trivial commutative ring R. Show

$$f(A_1 \oplus A_2 \oplus \cdots \oplus A_s) = f(A_1) \oplus f(A_2) \oplus \cdots \oplus f(A_s)$$

for all polynomials $f(x) \in R[x]$.

Hint: Consider first the case $s = 2$ and use induction on $\deg f(x)$.

(e) Let M be an R-module where R is a non-trivial commutative ring. Let $\alpha : M \to M$ and $\beta : M \to M$ be R-linear mappings and let $a \in R$. Show that $\alpha + \beta : M \to M$ and $\alpha\beta : M \to M$ are R-linear. Show that $a\alpha : M \to M$, defined by $(v)(a\alpha) = a((v)\alpha)$ for all $v \in M$, is R-linear. Does the set $\mathrm{End}\, M$ of all R-linear mappings $\alpha : M \to M$ have the structure of

(i) a ring (Yes/No),

(ii) an R-module (Yes/No)?

Following Definition 5.6, for each polynomial $f(x) = a_n x^n + a_{n-1}x^{n-1} + \cdots + a_1 x + a_0$ over R, write $f(\alpha) = a_n\alpha^n + a_{n-1}\alpha^{n-1} + \cdots + a_1\alpha + a_0\iota$ where $\iota : M \to M$ is the identity mapping. Show by induction that

(i) α^n is R-linear,

(ii) $f(\alpha)$ is R-linear.

Let $\alpha \in \mathrm{End}\, M$. Show that $\varepsilon_\alpha : R[x] \to \mathrm{End}\, M$ is a ring homomorphism where $(f(x))\varepsilon_\alpha = f(\alpha)$ for all $f(x) \in R[x]$, that is, ε_α is the *evaluation at α* homomorphism.

(f) Let V be a finite-dimensional vector space over the field F. Suppose $\dim V = t > 0$ and let $\mathcal{B}$ be a basis of V. Write $\mathrm{End}_F V$ for the ring of all F-linear mappings $\alpha : V \to V$. Show that $\theta : \mathrm{End}_F V \cong \mathfrak{M}_t(F)$ is

an F-linear ring isomorphism where $\theta(\alpha)$ is the matrix of α relative
to $\mathcal{B}$.

Hint: Model your proof on Theorem 3.15.

3. (a) Working in the $\mathbb{Q}[x]$-module $M(A)$, where $A = \left(\begin{smallmatrix} 1 & 0 \\ 0 & 0 \end{smallmatrix}\right)$, express each
    of xe_1, $x^2 e_1$, $(x+1)e_1$, $(x-1)e_1$ as a pair $(a_1, a_2) \in \mathbb{Q}^2$ where
    $e_1 = (1, 0)$. Express in the same way xe_2, $x(x+1)e_2$, $x(e_1 + e_2)$,
    $(x-1)(e_1 + e_2)$, $x(x-1)(e_1 + e_2)$ where $e_2 = (0, 1)$. Write down the
    orders of e_1, e_2 and $e_1 + e_2$ in $M(A)$. Is $e_1 + e_2$, $x(e_1 + e_2)$ a basis
    of $\mathbb{Q}^2$? Is $e_1 + e_2$ a generator of $M(A)$?
    Which of $\langle e_1 \rangle$, $\langle e_2 \rangle$ and $\langle e_1 + e_2 \rangle$ are submodules of $M(A)$? List the
    (four) submodules of $M(A)$. Which rational numbers a_1, a_2 are such
    that $v = a_1 e_1 + a_2 e_2$ generates $M(A)$?

    (b) Working in the $\mathbb{Q}[x]$-module $M(A)$, where $A = \left(\begin{smallmatrix} 4 & -2 \\ 5 & -3 \end{smallmatrix}\right)$, find the or-
    ders of $v_1 = (1, -1)$ and $v_2 = (5, -2)$. Are these vectors row eigen-
    vectors of A? (Yes/No). Write down the characteristic polynomial
    $\chi_A(x)$ of A. Verify that $\chi_A(A)$ is the zero matrix. Find the order of
    e_1 in $M(A)$. Does e_1 generate $M(A)$? Which (four) monic polynomi-
    als over $\mathbb{Q}$ arise as orders of elements in $M(A)$? Let $X = \left(\begin{smallmatrix} e_1 \\ xe_1 \end{smallmatrix}\right)$ and
    $Y = \left(\begin{smallmatrix} v_1 \\ v_2 \end{smallmatrix}\right)$. Verify that X and Y are invertible over $\mathbb{Q}$ and calculate the
    matrices C and D satisfying $XA = CX$ and $YA = DY$. Are C and D
    both similar to A? (Yes/No).

    (c) Let A be a $t \times t$ matrix over a field F. Suppose that the orders of
    $e_1, e_2, \ldots, e_t$ and $e_1 + e_2 + \cdots + e_t$ in $M(A)$ have degree 1. Show
    that A is a scalar matrix, that is, $A = \lambda I$ for some $\lambda \in F$. Let B be a
    2×2 matrix over F which is not a scalar matrix. Deduce that $M(B)$
    is cyclic being generated by one of e_1, e_2, $e_1 + e_2$.

    (d) Let V be a vector space over a field F and let $\alpha : V \to V$ be a linear
    mapping. Suppose $v \in V$. Show $K = \{f(x) \in F[x] : (v)f(\alpha) = 0\}$ to
    be an ideal of $F[x]$. K is the order ideal Definition 5.11 of v in $M(\alpha)$.
    Show V finite-dimensional implies K non-zero.

    (e) Let A be a $t \times t$ matrix over a field F and suppose $m(A) = 0$ where
    $m(x)$ is a monic polynomial over F. Show that F^t has the structure of
    an $F[x]/\langle m(x) \rangle$-module M' on defining $(f(x) + \langle m(x) \rangle)v = vf(A)$
    for all $f(x) \in F[x]$, $v \in F^t$.
    Hint: Start by showing that $\overline{f(x)}v$ is unambiguously defined by the
    r.h.s. of the above equation where $\overline{f(x)} = f(x) + \langle m(x) \rangle$.
    Suppose $m(x)$ is irreducible over F. Deduce that M' is a vector
    space over the field $F' = F[x]/\langle m(x) \rangle$. Let $v_1, v_2, \ldots, v_{t'}$ form a ba-
    sis of M' over F'. Show that the $\deg m(x) \times t'$ vectors $(x^i)v_j$ for
    $0 \le i < \deg m(x)$, $1 \le j \le t'$, form a basis of F^t. Deduce $t/t' = \deg m(x)$.

4. (a) Find xe_1 and x^2e_1 in the $\mathbb{Q}[x]$-module $M(A)$ where

$$A = \begin{pmatrix} 0 & 2 & -1 \\ 2 & 3 & -2 \\ 1 & 2 & -2 \end{pmatrix}.$$

Are e_1, xe_1 linearly dependent? Are e_1, xe_1, x^2e_1 linearly dependent? Show that e_1 has order $(x+1)(x-3)$ in $M(A)$. Are $(x+1)e_1$ and $(x-3)e_1$ row eigenvectors of A? Which elements v of $M(A)$ satisfy $(x+1)(x-3)v = 0$? Is $M(A)$ a cyclic $\mathbb{Q}[x]$-module? Determine the order of $(1, 0, -1)$ in $M(A)$ and hence construct an invertible matrix X over $\mathbb{Q}$ such that XAX^{-1} is diagonal.

Hint: The rows of X are linearly independent (row) eigenvectors of A.

(b) Find xe_1 and x^2e_1 in the $\mathbb{Q}[x]$-module $M(A)$ where

$$A = \begin{pmatrix} 2 & 2 & 1 \\ -1 & -1 & -1 \\ 1 & 2 & 2 \end{pmatrix}.$$

Show that e_1 has order $(x-1)^2$ in $M(A)$. Find the order of $e_1 + e_2$ in $M(A)$. Show that

$$X = \begin{pmatrix} e_1 + e_2 \\ e_1 \\ xe_1 \end{pmatrix}$$

is invertible over $\mathbb{Q}$ and satisfies

$$XAX^{-1} = \left(\begin{array}{c|cc} 1 & 0 & 0 \\ \hline 0 & 0 & 1 \\ 0 & -1 & 2 \end{array} \right).$$

(c) Let $\alpha : V \to V$ be a linear mapping of the vector space V over the field F. Use Lemma 5.15 to show that all subspaces N of V with $N \subseteq \ker \alpha$ are α-invariant. The subspace N' of V satisfies $\operatorname{im} \alpha \subseteq N'$. Show that N' is α-invariant.

Suppose $\operatorname{rank} \alpha = 1$. Show that there are no further α-invariant subspaces.

Let $\lambda \in F$ and let N be a subspace of V. Show that N is α-invariant if and only if N is $(\alpha - \lambda \iota)$-invariant where ι is the identity mapping of V.

(d) Let $\alpha : F^3 \to F^3$ be the linear mapping determined by

$$A = \begin{pmatrix} 1 & 0 & 0 \\ 0 & 0 & 0 \\ 0 & 0 & 0 \end{pmatrix}$$

over an arbitrary field F. Show that every α-invariant subspace N of F^3 satisfies either $\langle e_1 \rangle \subseteq N$ or $N \subseteq \langle e_2, e_3 \rangle$. List the ten α-invariant subspaces of F^3 in the case $F = \mathbb{Z}_2$.

(e) Let $\alpha : F^3 \to F^3$ be the linear mapping determined by

$$A = \begin{pmatrix} 0 & 1 & 0 \\ 0 & 0 & 0 \\ 0 & 0 & 0 \end{pmatrix}$$

over the arbitrary field F. Determine the α-invariant subspaces N of F^3 and list these subspaces in the case $F = \mathbb{Z}_2$.

(f) Specify the α-invariant subspaces of $\mathbb{Q}^3$ where α is the linear mapping determined by the matrix A of Question 4(a) above.
Hint: Consider $\alpha + \iota$.
Specify the α-invariant subspaces of $\mathbb{Q}^3$ where α is the linear mapping determined by the matrix A of Question 4(b) above.

5. Let A be a $t \times t$ matrix over a field F and let $\alpha : F^t \to F^t$ be given by $(v)\alpha = vA$ for all $v \in F^t$. Let N be a submodule of the $F[x]$-module $M(A)$. Let $u_1, u_2, \ldots, u_s$ be a basis of N and let B be the $s \times s$ matrix of $\alpha|_N$ relative to this basis. For $u \in N$ write $(u)\beta = (a_1, a_2, \ldots, a_s) \in F^s$ where $u = a_1 u_1 + a_2 u_2 + \cdots + a_s u_s$. Using Lemma 5.15 show $\beta : N \cong M(B)$, that is, show β to be an isomorphism between the $F[x]$-modules N and $M(B)$.

5.2 Cyclic Modules and Companion Matrices

Just as every finite abelian group decomposes into a direct sum of cyclic subgroups Theorem 3.4, so every $F[x]$-module $M(A)$ as in Definition 5.8 decomposes into a direct sum of cyclic submodules. As preparation for the fundamental theorem 6.5 we analyse cyclic $F[x]$-modules and the corresponding matrices, companion matrices, which explicitly determine cyclic modules. Companion matrices have many agreeable properties: their characteristic polynomials can be read off from the entries in the last row and they provide standard examples of cyclic modules. Submodules of cyclic $F[x]$-modules are also cyclic and easy to 'pin down', compared to those of a general

$F[x]$-module $M(A)$, and knowledge of them will help our study of variants of the rational canonical form in Section 6.2.

Definition 5.21

Let M be an R-module where R is a non-trivial commutative ring. Suppose M contains an element v_0 such that each element of M is expressible as $r v_0$ for some $r \in R$. Then M is said to be *cyclic* with *generator* v_0.

Taking $R = \mathbb{Z}$ we obtain the concept (Definition 2.1) of a cyclic abelian group. Taking $R = F[x]$ we obtain the important concept of a cyclic $F[x]$-module, examples of which we have already met in Examples 5.9a and 5.9b.

Let R be a non-trivial commutative ring. Then R is a free R-module of rank 1, that is, R is a cyclic R-module being generated by its 1-element 1 of order ideal $\{0\}$. The submodules of R are precisely the ideals K of R, but K may not be a cyclic R-module unless R is a PID. However R/K is a cyclic R-module being generated by its 1-element $K + 1$, where K is an ideal of R.

By Theorem 2.5 every cyclic $\mathbb{Z}$-module is isomorphic to the additive group of the ring $\mathbb{Z}/\langle n \rangle$ where $n \geq 0$. Our first lemma generalises Theorem 2.5 to include cyclic R-modules and introduces the standard examples of cyclic $F[x]$-modules.

Lemma 5.22

Let M be a cyclic R-module with generator v_0 where R is a non-trivial commutative ring. The mapping $\theta : R \to M$, where $(r)\theta = r v_0$ for all $r \in R$, is R-linear and surjective. Write $K = \ker \theta$. The mapping $\tilde{\theta} : R/K \cong M$, where $(K + r)\tilde{\theta} = (r)\theta$ for all $r \in R$, is an isomorphism of R-modules.

Suppose $R = F[x]$ where F is a field. Let v_0 have order $d_0(x)$ in M. Then $\tilde{\theta} : F[x]/\langle d_0(x) \rangle \cong M$.

Proof

As module laws 5 and 6 are obeyed in M we see $(r_1 + r_2)\theta = (r_1 + r_2)v_0 = r_1 v_0 + r_2 v_0 = (r_1)\theta + (r_2)\theta$ and $(rr_1)\theta = (rr_1)v_0 = r(r_1 v_0) = r((r_1)\theta)$ for all $r, r_1, r_2 \in R$. So θ is R-linear. As v_0 is a generator (Definition 5.21) of M, every element of M is $r v_0 = (r)\theta$ for some $r \in R$ showing $\operatorname{im} \theta = M$, that is, θ is surjective. From the first isomorphism theorem for R-modules (Corollary 2.28) we deduce $\tilde{\theta} : R/K \cong M$ where $K = \ker \theta$.

Now take $R = F[x]$. Then $K = \{f(x) \in F[x] : f(x)v_0 = 0\}$ is the order ideal (Definition 5.11) of v_0 in M. By hypothesis $K = \langle d_0(x) \rangle$ where $d_0(x)$ is either monic or zero. In any case $\tilde{\theta} : F[x]/\langle d_0(x) \rangle \cong M$. $\qquad\square$

Let us look at a few examples. For instance $M = \mathbb{Q}[x]/\langle x^3 + 1\rangle$ is a cyclic $\mathbb{Q}[x]$-module having generator $K + 1$ of order $x^3 + 1$ by Lemma 5.22 where $K = \langle x^3 + 1\rangle$. By Theorem 4.1 with $g(x) = x^3 + 1$ we see that each element of M is of the form $K + r(x)$ where $r(x) \in \mathbb{Q}[x]$, $\deg r(x) < 3$. In fact M is a 3-dimensional vector space over $\mathbb{Q}$ with basis $K + 1$, $K + x$, $K + x^2$. In the same way $M' = \mathbb{Z}_3[x]/\langle x^4\rangle$ is a cyclic $\mathbb{Z}_3[x]$-module being generated by $v_0 = \langle x^4\rangle + \overline{1}$ of order x^4; also M' is a 4-dimensional vector space over $\mathbb{Z}_3$ with basis $v_0, xv_0, x^2v_0, x^3v_0$ and $|M'| = 3^4 = 81$. The general case is discussed in Theorem 5.24. However we note in passing that $F[x]/\langle 0(x)\rangle$ is exceptional: it is the polynomial analogue of an infinite cyclic group, being a free $F[x]$-module of rank 1, and also it is a *countably infinite*-dimensional vector space over F: the elements of $F[x]/\langle 0(x)\rangle$ are singletons (subsets $\langle 0(x)\rangle + f(x) = \{f(x)\}$ of $F[x]$ having exactly one element) and $\{1\}, \{x\}, \{x^2\}, \ldots, \{x^n\}, \ldots$ is a countable F-basis of $F[x]/\langle 0(x)\rangle$.

The theory of finite abelian groups depends crucially on Lemma 2.7. The analogous lemma for $F[x]$-modules is stated next; it is equally important and can be proved in the same way (Exercises 5.2, Question 3(a)).

Lemma 5.23

Let v be an element of an $F[x]$-module M having monic order $d(x)$. Then $f(x)v$ has order $d(x)/\gcd\{f(x), d(x)\}$ in M where $f(x) \in F[x]$.

For example we know $v_0 = \langle x^4\rangle + \overline{1}$ has order x^4 in the $\mathbb{Z}_3[x]$-module M' above. So xv_0 has order $x^4/\gcd\{x, x^4\} = x^4/x = x^3$ in M' and $(x^2 + \overline{1})v_0$ has order $x^4/\gcd\{x^2 + \overline{1}, x^4\} = x^4/\overline{1} = x^4$ in M' using Lemma 5.23. This means that $(x^2 + \overline{1})v_0$ is also a generator of M' simply because it has the 'right' order as we show next.

Theorem 5.24

Let F be a field and let M be a cyclic $F[x]$-module with generator v_0 of monic order $d_0(x)$. Then each v in M has order $d(x)$ where $d(x)|d_0(x)$. Further v is a generator of M if and only if v has order $d_0(x)$. Also M is a t-dimensional vector space over F with basis $\mathcal{B}_{v_0}$ consisting of $v_0, xv_0, x^2v_0, \ldots, x^{t-1}v_0$ where $t = \deg d_0(x)$.

Proof

By Definition 5.21 there is $f(x) \in F[x]$ with $v = f(x)v_0$. So $d_0(x)v = d_0(x)f(x)v_0 = f(x)d_0(x)v_0 = f(x)0 = 0$ showing that $d_0(x)$ is in the order ideal K of v in M. So $K = \langle d(x)\rangle$ for some monic polynomial $d(x)$ over F by Theorem 4.4, that is, v has order $d(x)$ by Definition 5.11, and $d(x)|d_0(x)$.

Suppose v is a generator of M. Interchanging the roles of v and v_0 in the preceding paragraph gives $d_0(x)|d(x)$. So $d(x) = d_0(x)$ as the monic polynomials $d(x)$ and $d_0(x)$ are such that each is a divisor of the other.

Suppose $d(x) = d_0(x)$, that is, v has the same order as the generator v_0 of M. As $v = f(x)v_0$, by Lemma 5.23 we see $d_0(x) = d_0(x)/\gcd\{f(x), d_0(x)\}$. Therefore $\gcd\{f(x), d_0(x)\} = 1$. By Corollary 4.6 there are polynomials $a_1(x)$ and $a_2(x)$ over F with $a_1(x)f(x) + a_2(x)d_0(x) = 1$. So

$$v_0 = 1v_0 = (a_1(x)f(x) + a_2(x)d_0(x))v_0$$

$$= a_1(x)f(x)v_0 + a_2(x)d_0(x)v_0 = a_1(x)v$$

as $d_0(x)v_0 = 0$. Therefore v_0 is a polynomial multiple of v. This is good news as it is now only a small step to show that v generates M. Consider $v' \in M$. As v_0 generates M, by Definition 5.21 there is $f'(x) \in F[x]$ with $v' = f'(x)v_0$. So $v' = f'(x)a_1(x)v$ showing that v is a generator of M as $f'(x)a_1(x) \in F[x]$.

The vectors $v_0, xv_0, x^2v_0, \ldots, x^{t-1}v_0$ belong to the vector space M over F. Could these vectors be linearly dependent? If so there are $a_0, a_1, \ldots, a_{t-1} \in F$, not all zero, with $a_0v_0 + a_1xv_0 + a_2x^2v_0 + \cdots + a_{t-1}x^{t-1}v_0 = 0$. Write $a(x) = a_0 + a_1x + a_2x^2 + \cdots + a_{t-1}x^{t-1}$. Then $a(x)v_0 = 0$ and so $a(x) \in \langle d_0(x)\rangle$, that is, $a(x)$ belongs to the order ideal of v_0. Therefore $d_0(x)|a(x)$ which gives $t = \deg d_0(x) \le \deg a(x)$ as $a(x) \ne 0(x)$. But $\deg a(x) < t$. This contradiction shows that $v_0, xv_0, x^2v_0, \ldots, x^{t-1}v_0$ are linearly independent and we denote this ordered set of t vectors by $\mathcal{B}_{v_0}$. Does $\mathcal{B}_{v_0}$ span M? Each v in M is expressible as $v = f(x)v_0$. Dividing $f(x)$ by $d_0(x)$ there are $q(x)$ and $r(x)$ in $F[x]$ with $f(x) = q(x)d_0(x) + r(x)$ where $\deg r(x) < t$. Write $r(x) = r_0 + r_1x + r_2x^2 + \cdots + r_{t-1}x^{t-1}$. Using $d_0(x)v_0 = 0$ we obtain

$$v = f(x)v_0 = (q(x)d_0(x) + r(x))v_0 = q(x)d_0(x)v_0 + r(x)v_0$$

$$= r(x)v_0 = r_0v_0 + r_1xv_0 + r_2x^2v_0 + \cdots + r_{t-1}x^{t-1}v_0$$

showing that each vector v in M is expressible as a linear combination of the vectors in $\mathcal{B}_{v_0}$, that is, $\mathcal{B}_{v_0}$ spans M. Therefore $\mathcal{B}_{v_0}$ is an F-basis of M and $\dim M = t$. $\square$

We have used the basis $\mathcal{B}_{e_1}$ of $M(A)$ with $t = 2$ in Examples 5.9a and 5.9b. In Section 6.1 bases of the type $\mathcal{B}_{v_0}$ are exactly what is needed to construct an invertible matrix X with XAX^{-1} in rational canonical form Definition 6.4.

Suppose that the cyclic $F[x]$-module M is also a t-dimensional vector space over F. Let v_0 of order $d_0(x)$ be a generator of M. Then $d_0(x)$ is monic of degree t by Theorem 5.24, that is, $d_0(x) = x^t + b_{t-1}x^{t-1} + \cdots + b_1x + b_0$. Let $\alpha : M \to M$ be the F-linear mapping defined by $(v)\alpha = xv$ for all $v \in M$. What is the matrix of α relative to $\mathcal{B}_{v_0}$? Write $v_i = x^i v_0$ for $1 \leq i < t$. Then $\mathcal{B}_{v_0}$ is the F-basis $v_0, v_1, v_2, \ldots, v_{t-1}$ of the vector space M. As

$$(v)\alpha = xv, \qquad (xv)\alpha = x^2 v, \qquad (x^2 v)\alpha = x^3 v, \qquad \ldots, \qquad (x^{t-2} v)\alpha = x^{t-1} v$$

we see $(v_i)\alpha = v_{i+1}$ for $0 \leq i < t$, that is, α maps each vector in $\mathcal{B}_{v_0}$ into the next vector in $\mathcal{B}_{v_0}$ (when there is a next one). The only thing outstanding is: what does α 'do' to v_{t-1}? To answer this question we must express $(v_{t-1})\alpha = (x^{t-1} v_0)\alpha = x(x^{t-1} v_0) = x^t v_0$ as a linear combination of $v_0, v_1, v_2 \ldots, v_{t-1}$, that is, of $v_0, xv_0, x^2 v_0, \ldots, x^{t-1} v_0$. As v_0 has order $d_0(x)$ we know $d_0(x)v_0 = 0$, that is, $x^t v_0 + b_{t-1} x^{t-1} v_0 + \cdots + b_1 x v_0 + b_0 v_0 = 0$ which on rearranging gives

$$(v_{t-1})\alpha = -b_0 v_0 - b_1 v_1 - \cdots - b_{t-1} v_{t-1}$$

and this is the equation we are looking for.

Definition 5.25

The $t \times t$ matrix over the field F

$$C(d(x)) = \begin{pmatrix} 0 & 1 & 0 & 0 & \cdots & 0 \\ 0 & 0 & 1 & 0 & \cdots & 0 \\ 0 & 0 & 0 & 1 & \cdots & 0 \\ \vdots & \vdots & \vdots & \vdots & \ddots & \vdots \\ 0 & 0 & 0 & 0 & \cdots & 1 \\ -b_0 & -b_1 & -b_2 & -b_3 & \cdots & -b_{t-1} \end{pmatrix}$$

is called *the companion matrix* of $d_0(x) = x^t + b_{t-1} x^{t-1} + \cdots + b_1 x + b_0$ over F.

Companion matrices are the building blocks for the *rational canonical form* of a general square matrix A over F. Notice that a polynomial which isn't monic doesn't have a companion matrix. It is worth pointing out that, in spite of the notation, $C(d_0(x))$ is a matrix over F. Specifically note

$$C(d_0(x)) \text{ is the matrix of } \alpha \text{ relative to the basis } \mathcal{B}_{v_0} \text{ of the cyclic}$$
$$F[x]\text{-module } M \text{ with generator } v_0 \text{ of order } d_0(x)$$

on referring back to the discussion preceding Definition 5.25.

We have already seen in Examples 5.9a and 5.9b that companion matrices provide a ready source of cyclic modules. The main properties of companion matrices are established in the following theorem which involves the $F[x]$-module $M(C)$ (see Definition 5.8) determined by the companion matrix C.

Theorem 5.26

Write $C = C(d_0(x))$ where $d_0(x)$ is a monic polynomial of positive degree t over F. Then $M(C)$ is a cyclic $F[x]$-module with generator e_1 of order $d_0(x)$. Further $d_0(x)$ is the characteristic polynomial of C and $d_0(C) = 0$.

Proof

Let $d_0(x) = x^t + b_{t-1}x^{t-1} + \cdots + b_1 x + b_0$ where $b_i \in F$ for $0 \le i < t$. As $e_i C = e_{i+1}$ for $1 \le i < t$ we see $xe_i = e_{i+1}$ in $M(C)$. By induction $e_{i+1} = x^i e_1$ for $0 \le i < t$. So $\mathcal{B}_{e_1}$ is the standard basis $e_1, xe_1, x^2 e_1, \ldots, x^{t-1}e_1$ of F^t. As $\mathcal{B}_{e_1}$ spans F^t we see that $M(C)$ is cyclic with generator e_1. From Definition 5.25 the last row of the $t \times t$ matrix C is $e_t C = -(b_0, b_1, \ldots, b_{t-1})$, that is, $x^t e_1 = x(x^{t-1}e_1) = xe_t = e_t C = -(b_0 + b_1 x + \cdots + b_{t-1}x^{t-1})e_1$ which rearranges to produce $d_0(x)e_1 = 0$. Let K be the order ideal of e_1 in $M(C)$. Then $d_0(x) \in K$ and e_1 has order $d(x)$ in $M(C)$ where $d(x) | d_0(x)$ by Definition 5.11. As $M(C)$ is a t-dimensional vector space over F we see $\deg d(x) = t$ by Theorem 5.24. Therefore $d(x) = d_0(x)$ and so $K = \langle d_0(x) \rangle$ showing that e_1 has order $d_0(x)$ in $M(C)$. The characteristic polynomial of C is

$$\chi_C(x) = \det(xI - C) = \begin{vmatrix} x & -1 & 0 & \cdots & 0 & 0 \\ 0 & x & -1 & \cdots & 0 & 0 \\ 0 & 0 & x & \cdots & 0 & 0 \\ \vdots & \vdots & \vdots & \ddots & \vdots & \vdots \\ 0 & 0 & 0 & \cdots & x & -1 \\ b_0 & b_1 & b_2 & \cdots & b_{t-2} & x+b_{t-1} \end{vmatrix}.$$

We perform the column operation: $c_1 + (xc_2 + x^2 c_3 + \cdots + x^{t-1}c_t)$, that is, to col 1 add x^{i-1} col i for $1 < i \le t$ to the matrix $xI - C$. This column operation, being the composition of $t - 1$ ecos of type (iii), leaves the determinant unchanged and produces a new col 1 with only one non-zero entry, namely $d_0(x)$ in row t. So

$$\chi_C(x) = \begin{vmatrix} 0 & -1 & 0 & \cdots & 0 & 0 \\ 0 & x & -1 & \cdots & 0 & 0 \\ 0 & 0 & x & \cdots & 0 & 0 \\ \vdots & \vdots & \vdots & \ddots & \vdots & \vdots \\ 0 & 0 & 0 & \cdots & x & -1 \\ d_0(x) & b_1 & b_2 & \cdots & b_{t-2} & x+b_{t-1} \end{vmatrix}$$

$$= (-1)^{t+1} d_0(x) \begin{vmatrix} -1 & 0 & \cdots & 0 & 0 \\ x & -1 & \cdots & 0 & 0 \\ 0 & x & \cdots & 0 & 0 \\ \vdots & \vdots & \ddots & \vdots & \vdots \\ 0 & 0 & \cdots & x & -1 \end{vmatrix}$$

on expanding along col 1. The above determinant is $(-1)^{t-1}$ and so

$$\chi_C(x) = (-1)^{t+1} d_0(x)(-1)^{t-1} = (-1)^{2t} d_0(x) = d_0(x)$$

showing that $C(d_0(x))$ has characteristic polynomial $d_0(x)$.

In the $F[x]$-module $M(C)$ we know $d_0(x)e_i = e_i d_0(C)$ which is row i of the $t \times t$ matrix $d_0(C)$ for $1 \leq i \leq t$. But $e_i = x^{i-1} e_1$ and $d_0(x) e_1 = 0$ in $M(C)$. Therefore $d_0(x)e_i = d_0(x)x^{i-1} e_1 = x^{i-1} d_0(x) e_1 = x^{i-1} 0 = 0$ and so $e_i d_0(C) = 0$ for $1 \leq i \leq t$. We have shown $d_0(C) = 0$, that is, $d_0(C)$ is the $t \times t$ zero matrix as each of its rows is the zero vector of F^t.   □

Theorem 5.26 is brimming with useful information. First the $F[x]$-module $M(C(d_0(x)))$ is cyclic with generator e_1 and Corollary 5.27 below shows that every cyclic $F[x]$-module $M(A)$ is isomorphic to $M(C(\chi_A(x)))$.

Second should we wish to construct a matrix with a given characteristic polynomial then the companion matrix gives an immediate answer. For example find a matrix C over $\mathbb{Q}$ with characteristic polynomial $x^2 + (2/3)x - (4/5)$. An answer is $C = \begin{pmatrix} 0 & 1 \\ 4/5 & -2/3 \end{pmatrix}$.

Thirdly every companion matrix 'satisfies' its own characteristic polynomial. The reader can verify that C above satisfies

$$C^2 + (2/3)C - (4/5)I = \begin{pmatrix} 0 & 0 \\ 0 & 0 \end{pmatrix}.$$

By Theorem 5.26 *the Cayley–Hamilton Theorem* (Corollary 6.11) is valid for companion matrices, that is,

$$\chi_C(C) = 0 \quad \text{for } C = C(d_0(x)).$$

Conversely, as we will see, the general case of Corollary 6.11 can be reduced to that of a single companion matrix and so Theorem 5.26 'does the trick'.

Our next corollary lays bare the structure of every cyclic $F[x]$-module $M(A)$ provided we know a generator v_0 of $M(A)$ and the characteristic polynomial $\chi_A(x)$ of A.

Corollary 5.27

Let the $F[x]$-module $M(A)$ be cyclic with generator v_0 where A is a $t \times t$ matrix over a field F. The matrix X, having the vectors of the basis $\mathcal{B}_{v_0}$ of F^t as its rows, satisfies $XAX^{-1} = C(\chi_A(x))$. The order of v_0 in $M(A)$ is $\chi_A(x)$. The $F[x]$-modules $M(A)$ and $M(C(\chi_A(x)))$ are isomorphic.

Proof

Let $d_0(x)$ denote the order of v_0 in $M(A)$. Then $\deg d_0(x) = t$ by Theorem 5.24 as $M(A)$ has dimension t over F. Let $\alpha : F^t \to F^t$ be the linear mapping determined by A. Then α has matrix A relative to the standard basis $\mathcal{B}_0$ of F^t and, by the discussion preceding Definition 5.25, α has matrix $C(d_0(x))$ relative to $\mathcal{B}_{v_0}$. By Lemma 5.2 the matrix X, which relates $\mathcal{B}_{v_0}$ to $\mathcal{B}_0$, that is, $e_i X = x^{i-1} v_0$ for $1 \le i \le t$, satisfies $XAX^{-1} = C(d_0(x))$. By Theorem 5.26 the characteristic polynomial of $C(d_0(x))$ is $d_0(x)$ and so $d_0(x) = \chi_A(x)$ using Lemma 5.5. The order of v_0 in $M(A)$ is therefore $\chi_A(x)$ and $XAX^{-1} = C(\chi_A(x))$, that is, A and $C(\chi_A(x))$ are similar. By Theorem 5.13 the $F[x]$-modules $M(A)$ and $M(C(\chi_A(x)))$ are isomorphic.   □

So $M(A)$ is cyclic if and only if A is similar to the companion matrix of its characteristic polynomial.

As an example let

$$A = \begin{pmatrix} 1 & 1 & -1 \\ 1 & -1 & 2 \\ 0 & 2 & -1 \end{pmatrix}$$

over $\mathbb{Q}$. In Section 6.1 we describe a method leading to a generator of $M(A)$ should it have one. Also we will see shortly that cyclic modules have many generators. For the moment we adopt a 'trial and error' approach. Does e_1 generate $M(A)$? As $xe_1 = (1, 1, -1)$, $x^2 e_1 = (2, -2, 2)$ and

$$\begin{vmatrix} e_1 \\ xe_1 \\ x^2 e_1 \end{vmatrix} = \begin{vmatrix} 1 & 0 & 0 \\ 1 & 1 & -1 \\ 2 & -2 & 2 \end{vmatrix} = 0$$

we see that $e_1, xe_1, x^2 e_1$ are linearly dependent. So e_1 does not generate $M(A)$.

Does e_2 generate $M(A)$? As

$$\begin{vmatrix} e_2 \\ xe_2 \\ x^2 e_2 \end{vmatrix} = \begin{vmatrix} 0 & 1 & 0 \\ 1 & -1 & 2 \\ 0 & 6 & -5 \end{vmatrix} = 5 \ne 0$$

we see that $\mathcal{B}_{e_2}$ consisting of e_2, xe_2, x^2e_2 is a basis of $\mathbb{Q}^3$. So $M(A)$ is cyclic with generator e_2. We leave the reader to check that the characteristic polynomial of A is

$$\chi_A(x) = |xI - A| = \begin{vmatrix} x-1 & -1 & 1 \\ -1 & x+1 & -2 \\ 0 & -2 & x+1 \end{vmatrix}$$

$$= x^3 + x^2 - 6x + 4 = (x-1)(x^2 + 2x - 4)$$

and

$$X = \begin{pmatrix} e_2 \\ xe_2 \\ x^2e_2 \end{pmatrix} = \begin{pmatrix} 0 & 1 & 0 \\ 1 & -1 & 2 \\ 0 & 6 & -5 \end{pmatrix}$$

satisfies

$$XAX^{-1} = C(x^3 + x^2 - 6x + 4) = \begin{pmatrix} 0 & 1 & 0 \\ 0 & 0 & 1 \\ -4 & 6 & -1 \end{pmatrix}.$$

As the polynomial $x^2 + 2x - 4$ is irreducible over $\mathbb{Q}$, by Theorem 5.28 below the $\mathbb{Q}[x]$-module $M(A)$ has only four submodules: 0, N_1, N_2, $M(A)$ where $N_1 = \langle(2, 0, -1)\rangle$ and $N_2 = \langle e_1, xe_1 \rangle$. Also $M(A) = N_1 \oplus N_2$, that is, $M(A)$ is the internal direct sum of its submodules N_1 and N_2 as shown below:

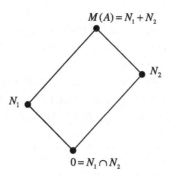

Any vector v of $\mathbb{Q}^3$ not in either N_1 or N_2 is a generator of $M(A)$, because $\langle v, xv, x^2v \rangle$ is a submodule N of $M(A)$ containing v and there is only one such submodule, namely $M(A)$ itself. So $\langle v, xv, x^2v \rangle = M(A)$.

From Lemma 2.2 all subgroups of $(\mathbb{Z}_n, +)$ are cyclic and correspond to the positive divisors of n. Our next theorem, which has already used in Examples 5.9a–5.9c, contains the polynomial analogue.

Theorem 5.28

Let M be a cyclic $F[x]$-module with generator v_0 having non-zero order $d_0(x)$ in M. Let N be a submodule of M. Then N is cyclic with generator $d(x)v_0$ where $d(x)$ is a monic divisor of $d_0(x)$. The order of $d(x)v_0$ in M is $d_0(x)/d(x)$. The quotient module M/N is cyclic with generator $N + v_0$ of order $d(x)$ in M/N.

Proof

Consider $K_N = \{f(x) \in F[x] : f(x)v_0 \in N\}$. It is straightforward to verify that K_N is an ideal of $F[x]$ (see Exercises 5.2, Question 4(a)). As $d_0(x)v_0 = 0 \in N$ we deduce $d_0(x) \in K_N$ and so K_N is non-zero. By Theorem 4.4 there is a unique monic polynomial $d(x)$ over F with $\langle d(x) \rangle = K_N$. As $d_0(x) \in K_N$ we see $d(x)|d_0(x)$. As $d(x) \in K_N$ we see $d(x)v_0 \in N$. We show that $w_0 = d(x)v_0$ is a generator of N. Consider $w \in N$. Since v_0 is a generator of M and $w \in M$ there is $g(x) \in F[x]$ with $w = g(x)v_0$ by Definition 5.21. So $g(x)v_0 = w \in N$ which gives $g(x) \in K_N$. Therefore $g(x) = h(x)d(x)$ for some $h(x) \in F[x]$ and so $w = h(x)d(x)v_0 = h(x)w_0$ showing that N is indeed cyclic with generator w_0.

Let K' denote the order ideal Definition 5.11 of w_0 in M. Then $d_0(x)/d(x) \in K'$ as

$$(d_0(x)/d(x))w_0 = (d_0(x)/d(x))d(x)v_0 = d_0(x)v_0 = 0$$

showing K' to be non-zero. By Theorem 4.4 there is a monic polynomial $d'(x)$ over F with $K' = \langle d'(x) \rangle$ and $d'(x)|d_0(x)/d(x)$. Then $d'(x)d(x)v_0 = d'(x)w_0 = 0$ showing $d'(x)d(x) \in \langle d_0(x) \rangle$, the order ideal of v_0 in M. Therefore $d_0(x)|d'(x)d(x)$ and so $d_0(x)/d(x)|d'(x)$. Each of the monic polynomials $d_0(x)/d(x)$ and $d'(x)$ is a divisor of the other and so they are equal: $d_0(x)/d(x) = d'(x)$. So $K' = \langle d_0(x)/d(x) \rangle$ showing that w_0 has order $d_0(x)/d(x)$ in M.

Using Lemma 2.27 each element $N + v$ of M/N is expressible as $N + f(x)v_0 = f(x)(N + v_0)$ where $f(x) \in F[x]$ showing that $N + v_0$ is a generator of M/N. Let K'' denote the order ideal of $N + v_0$ in M/N. Then $d(x) \in K''$ as

$$d(x)(N + v_0) = N + d(x)v_0 = N + w_0 = N$$

the zero element of M/N. So $K'' = \langle d''(x) \rangle$ where $d''(x)$ is a monic polynomial over F and $d''(x)|d(x)$ by Theorem 4.4. But $d''(x)(N + v_0) = N$ leads to $d''(x)v_0 \in N$ and so $d''(x) \in K_N = \langle d(x) \rangle$. Therefore $d(x)|d''(x)$ which gives $d''(x) = d(x)$. We have shown $K'' = K_N$ and so $N + v_0$ has order $d(x)$ in M/N. $\qquad\square$

Let M be as in Theorem 5.28. Then $N \to K_N$ is a bijection from the set of submodules N of M to the set of ideals K of $F[x]$ with $\langle d_0(x) \rangle \subseteq K$. The inverse bijection is $K \to N_K$ where $N_K = \{q(x)v_0 : q(x) \in K\}$. Further

$$N_1 \subseteq N_2 \quad \Leftrightarrow \quad K_{N_1} \subseteq K_{N_2}$$

that is, these bijections are inclusion-preserving (see Exercises 5.2, Question 4(a)).

Let v_0 be an element of the $F[x]$-module M. Then

$$N = \{f(x)v_0 : f(x) \in F[x]\} \quad \text{is the cyclic submodule of } M \text{ with generator } v_0.$$

It is straightforward to verify that N is a submodule Definition 2.26 of M and it is then immediate that v_0 is a generator Definition 5.21 of N. In the context of $F[x]$-modules we use *bold* brackets writing $N = \langle v_0 \rangle$ to indicate that v_0 is a generator of the submodule N, whereas $\langle v_0 \rangle$ denotes the *subspace* spanned by v_0. Therefore $\langle v_0 \rangle = \{f(x)v_0 : f(x) \in F[x]\}$ and $\langle v_0 \rangle = \{av_0 : a \in F\}$.

Corollary 5.29

Let the element v_0 of the $F[x]$-module M have order $d_0(x) \in F[x]$. Then the cyclic submodule $\langle v_0 \rangle$ is a t-dimensional vector space of M with basis $\mathcal{B}_{v_0}$ consisting of $v_0, xv_0, x^2v_0, \ldots, x^{t-1}v_0$ where $t = \deg d_0(x)$.

Proof

This is short and sweet as Corollary 5.29 is simply Theorem 5.24 applied to $\langle v_0 \rangle$. $\square$

With v_0 as in Corollary 5.29 we see $\langle v_0 \rangle = \langle v_0, xv_0, x^2v_0, \ldots, x^{t-1}v_0 \rangle$.

Example 5.30a

Let $A = \operatorname{diag}(-1, 0, 1)$ over the real field $\mathbb{R}$. We show that $v_0 = (1, 1, 1)$ generates the $\mathbb{R}[x]$-module $M(A)$. Since $xv_0 = (-1, 0, 1), x^2v_0 = (1, 0, 1)$ and

$$\begin{vmatrix} v_0 \\ xv_0 \\ x^2v_0 \end{vmatrix} = \begin{vmatrix} 1 & 1 & 1 \\ -1 & 0 & 1 \\ 1 & 0 & 1 \end{vmatrix} = 2 \neq 0,$$

the vectors v_0, xv_0, x^2v_0 of $\mathcal{B}_{v_0}$ are linearly independent. Hence $\mathcal{B}_{v_0}$ is a basis of $\mathbb{R}^3$ and so $v = a_0v_0 + a_1xv_0 + a_2x^2v_0 = (a_0 + a_1x + a_2x^2)v_0$ where $a_0, a_1, a_2 \in \mathbb{R}$ and $v \in \mathbb{R}^3$. Therefore $M(A) = \langle v_0 \rangle$ is cyclic with generator v_0. As $x^3v_0 = (-1, 0, 1) = xv_0$ we see $(x^3 - x)v_0 = 0$ and so v_0 has order $d_0(x) = x^3 - x = (x + 1)x(x - 1)$ in

$M(A)$. By Corollary 5.27 the matrix

$$X = \begin{pmatrix} v_0 \\ x v_0 \\ x^2 v_0 \end{pmatrix} = \begin{pmatrix} 1 & 1 & 1 \\ -1 & 0 & 1 \\ 1 & 0 & 1 \end{pmatrix}$$

having the vectors of $\mathcal{B}_{v_0}$ as its rows is invertible over $\mathbb{R}$ and satisfies

$$X A X^{-1} = C(x^3 - x) = \begin{pmatrix} 0 & 1 & 0 \\ 0 & 0 & 1 \\ 0 & 1 & 0 \end{pmatrix}.$$

The reader may think that we've taken a step backwards by undiagonalising a perfectly good diagonal matrix! But it is instructive to know that diagonal matrices with distinct diagonal entries determine cyclic modules (Exercises 5.2, Question 1(d)) and computations in these modules are easily carried out. Here the 2^3 monic divisors $d(x)$ of $d_0(x)$ can be arranged in a 'cubical' lattice:

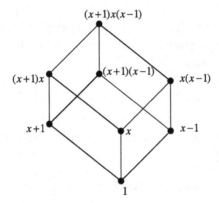

By Theorem 5.28 the 2^3 submodules of $M(A)$ are each of the form $\langle d(x)v_0 \rangle$ where $d(x)$ is a monic divisor of $(x + 1)x(x - 1)$ and these submodules fit together in the same way as shown in the above diagram: notice that $(x + 1)x(x - 1)/d(x)$ is the order of $d(x)v_0$ in $M(A)$ and the two lattices are related by the correspondence $(x + 1)x(x - 1)/d(x) \leftrightarrow \langle d(x)v_0 \rangle$ for all monic divisors $d(x)$ of $(x + 1)x(x - 1)$.

For instance

$$(x + 1)x(x - 1) \leftrightarrow \langle v_0 \rangle = M(A),$$

$$1 \leftrightarrow \langle (x + 1)x(x - 1)v_0 \rangle = \{(0, 0, 0)\} = 0,$$

$$x(x - 1) \leftrightarrow \langle (x + 1)v_0 \rangle = \langle (2, 1, 0) \rangle = \langle (x + 1)v_0, x(x + 1)v_0 \rangle$$

$$= \langle (2, 1, 0), (2, 0, 0) \rangle$$

which is the x_1x_2-plane (with equation $x_3 = 0$) in $\mathbb{R}^3$ and

$$x \leftrightarrow \langle (x+1)(x-1)v_0 \rangle = \langle (0,-1,0) \rangle = \langle -e_2 \rangle = \langle e_2 \rangle$$

which is the x_2-axis in $\mathbb{R}^3$. In fact the 1- and 2-dimensional submodules of $M(A)$ are respectively the coordinate axes and coordinate planes in $\mathbb{R}^3$.

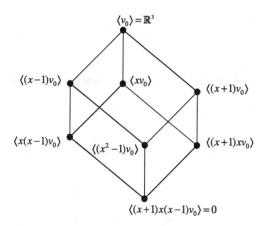

Consider $v = (a_1, a_2, a_3) \in \mathbb{R}^3$ with 3 non-zero entries. Which of the 8 submodules of $M(A)$ could $\langle v \rangle$ be? As v is not in any of the coordinate planes $\langle (x+1)v_0 \rangle$, $\langle xv_0 \rangle$, $\langle (x-1)v_0 \rangle$ we see $\langle v \rangle = M(A)$ is the only possibility. So v generates $M(A)$.

The reader can check that the theory of this example remains unchanged on replacing $\mathbb{R}$ by any field F of characteristic $\neq 2$ but retaining $A = \mathrm{diag}(-1, 0, 1)$ as before where 1 is the 1-element of F. In particular, for F a finite field with $|F| = q$ odd, there are $(q-1)^3$ vectors $v = (a_1, a_2, a_3) \in F^3$ with $M(A) = \langle v \rangle$, there being $q-1$ choices for $a_i \in F^*$, $i = 1, 2, 3$.

Example 5.30b

Let F be any field and let

$$A = C(x^2(x-1)) = \begin{pmatrix} 0 & 1 & 0 \\ 0 & 0 & 1 \\ 0 & 0 & 1 \end{pmatrix}$$

over F. By Theorem 5.26 the $F[x]$-module $M(A)$ is cyclic with generator e_1 of order $d_0(x) = x^2(x-1)$.

The six monic divisors of $d_0(x)$ can be arranged as shown.

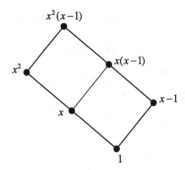

The submodules of $M(A)$, which are six in number by Theorem 5.28, fit together in the same way:

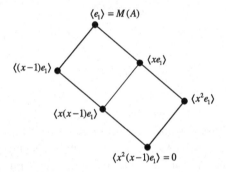

Here $\langle xe_1 \rangle = \langle e_2 \rangle = \langle e_2, xe_2 \rangle = \langle e_2, e_3 \rangle$ and $\langle (x-1)e_1 \rangle = \langle e_2 - e_1, e_3 - e_2 \rangle$ are subspaces of dimension 2, whereas $\langle x^2 e_1 \rangle = \langle e_3 \rangle = \langle e_3 \rangle$ and $\langle x(x-1)e_1 \rangle = \langle e_3 - e_2 \rangle = \langle e_3 - e_2 \rangle$ are 1-dimensional eigenspaces of A corresponding to the eigenvalues 0 and 1. This module decomposes as the internal direct sum

$$M(A) = \langle (x-1)e_1 \rangle \oplus \langle x^2 e_1 \rangle$$

using the submodules on the extreme left and right in the above diagram. We construct a basis of F^3 using bases of the submodules in this decomposition: specifically let $\mathcal{B} = \mathcal{B}_{(x-1)e_1} \cup \mathcal{B}_{x^2 e_1}$, that is,

$$\mathcal{B} \text{ consists of } (x-1)e_1, x(x-1)e_1, x^2 e_1.$$

This method (of obtaining a basis from bases of components in a direct sum) is universally applicable and we will use it time after time in Chapter 6. Let X denote the invertible matrix over F having the vectors of $\mathcal{B}$ as its rows. So

$$X = \begin{pmatrix} (x-1)e_1 \\ x(x-1)e_1 \\ x^2 e_1 \end{pmatrix} = \begin{pmatrix} -1 & 1 & 0 \\ 0 & -1 & 1 \\ 0 & 0 & 1 \end{pmatrix}$$

satisfies

$$XAX^{-1} = \left(\begin{array}{cc|c} 0 & 1 & 0 \\ 0 & 0 & 0 \\ \hline 0 & 0 & 1 \end{array}\right) = \left(\begin{array}{c|c} C(x^2) & 0 \\ \hline 0 & C(x-1) \end{array}\right).$$

As $A = C(x^2(x-1))$ we obtain

$$C(x^2(x-1)) \sim C(x^2) \oplus C(x-1)$$

which is a particular case of Theorem 5.31 below.

Finally note that any v in $M(A)$ which does not belong to either $N_1 = \langle (x-1)e_1 \rangle$ or $N_2 = \langle xe_1 \rangle$, the largest submodules $\neq M(A)$, is a generator of $M(A)$ by Theorem 5.28. In the case of a finite field F of order q, then $|M(A)| = q^3$, $|N_1| = |N_2| = q^2$ and $|N_1 \cap N_2| = |\langle(x-1)e_1\rangle \cap \langle xe_1 \rangle| = |\langle x(x-1)e_1 \rangle| = q$. Therefore $|N_1 \cup N_2| = |N_1| + |N_2| - |N_1 \cap N_2| = q^2 + q^2 - q$. So the number of vectors v with $M(A) = \langle v \rangle$ is $|M(A)| - |N_1 \cup N_2| = q^3 - q^2 - q^2 + q = q(q-1)^2$.

The next theorem will help our manipulation of direct sums of companion matrices and establish the connection between the rational and primary canonical forms.

Theorem 5.31 (The Chinese remainder theorem for companion matrices)

Let $f(x)$ and $g(x)$ be monic polynomials of positive degrees over a field F with $\gcd\{f(x), g(x)\} = 1$. Then $C(f(x)g(x)) \sim C(f(x)) \oplus C(g(x))$.

Proof

Write

$$f(x) = x^s + a_{s-1}x^{s-1} + \cdots + a_1 x + a_0,$$

$$g(x) = x^t + b_{t-1}x^{t-1} + \cdots + b_1 x + b_0$$

and so $\deg f(x) = s$, $\deg g(x) = t$. Working in the $F[x]$-module $M = M(C(f(x)g(x)))$ we describe a particular invertible $(s+t) \times (s+t)$ matrix Y over F satisfying $YC(f(x)g(x))Y^{-1} = C(f(x)) \oplus C(g(x))$. Now M is cyclic with generator $e_1 \in F^{s+t}$ of order $f(x)g(x)$ by Theorem 5.26. Then $v_1 = g(x)e_1$ has order $f(x)$ in M by Lemma 5.23. By Corollary 5.29 the submodule $N_1 = \langle v_1 \rangle$ of M has F-basis $\mathcal{B}_{v_1}$ consisting of $v_1, xv_1, \ldots, x^{s-1}v_1$. As $x^{k-1}e_1 = e_k$ in M for $1 \le k \le s + t$ we see $v_1 = g(x)e_1 = (b_0, b_1, \ldots, b_{t-1}, 1, 0, 0, \ldots, 0) \in F^{s+t}$, $xv_1 = xg(x)e_1 = (0, b_0, b_1, \ldots, b_{t-1}, 1, 0, 0, \ldots, 0) \in F^{s+t}$. This pattern (moving the coefficients of $g(x)$ one place right) continues until the last vector of $\mathcal{B}_{v_1}$ is obtained namely $x^{s-1}v_1 = x^{s-1}g(x)e_1 = (0, \ldots, 0, b_0, b_1, \ldots, b_{t-1}, 1) \in F^{s+t}$. In the same way, by Lemma 5.23 and Corollary 5.29, the order of $v_2 = f(x)e_1$ in M is $g(x)$ and the submodule $N_2 = \langle v_2 \rangle$ of M has F-basis $\mathcal{B}_{v_2}$ consisting of $v_2, xv_2, \ldots, x^{t-1}v_2$. As before these t elements of F^{s+t} are

$$v_2 = f(x)e_1 = (a_0, a_1, \ldots, a_{s-1}, 1, 0, \ldots, 0),$$

$$xv_2 = xf(x)e_1 = (0, a_0, a_1, \ldots, a_{s-1}, 1, 0, \ldots, 0), \quad \ldots,$$

until finally

$$x^{t-1}v_2 = x^{t-1}f(x)e_1 = (0, \ldots, 0, a_0, a_1, \ldots, a_{s-1}, 1).$$

The condition $\gcd\{f(x), g(x)\} = 1$ has not been used yet but it comes into play now to show that the vectors in $\mathcal{B} = \mathcal{B}_{v_1} \cup \mathcal{B}_{v_2}$ are linearly independent. Suppose there are scalars $c_i, d_j \in F$ for $0 \le i < s, 0 \le j < t$ satisfying

$$c_0 v_1 + c_1 x v_1 + \cdots + c_{s-1} x^{s-1} v_1 + d_0 v_2 + d_1 x v_2 + \cdots + d_{t-1} x^{t-1} v_2 = 0.$$

Write $c(x) = c_0 + c_1 x + \cdots + c_{s-1} x^{s-1}$ and $d(x) = d_0 + d_1 x + \cdots + d_{t-1} x^{t-1}$. On substituting $v_1 = g(x)e_1$ and $v_2 = f(x)e_1$ in the above equation, the polynomials $c(x)$ and $d(x)$ over F are seen to satisfy

$$(c(x)g(x) + d(x)f(x))e_1 = 0.$$

Therefore $c(x)g(x) + d(x)f(x)$ belongs to the order ideal $\langle f(x)g(x) \rangle$ of e_1 in M, that is, $f(x)g(x) | (c(x)g(x) + d(x)f(x))$. So $f(x) | c(x)g(x)$ which gives $f(x) | c(x)$ as $\gcd\{f(x), g(x)\} = 1$. As $\deg c(x) < s = \deg f(x)$ we deduce $c(x) = 0(x)$, that is, $c_i = 0$ for $0 \le i < s$. In the same way $g(x) | d(x)f(x)$ which leads to $g(x) | d(x)$ and so $d(x) = 0(x)$, that is, $d_j = 0$ for $0 \le j < t$ as $\deg d(x) < t = \deg f(x)$. The conclusion is: the $s + t$ vectors in $\mathcal{B}$ are linearly independent and so $\mathcal{B}$ is a basis of F^{s+t}, as these $s + t$ vectors necessarily span F^{s+t}. From Lemmas 5.18 and 5.19 we deduce $M = N_1 \oplus N_2$. The linear mapping $\alpha : F^{s+t} \to F^{s+t}$ determined by $C(f(x)g(x))$ has matrix $C(f(x)g(x))$ relative to the standard basis $\mathcal{B}_0$ of F^{s+t}. By Corollary 5.20 the matrix of α relative to $\mathcal{B}$ is $A_1 \oplus A_2$ where A_k is the matrix of the restriction $\alpha|_{N_k}$

of α to N_k relative to $\mathcal{B}_{v_k}$ for $k = 1, 2$. Now $(v)\alpha = xv$ for all $v \in M$ and so using the theory preceding Definition 5.25 we see $A_1 = C(f(x))$ and $A_2 = C(g(x))$. Let Y denote the invertible $(s + t) \times (s + t)$ matrix over F having the vectors of $\mathcal{B}$ as its rows. Therefore

$$
Y = \left(
\begin{array}{ccccccccc}
b_0 & b_1 & \cdots & b_{t-1} & 1 & 0 & 0 & \cdots & 0 \\
0 & b_0 & b_1 & \cdots & b_{t-1} & 1 & 0 & \cdots & 0 \\
\vdots & \ddots & \ddots & \ddots & & \ddots & \ddots & \ddots & \vdots \\
\vdots & & \ddots & \ddots & \ddots & & \ddots & \ddots & 0 \\
0 & 0 & \cdots & 0 & b_0 & b_1 & \cdots & b_{t-1} & 1 \\
\hline
a_0 & a_1 & \cdots & a_{s-1} & 1 & 0 & 0 & \cdots & 0 \\
0 & a_0 & a_1 & \cdots & a_{s-1} & 1 & 0 & \cdots & 0 \\
\vdots & \ddots & \ddots & \ddots & & \ddots & \ddots & \ddots & \vdots \\
\vdots & & \ddots & \ddots & \ddots & & \ddots & \ddots & 0 \\
0 & 0 & \cdots & 0 & a_0 & a_1 & \cdots & a_{s-1} & 1
\end{array}
\right)
$$

being partitioned by its first s rows and last t rows as indicated. By Lemma 5.2 we see that Y satisfies

$$ Y C(f(x)g(x)) Y^{-1} = C(f(x)) \oplus C(g(x)) $$

and so $C(f(x)g(x)) \sim C(f(x)) \oplus C(g(x))$. $\square$

We continue to use the notation

$$ f(x) = x^s + a_{s-1} x^{s-1} + \cdots + a_1 x + a_0, $$

$$ g(x) = x^t + b_{t-1} x^{t-1} + \cdots + b_1 x + b_0 $$

for monic polynomials over a field F although we do *not* assume $\gcd\{f(x), g(x)\} = 1$. Working in $M = M(C(f(x)g(x)))$ as in Theorem 5.31, let Y' denote the $(s + t) \times (s + t)$ matrix over F having the vectors of the ordered set $\mathcal{B}_{v_1} \cup \mathcal{B}_{v_2}$ as its rows. Let T denote the $(s + t) \times (s + t)$ matrix over F with $e_k T = e_{s+t+1-k}$ for $1 \le k \le s + t$. So the rows of T are the rows of the identity matrix I but appearing in the opposite order: the first row of T is the last row of I, the second row of T is the last-but-one row of I and so on. The reader can check $T^2 = I$ and so $T^{-1} = T$. Also $\det T = \pm 1$ (see Exercises 5.2, Question 6(d)). The $(s + t) \times (s + t)$ matrix $TY'T$ is obtained from Y' by reversing the order of its rows and at the same time reversing the order of its columns. Thus

$$
TY'T = \begin{pmatrix}
1 & a_{s-1} & \cdots & a_1 & a_0 & 0 & 0 & \cdots & 0 \\
0 & 1 & a_{s-1} & \cdots & a_1 & a_0 & 0 & \cdots & 0 \\
\vdots & \ddots & \ddots & \ddots & & \ddots & \ddots & \ddots & \\
\vdots & & \ddots & \ddots & \ddots & & \ddots & \ddots & 0 \\
0 & 0 & \cdots & 0 & 1 & a_{s-1} & \cdots & a_1 & a_0 \\
1 & b_{t-1} & \cdots & b_1 & b_0 & 0 & 0 & \cdots & 0 \\
0 & 1 & b_{t-1} & \cdots & b_1 & b_0 & 0 & \cdots & 0 \\
\vdots & \ddots & \ddots & \ddots & & \ddots & \ddots & & \vdots \\
\vdots & & \ddots & \ddots & \ddots & & \ddots & \ddots & 0 \\
0 & 0 & \cdots & 0 & 1 & b_{t-1} & \cdots & b_1 & b_0
\end{pmatrix}
$$

is partitioned into its first t rows and its last s rows. It is customary to write

$$R(f(x), g(x)) = \det TY'T \text{ which is called } \textit{the resultant} \text{ of } f(x) \text{ and } g(x).$$

For instance

$$
R(x - r, x^2 + ax + b) = \begin{vmatrix}
1 & -r & 0 \\
0 & 1 & -r \\
1 & a & b
\end{vmatrix} = r^2 + ar + b.
$$

We apply the *ecos* $c_3 + xc_2$, $c_3 + x^2 c_1$ to the underlying matrix and expand along col 3 to get

$$
\begin{vmatrix}
1 & -r & x(x - r) \\
0 & 1 & x - r \\
1 & a & x^2 + ax + b
\end{vmatrix} = -(x + a + r)(x - r) + x^2 + ax + b = r^2 + ar + b.
$$

We have expressed the resultant as a polynomial linear combination of $x - r$ and $x^2 + ax + b$. This method is used in the proof below of the main property Corollary 5.32 of resultants.

Corollary 5.32

Let $f(x)$, $g(x)$ be monic polynomials of positive degree over a field F. Then $\gcd\{f(x), g(x)\} = 1$ if and only if their resultant $R(f(x), g(x))$ is non-zero.

Proof

Suppose $\gcd\{f(x), g(x)\} = 1$. The matrix Y of Theorem 5.31 is invertible over F and as $Y = Y'$ we obtain $R(f(x), g(x)) = \det TYT \neq 0$. Conversely suppose $R(f(x), g(x)) \neq 0$. We apply the $s + t - 1$ *ecos* $c_{s+t} + x^{s+t-k}c_k$ over $F[x]$ to $TY'T$ for $1 \leq k < s + t$ where $\deg f(x) = s$, $\deg g(x) = t$. These *ecos* leave $\det TY'T$ unchanged and all columns of $TY'T$ except the last unchanged. However the new column $s + t$ is

$$(x^{t-1}f(x), \ldots, xf(x), f(x), x^{s-1}g(x), \ldots, xg(x), g(x))^T.$$

Expanding the resulting determinant along this column produces polynomials $a(x)$ and $b(x)$ of degrees at most $t - 1$ and $s - 1$ respectively over F satisfying $a(x)f(x) + b(x)g(x) = R(f(x), g(x))$. Therefore $\gcd\{f(x), g(x)\}$ is a divisor of the non-zero scalar $R(f(x), g(x))$. So $\gcd\{f(x), g(x)\} = 1$. □

The matrix Y satisfying $YC(f(x)g(x))Y^{-1} = C(f(x)) \oplus C(g(x))$ as in Theorem 5.31 is such that $\det Y = \det TYT = R(f(x), g(x))$. Because of the close connection between Y and $R(f(x), g(x))$ in the case $\gcd\{f(x), g(x)\} = 1$ we have allowed ourselves the above small digression on resultants in general.

EXERCISES 5.2

1. (a) Write down the companion matrix $C = C(x^3 - x)$ of $x^3 - x$ over $\mathbb{Q}$. List the eigenvalues of C and find a row eigenvector corresponding to each of them. Specify an invertible 3×3 matrix X over $\mathbb{Q}$ such that XCX^{-1} is diagonal.

   (b) Show $|C(f(x))| = (-1)^t a_0$ where $f(x) = x^t + a_{t-1}x^{t-1} + \cdots + a_1 x + a_0$ over a field F. Deduce $\operatorname{rank} C(f(x)) = t$ or $t - 1$ according as $a_0 \neq 0$ or $a_0 = 0$.

   (c) Let $C = C(x^2 + ax + b)$ over a field F. Verify $C^2 + aC + bI = 0$ and $|xI - C| = x^2 + ax + b$. Show that $D = C^2$ satisfies $D^2 + (2b - a^2)D + b^2 I = 0$.

   (d) Let $f(x)$ be a monic polynomial of degree t over a field F. Show that $C(f(x))$ is similar to a diagonal matrix over F if and only if $f(x)$ has t distinct zeros in F. Working over

   $$\mathbb{F}_4 = \{0, 1, c, c + 1 : 1 + 1 = 0, c^2 + c + 1 = 0\}$$

   find an invertible 4×4 matrix X with $XC(x^4 + x)X^{-1}$ diagonal.

2. (a) Verify that e_1 generates the $\mathbb{Q}[x]$-module $M(A)$ where

$$A = \begin{pmatrix} 1 & 0 & 1 \\ 1 & 1 & 0 \\ 1 & 1 & 0 \end{pmatrix}$$

over $\mathbb{Q}$. Calculate the order of e_1 in $M(A)$. Without further calculation write down the characteristic polynomial $\chi_A(x)$ of A and verify that

$$X = \begin{pmatrix} e_1 \\ xe_1 \\ x^2 e_1 \end{pmatrix}$$

is invertible over $\mathbb{Q}$ and satisfies $XAX^{-1} = C(\chi_A(x))$. Specify two linearly independent row eigenvectors of A.

(b) Let

$$A = \begin{pmatrix} 1 & 0 & 1 \\ 1 & 1 & 1 \\ 1 & 0 & 1 \end{pmatrix}$$

over $\mathbb{Q}$. Find the orders of e_1, e_2, e_3 and $e_1 + e_2 + e_3$ in the $\mathbb{Q}[x]$-module $M(A)$. Do any of these vectors generate $M(A)$? (Yes/No.) Verify that $e_1 + e_2$ generates $M(A)$ and find an invertible matrix X over $\mathbb{Q}$ satisfying $XAX^{-1} = C(\chi_A(x))$. Specify generators of the eight submodules of $M(A)$ and draw their lattice diagram.

(c) Let

$$A = \begin{pmatrix} 2 & 2 & 2 \\ -1 & -1 & 0 \\ -1 & -1 & -1 \end{pmatrix}$$

over $\mathbb{Q}$. Show that the $\mathbb{Q}[x]$-module $M(A)$ is cyclic and specify an invertible matrix X over $\mathbb{Q}$ satisfying $XAX^{-1} = C(\chi_A(x))$. Specify non-zero submodules N_1 and N_2 with $M(A) = N_1 \oplus N_2$.

(d) Write $C = C(f(x))$ where $f(x)$ is a monic polynomial of degree 3 over a field F. Use the factorisation of $f(x)$ into irreducible polynomials over F and Theorem 5.28 to show that the number of submodules of $M(C)$ is 2, 4, 6 or 8. Sketch the (five) possible lattice diagrams and state which of them cannot occur in the cases

(i) $F = \mathbb{R}$;   (ii) $F = \mathbb{C}$;   (iii) $F = \mathbb{Z}_2$.

(e) Let $A = (a_{ij})$ be a $t \times t$ matrix over a field F such that $a_{ii+1} \neq 0$ for $1 \leq i < t$ and $a_{ij} = 0$ whenever $j > i + 1$. Show that the $F[x]$-module $M(A)$ is cyclic with generator e_1.

Hint: Establish $\langle e_1, e_2, \ldots, e_s \rangle \subseteq \langle e_1, xe_1, \ldots, x^{s-1}e_1 \rangle$ for $1 \leq s \leq t$ by induction on s. Find the order of e_1 in the $\mathbb{Q}[x]$-module $M(A)$ where

$$A = \begin{pmatrix} 0 & 1 & 0 & 0 \\ 2 & 0 & 1 & 0 \\ 0 & 0 & 0 & 1 \\ 0 & 0 & 2 & 0 \end{pmatrix}.$$

Specify a generator of the submodule N of $M(A)$ satisfying $0 \neq N \neq M(A)$. Is every non-zero element of N a generator of N? Locate the elements v with $\langle v \rangle = M(A)$.

(f) Let

$$A = \begin{pmatrix} 0 & a & b \\ 0 & c & d \\ 0 & 0 & 0 \end{pmatrix}$$

be a matrix over a field F. Find a necessary and sufficient condition on a, b, c, d for the $F[x]$-module $M(A)$ to be cyclic with generator e_1.

3. (a) Write down a proof of Lemma 5.23 using Lemma 2.7 as guide.

(b) The element v_0 of the $\mathbb{Q}[x]$-module M has order $x^3 + x^2 + x + 1$. Find the orders in M of:

$$(x^2 + x)v_0, \qquad (x^4 + 2x^2 + 1)v_0, \qquad (2x^2 + x - 1)v_0,$$

$$(x^9 - x)v_0, \qquad (x^7 - x^6 + x^5 - x^4)v_0.$$

(c) Let $M = \mathbb{Z}_2[x]/K$ where $K = \langle x^4 + x \rangle$. Find the orders in M of: $K + x$, $K + x^2$, $K + \bar{1} + x + x^2$. Find those elements $v \in M$ such that $M = \langle v \rangle$.

(d) The cyclic $F[x]$-module M is generated by v_0 having monic order $d_0(x)$. Let v in M have order $d(x)$. Show

$$M = \langle v \rangle \quad \Leftrightarrow \quad d(x) = d_0(x).$$

The cyclic $F[x]$-module M' contains two elements v_0 and v_1 having equal orders such that $M' = \langle v_0 \rangle$, $M' \neq \langle v_1 \rangle$. What type of module must M' be? Specify an example of M', v_0, v_1 with this property.

(e) Let F be a field and let M be a cyclic $F[x]$-module with generator v_0 of monic order $f_1(x)f_2(x)$ where $\gcd\{f_1(x), f_2(x)\} = 1$ in M. Show that $M = N_1 \oplus N_2$ where $N_1 = \langle f_1(x)v_0 \rangle$ and $N_2 = \langle f_2(x)v_0 \rangle$. Suppose $v = v_1 + v_2$ where $v_1 \in N_1$, $v_2 \in N_2$. Show that the order of v in M is $g_1(x)g_2(x)$ where $g_i(x)$ is the order of v_i in N_i for $i = 1, 2$. Deduce $M = \langle v \rangle \Leftrightarrow N_1 = \langle v_1 \rangle$ and $N_2 = \langle v_2 \rangle$.

(f) Let $M = \mathbb{Z}_3[x]/K$ where $K = \langle x^4 + x^2 \rangle$ and write $v_0 = K + \bar{1}$. Write $N_1 = \langle x^2 v_0 \rangle$ and $N_2 = \langle (x^2 + \bar{1})v_0 \rangle$. What are the orders in M of

$$\text{(i)} \quad x^2 v_0 \quad \text{and} \quad \text{(ii)} \quad (x^2 + \bar{1})v_0?$$

Is $M = N_1 \oplus N_2$? (Yes/No). Show that N_1 contains 8 vectors v_1 with $N_1 = \langle v_1 \rangle$. How many vectors $v_2 \in N_2$ satisfy $N_2 = \langle v_2 \rangle$? How many vectors $v \in M$ satisfy $M = \langle v \rangle$?

4. The $F[x]$-module M contains an element v_0 of order $d_0(x)$ where F is a field.

(a) Let N be a submodule of M. Show that

$$K_N = \{f(x) \in F[x] : f(x)v_0 \in N\}$$

is an ideal of $F[x]$ with $\langle d_0(x) \rangle \subseteq K_N$. Show $K_N = \langle d_0(x) \rangle \Leftrightarrow \langle v_0 \rangle \cap N = 0$.
Let N_1 and N_2 be submodules of M. Show $N_1 \subseteq N_2 \Rightarrow K_{N_1} \subseteq K_{N_2}$. Show that the reverse implication is true in the case $M = \langle v_0 \rangle$.

(b) Let K be an ideal of $F[x]$. Show that $N_K = \{f(x)v_0 : f(x) \in K\}$ is a submodule of M. Specify a generator of the cyclic $F[x]$-module N_K. Let K_1 and K_2 be ideals of $F[x]$ containing $\langle d_0(x) \rangle$. Show $K_1 \subseteq K_2 \Leftrightarrow N_{K_1} \subseteq N_{K_2}$.

(c) Suppose $M = \langle v_0 \rangle$. Let $\mathbb{L}$ denote the set of submodules N of M and let $\mathbb{L}'$ denote the set of ideals K of $F[x]$ with $\langle d_0(x) \rangle \subseteq K$. Show $N = N_{K_N}$ for all $N \in \mathbb{L}$. Show also $K = K_{N_K}$ for all $K \in \mathbb{L}'$. Deduce that $N \to K_N$ is an inclusion-preserving bijection: $\mathbb{L} \to \mathbb{L}'$ with an inclusion-preserving inverse.

(d) Let $M = \mathbb{Z}_3[x]/\langle x^3 - x \rangle$ and $v_0 = \langle x^3 - x \rangle + \bar{1}$ (so $F = \mathbb{Z}_3$, $d_0(x) = x^3 - x$). Specify generators $d(x)v_0$ of each of the eight submodules N of M where $d(x)|d_0(x)$. For each N specify a generator of K_N. Arrange the submodules of M in their lattice diagram. Use the sieve formula (Exercises 4.1, Question 7(c)) to find the number of $v \in M$ with $M = \langle v \rangle$.

(e) Answer (d) above in the case of

$$M = \mathbb{Z}_3[x]/\langle x^5 + x \rangle \quad \text{and} \quad v_0 = \langle x^5 + x \rangle + \bar{1}.$$

Hint: The irreducible factors of $x^5 + x$ over $\mathbb{Z}_3$ have degrees 1 and 2.

5. (a) Let $d_0(x)$ be a monic polynomial of positive degree t over a field F. Write $M = F[x]/K$ where $K = \langle d_0(x) \rangle$ and so M is a cyclic $F[x]$-module with generator $K + 1$. Let $f(x)$ be a polynomial over F and let $N = \{K + g(x) : f(x)g(x) \in K\}$. Show that N is a submodule of M. Let $d(x) = \gcd\{f(x), d_0(x)\}$ and write $q(x) = d_0(x)/d(x)$. Show that $K + q(x)$ generates N.

   (b) Let A be a $t \times t$ matrix over a field F. Suppose that the $F[x]$-module $M(A)$ is cyclic with generator v_0 of order $\chi_A(x)$. Let $f(x)$ be a polynomial over F. Show that $N = \{v \in F^t : vf(A) = 0\}$ is a submodule of $M(A)$. (You should know, from the theory of linear homogeneous equations, that N is a subspace of F^t and $\dim N = t - \operatorname{rank} f(A)$.) As above let $d(x) = \gcd\{f(x), \chi_A(x)\}$ and $q(x) = \chi_A(x)/d(x)$. Show that N is cyclic with generator $v_0 q(A)$ of order $d(x)$ in $M(A)$. Deduce $\operatorname{rank} f(A) = t - \deg d(x)$.

   (c) Let $C = C((x^2 + 1)(x - 1))$ over $\mathbb{Q}$. Verify $\operatorname{rank} C - I = 2$ and $\operatorname{rank} C^2 + I = 1$. Determine the ranks of C^{100}, $C^{100} - C^{50}$, $C^{100} + C^{50}$ and $C^{100} - C^{52}$.

6. (a) Working over an arbitrary field F, use Theorem 5.31 to find invertible matrices Y_1 and Y_2 over F satisfying $Y_1 C(x(x - 1)^2) Y_1^{-1} = C(x) \oplus C((x - 1)^2)$, $Y_2 C(x^2(x + 1)^2) Y_2^{-1} = C(x^2) \oplus C((x + 1)^2)$.

   (b) Verify $R(x - r, x^3 + ax^2 + bx + c) = r^3 + ar^2 + br + c$ and evaluate $R(x^2 + ax + b, x^2 + cx + d)$ in the same way.

   (c) Let $f(x)$ and $g(x)$ be monic polynomials of positive degrees s and t over a field F. Show $R(f(x), g(x)) = (-1)^{st} R(g(x), f(x))$.

   (d) Let T denote the $t \times t$ matrix over a field F with $e_i T = e_{t+1-i}$ for $1 \leq i \leq t$. Show $\det T = (-1)^{t(t-1)/2}$.

Canonical Forms and Similarity Classes of Square Matrices over a Field

We are now ready to combine the theory developed in Chapters 4 and 5. Suppose given a $t \times t$ matrix A over a field F. We show how to find an invertible $t \times t$ matrix X over F such that the structure of XAX^{-1} is revealed for all to see. It is too much to expect XAX^{-1} to be diagonal although this can almost be achieved (any non-zero (i, j)-entries in the *Jordan normal form* of A have $i = j$ or $i + 1 = j$) should all eigenvalues of A belong to F. Rather we adopt an approach which works unconditionally for all square matrices: just as each finite abelian group G can be expressed as a direct sum of cyclic subgroups, so X can always be found such that XAX^{-1} is a direct sum of companion matrices. We will see that the analogy between $\mathbb{Z}$ and $F[x]$ extends to one between the non-zero finite $\mathbb{Z}$-module G and the $F[x]$-module $M(A)$ determined by A. This analogy is in no way hindered by the serendipitous fact that Cyclic and Companion begin with the same letter! Thus two finite non-trivial $\mathbb{Z}$-modules are isomorphic if and only if they have the same isomorphism type

$$C_{d_1} \oplus C_{d_2} \oplus \cdots \oplus C_{d_s}$$

where $d_j | d_{j+1}$ for $1 \leq j < s$, $d_1 > 1$, $s \geq 1$, whereas two $t \times t$ matrices over F are similar if and only if they have the same *rational canonical form* (rcf)

$$C(d_1(x)) \oplus C(d_2(x)) \oplus \cdots \oplus C(d_s(x))$$

where $d_j(x) | d_{j+1}(x)$ for $1 \leq j < s$, $\deg d_1(x) > 0$, $s \geq 1$. The reduction of the characteristic matrix $xI - A$ over $F[x]$ to its Smith normal form is the 'lion's share' of the calculation of X such that XAX^{-1} is the rcf of A. Consequently it is possible

C. Norman, *Finitely Generated Abelian Groups and Similarity of Matrices over a Field*, 251
Springer Undergraduate Mathematics Series,
DOI 10.1007/978-1-4471-2730-7_6, © Springer-Verlag London Limited 2012

to resolve the question of whether or not two square matrices over F are similar by an algorithmic method in which factorisation problems do not arise. Further it is essentially the Euclidean algorithm which does the trick. The monic polynomials $d_j(x)$ appearing in the rcf of A are unique and are called *the invariant factors of A*. This is the content of Section 6.1.

Should the factorisation of the characteristic polynomial $\chi_A(x) = |xI - A|$ into irreducible polynomials over F be known, it is relatively easy to derive *the primary canonical form (pcf)* of A and, where appropriate, *the Jordan normal form (Jnf)* of A. This is carried out in Section 6.2.

In Section 6.3 we study the endomorphisms and automorphisms of the $F[x]$-module $M(A)$, that is, the algebra of matrices in $\mathfrak{M}_t(F)$ which commute with A. Knowledge of the pcf of A will enable us to estimate the number of matrices similar to A and, in certain cases, find the partition of $\mathfrak{M}_t(F)$ into similarity classes.

6.1 The Rational Canonical Form of Square Matrices over a Field

Let F denote a field and let A be a $t \times t$ matrix over F. In order to analyse the $F[x]$-module $M(A)$ determined by A (see Definition 5.8), we begin with the $F[x]$-module

$$F[x]^t = F[x] \oplus F[x] \oplus \cdots \oplus F[x]$$

which is the direct sum of t copies of $F[x]$. The elements of $F[x]^t$ are t-tuples $(f_1(x), f_2(x), \ldots, f_t(x))$ of polynomials over F. Addition in $F[x]^t$ is carried out component-wise and the product of $f(x)$ in $F[x]$ by an element of $F[x]^t$ is given by

$$f(x)(f_1(x), f_2(x), \ldots, f_t(x)) = (f(x)f_1(x), f(x)f_2(x), \ldots, f(x)f_t(x)).$$

We will not make use of the fact that $F[x]^t$ is a ring as there is no need in this context to multiply two t-tuples together. However matrix multiplication will play a role: the elements of $F[x]^t$ are $1 \times t$ matrices over $F[x]$ and postmultiplication by matrices in $\mathfrak{M}_t(F[x])$ produce endomorphisms of $F[x]^t$.

Let $e_i(x)$ denote the element of $F[x]^t$ with $1(x)$ in position i and zeros elsewhere $(1 \leq i \leq t)$. Then $e_1(x), e_2(x), \ldots, e_t(x)$ is an $F[x]$-basis of $F[x]^t$ called the *standard basis* of $F[x]^t$. So $F[x]^t$ is a free $F[x]$-module of rank t by Definition 2.22. The connection between $F[x]^t$ and the $F[x]$-module $M(A)$ is provided by the *evaluation homomorphism*

$$\theta_A : F[x]^t \to M(A) \text{ defined by } (f_1(x), f_2(x), \ldots, f_t(x))\theta_A = \sum_{i=1}^{t} f_i(x)e_i.$$

The verification that θ_A is an $F[x]$-linear mapping is left to the reader (Exercises 6.1, Question 8(b)). The expression $\sum_{i=1}^{t} f_i(x)e_i$ on the right-hand side of the above equation is an element of the $F[x]$-module $M(A)$; so $f_i(x)e_i = e_i f_i(A)$ is row i of the $t \times t$ matrix $f_i(A)$ over F by Definition 5.8. Therefore

$$\sum_{i=1}^{t} f_i(x)e_i = \sum_{i=1}^{t} e_i f_i(A) \in F^t$$

showing that θ_A maps t-tuples of polynomials to t-tuples of scalars. In particular $(e_i(x))\theta_A = (0, \ldots, 0, 1(x), 0, \ldots, 0)\theta_A = e_i I = e_i$ for $1 \leq i \leq t$ as $1(A) = I$, showing that θ_A maps the standard basis of $F[x]^t$ to the standard basis of F^t. More generally

$$\left(\sum_{i=1}^{t} a_i e_i(x)\right)\theta_A = \sum_{i=1}^{t} a_i e_i,$$

that is, θ_A maps each t-tuple of constant polynomials over F into the corresponding vector in F^t for all $A \in \mathfrak{M}_t(F)$. As $F^t = \langle e_1, e_2, \ldots, e_t \rangle \subseteq \mathrm{im}\,\theta_A$ we see $\mathrm{im}\,\theta_A = F^t = M(A)$, that is, θ_A is surjective.

The next lemma helps us 'nail down' the kernel of θ_A.

Lemma 6.1

Let K denote the submodule of $F[x]^t$ generated by the rows of $xI - A$. Then $f(x)e_i(x) \equiv e_i f(A) \pmod{K}$ for $f(x) \in F[x]$ and $1 \leq i \leq t$.

Proof

Notice first that each element of K is expressible as the matrix product of an element $(f_1(x), f_2(x), \ldots, f_t(x))$ of $F[x]^t$ and $xI - A$, since

$$\sum_{i=1}^{t} f_i(x)(e_i(x)(xI - A)) = \left(\sum_{i=1}^{t} f_i(x)e_i(x)\right)(xI - A).$$

So K is a submodule of $F[x]^t$ being the image of the 'postmultiplication by $xI - A$' endomorphism of $F[x]^t$. Write $f(x) = \sum_{j=0}^{n} a_j x^j$ and let i satisfy $1 \leq i \leq t$. Then $a_0 e_i(x) = e_i(a_0 I)$ and so $a_0 e_i(x) \equiv e_i(a_0 I) \pmod{K}$. Also

$$e_i(x)(x^j I - A^j) = e_i(x)(x^{j-1}I + x^{j-2}A + \cdots + xA^{j-2} + A^{j-1})(xI - A) \in K$$

for $1 \leq j \leq n$ which shows $x^j e_i(x) = e_i(x)x^j I \equiv e_i(x)A^j \pmod{K}$. Hence $a_j x^j e_i(x) \equiv e_i(a_j A^j) \pmod{K}$ for $0 \leq j \leq n$. Adding up these $n + 1$ congruences gives $f(x)e_i(x) \equiv e_i f(A) \pmod{K}$ for $1 \leq i \leq t$. □

We show next that the kernel of θ_A is a free submodule of $F[x]^t$. In fact, by the polynomial analogue of Theorem 3.1, all submodules of $F[x]^t$ are free and of rank at most t (Exercises 6.1, Question 8(a)). Here rank $\ker \theta_A = t$ and what is more we are given an $F[x]$-basis of $\ker \theta_A$ 'on a plate' as we show next.

Theorem 6.2

Let A be a $t \times t$ matrix over a field F. The rows of the characteristic matrix $xI - A$ are the elements of an $F[x]$-basis of $\ker \theta_A$, where $\theta_A : F[x]^t \to M(A)$ is the evaluation homomorphism.

Proof

Let a_{ij} denote the (i, j)-entry in A for $1 \le i, j \le t$. We establish three properties of the rows of $xI - A$: first they belong to $\ker \theta_A$, secondly they generate $\ker \theta_A$, and thirdly they are $F[x]$-linearly independent.

Row i of $xI - A$ is $e_i(x)(xI - A) = e_i(x)xI - e_i A = xe_i(x) - (a_{i1}, a_{i2}, \ldots, a_{it})$ and so its image by θ_A is

$$(e_i(x)(xI - A))\theta_A = (xe_i(x))\theta_A - (a_{i1}, a_{i2}, \ldots, a_{it})\theta_A$$

$$= xe_i - (a_{i1}, a_{i2}, \ldots, a_{it}) = e_i A - e_i A = 0$$

for $1 \le i \le t$. Therefore the t rows of $xI - A$ belong to $\ker \theta_A$.

Using the notation of Lemma 6.1, let K denote the submodule of $F[x]^t$ generated by the rows of $xI - A$. By the preceding paragraph we obtain $K \subseteq \ker \theta_A$. Consider an element $\sum_{i=0}^{t} f_i(x)e_i(x)$ of $F[x]^t$. Taking $f(x) = f_i(x)$ in Lemma 6.1 gives $f_i(x)e_i(x) \equiv e_i f_i(A) \pmod{K}$ for $1 \le i \le t$. Adding these t congruences together produces

$$\sum_{i=0}^{t} f_i(x)e_i(x) \equiv \sum_{i=0}^{t} e_i f_i(A) \pmod{K}.$$

Now suppose $\sum_{i=0}^{t} f_i(x)e_i(x)$ belongs to $\ker \theta_A$ and so $\sum_{i=1}^{t} e_i f_i(A) = 0$. Therefore

$$\sum_{i=0}^{t} f_i(x)e_i(x) \equiv 0 \pmod{K}$$

which means $\sum_{i=0}^{t} f_i(x)e_i(x) \in K$. Therefore $\ker \theta_A \subseteq K$ which gives $\ker \theta_A = K$. So the rows of $xI - A$ generate $\ker \theta_A$.

Finally suppose the rows of $xI - A$ to be $F[x]$-linearly dependent (we are looking for a contradiction). There is a non-zero element $\sum_{i=1}^{t} g_i(x)e_i(x)$ of $F[x]^t$ with

$(g_1(x), g_2(x), \ldots, g_t(x))(xI - A) = 0$ which is an equation between t-tuples of polynomials over F. Let $g_{j_0}(x)$ have maximum degree among $g_i(x)$ for $1 \le i \le t$. Then $x g_{j_0}(x) = \sum_{i=1}^{t} a_{i j_0} g_i(x)$ on comparing j_0th entries. But

$$\deg x g_{j_0}(x) = 1 + \deg g_{j_0}(x) > \deg \sum_{i=1}^{t} a_{i j_0} g_i(x)$$

and so the above polynomial equality is false as the degrees of its sides are unequal. Therefore the rows of $xI - A$ are $F[x]$-linearly independent. So we've shown that the t rows of $xI - A$ form an $F[x]$-basis of $\ker \theta_A$. □

Example 6.3

Take

$$F = \mathbb{Q} \quad \text{and} \quad A = \begin{pmatrix} 1 & 1 & 1 \\ 0 & 2 & 1 \\ 0 & 0 & 1 \end{pmatrix}.$$

We find the images of a few elements of $\mathbb{Q}[x]^3$ by $\theta_A : \mathbb{Q}[x]^3 \to M(A)$. For instance

$$(2, x, x^2)\theta_A = (2e_1(x) + x e_2(x) + x^2 e_3(x))\theta_A = 2e_1 + x e_2 + x^2 e_3$$

$$= 2e_1 + e_2 A + e_3 A^2 = (2, 0, 0) + (0, 2, 1) + (0, 0, 1) = (2, 2, 2).$$

In the same way

$$(x^2, 1 - x, x^3)\theta_A = e_1 A^2 + e_2 - e_2 A + e_3 A^3$$

$$= (1, 3, 3) + (0, 1, 0) - (0, 2, 1) + (0, 0, 1) = (1, 2, 3).$$

Denote row i of

$$xI - A = \begin{pmatrix} x - 1 & -1 & -1 \\ 0 & x - 2 & -1 \\ 0 & 0 & x - 1 \end{pmatrix}$$

by $z_i(x)$ for $i = 1, 2, 3$.

Be careful to change the signs of all entries in A when constructing $xI - A$. In this case $z_1(x) = (x - 1, -1 - 1)$, $z_2(x) = (0, x - 2, -1)$, $z_3(x) = (0, 0, x - 1)$. By Theorem 6.2 we know that $z_1(x), z_2(x), z_3(x)$ is a $\mathbb{Q}[x]$-basis of $\ker \theta_A$. Does $(x^2 - x, -2, -2)$ belong to $\ker \theta_A$? To answer this question apply θ_A and see whether or not the resulting vector is zero. As

$$(x^2 - x, -2, -2)\theta_A = e_1 A^2 - e_1 A - 2e_2 - 2e_3 = (1, 3, 3) - (1, 1, 1) - (0, 2, 2)$$

$$= (0, 0, 0)$$

we see $(x^2 - x, -2, -2) \in \ker\theta_A$. As A is upper triangular (its (i, j)-entries are zero for $i > j$), it is easy to express elements of $\ker\theta_A$ as polynomial linear combinations of the rows of $xI - A$; here $(x^2 - x, -2, -2) = xz_1(x) + z_2(x) + z_3(x)$.

The equations $z_i(x) = e_i(x)(xI - A)$ for $i = 1, 2, 3$ show how the matrix $xI - A$ relates the $\mathbb{Q}[x]$-basis $z_1(x), z_2(x), z_3(x)$ of $\ker\theta_A$ and the $\mathbb{Q}[x]$-basis $e_1(x), e_2(x), e_3(x)$ of $\mathbb{Q}[x]^3$. Can we find $\mathbb{Q}[x]$-bases of $\ker\theta_A$ and $\mathbb{Q}[x]^3$ related in this way by a *diagonal* matrix? If so, then it is a small step, as we set out below, to express $M(A)$ as a direct sum of cyclic submodules and simultaneously find an invertible 3×3 matrix X over $\mathbb{Q}$ such that XAX^{-1} is a direct sum of companion matrices. Further, should the diagonal matrix referred to be in *Smith normal form* $\mathrm{diag}(1(x), d_1(x), d_2(x))$ where $d_1(x) \neq 1(x)$, then $XAX^{-1} = C(d_1(x)) \oplus C(d_2(x))$ where $d_1(x)|d_2(x)$, that is, XAX^{-1} is in rational canonical form. The reader should not be surprised that the answer to the above question is: Yes! A suitable $\mathbb{Q}[x]$-basis of $\mathbb{Q}[x]^3$ is provided by the rows of $Q(x)$ satisfying

$$P(x)(xI - A)Q(x)^{-1} = S(xI - A)$$

where $P(x)$, $Q(x)$ are invertible over $\mathbb{Q}[x]$ and $S(xI - A) = \mathrm{diag}(1, d_1(x), d_2(x))$ is the Smith normal form of $xI - A$. Using the method of Section 4.2, the sequence of elementary operations:

$$c_1 \leftrightarrow c_2, \qquad -c_1, \qquad c_2 - (x - 1)c_1, \qquad c_3 + c_1, \qquad r_2 + (x - 2)r_1,$$

$$r_2 \leftrightarrow r_3, \qquad c_2 \leftrightarrow c_3, \qquad r_3 + r_2$$

reduces $xI - A$ to its Smith normal form

$$S(xI - A) = \mathrm{diag}(1, x - 1, (x - 1)(x - 2))$$

showing $d_1(x) = x - 1$, $d_2(x) = (x - 1)(x - 2)$. Applying the *eros* in the above sequence to I gives

$$P(x) = \begin{pmatrix} 1 & 0 & 0 \\ 0 & 0 & 1 \\ x - 2 & 1 & 1 \end{pmatrix} \quad \text{and} \quad Q(x) = \begin{pmatrix} x - 1 & -1 & -1 \\ 0 & 0 & 1 \\ 1 & 0 & 0 \end{pmatrix}$$

on applying the conjugates of the *ecos* in the above sequence to I. Then

$$P(x)(xI - A) = S(xI - A)Q(x)$$

and so $P(x)(xI - A)Q(x)^{-1} = \mathrm{diag}(1, x - 1, (x - 1)(x - 2))$. Write $\rho_i(x) = e_i(x)Q(x)$ and so $\rho_i(x)$ is row i of $Q(x)$ for $i = 1, 2, 3$. By Corollary 2.23 we see that $\rho_1(x), \rho_2(x), \rho_3(x)$ form a $\mathbb{Q}[x]$-basis of $\mathbb{Q}[x]^3$ as $\det Q(x) = -1$. By Theorem 6.2 the rows of $xI - A$ form a $\mathbb{Q}[x]$-basis of $\ker\theta_A$ and as $\det P(x) = -1$ the same is true of the rows of $P(x)(xI - A)$. So the rows of $S(xI - A)Q(x)$, that is, $\rho_1(x), (x-1)\rho_2(x), (x-1)(x-2)\rho_3(x)$ form a $\mathbb{Q}[x]$-basis of $\ker\theta_A$. We have achieved the polynomial analogue of the $\mathbb{Z}$-bases ρ_1, ρ_2, ρ_3 of $\mathbb{Z}^3$ and $\rho_1, 2\rho_2, 26\rho_3$ of $\ker\theta$ used in Example 3.2. Here

$$(\rho_1(x))\theta_A = (x-1, -1, -1)\theta_A = e_1 A - e_1 - e_2 - e_3 = (0, 0, 0)$$

has order 1 in $M(A)$ and, being zero, it is a *redundant* generator. Write $v_1 = (\rho_2(x))\theta_A = (e_3(x))\theta_A = e_3$; then v_1 has order $x-1$ in $M(A)$ as $v_1 \neq 0$, but $(x-1)v_1 = e_3 A - e_3 = e_3 - e_3 = (0, 0, 0)$. Write $v_2 = (\rho_3(x))\theta_A = (e_1(x))\theta_A = e_1$; then v_2 has order $(x-1)(x-2)$ in $M(A)$ as $(x-1)v_2 = (x-1)e_1 = (0, 1, 1) \neq 0$, $(x-2)v_2 = (x-2)e_1 = (-1, 1, 1) \neq 0$ and $(x-1)(x-2)v_2 = (x-1)(-1, 1, 1) = (-1, 1, 1)(A - I) = 0$. Anticipating the theory in Theorem 6.5 below we conclude

$$M(A) = \langle v_1 \rangle \oplus \langle v_2 \rangle$$

that is, $M(A)$ is the direct sum of the cyclic submodule generated by v_1 and the cyclic submodule generated by v_2. Finally, combining Lemma 5.19, Corollary 5.20, Theorem 5.24 and Corollary 5.27, the vectors v_1, v_2, xv_2 make up the basis $\mathcal{B} = \mathcal{B}_{v_1} \cup \mathcal{B}_{v_2}$ of $\mathbb{Q}^3$ and

$$X = \begin{pmatrix} v_1 \\ v_2 \\ xv_2 \end{pmatrix} = \begin{pmatrix} 0 & 0 & 1 \\ 1 & 0 & 0 \\ 1 & 1 & 1 \end{pmatrix}$$

satisfies

$$XAX^{-1} = C(x-1) \oplus C((x-1)(x-2)) = \left(\begin{array}{c|cc} 1 & 0 & 0 \\ \hline 0 & 0 & 1 \\ 0 & -2 & 3 \end{array} \right)$$

which is the rational canonical form of A.

Definition 6.4

Let $d_1(x), d_2(x), \ldots, d_s(x)$ be non-constant monic polynomials over a field F with $d_j(x) | d_{j+1}(x)$ for $1 \leq j < s$. Let $t = \deg d_1(x) + \deg d_2(x) + \cdots + \deg d_s(x)$. Then the $t \times t$ matrix

$$C(d_1(x)) \oplus C(d_2(x)) \oplus \cdots \oplus C(d_s(x))$$

over F is said to be in *rational canonical form* (rcf).

So to be in rcf a matrix must be the direct sum of companion matrices of polynomials each being a divisor of the next.

Our next theorem is the culmination of the theory in Chapters 4 and 5. It is the analogue of Theorem 3.4 in the case of a *finite* abelian group G with $|G| > 1$.

Theorem 6.5 (The existence of the rational canonical form)

Let A be a $t \times t$ matrix over a field F. There are monic polynomials $d_1(x), d_2(x), \ldots,$ $d_s(x)$ of positive degree over F satisfying $d_i(x)|d_{i+1}(x)$ for $1 \le i < s$ where $s \le t$ and vectors v_i of order $d_i(x)$ in $M(A)$ for $1 \le i \le s$ such that

$$M(A) = \langle v_1 \rangle \oplus \langle v_2 \rangle \oplus \cdots \oplus \langle v_s \rangle.$$

Let X denote the invertible $t \times t$ matrix over F having as its rows the vectors of the basis $\mathcal{B} = \mathcal{B}_{v_1} \cup \mathcal{B}_{v_2} \cup \cdots \cup \mathcal{B}_{v_s}$ of F^t. Then

$$XAX^{-1} = C(d_1(x)) \oplus C(d_2(x)) \oplus \cdots \oplus C(d_s(x))$$

is in rational canonical form and $\chi_A(x) = d_1(x)d_2(x) \cdots d_s(x)$.

Proof

We use the evaluation homomorphism $\theta_A : F[x]^t \to M(A)$ which is surjective. By Theorem 6.2 the rows of $xI - A$ are the elements of an $F[x]$-basis of $\ker \theta_A$. Applying Theorem 4.16 with $A(x) = xI - A$, there are invertible $t \times t$ matrices $P(x)$ and $Q(x)$ over $F[x]$ satisfying

$$P(x)(xI - A) = S(xI - A)Q(x)$$

where $S(xI - A) = \mathrm{diag}(1, \ldots, 1, d_1(x), d_2(x), \ldots, d_s(x))$ is the Smith normal form of $xI - A$ and $1 \le s \le t$. The polynomials $d_1(x), d_2(x), \ldots, d_s(x)$ are the non-constant diagonal entries in $S(xI - A)$ and so are monic, of positive degree over F, and satisfy $d_i(x)|d_{i+1}(x)$ for $1 \le i < s$ by Definition 4.12. Write $\rho_i(x) = e_i(x)Q(x)$ for $1 \le i \le t$. As $P(x)$ is invertible over $F[x]$, the rows of $P(x)(xI - A)$ form an $F[x]$-basis of $\ker \theta_A$ by Corollary 2.21. Using the displayed equation above, the rows of $S(xI - A)Q(x)$ namely

$$\rho_1(x), \ldots, \rho_{t-s}(x), d_1(x)\rho_{t-s+1}(x), \ldots, d_s(x)\rho_t(x)$$

also form an (actually the same!) $F[x]$-basis of $\ker \theta_A$. As $Q(x)$ is invertible over $F[x]$, its rows

$$\rho_1(x), \quad \rho_2(x), \quad \ldots, \quad \rho_t(x)$$

form an $F[x]$-basis of $F[x]^t$ by Corollary 2.23. We now mimic the proof of Theorem 3.4 by using these closely related $F[x]$-bases to decompose $M(A) \cong F[x]^t/\ker \theta_A$. As $\rho_1(x), \rho_2(x), \ldots, \rho_t(x)$ generate the $F[x]$-module $F[x]^t$ and θ_A is surjective, we see that $(\rho_1(x))\theta_A, (\rho_2(x))\theta_A, \ldots, (\rho_t(x))\theta_A$ generate the $F[x]$-module $M(A) = \mathrm{im}\,\theta_A$. But $(\rho_i(x))\theta_A = 0$ for $1 \le i \le t - s$ as

$\rho_1(x), \ldots, \rho_{t-s}(x)$ belong to $\ker \theta_A$, being the first $t - s$ elements of the above $F[x]$-basis of $\ker \theta_A$. Discarding these $t - s$ redundant generators, all of which are zero, we see that the remaining s vectors $(\rho_{t-s+1}(x))\theta_A, \ldots, (\rho_t(x))\theta_A$ generate $M(A)$. Write $v_i = (\rho_{t-s+i}(x))\theta_A$ for $1 \le i \le s$. Then $M(A) = \langle v_1 \rangle + \langle v_2 \rangle + \cdots + \langle v_s \rangle$ showing $M(A)$ to be the sum of s non-trivial cyclic submodules. Next Lemma 2.15 is used to show that $M(A)$ is the internal direct sum of $\langle v_1 \rangle, \langle v_2 \rangle, \ldots, \langle v_s \rangle$ and to determine the order of each v_i in $M(A)$. Suppose

$$u_1 + u_2 + \cdots + u_s = 0 \qquad (\clubsuit)$$

where $u_i \in \langle v_i \rangle$ for $1 \le i \le s$. So $u_i = f_i(x)v_i$ where $f_i(x) \in F[x]$ for $1 \le i \le s$. Substituting for u_i and v_i equation $(\clubsuit)$ gives

$$(f_1(x)\rho_{t-s+1}(x) + f_2(x)\rho_{t-s+2}(x) + \cdots + f_s(x)\rho_t(x))\theta_A$$

$$= f_1(x)((\rho_{t-s+1}(x))\theta_A) + f_2(x)((\rho_{t-s+2}(x))\theta_A) + \cdots + f_s(x)((\rho_t(x))\theta_A)$$

$$= f_1(x)v_1 + f_2(x)v_2 + \cdots + f_s(x)v_s = u_1 + u_2 + \cdots + u_s = 0$$

which shows that $f_1(x)\rho_{t-s+1}(x) + f_2(x)\rho_{t-s+2}(x) + \cdots + f_s(x)\rho_t(x)$ belongs to $\ker \theta_A$. Using the $F[x]$-basis of $\ker \theta_A$ displayed above, there are polynomials $q_i(x)$ over F $(1 \le i \le t)$ with

$$f_1(x)\rho_{t-s+1}(x) + f_2(x)\rho_{t-s+2}(x) + \cdots + f_s(x)\rho_t(x)$$

$$= q_1(x)\rho_1(x) + \cdots + q_{t-s}(x)\rho_{t-s}(x) + q_{t-s+1}(x)d_1(x)\rho_{t-s+1}(x) + \cdots$$

$$+ q_t(x)d_s(x)\rho_t(x).$$

As $\rho_1(x), \rho_2(x), \ldots, \rho_t(x)$ are $F[x]$-linearly independent, the coefficients of $\rho_i(x)$ on opposite sides of the above equation are equal: so $q_i(x) = 0$ for $1 \le i \le t - s$ (since these $\rho_i(x)$ appear on one side only), and the $\rho_{t-s+i}(x)$ give

$$f_i(x) = q_{t-s+i}(x)d_i(x) \quad \text{for } 1 \le i \le s.$$

On substituting for $f_i(x)$ and v_i we obtain

$$u_i = f_i(x)v_i = q_{t-s+i}(x)d_i(x)(\rho_{t-s+i}(x))\theta_A$$

$$= q_{t-s+i}(x)((d_i(x)\rho_{t-s+i}(x))\theta_A) = 0$$

since $d_i(x)\rho_{t-s+i}(x) \in \ker \theta_A$ for $1 \le i \le s$. By Lemma 2.15 we conclude

$$M(A) = \langle v_1 \rangle \oplus \langle v_2 \rangle \oplus \cdots \oplus \langle v_s \rangle$$

showing that $M(A)$ is the internal direct sum of s non-zero cyclic submodules.

Let K_i be the order ideal Definition 5.11 of v_i in $M(A)$ for $1 \leq i \leq s$. As $d_i(x)\rho_{t-s+i}(x) \in \ker\theta_A$, we see

$$d_i(x)v_i(x) = d_i(x)((\rho_{t-s+i}(x))\theta_A) = (d_i(x)\rho_{t-s+i}(x))\theta_A = 0$$

showing $d_i(x) \in K_i$. Conversely suppose $g_i(x) \in K_i$. Then $g_i(x)v_i = 0$ which leads to $g_i(x)\rho_{t-s+i}(x) \in \ker\theta_A$. Using the above $F[x]$-basis of $\ker\theta_A$ consisting of the rows of $S(xI - A)Q(x)$ we deduce $d_i(x)|g_i(x)$. Therefore $K_i = \langle d_i(x)\rangle$ showing that v_i has order $d_i(x)$ in $M(A)$ for $1 \leq i \leq s$.

Finally we construct a matrix X such that XAX^{-1} is in rcf Definition 6.4. Let $\alpha : F^t \to F^t$ be the linear mapping determined by the given $t \times t$ matrix A over F and so $(v)\alpha = xv = vA$ for all $v \in F^t$. Write $N_i = \langle v_i\rangle$ for $1 \leq i \leq s$. By Theorem 5.24 the vectors $x^j v_i$ for $j = 0, 1, 2, \ldots, \deg d_i(x) - 1$, in that order, make up the basis $\mathcal{B}_{v_i}$ of N_i. By the discussion following Theorem 5.24, the matrix of $\alpha|_{N_i} : N_i \to N_i$, the restriction of α to the cyclic submodule N_i, relative to $\mathcal{B}_{v_i}$ is the companion matrix $C(d_i(x))$ for $1 \leq i \leq s$. By the first part of the proof $M(A) = N_1 \oplus N_2 \oplus \cdots \oplus N_s$. Therefore $\mathcal{B} = \mathcal{B}_{v_1} \cup \mathcal{B}_{v_2} \cup \cdots \cup \mathcal{B}_{v_s}$ is a basis of F^t by Lemma 5.18. Let X be the $t \times t$ matrix having the vectors of $\mathcal{B}$ as its rows. Using Lemma 5.2 and Corollary 5.20 we obtain

$$XAX^{-1} = C(d_1(x)) \oplus C(d_2(x)) \oplus \cdots \oplus C(d_s(x)).$$

The polynomials $d_i(x)$ for $1 \leq i \leq s$ satisfy Definition 6.4 and so XAX^{-1} is in rational canonical form. The above matrix equation leads to

$$X(xI - A)X^{-1} = (xI - C(d_1(x))) \oplus (xI - C(d_2(x))) \oplus \cdots \oplus (xI - C(d_s(x))).$$

Taking determinants:

$$\chi_A(x) = |xI - A| = |X||xI - A||X|^{-1} = |X(xI - A)X^{-1}|$$

$$= |(xI - C(d_1(x))) \oplus (xI - C(d_2(x))) \oplus \cdots \oplus (xI - C(d_s(x)))|$$

$$= |xI - C(d_1(x))||xI - C(d_2(x))| \cdots |xI - C(d_s(x))| = d_1(x)d_2(x)\cdots d_s(x)$$

on using the theory following Definition 5.17 and Theorem 5.26.   □

The 'end of the road' is in sight thanks to Theorem 6.5 and some comments are in order. First the $F[x]$-module $M(A)$ decomposes into a direct sum of cyclic modules for any $t \times t$ matrix A over any field F. The set-up is uncannily analogous to that of finite abelian groups Theorem 3.7: although the decomposition itself is not in general unique, the divisor sequence $(d_1(x), d_2(x), \ldots, d_s(x))$ as in Theorem 6.5 is uniquely determined by A (see Theorem 6.6 below) and characterises the similarity class of A.

Secondly the reader is now able to transform any $t \times t$ matrix A over any field F into its rcf XAX^{-1}. The 'stumbling-block' is the elementary but laborious process of calculating $S(xI - A)$ and finding a suitable matrix $Q(x)$ as in Theorem 4.16; the reader will be aware that all the foregoing examples of A are either 'nice' from a theoretical point of view or have $t \leq 3$. Nevertheless the rcf is as good as one can hope for: each similarity class contains a unique matrix in rcf and, being a direct sum of companion matrices, the Cayley–Hamilton theorem Corollary 6.11 is a direct consequence of its existence.

We now prepare to prove the polynomial analogue of Theorem 3.7: for each $t \times t$ matrix A over a field F the polynomials $d_1(x), d_2(x), \ldots, d_s(x)$ in Theorem 6.5 are *unique*. Let $d(x)$ be a polynomial over the field F and let M be an $F[x]$-module. Then $\mu_{d(x)} : M \rightarrow M$, defined by $(v)\mu_{d(x)} = d(x)v$ for all $v \in M$, is $F[x]$-linear. In other words the mapping $\mu_{d(x)}$, which multiplies each element v of M by $d(x)$, is an endomorphism of M. Write

$$\operatorname{im} \mu_{d(x)} = d(x)M \quad \text{and} \quad \ker \mu_{d(x)} = M_{(d(x))}$$

and so $d(x)M$ and $M_{(d(x))}$ are submodules of M.

For $F[x]$-modules M and M' we have

$$d(x)(M \oplus M') = (d(x)M) \oplus (d(x)M') \quad \text{and}$$

$$(M \oplus M')_{(d(x))} = M_{(d(x))} \oplus M'_{(d(x))}$$

showing that the external direct sum (Exercises 2.3, Question 7(f)) is respected in this context. We are duty bound to consider decompositions of modules into direct sums of cyclic modules as in Theorem 6.5 and it is convenient to use the most economic notation Lemma 5.22 for cyclic torsion $F[x]$-modules, namely

$$F[x]/\langle d_0(x)\rangle \quad \text{with } d_0(x) \text{ monic}$$

which is cyclic with generator

$$\langle d_0(x)\rangle + 1(x) \quad \text{of order } d_0(x).$$

From Lemmas 5.22 and 5.23 we deduce

$$d(x)(F[x]/\langle d_0(x)\rangle) \cong F[x]/\langle d_0(x)/\gcd\{d(x), d_0(x)\}\rangle$$

as the above module on the left is cyclic with generator $\langle d_0(x)\rangle + d(x)$ of order $d_0(x)/\gcd\{d(x), d_0(x)\}$. Using Lemma 5.23 again gives

$$(F[x]/\langle d_0(x)\rangle)_{(d(x))} \cong F[x]/\langle \gcd\{d(x), d_0(x)\}\rangle$$

as the coset $\langle d_0(x)\rangle + d_0(x)/\gcd\{d(x), d_0(x)\}$ of order $\gcd\{d(x), d_0(x)\}$ generates the above module on the left.

The analogy between Theorems 3.7 and 6.6 is so close that the reader is merely given a start on the proof of Theorem 6.6 and then encouraged to complete it by referring back to Theorem 3.7.

Theorem 6.6 (The invariance theorem for $F[x]$-modules $M(A)$)

Let M and M' be isomorphic $F[x]$-modules where F is a field. Suppose

$$M \cong F[x]/\langle d_1(x)\rangle \oplus F[x]/\langle d_2(x)\rangle \oplus \cdots \oplus F[x]/\langle d_s(x)\rangle$$

where $d_1(x), d_2(x), \ldots, d_s(x)$ are monic polynomials of positive degree satisfying $d_i(x)|d_{i+1}(x)$ for $1 \leq i < s$. Also suppose

$$M' \cong F[x]/\langle d_1'(x)\rangle \oplus F[x]/\langle d_2'(x)\rangle \oplus \cdots \oplus F[x]/\langle d_{s'}'(x)\rangle$$

where $d_1'(x), d_2'(x), \ldots, d_{s'}'(x)$ are monic polynomials of positive degree satisfying $d_i'(x)|d_{i+1}'(x)$ for $1 \leq i < s'$. Then $s = s'$ and $d_i(x) = d_i'(x)$ for $1 \leq i \leq s$.

Proof

By hypothesis there is an isomorphism $\alpha : M \cong M'$. Let $d(x) \in F[x]$. The $F[x]$-linearity of α gives $\mu_{d(x)}\alpha = \alpha\mu_{d(x)}$ as $(d(x)v)\alpha = d(x)(v)\alpha$ for all $v \in M$. Therefore $(d(x)M)\alpha = \{(d(x)v)\alpha : v \in M\} = \{d(x)(v)\alpha : v \in M\} \subseteq d(x)M'$ as $(v)\alpha \in M'$ for $v \in M$. Replacing α by α^{-1} gives $(d(x)M')\alpha^{-1} \subseteq d(x)M$. Applying α to this inclusion gives $d(x)M' \subseteq (d(x)M)\alpha$. Therefore $(d(x)M)\alpha = d(x)M'$ showing that the submodules $d(x)M$ and $d(x)M'$ correspond under α and so are isomorphic. We write $\alpha| : d(x)M \cong d(x)M'$ as the restriction $\alpha|$ is an isomorphism between $d(x)M$ and $d(x)M'$.

In the same way for $v \in M_{(d(x))}$ we have $(v)\mu_{d(x)} = 0$ and so $0 = (v)\mu_{d(x)}\alpha = (v)\alpha\mu_{d(x)}$ showing $(v)\alpha \in M'_{(d(x))}$. We've shown $(M_{(d(x))})\alpha \subseteq M'_{(d(x))}$. Replacing α by α^{-1} gives $(M'_{(d(x))})\alpha^{-1} \subseteq M_{(d(x))}$ and so on applying α we obtain $M'_{(d(x))} \subseteq (M_{(d(x))})\alpha$. Therefore $(M_{(d(x))})\alpha = M'_{(d(x))}$ showing that the submodules $M_{(d(x))}$ and $M'_{(d(x))}$ correspond under α and so are isomorphic. As before we write $\alpha| : M_{(d(x))} \cong M'_{(d(x))}$. Now take $d(x) = d_1(x)$. Mimicking the proof of Theorem 3.7 step-by-step (Exercises 6.1, Question 8(c)) leads to the conclusion $s = s'$ and $d_i(x) = d_i'(x)$ for $1 \leq i \leq s$. $\qquad\square$

The following corollary is a direct consequence of Theorem 6.6.

Corollary 6.7 (The uniqueness of the rational canonical form)

Let A be a $t \times t$ matrix over a field F. The monic polynomials $d_1(x), d_2(x), \ldots, d_s(x)$ of positive degree over F as in Theorem 6.5 are unique. Also A is similar to a unique matrix in rational canonical form Definition 6.4.

Proof

Suppose as well as the decomposition of $M(A)$ described in Theorem 6.5 we have $M(A) = \langle v_1' \rangle \oplus \langle v_2' \rangle \oplus \cdots \oplus \langle v_{s'}' \rangle$ where $v_i' \neq 0$ has order $d_i'(x)$ in $M(A)$ for $1 \leq i \leq s'$ and $d_i'(x) | d_{i+1}'(x)$ for $1 \leq i < s'$. Now $\langle v_i' \rangle \cong F[x]/\langle d_i'(x) \rangle$ by Lemma 5.22 for $1 \leq i < s'$. Applying Theorem 6.6 with $M = M' = M(A)$ and $\alpha = \iota$, the identity mapping of M, gives $s = s'$ and $d_i(x) = d_i'(x)$ for $1 \leq i \leq s$. So the polynomials $d_1(x), d_2(x), \ldots, d_s(x)$ are unique.

Suppose $A \sim C = C(d_1(x)) \oplus C(d_2(x)) \oplus \cdots \oplus C(d_s(x))$ and $A \sim C' = C(d_1'(x)) \oplus C(d_2'(x)) \oplus \cdots \oplus C(d_{s'}'(x))$ where C and C' are in rcf Definition 6.4. Then $C \sim C'$ and so $M(C) \cong M(C')$ by Theorem 5.13. Write $r(i) = 1 + \sum_{j<i} \deg d_j(x)$ for $1 \leq i \leq s$. Then $v_i = e_{r(i)}$ has order $d_i(x)$ in $M(C)$ for $1 \leq i \leq s$ by Theorem 5.26 and $M(C) = \langle v_1 \rangle \oplus \langle v_2 \rangle \oplus \cdots \oplus \langle v_s \rangle$. In the same way $M(C') = \langle v_1' \rangle \oplus \langle v_2' \rangle \oplus \cdots \oplus \langle v_{s'}' \rangle$ where v_i' has order $d_i'(x)$ in $M(C')$ for $1 \leq i \leq s'$. Applying Theorem 6.6 with $M = M(C)$ and $M' = M(C')$ we conclude $s = s'$ and $d_i(x) = d_i'(x)$ for $1 \leq i \leq s$. So A is similar to a unique matrix in rcf. $\qquad\square$

It is now legitimate to make the next definition.

Definition 6.8

Let A be a $t \times t$ matrix over a field F. The sequence $(d_1(x), d_2(x), \ldots, d_s(x))$ of polynomials as in Theorem 6.5 is called *the invariant factor sequence of A*. The unique matrix $C(d_1(x)) \oplus C(d_2(x)) \oplus \cdots \oplus C(d_s(x))$ as in Definition 6.4 which is similar to A is called *the rational canonical form of A*.

A number of important things are immediately apparent. Suppose we are presented with two $t \times t$ matrices A and A' over a field F. How can we determine whether or not A and A' are similar? The method should be almost second nature to the reader: reduce $xI - A$ and $xI - A'$ to their Smith normal forms $S(xI - A)$ and $S(xI - A')$ using elementary operations over $F[x]$. Then

$$A \sim A' \quad \Leftrightarrow \quad S(xI - A) = S(xI - A')$$

that is, A and A' are similar if and only if the Smith normal forms of their characteristic matrices are equal. Also

$$S(xI - A) = \operatorname{diag}(1, 1, \ldots, 1, d_1(x), d_2(x), \ldots, d_s(x))$$

that is, the invariant factors $d_i(x)$ of A are the last s diagonal entries in $S(xI - A)$ for $1 \leq i \leq s$ being preceded by $t - s$ constant polynomials $1 = 1(x)$ where $s \leq t$. So

$$A \text{ and } A' \text{ are similar} \quad \Leftrightarrow \quad \text{their invariant factor sequences are equal.}$$

Using Theorem 5.13 we see

$$M(A) \cong M(A') \quad \Leftrightarrow \quad A \text{ and } A' \text{ have equal invariant factor sequences.}$$

As an illustration consider the ring $\mathfrak{M}_3(\mathbb{Z}_2)$ of 3×3 matrices A over $\mathbb{Z}_2$. There are $2^9 = 512$ such matrices, there being 2 choices for each of the 9 entries in A. There are $2^3 = 8$ possibilities for the characteristic polynomial of A as $\chi_A(x) = |xI - A| = x^3 + a_2 x^2 + a_1 x + a_0$ there being 2 choices for each of a_0, a_1, a_2. So there are 8 similarity classes of matrices A with $M(A)$ cyclic by Corollary 5.27, such A having a single invariant factor, namely $\chi_A(x)$ and $S(xI - A) = \mathrm{diag}(1, 1, \chi_A(x))$. We know $d_1(x)d_2(x)\cdots d_s(x) = \chi_A(x)$ by Theorem 6.5, that is, the invariant factors of A are certain divisors of the characteristic polynomial of A. For $M(A)$ non-cyclic we have $s \geq 2$ and so $d_1(x)^2 | \chi_A(x)$ as $d_1(x) | d_2(x)$. As $2 \deg d_1(x) \leq \deg \chi_A(x) = 3$ we see $\deg d_1(x) = 1$ and $d_2(x)$ is reducible of degree at most 2. For $s = 3$ the possible invariant factor sequences are (x, x, x) and $(x + 1, x + 1, x + 1)$ arising from $A = 0$ and $A = I$ respectively. For $s = 2$ the possible invariant factor sequences are

$$(x, x^2), \qquad (x + 1, (x + 1)^2), \qquad (x, x(x + 1)), \qquad (x + 1, x(x + 1)).$$

So in all there are 14 similarity classes of 3×3 matrices over $\mathbb{Z}_2$ of which 6 arise from invertible matrices A, that is, those satisfying $\chi_A(0) \neq 0$. So the group $GL_3(\mathbb{Z}_2)$ of order 168 partitions into 6 *conjugacy classes* (in $GL_t(F)$ the terms 'similarity' and 'conjugacy' are interchangeable). By Theorem 5.13 there are 14 isomorphism classes of $\mathbb{Z}_2[x]$-modules $M(A)$ for $A \in \mathfrak{M}_3(\mathbb{Z}_2)$.

We now introduce the polynomial analogue of the exponent Definition 3.11 of a finite abelian group.

Definition 6.9

Let A be a $t \times t$ matrix over a field F. The monic polynomial $\mu_A(x)$ of least degree over F satisfying $\mu_A(A) = 0$ is called *the minimum polynomial of A*.

It is not clear at the outset that each $t \times t$ matrix A over F has a polynomial $\mu_A(x)$ as in Definition 6.9. However it *is* clear that the constant polynomial $1(x)$ cannot be the minimum polynomial of any A since $1(A) = I$ by Definition 5.8 and $I \neq 0$ (the identity matrix cannot equal the zero matrix). So if $\mu_A(x)$ exists then $\deg \mu(x) \geq 1$. So the minimum polynomial of the zero $t \times t$ matrix 0 over F is $\mu_0(x) = x$ since $\mu_0(0) = 0$ and no other monic polynomial of degree 1 has this property. In the same way the minimum polynomial of the identity $t \times t$ matrix I over F is $\mu_I(x) = x - 1$ since $\deg \mu_I(x) = 1$ and $\mu_I(I) = I - I = 0$. More generally $\mu_A(x) = x - a \Leftrightarrow A = aI$ for $a \in F$, that is, scalar matrices and only scalar matrices have minimum polynomials of degree 1. Does the 3×3 matrix

$$A = C(x) \oplus C(x^2) = \left(\begin{array}{c|cc} 0 & 0 & 0 \\ \hline 0 & 0 & 1 \\ 0 & 0 & 0 \end{array}\right)$$

over $\mathbb{Q}$ have a minimum polynomial? A likely candidate is x^2 as $A^2 = 0$, but having Definition 6.9 in mind we must ask: could there be another monic polynomial $f(x)$ of degree 2 over $\mathbb{Q}$ with $f(A) = 0$? If so then $r(x) = f(x) - x^2 = ax + b$ satisfies $r(A) = 0$ where $a, b \in \mathbb{Q}$, that is, $aA + bI = 0$. But

$$aA + bI = \left(\begin{array}{c|cc} b & 0 & 0 \\ \hline 0 & b & a \\ 0 & 0 & b \end{array}\right) = \left(\begin{array}{ccc} 0 & 0 & 0 \\ 0 & 0 & 0 \\ 0 & 0 & 0 \end{array}\right)$$

gives $a = b = 0$. Therefore $f(x) = x^2$ and we conclude $\mu_A(x) = x^2$ from Definition 6.9. Now (x, x^2) is the invariant factor sequence of the above 3×3 matrix A and so, in this case, the minimum polynomial and the *largest* (and last) invariant factor coincide. Our next corollary (analogous to part one of Corollary 3.12) shows that this is always true. It follows that each square matrix over a field *does* have a unique minimum polynomial as described in Definition 6.9.

Corollary 6.10

Let A be a $t \times t$ matrix over a field F. Let $(d_1(x), d_2(x), \ldots, d_s(x))$ denote the invariant factor sequence of A. Let $K_A = \{f(x) \in F[x] : f(A) = 0\}$. Then $K_A = \langle d_s(x) \rangle$, that is, K_A is the principal ideal of $F[x]$ with generator $d_s(x)$. Also $\mu_A(x) = d_s(x)$ and similar matrices have equal minimum polynomials.

Proof

Notice first $K_A = \ker \varepsilon_A$ where $\varepsilon_A : F[x] \to \mathfrak{M}_t(F)$ is the evaluation at A ring homomorphism Definition 5.8 given by $(f(x))\varepsilon_A = f(A)$ for all $f(x) \in F[x]$. Therefore K_A is an ideal of $F[x]$ by Exercises 2.3, Question 3(b) and is called *the annihilator ideal K_A of A*. Our first task is to show $d_s(x) \in K_A$ which we do using the theory of Chapter 5. Write $C = C(d_1(x)) \oplus C(d_2(x)) \oplus \cdots \oplus C(d_s(x))$ for the rcf Definition 6.8 of A. By Theorem 6.5 there is an invertible $t \times t$ matrix X over F with $XAX^{-1} = C$. By Lemma 5.12, with C in place of B, we have $Xf(A)X^{-1} = f(C)$ for all $f(x) \in F[x]$. Therefore $f(A) = 0 \Leftrightarrow f(C) = 0$ which shows $K_A = K_C$; in fact similar matrices have equal annihilator ideals. From the discussion of matrix direct sum after Definition 5.17 and Exercises 5.1, Question 2(d) we obtain

$$f(C) = f(C(d_1(x))) \oplus f(C(d_2(x))) \oplus \cdots \oplus f(C(d_s(x))) \quad \text{for all } f(x) \in F[x] \quad (\spadesuit)$$

which expresses $f(C)$ as a direct sum of matrices $f(C(d_i(x)))$ for $1 \leq i \leq s$. By Theorem 5.26 we know $d_i(C(d_i(x))) = 0$ for $1 \leq i \leq s$. However $d_i(x)|d_s(x)$ and so $d_i(C(d_i(x)))$ is a factor of $d_s(C(d_i(x)))$ for $1 \leq i \leq s$ using the evaluation homomorphism $\varepsilon_{C(d_i(x))}$ of Definition 5.8. So $d_s(C(d_i(x))) = 0$ for $1 \leq i \leq s$ and hence $d_s(C) = 0$ as each of the s matrices in $(\spadesuit)$, with $f(x) = d_s(x)$, is zero. So $d_s(x) \in K_A$.

We now show that $d_s(x)$ generates the ideal K_A. Consider $f(x) \in K_A$. Then $f(A) = 0$ and so $f(C) = Xf(A)X^{-1} = X0X^{-1} = 0$. From $(\spadesuit)$ we deduce $f(C(d_s(x))) = 0$ as a direct sum of matrices is zero if only if each summand (each individual matrix in the direct sum) is zero. The element e_1 has order $d_s(x)$ in the $F[x]$-module $M = M(C(d_s(x)))$ by Theorem 5.26. As $f(x)e_1 = e_1 f(C(d_s(x))) = e_1 0 = 0$ in M we deduce $d_s(x)|f(x)$ from Definition 5.11. From the discussion following Definition 4.3 we conclude $K_A = \langle d_s(x) \rangle$.

Now $d_s(x)$ is the unique monic generator of K_A by Theorem 4.4. Let $f(x)$ in K_A be monic. Then $d_s(x)|f(x)$ implies $\deg d_s(x) \leq \deg f(x)$ and $\deg d_s(x) = \deg f(x)$ implies $d_s(x) = f(x)$. Therefore $d_s(x)$ is the monic polynomial of least degree in K_A, that is, $d_s(x) = \mu_A(x)$ by Definition 6.9. By Corollary 6.7 similar matrices have equal sequences of invariant factors and so, on comparing the last polynomials in these sequences, their minimum polynomials are also equal. $\square$

The minimum polynomial $\mu_A(x)$ is a useful similarity invariant. In the 2×2 case it alone 'does the job': the 2×2 matrices A and B over a field F are similar if and only if $\mu_A(x) = \mu_B(x)$ (see Exercises 6.1, Question 6(a)).

The relationship between $\mu_A(x)$ and the characteristic polynomial $\chi_A(x)$ is the subject of the next corollary, which is analogous to the last part of Corollary 3.12.

Corollary 6.11

Let $\mu_A(x)$ and $\chi_A(x)$ be the minimum and characteristic polynomials of the $t \times t$ matrix A over the field F. Then $\mu_A(x)|\chi_A(x)$ and

$$\chi_A(A) = 0 \quad \text{(the Cayley–Hamilton theorem)}.$$

Also $\chi_A(x)|(\mu_A(x))^t$. The polynomials $\chi_A(x)$ and $\mu_A(x)$ have the same irreducible factors over F.

Proof

As usual let $(d_1(x), d_2(x), \ldots, d_s(x))$ denote the invariant factor sequence of A. Then $\chi_A(x) = \det(xI - A) = d_1(x)d_2(x) \cdots d_s(x)$ by Theorem 6.5. As $\mu_A(x) = d_s(x)$ by Corollary 6.10 we see $\chi_A(x) = d_1(x)d_2(x) \cdots d_{s-1}(x)\mu_A(x)$ showing $\mu_A(x)|\chi_A(x)$. Evaluation Definition 5.8 at A gives $\chi_A(A) = d_1(A)d_2(A) \cdots d_{s-1}(A)\mu_A(A) = 0$ since $\mu_A(A) = 0$ which establishes the Cayley–Hamilton theorem for square matrices over a field.

As $d_i(x)|d_{i+1}(x)$ there is a monic polynomial $q_i(x)$ over F with $d_i(x)q_i(x) = d_{i+1}(x)$ for $1 \leq i < s$. Hence $d_i(x)|d_s(x)$ for $1 \leq i \leq s$ as

$$d_i(x)q_i(x)q_{i+1}(x) \cdots q_{s-1}(x) = d_s(x).$$

Multiplying these s equations together and substituting $\chi_A(x) = d_1(x)d_2(x) \cdots d_s(x)$, $\mu_A(x) = d_s(x)$ gives $\chi_A(x)q_1(x)q_2(x)^2 \cdots q_{s-1}(x)^{s-1} = (\mu_A(x))^s$ which shows $\chi_A(x)|(\mu_A(x))^s$. As $s \leq t$ we also obtain $\chi_A(x)|(\mu_A(x))^t$.

Let $p(x)$ be irreducible Definition 4.7 over F. Suppose $p(x)|\chi_A(x)$. From $\chi_A(x)|(\mu_A(x))^t$ we deduce $p(x)|(\mu_A(x))^t$. Hence $p(x)|\mu_A(x)$ as an irreducible divisor of a product of polynomials must be a divisor of at least one of the polynomials. Conversely suppose $p(x)|\mu_A(x)$. From $\mu_A(x)|\chi_A(x)$ we deduce directly $p(x)|\chi_A(x)$. Therefore $\chi_A(x)$ and $\mu_A(x)$ have the same irreducible factors over F.   □

In Section 6.2 we assume that the factorisation

$$\chi_A(x) = p_1(x)^{n_1} p_2(x)^{n_2} \cdots p_k(x)^{n_k}$$

of $\chi_A(x)$ into positive powers n_j of monic irreducible polynomials $p_j(x)$ over F is known. By Corollary 6.11 the factorisation of $\mu_A(x)$ involves positive (but no larger) powers of the same monic irreducible polynomials, that is,

$$\mu_A(x) = p_1(x)^{n'_1} p_2(x)^{n'_2} \cdots p_k(x)^{n'_k} \quad \text{where } 1 \leq n'_j \leq n_j \text{ for } 1 \leq j \leq k.$$

One further comment: the Cayley–Hamilton theorem holds for square matrices over commutative rings (see P.M. Cohn: Algebra, Volume 1, Wiley (1974)). However the above proof does not 'work' in this general setting as it depends on the existence of the rcf Definition 6.4.

EXERCISES 6.1

1. (a) For each of the following 3×3 matrices A over the rational field $\mathbb{Q}$ reduce its characteristic matrix $xI - A$ to Smith normal form $S(xI - A)$

noting the *eros* and *ecos* used in the reduction. State the invariant factor sequence of A and find invertible 3×3 matrices $P(x)$ and $Q(x)$ over $\mathbb{Q}[x]$ satisfying $P(x)(xI - A) = S(xI - A)Q(x)$. Specify an invertible 3×3 matrix X over $\mathbb{Q}$ such that XAX^{-1} is in rational canonical form.

(i) $\begin{pmatrix} 1 & -1 & 1 \\ -1 & 1 & -1 \\ -2 & 2 & -2 \end{pmatrix}$;   (ii) $\begin{pmatrix} 1 & 1 & -1 \\ -1 & -1 & 1 \\ 2 & 2 & -2 \end{pmatrix}$;

(iii) $\begin{pmatrix} 1 & 1 & -1 \\ 1 & 0 & -1 \\ 5 & 3 & -4 \end{pmatrix}$.

(b) The $t \times t$ matrix A over a field F has sequence

$$(d_1(x), d_2(x), \ldots, d_s(x))$$

of invariant factors $(t \geq 2)$. Use the rcf of A to show rank $A \geq t - s$. Show rank $A = 1$ implies $s = t - 1$, $d_j(x) = x$ for $1 \leq j \leq t - 2$ and $d_{t-1}(x) = x(x - \text{trace } A)$.

(c) Let A be a $t \times t$ matrix over a field F and let $P(x)$, $Q(x)$ be invertible $t \times t$ matrices over $F[x]$ satisfying $P(x)(xI - A) = S(xI - A)Q(x)$. Show $\det P(x) = \det Q(x)$.

2. (a) The $t \times t$ matrix A over a field F has sequence

$$(d_1(x), d_2(x), \ldots, d_s(x))$$

of invariant factors $(t \geq 2)$. For $\lambda \in F$ show that $A - \lambda I$ has sequence of invariant factors $(d_1(x + \lambda), d_2(x + \lambda), \ldots, d_s(x + \lambda))$. *Hint*: Replace x by $x + \lambda$ in

$$P(x)(xI - A) = \text{diag}(1, 1, \ldots, 1, d_1(x), d_2(x), \ldots, d_s(x))Q(x).$$

(b) Find the invariant factor sequences of the following matrices over $\mathbb{Q}$:

(i) $\begin{pmatrix} 2 & -1 & 1 \\ -1 & 2 & -1 \\ -2 & 2 & -1 \end{pmatrix}$;   (ii) $\begin{pmatrix} 3 & 1 & -1 \\ -1 & 1 & 1 \\ 2 & 2 & 0 \end{pmatrix}$;

(iii) $\begin{pmatrix} 6 & 1 & -1 \\ 1 & 5 & -1 \\ 5 & 3 & 1 \end{pmatrix}$.

Hint: Use the answer to Question 1(a) above.

(c) Let $d(x)$ be a monic polynomial of positive degree over a field F. Show nullity $C(d(x)) \leq 1$. Show also nullity $C(d(x)) = 1$ if and only if $d(0) = 0$.

(Remember nullity $A = \dim\{v \in F^t : vA = 0\}$ where A is a $t \times t$ matrix over F.)

Let the $t \times t$ matrix A have invariant factors $d_1(x), d_2(x), \ldots, d_s(x)$. Write nullity $A = n$. Show $n \leq s$. Show also nullity $A = n$ if and only if $x | d_i(x)$ for $s - n < i \leq s$.

(d) Write $C_0 = C(x^4)$ over an arbitrary field F. Determine the invariant factors of C_0^2, C_0^3 and C_0^4. Write $C_1 = C((x^2+1)^2)$ over F. Determine the invariant factors of $C_1^2 + I$. Are C_0^2 and $C_1^2 + I$ similar? Specify invertible 4×4 matrices X_0 and X_1 over F such that $X_0 C_0^2 X_0^{-1}$ is in rcf and $X_1(C_1^2 + I)X_1^{-1}$ is in rcf.

Write $C = C(d(x)^n)$ where $d(x)$ is a monic polynomial of positive degree m over F and n is a positive integer. Use a basis of F^{mn} consisting of vectors, suitably ordered, of the type $x^i d(x)^j e_1$ in the $F[x]$-module $M(C)$ for $0 \leq i < m$, $0 \leq j < n$, to construct an invertible matrix X over F such that $Xd(C)X^{-1}$ is in rcf. Are all the invariant factors of $d(C)$ equal to each other? Is $d(C)$ similar to $C(x^{mn})^m$?

(e) Write $C = C(x^n)$ over an arbitrary field F where n is a positive integer. Let m be an integer with $1 \leq m \leq n$ and suppose $n = qm + r$ with $0 \leq r < m$. Show that the invariant factors of C^m are x^q ($m - r$ times), x^{q+1} (r times).

Hint: Transform C^m into rcf by using, suitably ordered, the basis vectors $x^{jm+i} e_1$ in the module $M(C)$ for $0 \leq jm + i < n$ where $0 \leq i < m$, $0 \leq j \leq q$.

(f) Write $C = C(x^4 + x^2 + 1)$ over an arbitrary field F. Find an invertible 4×4 matrix X over F such that

$$XC^2X^{-1} = C(x^2 + x + 1) \oplus C(x^2 + x + 1).$$

More generally let $d(x)$ be a monic polynomial of positive degree t over a field F. Write $C = C(d(x^2))$. Find an invertible $2t \times 2t$ matrix X over F satisfying

$$XC^2X^{-1} = C(d(x)) \oplus C(d(x)).$$

What is the rcf of $C(d(x^4))^4$?

Hint: Use the module $M(C^2)$ to find X.

3. (a) Let A be a $t \times t$ matrix over a field F where $\chi(F) \neq 2$ ($-1 \neq 1$ in F). Suppose v has order $f(x)$ in the $F[x]$-module $M(A)$. Show that $(-1)^{\deg f(x)} f(-x)$ is the order of v in the $F[x]$-module $M(-A)$.

Suppose that $(d_1(x), d_2(x), \ldots, d_s(x))$ is the invariant factor sequence of A. Describe the invariant factor sequence of $-A$ in terms of the polynomials $d_j(x)$ $(1 \le j \le s)$. State a necessary and sufficient condition on $d_j(x)$ $(1 \le j \le s)$ for $-A \sim A$.

(b) Decide whether or not $-A \sim A$ in the case of

$$A = \begin{pmatrix} 1 & 2 & 3 \\ 2 & 1 & 1 \\ 1 & -1 & -2 \end{pmatrix}$$

over $\mathbb{Q}$.

(c) Let A be a 3×3 matrix over a field F of characteristic not 2. Is the following statement true? $-A \sim A \Leftrightarrow |A| = \text{trace } A = 0$. (Either find a proof or produce a counter-example.)

4. (a) Let A be a $t \times t$ matrix over a field F. By transposing the equation $P(x)(xI - A) = S(xI - A)Q(x)$ show that A and A^T have the same sequence of invariant factors and deduce $A \sim A^T$.

(b) Let $f(x) = a_0 + a_1 x + a_2 x^2 + a_3 x^3 + x^4$ be a monic polynomial over a field F and write

$$R_f = \begin{pmatrix} a_1 & a_2 & a_3 & 1 \\ a_2 & a_3 & 1 & 0 \\ a_3 & 1 & 0 & 0 \\ 1 & 0 & 0 & 0 \end{pmatrix}.$$

Show that $R_f C(f(x))$ is symmetric and deduce $R_f C(f(x)) R_f^{-1} = C(f(x))^T$. More generally let $f(x)$ be a monic polynomial of positive degree t over F. Specify a symmetric invertible $t \times t$ matrix R_f satisfying $R_f C(f(x)) R_f^{-1} = C(f(x))^T$.

(c) Write $C = C(d_1(x)) \oplus C(d_2(x))$ where $d_1(x) = x^2 + a_1 x + a_0$, and $d_2(x) = x^2 + b_1 x + b_0$ are polynomials over F. Using the notation of (b) above, show that $R = R_{d_1} \oplus R_{d_2}$ is a symmetric invertible 4×4 matrix over F satisfying $RCR^{-1} = C^T$. More generally suppose the $t \times t$ matrix C over F to be in rcf. Specify a symmetric invertible $t \times t$ matrix R with $RCR^{-1} = C^T$.

(d) Let A be an arbitrary $t \times t$ matrix over F. Show that there is a symmetric invertible $t \times t$ matrix Y such that $YAY^{-1} = A^T$.
   Hint: Consider $Y = X^T R X$ where $XAX^{-1} = C$ is in rcf.

(e) Find Y as in (d) above in the case of the 3×3 matrix A of Question 1(i) above.

(f) Let F be a field and let U be a subspace of F^t. Write $s = \dim U$ and $U^o = \{v \in F^t : uv^T = 0 \text{ for all } u \in U\}$.

Let A be a $t \times t$ matrix over F and let Y, as in (d) above, be a symmetric and invertible $t \times t$ matrix over F with $YAY^{-1} = A^T$. Let $\gamma : F^t \cong F^t$ be the isomorphism determined Definition 5.8 by Y. Let N be a submodule of $M(A)$. Show that N^o is a submodule of $M(A^T)$ and $(N^o)\gamma$ is a submodule of $M(A)$. Write $(N)\pi = (N^o)\gamma$ and denote by $\mathbb{L}(M(A))$ the set of all submodules of $M(A)$. Show that $\pi : \mathbb{L}(M(A)) \to \mathbb{L}(M(A))$ is a *polarity*, that is,
   (i) $(N)\pi^2 = N$ for all $N \in \mathbb{L}(M(A))$,
   (ii) $N_1 \subseteq N_2 \Leftrightarrow (N_2)\pi \subseteq (N_1)\pi$ where $N_1, N_2 \in \mathbb{L}(M(A))$
      (π is inclusion-reversing).
   Hint: For (i) show $(N)\gamma^{-1} \subseteq ((N^o)\gamma)^o$ (see Exercises 3.1, Question 6(b)).
5. (a) Let $f(x) = a_0 + a_1 x + a_2 x^2 + \cdots + a_{t-1}x^{t-1} + x^t$ be a monic polynomial of positive degree t over a field F with $a_0 \neq 0$. Write

$$f(x)^* = (x^t/a_0)f(1/x)$$

$$= 1/a_0 + (a_{t-1}/a_0)x + (a_{t-2}/a_0)x^2 + \cdots + (a_1/a_0)x^{t-1} + x^t.$$

Let $g(x)$ be a monic polynomial of positive degree s over F with $g(0) \neq 0$. Show $(f(x)g(x))^* = f(x)^* g(x)^*$ and $f(x)^{**} = f(x)$. The polynomial $f(x)$ as above is called *palindromic* if $f(x) = f(x)^*$. Show that the product of palindromic polynomials is itself palindromic and $g(x)g(x)^*$ is palindromic.
Let $f(x)$ as above be palindromic. Show that $f(x)$ satisfies $f(0) = a_0 = \pm 1$. Hence show either $a_i = a_{t-i}$ for $0 \leq i \leq t$ or $a_i = -a_{t-i}$ for $0 \leq i \leq t$. List the palindromic polynomials of degrees 1, 2 and 3 over $\mathbb{Z}_3$.
   (b) Let $f(x) = a_0 + a_1 x + a_2 x^2 + x^3$ over a field F where $a_0 \neq 0$. Calculate (the entries in) $C(f(x))^{-1}$ and hence find an invertible 3×3 matrix X over F with $XC(f(x))^{-1} = C(f(x)^*)X$.
   Hint: Show e_3 generates the $F[x]$-module $M(C(f(x))^{-1})$ and has order $f(x)^*$ in this module.
   (c) Let $f(x)$ be a monic polynomial of positive degree t over the field F with $f(0) \neq 0$. Generalise (b) above to show that the invertible $t \times t$ matrix X with $e_i X = e_{t+1-i}$ for $1 \leq i \leq t$ satisfies $XC(f(x))^{-1} = C(f(x)^*)X$.
   (d) Let A be an invertible $t \times t$ matrix over F with rcf

$$C(d_1(x)) \oplus C(d_2(x)) \oplus \cdots \oplus C(d_s(x)).$$

Show $\chi_A(0) \neq 0$ and deduce $d_j(0) \neq 0$ for $1 \leq j \leq s$. Show also

$$A^{-1} \sim C(d_1(x))^{-1} \oplus C(d_2(x))^{-1} \oplus \cdots \oplus C(d_s(x))^{-1}.$$

Deduce from (c) above that A^{-1} has invariant factor sequence $(d_1(x)^*, d_2(x)^*, \ldots, d_s(x)^*)$. Hence show $A \sim A^{-1}$ if and only if $d_j(x)$ is palindromic for $1 \leq j \leq s$.

(e) List the 12 invariant factor sequences of 3×3 matrices A over $\mathbb{Z}_3$ with $A \sim A^{-1}$.

(f) Working over $\mathbb{Q}$ calculate $\chi_A(x)$ and decide whether or not $A \sim A^{-1}$ in the cases

(i) $A = \begin{pmatrix} 1 & 2 & 5 \\ 2 & 2 & 7 \\ -1 & -1 & -3 \end{pmatrix}$;

(ii) $A = \begin{pmatrix} 2 & 1 & -1 \\ -1 & 0 & 1 \\ 2 & 2 & -1 \end{pmatrix}$.

6. (a) Let A and B be 2×2 matrices over a field F. Show $\mu_A(x) = \mu_B(x)$ implies $A \sim B$.

   Hint: Consider the cases $\deg \mu_A(x) = 1, 2$ separately.

   Construct an example of 3×3 matrices A and B over $\mathbb{Q}$ with $\mu_A(x) = \mu_B(x)$ but A is not similar to B.

   Find a formula (in terms of q) for the number of similarity classes of 2×2 matrices over the finite field $\mathbb{F}_q$ having q elements.

   (b) True or false? The 3×3 matrices A and B over the field F are similar if and only if $\chi_A(x) = \chi_B(x)$ and $\mu_A(x) = \mu_B(x)$.

   (c) Let A be a $t \times t$ matrix over a field F. Let $\varepsilon_A : F[x] \to \mathfrak{M}_t(F)$ denote the evaluation at A ring homomorphism (Definition 5.8). Show $\tilde{\varepsilon}_A : F[x]/\langle \mu_A(x) \rangle \cong \operatorname{im} \varepsilon_A$.

   Hint: Use Exercises 2.3, Question 3(b).

   (d) The $t \times t$ matrix A over the field F is such that $I, A, A^2, \ldots, A^{n-1}$ are linearly independent but $I, A, A^2, \ldots, A^{n-1}, A^n$ are linearly dependent. Show $A^n = a_0 I + a_1 A + a_2 A^2 + \cdots + a_{n-1} A^{n-1}$ where $\mu_A(x) = x^n - a_{n-1} x^{n-1} - \cdots - a_1 x - a_0$ is the minimum polynomial of A.

   Calculate A^2 in the case of

   $$A = \begin{pmatrix} 2 & 1 & 1 \\ -2 & -1 & -2 \\ 1 & 1 & 2 \end{pmatrix}$$

   over $\mathbb{Q}$. Hence find $\mu_A(x)$ and the invariant factors of A.

   (e) Let A_i be a $t_i \times t_i$ matrix over a field F with minimum polynomial $\mu_{A_i}(x)$ for $i = 1, 2$. Show directly from Definition 6.9 that the minimum polynomial of $A_1 \oplus A_2$ is $\operatorname{lcm}\{\mu_{A_1}(x), \mu_{A_2}(x)\}$.

7. (a) By describing the possible invariant factor sequences show that there are $q^3 + q^2 + q$ similarity classes of 3×3 matrices over the finite field $\mathbb{F}_q$. Determine the number of similarity classes of 4×4 matrices over $\mathbb{F}_q$ (it isn't quite $q^4 + q^3 + q^2 + q$).

Show that the number of similarity classes of $n \times n$ matrices A over
$\mathbb{F}_q$ with minimum polynomial of degree m is $P(n, m) \, q^m$ where
$P(n, m)$ is the number of partitions (Definition 3.13) of n having
largest part m. Denote by $P(n, m, l)$ the number of partitions of n with
last part m having l distinct parts. Find a formula involving $P(n, m, l)$
for the number of conjugacy classes in $GL_n(\mathbb{F}_q)$. Determine $P(8, 3)$,
$P(8, 3, 1)$, $P(8, 3, 2)$, $P(8, 3, 3)$ and hence find the number of conjugacy classes in $GL_8(\mathbb{Z}_5)$ having minimum polynomial of degree 3.
Hint: Consider the partition $(t_1, t_2, \ldots, t_s)$ where $(d_1(x), d_2(x), \ldots,$
$d_s(x))$ is the sequence of invariant factors of A and $t_j = \deg d_j(x)$ for
$1 \le j \le s$.

(b) Determine the number of similarity classes of $t \times t$ matrices A over a
given field F satisfying $A^2 = 0$.
How many similarity classes of $t \times t$ matrices A over $\mathbb{Z}_2$ satisfying
$A^2 = I$, $A \ne I$ are there?
Hint: $\mu_A(x)$ is a divisor of $(x - 1)^2$. Does the answer change if $\mathbb{Z}_2$ is
replaced by an arbitrary field F of characteristic 2? How many conjugacy classes of involutions (elements of multiplicative order 2) are
there in the group $GL_t(F)$ where $\chi(F) = 2$?

(c) Let F be a field and let $N(t)$ denote the number of similarity
classes of $t \times t$ matrices A over F satisfying $A^3 = 0$. Show $N(t) =$
$\lfloor (t + 2)/2 \rfloor + N(t - 3)$ for $t > 3$.
Hint: There are $N(t - 3)$ similarity classes of $t \times t$ matrices A over F
with $\mu_A(x) = x^3$. Calculate $N(t)$ for $1 \le t \le 10$.
Write $N'(t) = (1/2)((t/2) + 1)((t/2) + 2) - (\lfloor t/6 \rfloor + 1)((t/2) -$
$(3/2)\lfloor t/6 \rfloor)$ for even t and $N'(t) = (1/2)(\lfloor t/2 \rfloor + 1)(\lfloor t/2 \rfloor + 2) -$
$(\lfloor (t - 3)/6 \rfloor + 1)((t - 3)/2 - (3/2)\lfloor (t - 3)/6 \rfloor)$ for t odd. Verify the
formula $N'(t) = \lfloor (t + 2)/2 \rfloor + N'(t - 3)$ for $t > 3$.
Hint: Treat the cases t odd and t even separately.
Using induction on t deduce $N'(t) = N(t)$ for $t \ge 1$. Find the number
of conjugacy classes of elements of order 3 in the group $GL_{100}(\mathbb{Z}_3)$.

8. (a) Let F be a field and t a positive integer. Show that each submodule K
of $F[x]^t$ is free with rank $K \le t$.
Hint: Generalise Theorem 3.1.

(b) The evaluation mapping $\theta_A : F[x]^t \to M(A)$ is defined by
$(f_1(x), f_2(x), \ldots, f_t(x))\theta_A = \sum_{i=1}^{t} f_i(x)e_i$ for all t-tuples of polynomials $(f_1(x), f_2(x), \ldots, f_t(x)) \in F[x]^t$. Show that θ_A is
$F[x]$-linear.

(c) Complete the proof of Theorem 6.6 by referring back to Theorem 3.7.

6.2 Primary Decomposition of $M(A)$ and Jordan Form

As usual let A denote a $t \times t$ matrix over a field F. Here we assume that the factorisation of the characteristic polynomial $\chi_A(x)$ of A into irreducible polynomials over F is known. The primary decomposition of the $F[x]$-module $M(A)$ is established in Theorem 6.12, the theory being analogous to that of Section 3.2. We obtain *the primary canonical form (pcf)* of A and using the partition function $p(n)$ find the number of similarity classes of matrices having a given characteristic polynomial. The pcf is then modified to give two versions of *the Jordan normal form (Jnf)*: first the 'non-split' case which becomes the 'usual' Jnf should $\chi_A(x)$ factorise into polynomials of degree 1 over F, and secondly the *separable Jordan form (sJf)* in case $\chi_A(x)$ factorises into irreducible polynomials over F none of which have repeated zeros in any extension field of F. The sJf is used in representation theory. Finally we discuss the *real Jordan form* which is important in the theory of dynamical systems.

Let A be a $t \times t$ matrix over a field F with $\chi_A(x) = p_1(x)^{n_1} p_2(x)^{n_2} \cdots p_k(x)^{n_k}$ being the factorisation of $\chi_A(x)$ into positive powers n_j of distinct monic irreducible polynomials $p_j(x)$ over F for $1 \leq j \leq k$.

The $p_j(x)$-component of $M(A)$ is

$$M(A)_{p_j(x)} = \{v \in M(A) : p_j(x)^{n_j} v = 0\} \text{ for } 1 \leq j \leq k.$$

Directly from the above definition we see that the order of each element v of $M(A)_{p_j(x)}$ is a divisor of $p_j(x)^{n_j}$. Conversely let u have order $p_j(x)^{l_j}$ in $M(A)$ where $l_j \geq 0$. So $p_j(x)^{l_j} u = 0$ and also $\chi_A(x)u = 0$ by Corollary 6.11. By Corollary 4.6 we see $p_j(x)^{\min\{l_j, n_j\}} = \gcd\{p_j(x)^{l_j}, \chi_A(x)\}$ satisfies $p_j(x)^{\min\{l_j, n_j\}} u = 0$ and so $u \in M(A)_{p_j(x)}$. Therefore

$M(A)_{p_j(x)}$ consists exactly of those elements u having order a power of $p_j(x)$.

It is straightforward to show that $M(A)_{p_j(x)}$ is a submodule of $M(A)$. Also $M(A)_{p_j(x)}$ is non-zero (Exercises 6.2, Question 3(a)). The k submodules $M(A)_{p_j(x)}$ for $1 \leq j \leq k$ are collectively referred to as *the primary components of* $M(A)$.

For example let

$$A = \begin{pmatrix} 1 & 1 & 1 \\ -1 & -1 & -1 \\ 2 & 2 & 1 \end{pmatrix}$$

over $\mathbb{Q}$. The reader can verify $\chi_A(x) = x^2(x - 1)$ and so $M(A)$ has primary components $M(A)_x$ and $M(A)_{x-1}$. We now find $\mathbb{Q}$-bases of these A-invariant subspaces of

$\mathbb{Q}^3$ in the same way as (row) eigenvectors are found:

$$A^2 = \begin{pmatrix} 1 & 1 & 1 \\ -1 & -1 & -1 \\ 2 & 2 & 1 \end{pmatrix} \begin{pmatrix} 1 & 1 & 1 \\ -1 & -1 & -1 \\ 2 & 2 & 1 \end{pmatrix} = \begin{pmatrix} 2 & 2 & 1 \\ -2 & -2 & -1 \\ 2 & 2 & 1 \end{pmatrix}$$

and therefore

$$M(A)_x = \{v \in \mathbb{Q}^3 : x^2 v = v A^2 = 0\}$$

$$= \{(a, b, c) \in \mathbb{Q}^3 : a - b + c = 0\} = \langle (0, 1, 1), (1, 1, 0) \rangle$$

where $v = (a, b, c)$. So $M(A)_x$ has $\mathbb{Q}$-basis $\mathcal{B}_1$ consisting of u_1, xu_1 where $u_1 = (0, 1, 1)$ and $x^2 u_1 = 0$. Therefore u_1 has order x^2 in $M(A)$ and xu_1 is a row eigenvector of A associated with the eigenvalue 0. The reader can check that

$$M(A)_{x-1} = \{v \in \mathbb{Q}^3 : (x-1)v = v(A-I) = 0\} = \langle (2, 2, 1) \rangle$$

is the row eigenspace of A associated with the eigenvalue 1. So $M(A)_{x-1}$ has $\mathbb{Q}$-basis $\mathcal{B}_2$ consisting of the single vector $u_2 = (2, 2, 1)$ of order $x - 1$ in $M(A)$. In fact $\mathcal{B}_1 \cup \mathcal{B}_2$ is a basis of $\mathbb{Q}^3$ (see the proof of Corollary 6.13) and

$$Y = \begin{pmatrix} u_1 \\ xu_1 \\ u_2 \end{pmatrix} = \begin{pmatrix} 0 & 1 & 1 \\ 1 & 1 & 0 \\ 2 & 2 & 1 \end{pmatrix} \text{ satisfies } YAY^{-1} = C(x^2) \oplus C(x-1).$$

As a second example let $A = C(x^2(x+1)) \oplus C(x(x+1)^2(x-1))$ over $\mathbb{Q}$. So A is a 7×7 partitioned matrix Definition 5.17, the leading entries in the two diagonal blocks (submatrices) being in the $(1, 1)$- and $(4, 4)$-positions. Then $\chi_A(x) = x^3(x+1)^3(x-1)$ and

$$M(A) = \langle e_1 \rangle \oplus \langle e_4 \rangle$$

by Theorem 5.26 where $e_1, e_4 \in \mathbb{Q}^7$. Now e_1 has order $x^2(x+1)$ and e_4 has order $x(x+1)^2(x-1)$ in $M(A)$. From these generators with hybrid orders (orders divisible by two or more irreducible polynomials) we construct vectors having 'pure' orders (orders which are powers of a single irreducible polynomial) which generate the primary components. In this case the primary components of $M(A)$ are:

$$M(A)_x = \langle (x+1)e_1 \rangle \oplus \langle (x+1)^2(x-1)e_4 \rangle \quad \text{of dimension } 2 + 1 = 3,$$

$$M(A)_{x+1} = \langle x^2 e_1 \rangle \oplus \langle x(x-1)e_4 \rangle \quad \text{of dimension } 1 + 2 = 3,$$

$$M(A)_{x-1} = \langle x(x+1)^2 e_4 \rangle \quad \text{of dimension } 1.$$

In fact $M(A) = M(A)_x \oplus M(A)_{x+1} \oplus M(A)_{x-1}$, that is, $M(A)$ is the internal direct sum of its primary components (see Theorem 6.12).

Returning to the general case let A and B be $t \times t$ matrices over a field F and let $\alpha : M(A) \cong M(B)$ be an isomorphism of $F[x]$-modules. Then A and B are similar by Theorem 5.13 and so $\chi_A(x) = \chi_B(x)$ from Lemma 5.5. We proceed to show that restrictions of α give rise to isomorphisms between the primary components of $M(A)$ and those of $M(B)$. Suppose $v \in M(A)_{p_j(x)}$. Applying α to the equation $p_j(x)^{n_j} v = 0$ gives $p_j(x)^{n_j}(v)\alpha = 0$, since α is $F[x]$-linear, which shows $(v)\alpha \in M(B)_{p_j(x)}$. In the same way $w \in M(B)_{p_j(x)}$ implies $(w)\alpha^{-1} \in M(A)_{p_j(x)}$. So the restriction (Definition 5.16) of α to $M(A)_{p_j(x)}$ is an isomorphism $\alpha| : M(A)_{p_j(x)} \cong M(B)_{p_j(x)}$ for $1 \leq j \leq k$.

Our next theorem is the polynomial analogue of Theorem 3.10.

Theorem 6.12 (The primary decomposition of the $F[x]$-module $M(A)$)

Let A be a $t \times t$ matrix over a field F with $\chi_A(x) = p_1(x)^{n_1} p_2(x)^{n_2} \cdots p_k(x)^{n_k}$ where $p_1(x), p_2(x), \ldots, p_k(x)$ are k different monic irreducible polynomials over F and $n_1, n_2, \ldots, n_k$ are positive integers. Then

$$M(A) = M(A)_{p_1(x)} \oplus M(A)_{p_2(x)} \oplus \cdots \oplus M(A)_{p_k(x)}.$$

We omit the proof of Theorem 6.12 as it is the analogous to the proof of Theorem 3.10 (Exercises 6.2, Question 3(d)). Notice that the notation $p_j(x)$ for $1 \leq j \leq k$ imposes an arbitrary ordering on the k monic irreducible factors of $\chi_A(x)$ and determines the order in which the primary components appear in the above decomposition.

Let A and B be $t \times t$ matrices over a field F. Then

$$A \sim B \quad \Leftrightarrow \quad M(A) \cong M(B)$$

$$\Leftrightarrow \quad M(A)_{p_j(x)} \cong M(B)_{p_j(x)} \quad \text{for all } j \text{ with } 1 \leq j \leq k$$

by Theorems 5.13, 6.12 and the preceding discussion. So the similarity class of A depends only on the isomorphism classes of the primary components of $M(A)$. This fact will help in the enumeration of similarity classes discussed after Definition 6.14.

Corollary 6.13

Let A and $\chi_A(x)$ be as in Theorem 6.12 and let α be the linear mapping determined by A. Let $\mathcal{B}_j$ be a basis of $M(A)_{p_j(x)}$ and let A_j be the matrix of $\alpha| : M(A)_{p_j(x)} \cong M(A)_{p_j(x)}$ relative to $\mathcal{B}_j$ for $1 \leq j \leq k$. Then

$$\chi_{A_j}(x) = p_j(x)^{n_j} \quad \text{and} \quad \dim M(A)_{p_j(x)} = \deg p_j(x)^{n_j} \quad \text{for } 1 \leq j \leq k.$$

Further $M(A)_{p_j(x)} \cong M(A_j)$ for $1 \leq j \leq k$.

Proof

Note that $\mathcal{B} = \mathcal{B}_1 \cup \mathcal{B}_2 \cup \cdots \cup \mathcal{B}_k$ is a basis of F^t by Lemma 5.18 and Theorem 6.12. As usual $\chi_{A_j}(x) = \det(xI - A_j)$ and A_j has minimum polynomial $\mu_{A_j}(x)$. From the definition of primary component we deduce $\mu_{A_j}(x)|p_j(x)^{n_j}$ and so $\chi_{A_j}(x) = p_j(x)^{m_j}$ where m_j is a positive integer for $1 \le j \le k$ by Corollary 6.11. By Corollary 5.20 we know $XAX^{-1} = A_1 \oplus A_2 \oplus \cdots \oplus A_k$ where the vectors in $\mathcal{B}$ are the rows of X. By Lemma 5.5 and the discussion of direct sums following Definition 5.17 we obtain $\chi_A(x) = \chi_{A_1}(x)\chi_{A_2}(x) \cdots \chi_{A_k}(x)$, that is,

$$p_1(x)^{n_1} p_2(x)^{n_2} \cdots p_k(x)^{n_k} = p_1(x)^{m_1} p_2(x)^{m_2} \cdots p_k(x)^{m_k}.$$

As $p_1(x), p_2(x), \ldots, p_k(x)$ are distinct monic irreducible polynomials we deduce $n_j = m_j$ for $1 \le j \le k$ from the polynomial analogue of the fundamental theorem of arithmetic. Therefore $\dim M(A)_{p_j(x)} = |\mathcal{B}_j| = \deg \chi_{A_j}(x) = \deg p_j(x)^{n_j}$ for $1 \le j \le k$. Taking $s = k$ and $N_j = M(A)_{p_j(x)}$ in Corollary 5.20 and $N = N_j, \mathcal{B} = \mathcal{B}_j$ in Exercises 5.1, Question 5, gives the stated module isomorphism. $\qquad\square$

The primary decomposition Theorem 3.10 of a finite abelian group G corresponds to the factorisation of $|G|$ into powers of distinct primes. From Theorem 6.12 and Corollary 6.13 we obtain the module analogue: the primary decomposition of $M(A)$ corresponds to the factorisation of $\chi_A(x)$ into powers of distinct monic irreducible polynomials.

For example let A be an 8×8 matrix over $\mathbb{Q}$ with $\chi_A(x) = p_1(x)^2 p_2(x)^3$ where $p_1(x) = x + 1$ and $p_2(x) = x^2 + 1$. We will see shortly that there are 6 similarity classes of such matrices A. The $\mathbb{Q}[x]$-modules $M(A)$ have a common feature: in each case

$$M(A) = M(A)_{p_1(x)} \oplus M(A)_{p_2(x)} \cong M(A_1) \oplus M(A_2)$$

where $\chi_{A_1}(x) = p_1(x)^2$, $\chi_{A_2}(x) = p_2(x)^3$. So $\chi_A(x) = \chi_{A_1}(x)\chi_{A_2}(x)$ is the factorisation of $\chi_A(x)$ into powers of irreducible polynomials over $\mathbb{Q}$.

We now return to the general case of a $t \times t$ matrix A over a field F. Suppose $M(A) = N_1 \oplus N_2 \oplus \cdots \oplus N_s$ where $N_i = \langle v_i \rangle$, for $1 \le i \le s$ v_i has order $d_i(x)$ in $M(A)$ and $d_i(x)|d_{i+1}(x)$ for $1 \le i < s$ as in Theorem 6.5. As before let $\chi_A(x) = p_1(x)^{n_1} p_2(x)^{n_2} \cdots p_k(x)^{n_k}$ be the factorisation of $\chi_A(x)$ into positive powers n_j of distinct monic irreducible polynomials $p_j(x)$ over F for $1 \le j \le k$. As $\chi_A(x) = d_1(x)d_2(x) \cdots d_s(x)$ by Theorem 6.5, each invariant factor $d_i(x)$ has factorisation $d_i(x) = p_1(x)^{t_{i1}} p_2(x)^{t_{i2}} \cdots p_k(x)^{t_{ik}}$ where $0 \le t_{ij} \le n_j$. On comparing powers of $p_j(x)$ we obtain $n_j = t_{1j} + t_{2j} + \cdots + t_{sj}$ where $0 \le t_{1j} \le t_{2j} \le \cdots \le t_{sj}$ for $1 \le j \le k$. By Corollary 6.11 each $t_{sj} > 0$. Let l_j denote the smallest i with $t_{ij} > 0$ for

$1 \leq j \leq k$. Therefore

$(t_{l_j j}, t_{l_j+1 j}, \ldots, t_{sj})$ is a partition of n_j with at most s parts for $1 \leq j \leq k$.

As $l_j = 1 \Leftrightarrow p_j(x) | d_1(x)$, one at least of these partitions has s parts.

Write $m_{ij}(x) = d_i(x)/p_j(x)^{t_{ij}}$ and $v_{ij} = m_{ij}(x)v_i$ for $1 \leq i \leq s, 1 \leq j \leq k$.

The vector v_i generates N_i and has order $d_i(x)$ in $M(A)$. So v_{ij} has order $d_i(x)/\gcd\{m_{ij}(x), d_i(x)\} = d_i(x)/m_{ij}(x) = p_j(x)^{t_{ij}}$ by Lemma 5.23. Let k_i denote the number of positive exponents t_{ij} for $1 \leq i \leq s$. As $d_i(x) | d_{i+1}(x)$ for $1 \leq i < s$ we see $k_1 \leq k_2 \leq \cdots \leq k_s$ and $k_s = k$ by Corollary 6.11. On omitting trivial terms the primary decomposition of N_i is

$$N_i = \sum_j \oplus \langle v_{ij} \rangle \quad \text{for } 1 \leq i \leq s$$

as $d_i(x) = p_1(x)^{t_{i1}} p_2(x)^{t_{i2}} \cdots p_k(x)^{t_{ik}}$, that is, N_i is the internal direct sum of its k_i non-trivial primary components $\langle v_{ij} \rangle$ for $1 \leq i \leq s$. So $N_i \cap M(A)_{p_j(x)} = \langle v_{ij} \rangle$ for $1 \leq i \leq s, 1 \leq j \leq k$. On omitting zero terms in

$$M(A)_{p_j(x)} = \sum_i \oplus \langle v_{ij} \rangle$$

gives the invariant factor decomposition of $M(A)_{p_j(x)}$.

The non-constant monic polynomials $p_j(x)^{t_{ij}}$ are called *the elementary divisors of A*. In other words the elementary divisors of A are the invariant factors of the primary components of A. From the discussion following Corollary 6.7 and Theorem 6.12 we deduce:

two $t \times t$ matrices over a field are similar if and only if their elementary divisors (taking repetitions into account) are equal.

For example consider $A = C((x+1)(x^2+1)) \oplus C((x+1)(x^2+1)^2)$ over $\mathbb{Q}$. So A is in rcf. Write $p_1(x) = x+1$ and $p_2(x) = x^2+1$. Then $\chi_A(x) = p_1(x)^2 p_2(x)^3$ giving $n_1 = 2, n_2 = 3$. As A is a direct sum we obtain $M(A) = N_1 \oplus N_2$ where $N_1 = \langle e_1 \rangle$ and $N_2 = \langle e_4 \rangle$ are cyclic submodules with generators $v_1 = e_1$ and $v_2 = e_4$ of orders $(x+1)(x^2+1)$ and $(x+1)(x^2+1)^2$ respectively from Theorem 5.26. Using Lemma 5.23 we see $v_{11} = (x^2+1)e_1 = e_1 + e_3$ has order $x+1$ and $v_{21} = (x^2+1)^2 e_4 = (x^4+2x^2+1)e_4 = e_4 + 2e_6 + e_8$ also has order $x+1$ in $M(A)$. So these vectors are row eigenvectors of A corresponding to the eigenvalue -1 and the primary component $M(A)_{x+1} = \langle v_{11}, v_{21} \rangle$ is in this case the eigenspace of A corresponding to -1. By Lemma 5.23 the vectors $v_{12} = (x+1)e_1 = e_1 + e_2$ and $v_{22} = (x+1)e_4 = e_4 + e_5$ have orders x^2+1 and $(x^2+1)^2$ respectively in $M(A)$. The primary component $M(A)_{x^2+1} = \langle v_{12} \rangle \oplus \langle v_{22} \rangle$ is a subspace of $\mathbb{Q}^8$ having dimension $2 + 4 = 6$.

In this case $t_{11} = 1$, $t_{21} = 1$, $t_{12} = 1$, $t_{22} = 2$ and the elementary divisors of $M(A)$ are $x + 1, x + 1; x^2 + 1, (x^2 + 1)^2$. Looking ahead, the matrix

$$Y = \begin{pmatrix} v_{11} \\ \hline v_{21} \\ \hline v_{12} \\ xv_{12} \\ \hline v_{22} \\ xv_{22} \\ x^2 v_{22} \\ x^3 v_{22} \end{pmatrix} = \left(\begin{array}{cc|cc|cccc} 1 & 0 & 1 & 0 & 0 & 0 & 0 & 0 \\ \hline 0 & 0 & 0 & 1 & 0 & 2 & 0 & 1 \\ \hline 1 & 1 & 0 & 0 & 0 & 0 & 0 & 0 \\ 0 & 1 & 1 & 0 & 0 & 0 & 0 & 0 \\ \hline 0 & 0 & 0 & 1 & 1 & 0 & 0 & 0 \\ 0 & 0 & 0 & 0 & 1 & 1 & 0 & 0 \\ 0 & 0 & 0 & 0 & 0 & 1 & 1 & 0 \\ 0 & 0 & 0 & 0 & 0 & 0 & 1 & 1 \end{array}\right)$$

is invertible over $\mathbb{Q}$ and satisfies

$$YAY^{-1} = \left(\begin{array}{cc|cc|cc|cc} -1 & 0 & 0 & 0 & 0 & 0 & 0 & 0 \\ 0 & -1 & 0 & 0 & 0 & 0 & 0 & 0 \\ \hline 0 & 0 & 0 & 1 & 0 & 0 & 0 & 0 \\ 0 & 0 & -1 & 0 & 0 & 0 & 0 & 0 \\ \hline 0 & 0 & 0 & 0 & 0 & 1 & 0 & 0 \\ 0 & 0 & 0 & 0 & 0 & 0 & 1 & 0 \\ 0 & 0 & 0 & 0 & 0 & 0 & 0 & 1 \\ 0 & 0 & 0 & 0 & -1 & 0 & -2 & 0 \end{array}\right)$$

$$= C(x + 1) \oplus C(x + 1) \oplus C(x^2 + 1) \oplus C((x^2 + 1)^2)$$

which is in primary canonical form according to the following definition.

Definition 6.14

Let A be a $t \times t$ matrix over a field F as in Theorem 6.12. Let P_j denote the rcf of A_j where $M(A)_{p_j(x)} \cong M(A_j)$ as in the proof of Corollary 6.13 for $1 \le j \le k$. The matrix $P_1 \oplus P_2 \oplus \cdots \oplus P_k$ is said to be in *primary canonical form (pcf)*.

As $A \sim A_1 \oplus A_2 \oplus \cdots \oplus A_k$ and $A_j \sim P_j$ for $1 \le j \le k$ we see

$$A \sim P_1 \oplus P_2 \oplus \cdots \oplus P_k$$

where $\sim$ denotes similarity Definition 5.3. Apart from the ordering of the primary components each A is similar to a unique matrix in pcf.

We proceed to construct an invertible matrix Y over F such that YAY^{-1} is in pcf. For each i and j with $1 \le i \le s$, $1 \le j \le k$ and $t_{ij} > 0$, the vector v_{ij} has order $p_j(x)^{t_{ij}}$ and so the basis $\mathcal{B}_{v_{ij}}$ of $N_i \cap M(A)_{p_j(x)}$ has $\deg p_j(x)^{t_{ij}}$ elements by Corollary 5.29. As $N_1, N_2, \ldots, N_s$ are independent we see that the vectors in $\bigcup_i \mathcal{B}_{v_{ij}}$ are linearly independent and are $\sum_i \deg p_j(x)^{t_{ij}} = \deg p_j(x)^{n_j}$ in number. Therefore $\mathcal{B}_j = \bigcup_i \mathcal{B}_{v_{ij}}$ is a basis of the $p_j(x)$-component of $M(A)$ for $1 \leq j \leq k$. Also P_j is the matrix relative to $\mathcal{B}_j$ of the restriction to $M(A)_{p_j(x)}$ of the linear mapping determined by A for $1 \leq j \leq k$ by Corollaries 5.20 and 5.27. Finally $\mathcal{B} = \bigcup_{i,j} \mathcal{B}_{v_{ij}} = \mathcal{B}_1 \cup \mathcal{B}_2 \cup \cdots \cup \mathcal{B}_k$ is a basis of F^t and the vectors in $\mathcal{B}$ are the rows of an invertible matrix Y satisfying $YAY^{-1} = P_1 \oplus P_2 \oplus \cdots \oplus P_k$ in pcf by Corollary 5.20. Now $P_j = \sum_{i=1}^{s} \oplus C(p_j(x)^{t_{ij}})$ for $1 \leq j \leq k$ and so YAY^{-1} is the direct sum of $k_1 + k_2 + \cdots + k_s$ companion matrices $C(p_j(x)^{t_{ij}})$, one for each elementary divisor $p_j(x)^{t_{ij}}$ of A. The non-zero $F[x]$-modules $M(C(p_j(x)^{t_{ij}}))$ are *indecomposable*, that is, none can be expressed as the direct sum of two non-zero submodules (Exercises 6.2, Question 3(b)). So the above decomposition of $M(A)$ into $k_1 + k_2 + \cdots + k_s$ non-trivial submodules cannot be 'improved', that is, $M(A)$ cannot be expressed as the direct sum of more than $k_1 + k_2 + \cdots + k_s$ non-zero submodules (Exercises 6.2, Question 3(c)).

We return briefly to Example 6.3. With

$$A = \begin{pmatrix} 1 & 1 & 1 \\ 0 & 2 & 1 \\ 0 & 0 & 1 \end{pmatrix}$$

over $\mathbb{Q}$ we found $M(A) = \langle v_1 \rangle \oplus \langle v_2 \rangle$ where $v_1 = (0, 0, 1)$ and $v_2 = (1, 0, 0)$ have orders $d_1(x) = x - 1$ and $d_2(x) = (x - 1)(x - 2)$ respectively in $M(A)$. Here $s = k = 2$. Write $p_1(x) = x - 1$ and $p_2(x) = x - 2$. Then $\chi_A(x) = (x - 1)^2(x - 2) = p_1(x)^2 p_2(x)$ and so $n_1 = 2$, $n_2 = 1$. Also $t_{11} = 1$, $t_{12} = 0$, $t_{21} = 1$, $t_{22} = 1$ and so $m_{11}(x) = 1$, $m_{21}(x) = x - 2$, $m_{22}(x) = x - 1$ giving $v_{11} = v_1$, $v_{21} = (x - 2)v_2 = (-1, 1, 1)$, $v_{22} = (x - 1)v_1 = (0, 1, 1)$ of orders $x - 1$, $x - 1$, $x - 2$ respectively in $M(A)$. In this case the rows of Y are eigenvectors of A and

$$Y = \begin{pmatrix} v_{11} \\ v_{21} \\ v_{22} \end{pmatrix} = \begin{pmatrix} 0 & 0 & 1 \\ -1 & 1 & 1 \\ 0 & 1 & 1 \end{pmatrix} \text{ satisfies } YAY^{-1} = \operatorname{diag}(1, 1, 2) = P_1 \oplus P_2 \text{ in pcf}$$

where $P_1 = C(x - 1) \oplus C(x - 1)$ and $P_2 = C(x - 2)$. We've modified X with XAX^{-1} in rcf to get Y with YAY^{-1} in pcf. It is always possible to carry out this procedure provided the irreducible factorisation of $\chi_A(x)$ is known.

We now ask: what is the number of similarity classes of $t \times t$ matrices A over F all having the same $\chi_A(x)$? In other words by Theorem 5.13 how many isomorphism classes of $F[x]$-modules $M(A)$ are there where A has a given characteristic polynomial? As before it is necessary to know the factorisation

$$\chi_A(x) = p_1(x)^{n_1} p_2(x)^{n_2} \cdots p_k(x)^{n_k}.$$

The analogous problem for finite abelian groups was solved at the end of Section 3.2 using the partition function $p(n)$. We use the same approach here.

The relationship between the invariant factors $d_i(x)$ and the elementary divisors $p_j(x)^{t_{ij}}$, $t_{ij} > 0$, of A is expressed by the entries in the following bordered $s \times k$ table:

$\cong M(A)$	$p_1(x)$	$p_2(x)$	$\cdots$	$p_k(x)$
$d_1(x)$	t_{11}	t_{12}	$\cdots$	t_{1k}
$d_2(x)$	t_{21}	t_{22}	$\cdots$	t_{2k}
$\vdots$	$\vdots$	$\vdots$		$\vdots$
$d_s(x)$	t_{s1}	t_{s2}	$\cdots$	t_{sk}

where $\cong M(A)$ denotes the isomorphism class of the $F[x]$-module $M(A)$. The order in which the k columns appear in the table is arbitrary, but otherwise different tables correspond to different isomorphism classes of $F[x]$-modules $M(A)$. As noted above the non-zero entries in column j are the parts of a partition $(t_{l_j j}, t_{l_j+1 j}, \ldots, t_{sj})$ of n_j for $1 \leq j \leq k$. Notice $t_{1j} > 0$ for some j as $\deg d_1(x) \geq 1$ and so s is the maximum number of parts in these k partitions.

Conversely suppose given a partition of n_j for each j with $1 \leq j \leq k$. Denote by s the largest number of parts in any one of these partitions. Then the partition of n_j has $s - l_j + 1$ parts where l_j is a positive integer and so can be written $(t_{l_j j}, t_{l_j+1 j}, \ldots, t_{sj})$ as above. Write $t_{ij} = 0$ for $i < l_j$. Then t_{ij} is the (i, j)-entry in the table of the isomorphism class of $M(A)$ where A is a $t \times t$ matrix with $\chi_A(x) = p_1(x)^{n_1} p_2(x)^{n_2} \cdots p_k(x)^{n_k}$ and invariant factor sequence $(d_1(x), d_2(x), \ldots, d_s(x))$ where $d_i(x) = p_1(x)^{t_{i1}} p_2(x)^{t_{i2}} \cdots p_k(x)^{t_{ik}}$ for $1 \leq i \leq s$. Using the partition function $p(n)$ of Definition 3.13 the conclusion is:

there are $p(n_1)p(n_2) \cdots p(n_k)$ isomorphism classes of $F[x]$-modules $M(A)$ where A is a $t \times t$ matrix over F with $\chi_A(x) = p_1(x)^{n_1} p_2(x)^{n_2} \cdots p_k(x)^{n_k}$

since there are $p(n_j)$ choices for each partition of n_j for $1 \leq j \leq k$.

For example consider 25×25 matrices A over $\mathbb{Q}$ with $\chi_A(x) = x^5(x + 1)^6 \cdot (x^2 + 1)^7$. From the discussion following Definition 3.13 we know $p(5) = 7$, $p(6) = 11$, $p(7) = 15$. The number of isomorphism classes of $\mathbb{Q}[x]$-modules $M(A)$ is $7 \times 11 \times 15 = 1155$. One such class is specified by the partitions $(2, 3)$, $(2, 2, 2)$, $(3, 4)$ of 5, 6, 7 respectively. This class has elementary divisors x^2, x^3; $(x + 1)^2$, $(x + 1)^2$, $(x + 1)^2$; $(x^2 + 1)^3$, $(x^2 + 1)^4$ and representative matrix C in pcf:

$$C(x^2) \oplus C(x^3) \oplus C((x + 1)^2) \oplus C((x + 1)^2) \oplus C((x + 1)^2)$$

$$\oplus C((x^2 + 1)^3) \oplus C((x^2 + 1)^4).$$

The table of $\cong M(C)$ is

$\cong M(C)$	x	$x+1$	x^2+1
$(x+1)^2$	0	2	0
$x^2(x+1)^2(x^2+1)^3$	2	2	3
$x^3(x+1)^2(x^2+1)^4$	3	2	4

The invariant factors of C appear in the rows of the table. So

$$C \sim C((x+1)^2) \oplus C(x^2(x+1)^2(x^2+1)^3) \oplus C(x^3(x+1)^2(x^2+1)^4)$$

the matrix on the right being the rcf of C and $\mu_C(x) = x^3(x+1)^2(x^2+1)^4$ being the minimum polynomial of C. The invariant factor decompositions of the primary components of $M(C)$ appear in the columns of the table. So

$$M(C)_x \cong M(C(x^2)) \oplus M(C(x^3)),$$

$$M(C)_{x+1} \cong M(C((x+1)^2)) \oplus M(C((x+1)^2)) \oplus M(C((x+1)^2)) \quad \text{and}$$

$$M(C)_{x^2+1} \cong M(C((x^2+1)^3)) \oplus M(C((x^2+1)^4)).$$

We have now completed our counting of the number of similarity classes of $t \times t$ matrices A over a field F having a given characteristic polynomial $\chi_A(x)$. The pcf of A plays a crucial role in this theory. We next obtain *the Jordan normal form* of A by modifying the companion matrices appearing in the pcf of A.

Definition 6.15

Let $f(x)$ be a monic polynomial of positive degree n over a field F and let l be a positive integer.

The *Jordan block matrix*

$$J(f(x), l) = \begin{pmatrix} C & L & 0 & 0 & \cdots & 0 \\ 0 & C & L & 0 & \cdots & 0 \\ 0 & 0 & C & L & \ddots & \vdots \\ \vdots & \vdots & \ddots & \ddots & \ddots & 0 \\ 0 & 0 & \cdots & 0 & C & L \\ 0 & 0 & \cdots & 0 & 0 & C \end{pmatrix}$$

is the $ln \times ln$ matrix over F partitioned into $n \times n$ submatrices where $C = C(f(x))$, L denotes the $n \times n$ matrix with $(n, 1)$-entry 1 all other entries being zero, and 0 denotes the $n \times n$ zero matrix.

A direct sum of matrices of the type $J(p(x), l)$, where $p(x)$ is monic and irreducible over F, is said to be in *Jordan normal form (Jnf)* over F.

For example

$$J(x-2,3) = \left(\begin{array}{cc|c} 2 & 1 & 0 \\ 0 & 2 & 1 \\ \hline 0 & 0 & 2 \end{array}\right) \quad \text{and} \quad J(x^2+1,2) = \left(\begin{array}{cc|cc} 0 & 1 & 0 & 0 \\ -1 & 0 & 1 & 0 \\ \hline 0 & 0 & 0 & 1 \\ 0 & 0 & -1 & 0 \end{array}\right).$$

Also $J(x-2,3) \oplus J(x^2+1,2)$ is in Jnf over $\mathbb{Q}$.

The Jordan block matrix $J(f(x),l)$ and the companion matrix $C(f(x)^l)$ are similar as we'll see shortly. These matrices have a number of entries in common: in both cases the $(i,i+1)$-entry is 1 for $1 \leq i < ln$ and all (i,j)-entries are zero for $i+1 < j$. The presence of the somewhat strange matrices L achieves this as far as $J(f(x),l)$ is concerned. The structure of $J(f(x),l)$ exploits the fact that the module $M(C(f(x)^l))$ has a chain Theorem 5.28 of submodules:

$$\langle e_1 \rangle \supset \langle f(x)e_1 \rangle \supset \langle f(x)^2 e_1 \rangle \supset \cdots \supset \langle f(x)^{l-1} e_1 \rangle \supset 0$$

whereas the structure of $C(f(x)^l)$ ignores its existence! The precise connection between $C(f(x)^l)$ and $J(f(x),l)$ is established in Theorem 6.16. We first look at the case $l=2$, $n=2$. Write $C' = C(f(x)^2)$ where $f(x) = a_0 + a_1 x + x^2$. Then e_1 has order $f(x)^2$ in $M(C')$. We are looking for an invertible 4×4 matrix Z with

$$ZC'Z^{-1} = J(f(x),2) = \left(\begin{array}{cc|cc} 0 & 1 & 0 & 0 \\ -a_0 & -a_1 & 1 & 0 \\ \hline 0 & 0 & 0 & 1 \\ 0 & 0 & -a_0 & -a_1 \end{array}\right).$$

On equating rows in $ZC' = J(f(x),2)Z$ we see that the rows z_1, z_2, z_3, z_4 of Z are linearly independent and satisfy

$$xz_1 = z_2, \qquad xz_2 = -a_0 z_1 - a_1 z_2 + z_3, \qquad xz_3 = z_4, \qquad xz_4 = -a_0 z_3 - a_1 z_4$$

working in the module $M(C')$. The first two equations rearrange to give $z_3 = (a_0 + a_1 x + x^2)z_1 = f(x)z_1$ and the last two equations give $f(x)z_3 = 0$. So $f(x)^2 z_1 = f(x)z_3 = 0$. The above four equations and the linear independence of z_1, z_2, z_3, z_4 are expressed by the single statement: z_1 has order $f(x)^2$ in $M(C')$. We take $z_1 = e_1$ (it would be perverse to do otherwise as e_1 has the property required of z_1). Knowing $f(x)^2 = a_0^2 + 2a_0 a_1 x + (2a_0 + a_1^2)x^2 + 2a_1 x^3 + x^4$ the reader can now check

$$Z = \begin{pmatrix} z_1 \\ z_2 \\ z_3 \\ z_4 \end{pmatrix} = \begin{pmatrix} z_1 \\ xz_1 \\ f(x)z_1 \\ xf(x)z_1 \end{pmatrix} = \begin{pmatrix} e_1 \\ xe_1 \\ (a_0 + a_1 x + x^2)e_1 \\ (a_0 x + a_1 x^2 + x^3)e_1 \end{pmatrix} = \begin{pmatrix} 1 & 0 & 0 & 0 \\ 0 & 1 & 0 & 0 \\ a_0 & a_1 & 1 & 0 \\ 0 & a_0 & a_1 & 1 \end{pmatrix}$$

is invertible over F and does satisfy $ZC' = J(f(x), 2)Z$. We are now ready for the general case.

Theorem 6.16

Let $f(x)$ be a monic polynomial of positive degree n over a field F and let l be a positive integer. Let Z be the invertible $ln \times ln$ matrix over F with $e_i Z = x^r f(x)^q e_1$ evaluated in $M(C(f(x)^l))$ for $1 \le i \le ln$ where $i - 1 = qn + r$, $0 \le r < n$. Then

$$ZC(f(x)^l)Z^{-1} = J(f(x), l) \quad \text{and} \quad M(C(f(x)^l)) \cong M(J(f(x), l)).$$

Proof

Notice first that q and r are respectively the quotient and remainder on dividing $i - 1$ by n. As $x^{j-1}e_1 = e_j$ in $M(C(f(x)^l))$ for $1 \le j \le ln$, the jth entry in $e_i Z = x^r f(x)^q e_1$ is the coefficient of x^{j-1} in $x^r f(x)^q$ which is monic of degree $qn + r = i - 1$. So the (i, i)-entry in Z is 1 and the (i, j)-entry in Z is 0 for $i < j$, that is, Z is a lower triangular matrix (all its entries above the diagonal are zero). Hence $|Z| = 1$ and so Z is invertible over F. It is enough to verify $ZC(f(x)^l) = J(f(x), l)Z$ which we now carry out row by row.

Suppose first $i \not\equiv 0 \pmod{n}$, that is, $r \ne n - 1$ where $i - 1 = qn + r$ and $1 \le i < ln$. Then $e_i J(f(x), l) = e_{i+1}$ and so

$$e_i J(f(x), l)Y = e_{i+1}Y = x^{r+1} f(x)^q e_1 = x(x^r f(x)^q e_1)$$

$$= x(e_i Z) = e_i ZC(f(x)^l)$$

as multiplication by x on the left means multiplication by $C(f(x)^l)$ on the right by Lemma 5.7 and Definition 5.8. The above equation says that $J(f(x), l)Z$ and $ZC(f(x)^l)$ have the same row i.

Suppose now $i \equiv 0 \pmod{n}$ and $i < ln$. So $i - 1 = qn + (n - 1)$, that is, $r = n - 1$ and $i = (q + 1)n$ where $0 \le q + 1 < l$. The last row of $C = C(f(x))$ is $(x^n - f(x))e_1$ and so row i of $J(f(x), l)$ is $e_i J(f(x), l) = (x^n - f(x))e_{i-n+1} + e_{i+1}$. As $i - n = qn$ we see $e_{i-n+1}Z = f(x)^q e_1$ and $e_{i+1}Z = f(x)^{q+1}e_1$. Therefore

$$e_i J(f(x), l)Z = (x^n - f(x))f(x)^q e_1 + f(x)^{q+1}e_1$$

$$= x^n f(x)^q e_1 = x(x^{n-1} f(x)^q e_1) = x(e_i Z) = e_i ZC(f(x)^l)$$

showing that $J(f(x), l)Z$ and $ZC(f(x)^l)$ have the same row i.

Lastly suppose $i = ln$. Then $e_i J(f(x), l) = (x^n - f(x))e_{i-n+1}$. As $i - n = (l - 1)n$ we have $e_{i-n+1}Z = f(x)^{l-1}e_1$. Therefore

$$e_i J(f(x), l)Z = (x^n - f(x))f(x)^{l-1}e_1 = x^n f(x)^{l-1}e_1$$

since $f(x)^l e_1 = 0$ in the $F[x]$-module $M(C(f(x)^l))$. As $i - 1 = ln - 1 = (l - 1)n + n - 1$ the last row of Z is $e_i Z = x^{n-1} f(x)^{l-1} e_1$. Hence $e_i J(f(x), l) Z = x(x^{n-1} f(x)^{l-1} e_1) = x(e_i Z) = e_i Z C(f(x)^l)$. The conclusion is: $J(f(x), l) Z = Z C(f(x)^l)$ as row by row these matrices are equal. The matrices $C(f(x)^l)$ and $J(f(x), l)$ are similar and so the $F[x]$-modules $M(C(f(x)^l))$ and $M(J(f(x), l))$ are isomorphic by Theorem 5.13.   □

Notice $l = 1$ in Theorem 6.16 gives $q = 0$, $Z = I$ and $J(f(x), 1) = C(f(x))$. Also $n = 1$ in Theorem 6.16 gives $r = 0$, $f(x) = x - \lambda$,

$$Z = \begin{pmatrix} e_1 \\ (x - \lambda)e_1 \\ (x - \lambda)^2 e_1 \\ \vdots \\ (x - \lambda)^{l-1} e_1 \end{pmatrix} \quad \text{and}$$

$$J(x - \lambda, l) = \lambda I + C(x^l) = \begin{pmatrix} \lambda & 1 & 0 & \cdots & 0 \\ 0 & \lambda & 1 & \ddots & \vdots \\ 0 & 0 & \lambda & \ddots & 0 \\ \vdots & \vdots & \ddots & \ddots & 1 \\ 0 & 0 & \cdots & 0 & \lambda \end{pmatrix},$$

the 'usual' $l \times l$ Jordan block matrix which the reader may already have met.

Let A be a $t \times t$ matrix over a field F having m elementary divisors. There is an invertible matrix Y over F with $YAY^{-1} = C_1 \oplus C_2 \oplus \cdots \oplus C_m$ where each C_i is the companion matrix of a power of an irreducible polynomial over F. By Theorem 6.16 there is an invertible matrix Z_i over F such that $Z_i C_i Z_i^{-1} = J_i$ is a Jordan block matrix for $1 \le i \le m$. Then $Z = Z_1 \oplus Z_2 \oplus \cdots \oplus Z_m$ is invertible over F and

$$(ZY)A(ZY)^{-1} = Z(YAY^{-1})Z^{-1} = Z(C_1 \oplus C_2 \oplus \cdots \oplus C_m)Z^{-1}$$

$$= J_1 \oplus J_2 \oplus \cdots \oplus J_m$$

is in Jnf Definition 6.15.

Suppose $\chi_A(x)$ factorises as a product of polynomials $x - \lambda_j$ over F, that is, all the eigenvalues of A belong to F. Then A is similar to the 'almost diagonal' matrix $J_1 \oplus J_2 \oplus \cdots \oplus J_m$ in Jnf where $J_i = J(x - \lambda_j, l_i)$ for $1 \le i \le m$.

As an illustration consider $A = C((x + 1)^3 (x^2 - 2)^2)$ over $\mathbb{Q}$ which is in rcf. Then $M(A) = \langle v_{12} \rangle \oplus \langle v_{11} \rangle$ is the primary decomposition of the cyclic $\mathbb{Q}[x]$-module $M(A)$ where $v_{12} = (x^2 - 2)^2 e_1 = (4, 0, -4, 0, 1, 0, 0)$ and $v_{11} = (x + 1)^3 e_1 = (1, 3, 3, 1, 0, 0, 0)$. The matrix Y of Theorem 5.31 satisfies $YAY^{-1} =$

$C((x + 1)^3) \oplus C((x^2 - 2)^2)$ in pcf. We 'short-circuit' the above theory by directly constructing

$$ZY = \begin{pmatrix} v_{21} \\ (x+1)v_{21} \\ (x+1)^2 v_{21} \\ \hline v_{11} \\ x v_{11} \\ (x^2 - 2)v_{11} \\ x(x^2 - 2)v_{11} \end{pmatrix} = \begin{pmatrix} 4 & 0 & -4 & 0 & 1 & 0 & 0 \\ 4 & 4 & -4 & -4 & 1 & 1 & 0 \\ 4 & 8 & 0 & -8 & -3 & 2 & 1 \\ 1 & 3 & 3 & 1 & 0 & 0 & 0 \\ 0 & 1 & 3 & 3 & 1 & 0 & 0 \\ -2 & -6 & -5 & -1 & 3 & 1 & 0 \\ 0 & -2 & -6 & -5 & -1 & 3 & 1 \end{pmatrix}.$$

Then ZY is invertible over $\mathbb{Q}$ and satisfies

$$(ZY)A(ZY)^{-1} = J(x+1,3) \oplus J(x^2 - 2, 2)$$

$$= \left(\begin{array}{cc|c} -1 & 1 & 0 \\ \hline 0 & -1 & 1 \\ \hline 0 & 0 & -1 \end{array} \right) \oplus \begin{pmatrix} 0 & 1 & 0 & 0 \\ 2 & 0 & 1 & 0 \\ 0 & 0 & 0 & 1 \\ 0 & 0 & 2 & 0 \end{pmatrix}$$

which is in Jnf.

We now prepare to meet *separable* polynomials. This important concept leads to a useful modification of the Jordan normal form.

Let $f(x)$ be a polynomial of positive degree n over a field F with leading coefficient a_n. If we are 'lucky' then $f(x)$ *splits over* F, that is, there are elements $c_1, c_2, \ldots, c_n \in F$ with $f(x) = a_n(x - c_1)(x - c_2) \cdots (x - c_n)$; as F has no divisors of zero we see that $c_1, c_2, \ldots, c_n$ are the zeros of $f(x)$. If we are 'unlucky' then $f(x)$ has an irreducible factor $p_1(x)$ over F with $2 \leq \deg p_1(x) \leq n$. Using Theorem 4.9 there is an extension field E_1 of F which contains a zero c_1 of $p_1(x)$. Either $f(x)$ splits over E_1 or the process is repeated. In the latter case $f(x)$ has irreducible factor $p_2(x)$ over E_1 with $2 \leq \deg p_2(x) \leq n - 1$; also there is an extension field E_2 of E_1 containing a zero c_2 of $p_2(x)$. After at most $n - 1$ steps this process terminates in an extension field E of F such that $f(x)$ splits over E. Assuming that the construction of E has been carried out as economically as possible (there is no subfield E' of E which is an extension of F such that $E' \neq E$ and $f(x)$ splits over E') then E is called *a splitting field* of $f(x)$ over F.

The theory of splitting fields is fundamental in Galois Theory but is not carried further here. Notice however that $E = \mathbb{F}_{32}$ is a splitting field of $f(x) = x^{32} - x$ over $F = \mathbb{Z}_2$ and provides an unbiased view of $\mathbb{F}_{32}$, that is, no one of the six irreducible quintics over $\mathbb{Z}_2$ is given preferential treatment (Exercises 4.1, Question 3(c)). No two of the 32 zeros of $x^{32} - x$ in $\mathbb{F}_{32}$ are equal. We now investigate this important property in general.

Definition 6.17

An irreducible polynomial $p_0(x)$ of degree n over a field F is called *separable over F* if there is a field E with $F \subseteq E$ containing n distinct zeros $c_1, c_2, \ldots, c_n$ of $p_0(x)$. A polynomial $f(x)$ over F is called *separable over F* if all its irreducible factors over F are separable over F.

The reader might expect all polynomials over certain fields to be separable. This is true for F a finite field (Exercises 4.1, Question 3(c)) and in the case $\chi(F) = 0$, that is, F has characteristic zero as we show after Lemma 6.19. In particular all polynomials over the real field $\mathbb{R}$ are separable. However, as we now demonstrate, $x^2 - y$ over F is an example of an *inseparable* (not separable) polynomial over F, where

$$F = \mathbb{Z}_2(y) = \{f(y)/g(y) : f(y), g(y) \in \mathbb{Z}_2[y], \ g(y) \neq 0(y)\}$$

is *the field of fractions* of $\mathbb{Z}_2[y]$. Note first that $x^2 - y$ has no zeros in F: the equation $y = (f(y)/g(y))^2$ leads to $yg(y)^2 = f(y)^2$ which is impossible as $\deg yg(y)^2$ is odd and $\deg f(y)^2$ is even. So $x^2 - y$ is irreducible over F by Lemma 4.8(ii). Also any extension field E of F containing a zero c_1 of $x^2 - y$ does not contain another zero c_2, since $\chi(E) = 2$ giving

$$c_1^2 = c_2^2 = y \ \Rightarrow (c_1 - c_2)^2 = 0 \ \Rightarrow \ c_1 = c_2 \quad \text{and} \quad x^2 - y = (x - c_1)^2.$$

So $x^2 - y$ is irreducible and inseparable over F. The reducibility of a polynomial depends on its ground field and the same goes for separability: in this case $x^2 - y$ is reducible and separable over E.

We next introduce the derivative of a polynomial without appeal to any limiting process. The derivative in this context is closely related to separability.

Definition 6.18

Let $f(x) = a_n x^n + a_{n-1} x^{n-1} + \cdots + a_1 x + a_0$ be a polynomial over a field F. The polynomial $f'(x) = na_n x^{n-1} + (n-1)a_{n-1} x^{n-1} + \cdots + a_1$ over F is called *the (formal) derivative of $f(x)$.*

The coefficient $na_n = a_n + a_n + \cdots + a_n$ (n terms) of x^n in $f'(x)$ is an element of F as are all the coefficients in $f'(x)$. For example $f(x) = x^3 + x^2 + 1$ over $\mathbb{Z}_2$ has $f'(x) = x^2$ since $3x^2 = x^2$ and $2x = 0$; in this case $\gcd\{f(x), f'(x)\} = 1$.

Lemma 6.19

Let $f(x) = a_n x^n + a_{n-1} x^{n-1} + \cdots + a_1 x + a_0$ be a polynomial of positive degree n over a field F. Suppose F has an extension field E containing $c_1, c_2, \ldots, c_n$ such that $f(x) = a_n(x - c_1)(x - c_2) \cdots (x - c_n)$. Then

$$\gcd\{f(x), f'(x)\} = 1 \quad \Leftrightarrow \quad c_1, c_2, \ldots, c_n \text{ are distinct.}$$

Proof

Suppose $c_1, c_2, \ldots, c_n$ are not distinct. We arrange the notation so that $c_1 = c_2$. Therefore $f(x) = (x - c_1)^2 g(x)$ where $g(x) = a_n(x - c_3)(x - c_4) \cdots (x - c_n)$. The usual rules of differentiation are valid in this context (see Exercises 6.2, Question 8(a)) and so $f'(x) = 2(x - c_1)g(x) + (x - c_1)^2 g'(x)$ showing that $x - c_1$ is a divisor of both $f(x)$ and $f'(x)$. So $\gcd\{f(x), f'(x)\} = 1 \Rightarrow c_1, c_2, \ldots, c_n$ distinct.

Conversely suppose $c_1, c_2, \ldots, c_n$ to be distinct. Then $f(x)$ has n distinct monic irreducible factors over E namely $x - c_i$ for $1 \leq i \leq n$. Is it possible for $x - c_i$ to be a divisor of $f'(x)$ for some i? If this is the case then we choose the notation so that $i = 1$. Write $f(x) = (x - c_1)h(x)$ where $h(x) = a_n(x - c_2)(x - c_3) \cdots (x - c_n)$. Then $f'(x) = 1 \times h(x) + (x - c_1)h'(x)$ using the familiar rule for differentiating a product. As $(x - c_1) | f(x)$ we deduce $(x - c_1) | h(x)$ showing $h(x) = (x - c_1)q(x)$ for some $q(x) \in E[x]$. Evaluating $h(x) = (x - c_1)q(x)$ at c_1 gives $h(c_1) = (c_1 - c_1)q(c_1) = 0$. But $h(c_1) = a_n(c_1 - c_2)(c_1 - c_3) \cdots (c_1 - c_n) \neq 0$ as $a_n \neq 0$ and $c_1 - c_i \neq 0$ for $1 < i \leq n$. We conclude: none of the n factors $x - c_i$ of $f(x)$ is a divisor of $f'(x)$. So $c_1, c_2, \ldots, c_n$ distinct implies $\gcd\{f(x), f'(x)\} = 1$. □

There is a surprising consequence of Lemma 6.19: let $p_0(x)$ be a monic irreducible polynomial over a field F which is inseparable over F. Then $\gcd\{p_0(x), p_0'(x)\}$ is a non-constant monic divisor over F of $p_0(x)$. Therefore $\gcd\{p_0(x), p_0'(x)\} = p_0(x)$ by Definition 4.7 and so $p_0(x) | p_0'(x)$. As $\deg p_0(x) > \deg p_0'(x)$ there is only one way out of this apparent impasse, namely $p_0'(x) = 0(x)$, that is, the derivative of $p_0(x)$ is the zero polynomial. As $\deg p_0(x) \geq 1$ we cannot conclude that $p_0(x)$ is a constant polynomial! However it is correct to infer $\chi(F) \neq 0$ as all non-constant polynomials over fields of characteristic zero have non-zero derivatives. So $\chi(F) = p$ where p is prime and the only powers of x appearing in $p_0(x)$ with non-zero coefficients are powers of x^p, that is, $p_0(x) \in F[x^p]$. For example, as we saw following Definition 6.17, $p_0(x) = x^2 - y$ over $F = \mathbb{Z}_2(y)$ is inseparable; in this case $p_0(x)$ belongs to $F[x^2]$ and has zero derivative.

We now discuss a modification of the Jordan normal form involving separability.

Definition 6.20

Let B be an $n \times n$ matrix over a field F and let l be a positive integer. The $ln \times ln$ matrix

$$J_S(B, l) = \begin{pmatrix} B & I & 0 & \cdots & 0 & 0 \\ 0 & B & I & \ddots & 0 & 0 \\ \vdots & \ddots & \ddots & \ddots & \ddots & \vdots \\ \vdots & \vdots & \ddots & \ddots & I & 0 \\ 0 & 0 & \cdots & \ddots & B & I \\ 0 & 0 & \cdots & \cdots & 0 & B \end{pmatrix},$$

where I and 0 denote respectively the $n \times n$ identity and zero matrices over F, is called a *separable Jordan block over* F.

A direct sum of matrices of the type $J_S(C(p(x)), l)$, where each $p(x)$ is monic, separable over F and irreducible over F, is said to be in *separable Jordan form (sJf)* *over* F.

The next theorem studies the connection between $J_S(B, l)$ and $J(\chi_B(x), l)$ in the case $\chi_B(x)$ irreducible. Computations involving $J_S(B, l)$, rather than $J(\chi_B(x), l)$, are easier to carry out.

Theorem 6.21

Let B be a $n \times n$ matrix over a field F with $\chi_B(x)$ irreducible over F and let l be an integer with $l \geq 2$. Then $J(\chi_B(x), l) \sim J_S(B, l)$ if and only if $\chi_B(x)$ is separable over F. In the separable case the $F[x]$-module $M(J_S(B, l))$ is cyclic with generator e_1.

Proof

For convenience we write $\chi_B(x) = a_n x^n + a_{n-1} x^{n-1} + \cdots + a_1 x + a_0$ where $a_n = 1$. Our first task is to determine the structure of the $ln \times ln$ matrix $\chi_B(J_S(B, l))$. Then $J_S(B, l) = \tilde{B} + \tilde{N}$ where

$$\tilde{B} = \begin{pmatrix} B & 0 & 0 & \cdots & 0 & 0 \\ 0 & B & 0 & \ddots & 0 & 0 \\ \vdots & \ddots & \ddots & \ddots & \vdots & \vdots \\ \vdots & \vdots & \ddots & \ddots & 0 & 0 \\ 0 & 0 & 0 & \ddots & B & 0 \\ 0 & 0 & 0 & \cdots & 0 & B \end{pmatrix}, \quad \tilde{N} = \begin{pmatrix} 0 & I & 0 & \cdots & 0 & 0 \\ 0 & 0 & I & \ddots & 0 & 0 \\ \vdots & \vdots & \ddots & \ddots & \ddots & \vdots \\ \vdots & \vdots & \ddots & \ddots & I & 0 \\ 0 & 0 & 0 & \ddots & 0 & I \\ 0 & 0 & 0 & \cdots & 0 & 0 \end{pmatrix}$$

and

$$\tilde{B}\tilde{N} = \begin{pmatrix} 0 & B & 0 & \cdots & 0 & 0 \\ 0 & 0 & B & \ddots & 0 & 0 \\ \vdots & \vdots & \ddots & \ddots & \ddots & \vdots \\ \vdots & \vdots & \vdots & \ddots & B & 0 \\ 0 & 0 & 0 & \cdots & 0 & B \\ 0 & 0 & 0 & \cdots & 0 & 0 \end{pmatrix} = \tilde{N}\tilde{B},$$

that is, $\tilde{B}$ and $\tilde{N}$ commute – a big advantage of $J_S(B, l)$ over $J(\chi_B(x), l)$. In particular powers of $J_S(B, l)$ can be calculated using the binomial expansion

$$J_S(B, l)^j = (\tilde{B} + \tilde{N})^j = \tilde{B}^j + j\tilde{B}^{j-1}\tilde{N} + \binom{j}{2}\tilde{B}^{j-2}\tilde{N}^2 + \cdots + \binom{j}{j}\tilde{N}^j.$$

Notice that all matrices in this expansion are partitioned – they are $l \times l$ matrices having entries which are $n \times n$ matrices over F and so belong to the subring $\mathfrak{M}_l(\mathfrak{M}_n(F))$ of $\mathfrak{M}_{ln}(F)$. Further the $n \times n$ submatrices in $\tilde{B}^{j-i}\tilde{N}^i$ on the ith diagonal above (and parallel to) the main diagonal are all equal B^{j-i} for $1 \le i < l$, all other $n \times n$ submatrices in the partition being zero. For instance with $l \ge 3$,

$$J_S(B, l)^2 = \begin{pmatrix} B^2 & 2B & I & 0 & \cdots & 0 \\ 0 & B^2 & 2B & \ddots & \ddots & \vdots \\ \vdots & \ddots & \ddots & \ddots & I & 0 \\ \vdots & \vdots & \ddots & \ddots & 2B & I \\ 0 & 0 & \cdots & \ddots & B^2 & 2B \\ 0 & 0 & \cdots & \cdots & 0 & B^2 \end{pmatrix}$$

which displays the above binomial expansion in diagonal 'stripes'. Now

$$
\tilde{N}^{l-1} =
\left(
\begin{array}{ccccc|cc}
0 & 0 & 0 & \cdots & 0 & I \\
0 & 0 & 0 & \cdots & 0 & 0 \\
\vdots & \vdots & \ddots & \ddots & \vdots & \vdots \\
\vdots & \vdots & \ddots & \ddots & \vdots & \vdots \\
0 & 0 & \cdots & 0 & 0 & 0 \\
0 & 0 & \cdots & 0 & 0 & 0
\end{array}
\right)
$$

and $\tilde{N}^l = 0$ showing that $\tilde{N}$ is a *nilpotent matrix* (some power of $\tilde{N}$ is the zero matrix). Therefore the above binomial expansion can be expressed

$$
J_S(B,l)^j = \sum_{i=0}^{l} \binom{j}{i} \tilde{B}^{j-i} \tilde{N}^i .
$$

Multiplying this equation by the coefficient a_j of x^j in $\chi_B(x)$ and summing over j gives

$$
\chi_B(J_S(B,l)) = \sum_{j=0}^{n} a_j J_S(B,l)^j
$$

$$
=
\left(
\begin{array}{cccccc}
\chi_B(B) & \chi_B'(B) & ? & ? & \cdots & ? \\
0 & \chi_B(B) & \chi_B'(B) & ? & \cdots & ? \\
0 & 0 & \chi_B(B) & \ddots & \ddots & \vdots \\
\vdots & \vdots & \ddots & \ddots & \ddots & ? \\
0 & 0 & \cdots & 0 & \chi_B(B) & \chi_B'(B) \\
0 & 0 & \cdots & 0 & 0 & \chi_B(B)
\end{array}
\right)
$$

where the question marks ? denote $n \times n$ matrices closely related to the higher derivatives $\chi_B^{(i)}(x)$ for $i \geq 2$ of $\chi_B(x)$ but which are not relevant here (Exercises 6.2, Question 7(c)). By Corollary 6.11 we know $\chi_B(B) = 0$ and so $\chi_B(J_S(B,l))$ is a partitioned upper triangular matrix with zero $n \times n$ matrices on the main diagonal and the same $n \times n$ matrix $\chi_B'(B)$ along the next diagonal. We have found the structure of

$\chi_B(J_S(B, l))$ we were looking for and leave the reader to verify

$$
(\chi_B(J_S(B, l)))^{l-1} = \begin{pmatrix}
0 & 0 & 0 & \ldots & 0 & (\chi'_B(B))^{l-1} \\
\hline
0 & 0 & 0 & \ldots & 0 & 0 \\
\hline
\vdots & \vdots & \ddots & \ddots & \ddots & \vdots \\
\hline
\vdots & \vdots & \ddots & \ddots & \ddots & \vdots \\
\hline
0 & 0 & \ldots & \ldots & 0 & 0 \\
\hline
0 & 0 & \ldots & \ldots & 0 & 0
\end{pmatrix} \qquad (\clubsuit\clubsuit)
$$

showing that this partitioned matrix has at most one non-zero $n \times n$ submatrix entry, namely $(\chi'_B(B))^{l-1}$ in the top right-hand corner. So far we have not used the irreducibility of $\chi_B(x)$ nor the separability (or otherwise) of $\chi_B(x)$. But things are about to change!

Suppose $\chi_B(x)$ is irreducible and separable over F. By Corollary 4.6 and Lemma 6.19 there are $a(x), b(x) \in F[x]$ with $a(x)\chi_B(x) + b(x)\chi'_B(x) = 1$. Evaluating this polynomial equation at B gives $b(B)\chi'_B(B) = I$ as $\chi_B(B) = 0$ by Corollary 6.11. Therefore $\chi'_B(B)$ is invertible over F. Using Exercises 5.1, Question 2(b) we see that the characteristic polynomial of $J_S(B, l)$ is $|xI - J_S(B, l)| = |xI - B|^l = \chi_B(x)^l$. What is the minimum polynomial $\mu(x)$ of $J_S(B, l)$? By Corollary 6.11 we know $\mu(x) = \chi_B(x)^{l'}$ where $1 \le l' \le l$ as $\chi_B(x)$ is irreducible over F. From $(\clubsuit\clubsuit)$ we see $\chi_B(J_S(B, l))^{l-1} \ne 0$. So $\mu(x)$ is not a divisor of $\chi_B(x)^{l-1}$. Therefore $l' = l$ and $\mu(x) = \chi_B(x)^l$. So $J_S(B, l)$ has only one invariant factor, namely $\chi_B(x)^l$ by Corollary 6.10, which means $J_S(B, l) \sim C(\chi_B(x)^l)$ by Corollary 6.7. But $C(\chi_B(x)^l) \sim J(\chi_B(x), l)$ by Theorem 6.16 and so $J(\chi_B(x), l) \sim J_S(B, l)$. Premultiplying $(\clubsuit\clubsuit)$ by e_1 gives $\chi_B(x)^{l-1}e_1 = e_1\chi_B(J_S(B, l)) \ne 0$ in the $F[x]$-module $M(J_S(B, l))$. So e_1 has order $\chi_B(x)^l$ in $M(J_S(B, l))$, that is, $M(J_S(B, l)) = \langle e_1 \rangle$.

Suppose now that $\chi_B(x)$ is irreducible over F but not separable over F. In this case $\chi'_B(x) = 0$ and so $(\chi_B(J_S(B, l)))^{l-1} = 0$ from $(\clubsuit\clubsuit)$. Therefore $\mu(x)|\chi_B(x)^{l-1}$ giving $\mu(x) \ne \chi_B(x)^l$, that is, the minimum and characteristic polynomials of $J_S(B, l)$ are different. Combining Theorems 5.13, 5.26 and 6.16 we see that the $F[x]$-module $M(J(\chi_B(x), l))$ is cyclic but the $F[x]$-module $M(J_S(B, l))$ is not cyclic. So $M(J(\chi_B(x), l))$ and $M(J_S(B, l))$ are not isomorphic. From Theorem 5.13 we conclude that $J(\chi_B(x), l)$ and $J_S(B, l)$ are not similar. $\qquad \square$

Corollary 6.22

Let A be a $t \times t$ matrix over a field F. Then A is similar to a matrix in separable Jordan form over F if and only if $\chi_A(x)$ is separable over F.

Proof

Suppose $A \sim J_S(C(p_1(x)), l_1) \oplus J_S(C(p_2(x), l_2)) \oplus \cdots \oplus J_S(C(p_m(x), l_m))$ where $p_i(x)$ is monic, separable and irreducible over F for $1 \leq i \leq m$. As $p_i(x)^{l_i}$ is the characteristic polynomial of $J_S(C(p_i(x)), l_i)$ for $1 \leq i \leq m$ we deduce $\chi_A(x) = \prod_{i=1}^{m} p_i(x)^{l_i}$ from Lemma 5.5 and Exercises 5.1, Question 2(b). There may be repetitions among the polynomials $p_i(x)$ but all are separable over F. So $\chi_A(x)$ is separable over F by Definition 6.17.

Suppose $\chi_A(x)$ is separable over F. By Theorem 6.16 and the discussion after it we know $A \sim J_1 \oplus J_2 \oplus \cdots \oplus J_m$ where $J_i = J(p_i(x), l_i)$ and $p_i(x)$ is irreducible over F for $1 \leq i \leq m$. As $p_i(x)^{l_i}$ is the characteristic polynomial of $J(p_i(x), l_i)$ we obtain $\chi_A(x) = \prod_{i=1}^{m} p_i(x)^{l_i}$ as above. Therefore each $p_i(x)$ is separable over F by Definition 6.17. As $J(p_i(x), 1) = C(p_i(x)) = J_S(C(p_i(x), 1))$ we see $J_i = J(p_i(x), l_i) \sim J_S(C(p_i(x), l_i))$ by Theorem 6.21 for $l_i \geq 1$ and $1 \leq i \leq m$. Therefore

$$A \sim J_S(C(p_1(x)), l_1) \oplus J_S(C(p_2(x), l_2)) \oplus \cdots \oplus J_S(C(p_m(x), l_m))$$

showing that A is similar to a matrix in sJf over F. $\qquad\square$

Let $p_0(x) = x^2 + a_1 x + a_0$ be separable and irreducible over a field F. Our next theorem *transforms* the companion matrix $C(p_0(x)^l)$ into the separable Jordan matrix $J_S(C(p_0(x))^T, l)$, that is, a specific invertible matrix Z_S over F is constructed satisfying $Z_S C(p_0(x)^l) Z_S^{-1} = J_S(C(p_0(x))^T, l)$. From Theorem 6.21 we know that Z_S exists but Theorem 6.21 does not help us find it. Keep the case $F = \mathbb{R}$ and the discussion after Lemma 4.8 in mind as this is the motivation for our next theorem. Every square matrix A over $\mathbb{R}$ in pcf Definition 6.14 can be transformed into sJf Definition 6.20 and a final 'tweak' Corollary 6.25 gives a matrix in *real Jordan form* Definition 6.24.

Theorem 6.23

Let $p_0(x) = x^2 + a_1 x + a_0$ be separable and irreducible over a field F and let l be a positive integer. Let $E = F(c)$ be an extension field of F where $p_0(x) = (x - c)(x - c')$ and so $c' \in E$. Regarding $p_0(x)^l$ as a polynomial over E, denote by M the $E[x]$-module (Definition 5.8) determined by $C(p_0(x)^l)$. Write $w = (x - c')^l e_1$. Then w has order $(x - c)^l$ in M. Also $(x - c)^{j-1} w = v_{j0} + c v_{j1}$ where $v_{j0}, v_{j1} \in F^{2l}$ for $1 \leq j \leq l$. Let Z_S be the $2l \times 2l$ matrix over F with $e_i Z = v_{jr}$ where $i - 1 = 2(j - 1) + r, 0 \leq r < 2$ for $1 \leq i \leq 2l$. Then Z_S is invertible over F and satisfies

$$Z_S C(p_0(x)^l) Z_S^{-1} = J_S(C(p_0(x))^T, l).$$

Proof

The vector $e_1 \in E^{2l}$ generates M by Theorem 5.26 and so has order $p_0(x)^l = (x-c)^l(x-c')^l$ in M. Therefore

$$w = (x-c')^l e_1$$

has order $p_0(x)^l / \gcd\{(x-c')^l, p_0(x)^l\} = p_0(x)^l/(x-c')^l = (x-c)^l$ in M by Lemma 5.23. Now $E = F \oplus cF$, that is, the additive group of E is the direct sum of its subgroups F and $cF = \{cv : v \in F\}$. In the same way $E^{2l} = F^{2l} \oplus cF^{2l}$, that is, the vector space E^{2l}, which is $4l$-dimensional over F, is the direct sum of its subspaces F^{2l} and $cF^{2l} = \{cv : v \in F^{2l}\}$ both of which have dimension $2l$ over F. Note that E^{2l} is the set of elements of the $E[x]$-module M. As $(x-c)^{j-1}w \in M$ there are unique vectors $v_{j0}, v_{j1} \in F^{2l}$ with $(x-c)^{j-1}w = v_{j0} + cv_{j1}$ for $1 \le j \le l$.

As $p_0(x)$ is separable over F we know $c \ne c'$ by Definition 6.17. Let $w' = (x-c)^l e_1$. Then w' has order $(x-c')^l$ in M. Also $M = \langle w \rangle \oplus \langle w' \rangle$ by the primary decomposition theorem 6.12. By Exercises 4.1, Question 4(c) there is a self-inverse automorphism θ of the field E with $(c)\theta = c'$, $(c')\theta = c$ and $(v)\theta = v$ for all $v \in F$; notice that θ is complex conjugation in the case $F = \mathbb{R}$, $E = \mathbb{C}$ as c and c' are complex conjugates. The automorphism θ of E can be extended to a ring automorphism $\widehat{\theta}$ of $E[x]$ by defining

$$(a_0 + a_1x + \cdots + a_sx^s)\widehat{\theta} = (a_0)\theta + (a_1)\theta x + \cdots + (a_s)\theta x^s$$

for all $a_0 + a_1x + \cdots + a_sx^s \in E[x]$. So

$$((x-c)^j)\widehat{\theta} = ((x-c)\theta)^j = (x-(c)\theta)^j = (x-c')^j \quad \text{for } 1 \le j \le l.$$

We introduce the invertible F-linear mapping $\hat{\theta} : M \to M$ defined by

$$(x_1, x_2, \ldots, x_{2l})\hat{\theta} = ((x_1)\theta, (x_2)\theta, \ldots, (x_{2l})\theta) \quad \text{for all } x_i \in E, \ 1 \le i \le 2l.$$

Then $\hat{\theta}$ fixes all vectors in F^{2l} since θ fixes all elements of F. The mapping $\hat{\theta}$ is not $E[x]$-linear, but it does have *the semi-linearity property*

$$(f(x)v)\hat{\theta} = ((f(x))\widehat{\theta})((v)\hat{\theta}) \quad \text{for all } f(x) \in E[x], v \in M \tag{$\blacklozenge$}$$

(Exercises 6.2, Question 9(a)). As θ is self-inverse so also are $\widehat{\theta}$ and $\hat{\theta}$. Now θ interchanges c and c'. Using ($\blacklozenge$) we see that $\hat{\theta}$ interchanges w and w' because

$$(w)\hat{\theta} = ((x-c')^l e_1)\hat{\theta} = (x-c')^l \widehat{\theta}(e_1)\hat{\theta} = (x-c)^l e_1 = w'$$

and $(w')\hat{\theta} = w$ as $(\hat{\theta})^{-1} = \hat{\theta}$. Now

$$w' = (w)\hat{\theta} = (v_{10} + cv_{11})\hat{\theta} = (v_{10})\hat{\theta} + (c)\theta(v_{11})\hat{\theta} = v_{10} + c'v_{11}$$

using ($\blacklozenge$) again and more generally for $1 \leq j \leq l$

$$(x-c)^{j-1}w' = ((x-c')^{j-1})\widehat{\theta}(w)\widehat{\theta} = ((x-c')^{j-1}w)\widehat{\theta}$$

$$= (v_{j0} + cv_{j1})\widehat{\theta} = (v_{j0})\widehat{\theta} + (c)\theta(v_{j1})\theta = v_{j0} + c'v_{j1}.$$

We can now construct the $2l \times 2l$ matrices

$$Y = \begin{pmatrix} w \\ (x-c)w \\ \vdots \\ (x-c)^{l-1}w \\ \hline w' \\ (x-c')w' \\ \vdots \\ (x-c')^{l-1}w' \end{pmatrix} = \begin{pmatrix} v_{10}+cv_{11} \\ v_{20}+cv_{21} \\ \vdots \\ v_{l0}+cv_{l1} \\ v_{10}+c'v_{11} \\ v_{20}+c'v_{21} \\ \vdots \\ v_{l0}+c'v_{l1} \end{pmatrix} \quad \text{and} \quad Z_S = \begin{pmatrix} v_{10} \\ v_{11} \\ v_{20} \\ v_{21} \\ \vdots \\ \vdots \\ v_{l0} \\ v_{l1} \end{pmatrix}.$$

By Theorem 6.16 and the discussion following it Y is invertible over E and satisfies $YC(p_0(x)^l)Y^{-1} = J(x-c,l) \oplus J(x-c',l)$. Now $\det Y$ and $\det Z_S$ are closely related: applying *eros* to Y we obtain $\det Y = (-1)^{l(l+1)/2}(c-c')^l \det Z_S$ (Exercises 6.2, Question 9(b)). So Z_S is invertible over F and its rows form a basis $\mathcal{B}$ of F^{2l} by Corollary 2.23.

Regarding $p_0(x)^l$ as a polynomial over F, let M_0 denote the $F[x]$-module determined by $C(p_0(x)^l)$ and let $\alpha : M_0 \to M_0$ denote the F-linear mapping determined by $C(p_0(x)^l)$ as in Definition 5.8. Then $(v)\alpha = xv = vC(p_0(x)^l)$ for all $v \in F^{2l}$. We now use the decomposition $E^{2l} = F^{2l} \oplus cF^{2l}$ above to find the matrix of α relative to $\mathcal{B}$.

For $1 \leq j \leq l$ the equation $x(x-c)^{j-1}w = c(x-c)^{j-1}w + (x-c)^j w$ in M gives

$$x(v_{j0} + cv_{j1}) = c(v_{j0} + cv_{j1}) + v_{j+10} + cv_{j+11}$$

$$= -a_0v_{j1} + v_{j+10} + c(v_{j0} - a_1v_{j1} + v_{j+11})$$

as $c^2 = -a_1c - a_0$ and where $v_{l+10} = v_{l+11} = 0$ since $(x-c)^l w = 0$. Comparing 'real' and 'imaginary' parts we obtain for $1 \leq j < l$

$$\begin{array}{ll} xv_{j0} = 0 \times v_{j0} - a_0v_{j1} + 1 \times v_{j+10} + 0 \times v_{j+11} \\ xv_{j1} = 1 \times v_{j0} - a_1v_{j1} + 0 \times v_{j+10} + 1 \times v_{j+11} \end{array} \quad \text{and} \quad \begin{array}{l} xv_{l0} = 0 \times v_{l0} - a_0v_{l1} \\ xv_{l1} = 1 \times v_{l0} - a_1v_{l1} \end{array}$$

and these equations in M_0 tell us by Definition 6.20 that $J_S(C(p_0(x))^T, l)$ is the matrix of α relative to $\mathcal{B}$. As $C(p_0(x)^l)$ is the matrix of α relative to the standard basis $\mathcal{B}_0$ of F^{2l} we conclude $Z_S C(p_0(x)^l)Z_S^{-1} = J_S(C(p_0(x))^T, l)$ by Lemma 5.2. $\qquad\square$

Although the details of Theorem 6.23 are lengthy, the construction of the invertible matrices Y and Z_S is relatively straightforward. Remember that $x^{i-1}e_1 = e_i$ in M as in Theorem 6.23 for $1 \le i \le 2l$ and so the ith entry in $w = (x - c')^l e_1$ is the coefficient of x^{i-1} in $(x - c')^l = (x + a_1 + c)^l$ namely $\binom{l}{i-1}(a_1 + c)^{l-i+1}$ and the ith entry in $w' = (x - c)^l e_1$ is the coefficient of x^{i-1} in $(x - c)^l$ namely $\binom{l}{i-1}(-c)^{l-i+1}$ for $1 \le i \le 2l$.

We work through the case $p_0(x) = x^2 + x + 1$, $F = \mathbb{Q}$, $l = 3$. The zeros of $p_0(x)$ are $c = -1/2 + (\sqrt{3}/2)i$, $c' = -1/2 - (\sqrt{3}/2)i$ and so $E = \mathbb{Q}(c) = \mathbb{Q}(\sqrt{3}i)$. In this case $w = (x - c')^3 e_1 = (x + 1 + c)^3 e_1 = ((1 + c)^3, 3(1 + c)^2, 3(1 + c), 1, 0, 0)$ and $w' = (x - c)^3 e_1 = (-c^3, 3c^2, -3c, 1, 0, 0)$ have orders $(x - c)^3$ and $(x + 1 + c)^3$ respectively in the $E[x]$-module M determined by the companion matrix $C((x^2 + x + 1)^3)$. Then

$$
X = \begin{pmatrix} w \\ xw \\ x^2w \\ \hline w' \\ xw' \\ x^2w' \end{pmatrix} = \left(\begin{array}{cccccc} (1+c)^3 & 3(1+c)^2 & 3(1+c) & 1 & 0 & 0 \\ 0 & (1+c)^3 & 3(1+c)^2 & 3(1+c) & 1 & 0 \\ 0 & 0 & (1+c)^3 & 3(1+c)^2 & 3(1+c) & 1 \\ \hline -c^3 & 3c^2 & -3c & 1 & 0 & 0 \\ 0 & -c^3 & 3c^2 & -3c & 1 & 0 \\ 0 & 0 & -c^3 & 3c^2 & -3c & 1 \end{array} \right)
$$

satisfies $XC((x^2 + x + 1)^3)X^{-1} = C((x - c)^3) \oplus C((x + c + 1)^3)$ by Theorem 5.31. Eliminating c^2 and higher powers of c using $c^2 = -c - 1$ produces

$$
X = \left(\begin{array}{cccccc} -1 & 3c & 3+3c & 1 & 0 & 0 \\ 0 & -1 & 3c & 3+3c & 1 & 0 \\ 0 & 0 & -1 & 3c & 3+3c & 1 \\ \hline -1 & -3-3c & -3c & 1 & 0 & 0 \\ 0 & -1 & -3-3c & -3c & 1 & 0 \\ 0 & 0 & -1 & -3-3c & -3c & 1 \end{array} \right).
$$

Applying *eros* of type (iii) over $\mathbb{Q}(c)$ to X and eliminating c^2 produces

$$
Y = \begin{pmatrix} w \\ (x-c)w \\ (x-c)^2w \\ \hline w' \\ (x+c+1)w' \\ (x+c+1)^2w' \end{pmatrix} = \left(\begin{array}{cccccc} -1 & 3c & 3+3c & 1 & 0 & 0 \\ c & 2+3c & 3+3c & 3+2c & 1 & 0 \\ 1+c & 3+2c & 5+3c & 5+2c & 3+c & 1 \\ \hline -1 & -3-3c & -3c & 1 & 0 & 0 \\ -1-c & -1-3c & -3c & 1-2c & 1 & 0 \\ -c & 1-2c & 2-3c & 3-2c & 2-c & 1 \end{array} \right)
$$

$$= \begin{pmatrix} v_{10} + cv_{11} \\ v_{20} + cv_{21} \\ v_{30} + cv_{31} \\ \hline v_{10} - (c+1)v_{11} \\ v_{20} - (c+1)v_{21} \\ v_{30} - (c+1)v_{31} \end{pmatrix}.$$

The matrix Y transforms $C((x^2 + x + 1)^3)$ into $C((x - c)^3) \oplus C((x - c')^3)$ by Theorem 5.31. Notice $\det Y = \det X = R((x-c)^3, (x-c')^3) \neq 0$ by Corollary 5.32. Finally

$$Z_S = \begin{pmatrix} v_{10} \\ v_{11} \\ v_{20} \\ v_{21} \\ v_{30} \\ v_{31} \end{pmatrix} = \begin{pmatrix} -1 & 0 & 3 & 1 & 0 & 0 \\ 0 & 3 & 3 & 0 & 0 & 0 \\ 0 & 2 & 3 & 3 & 1 & 0 \\ 1 & 3 & 3 & 2 & 0 & 0 \\ 1 & 3 & 5 & 5 & 3 & 1 \\ 1 & 2 & 3 & 2 & 1 & 0 \end{pmatrix}$$

is invertible over $\mathbb{Q}$ as $\det Z_S = -27$ by direct computation. Also Z_S satisfies

$$Z_S \begin{pmatrix} 0 & 1 & 0 & 0 & 0 & 0 \\ 0 & 0 & 1 & 0 & 0 & 0 \\ 0 & 0 & 0 & 1 & 0 & 0 \\ 0 & 0 & 0 & 0 & 1 & 0 \\ 0 & 0 & 0 & 0 & 0 & 1 \\ -1 & -3 & -6 & -7 & -6 & -3 \end{pmatrix} = \left(\begin{array}{cc|cc|cc} 0 & -1 & 1 & 0 & 0 & 0 \\ 1 & -1 & 0 & 1 & 0 & 0 \\ 0 & 0 & 0 & -1 & 1 & 0 \\ 0 & 0 & 1 & -1 & 0 & 1 \\ 0 & 0 & 0 & 0 & 0 & -1 \\ 0 & 0 & 0 & 0 & 1 & -1 \end{array} \right) Z_S$$

as the reader may verify. So $Z_S C((x^2 + x + 1)^3) Z_S^{-1} = J_S(C(x^2 + x + 1)^T, 3)$.

As an example of our next corollary notice $C(x^2 + x + 1)^T = \begin{pmatrix} 0 & -1 \\ 1 & -1 \end{pmatrix}$ is similar over $\mathbb{R}$ to the rotation matrix

$$R_{2\pi/3} = \begin{pmatrix} -1/2 & \sqrt{3}/2 \\ -\sqrt{3}/2 & -1/2 \end{pmatrix}$$

as $Z_0 C(x^2 + x + 1)^T Z_0^{-1} = R_{2\pi/3}$ where

$$Z_0 = \begin{pmatrix} 1 & -(1 - \sqrt{3})/2 \\ 1 & -(1 + \sqrt{3})/2 \end{pmatrix}.$$

The 6×6 matrix $Z_1 = Z_0 \oplus Z_0 \oplus Z_0$ is invertible over $\mathbb{R}$ and transforms

$J_S(C(x^2 + x + 1)^T, 3)$ into

$$J_S(R_{2\pi/3}, 3) = \begin{pmatrix} -1/2 & \sqrt{3}/2 & 1 & 0 & 0 & 0 \\ -\sqrt{3}/2 & -1/2 & 0 & 1 & 0 & 0 \\ 0 & 0 & -1/2 & \sqrt{3}/2 & 1 & 0 \\ 0 & 0 & -\sqrt{3}/2 & -1/2 & 0 & 1 \\ 0 & 0 & 0 & 0 & -1/2 & \sqrt{3}/2 \\ 0 & 0 & 0 & 0 & -\sqrt{3}/2 & -1/2 \end{pmatrix}$$

which is in real Jordan form according to the next definition.

Definition 6.24

A direct sum of matrices, each being either of the type $J(x - \lambda, k)$ or $J_S(B, l)$ (see Definitions 6.15 and 6.20) where $B = \begin{pmatrix} a & b \\ -b & a \end{pmatrix}$ for $\lambda, a, b \in \mathbb{R}$, $b > 0$, is said to be in *real Jordan form*.

So a matrix in real Jordan form is the direct sum of certain separable Jordan blocks over $\mathbb{R}$. For example

$$J(x - 3, 1) \oplus J(x - 4, 2) \oplus J_S(2R_{2\pi/3}, 2)$$

$$= \begin{pmatrix} 3 & 0 & 0 & 0 & 0 & 0 & 0 \\ 0 & 4 & 1 & 0 & 0 & 0 & 0 \\ 0 & 0 & 4 & 0 & 0 & 0 & 0 \\ 0 & 0 & 0 & -1 & \sqrt{3} & 1 & 0 \\ 0 & 0 & 0 & -\sqrt{3} & -1 & 0 & 1 \\ 0 & 0 & 0 & 0 & 0 & -1 & \sqrt{3} \\ 0 & 0 & 0 & 0 & 0 & -\sqrt{3} & -1 \end{pmatrix}$$

is in real Jordan form. Notice that $(x - 3)(x - 4)^2(x^2 + 2x + 4)^2$ is the factorisation into irreducible polynomials over $\mathbb{R}$ of the characteristic polynomial of the above 7×7 matrix.

Note also that $\begin{pmatrix} a & b \\ -b & a \end{pmatrix}$ is the matrix of $\alpha : \mathbb{C} \to \mathbb{C}$ relative to the basis $1, i$ of the real vector space $\mathbb{C}$ where α is the $\mathbb{R}$-linear mapping 'multiplication by $z_0 = a + ib$', that is, $(z)\alpha = z_0 z$ for all $z \in \mathbb{C}$ and so $(1)\alpha = a + ib$, $(i)\alpha = -b + ia$. In geometric terms α is the composition of the commuting mappings 'radial expansion by $|z_0| = \sqrt{a^2 + b^2}$' and '(anticlockwise) rotation through $\arg z_0$', both fixing the origin.

Corollary 6.25

Let A be a $t \times t$ matrix over the real field $\mathbb{R}$. Then A is similar to a matrix in real Jordan form.

Proof

We construct a matrix which transforms the pcf Definition 6.14 of A into a matrix in real Jordan form. By the discussion after Lemma 4.8 the monic factors of $\chi_A(x)$ which are irreducible over $\mathbb{R}$ are of at most two types: $x - \lambda$ where $\lambda \in \mathbb{R}$ and $p_0(x) = x^2 + a_1 x + a_0$ where $a_1^2 < 4a_0$. Let m be the number of elementary divisors of A. Then the pcf of A is the direct sum of m matrices each being either $C((x - \lambda)^l)$ or $C((p_0(x))^l)$ as above. There is an invertible $l \times l$ matrix Z over $\mathbb{R}$ with $ZC((x - \lambda)^l)Z^{-1} = J(x - \lambda, l)$ by Theorem 6.16.

By Theorem 6.23 there is an invertible $2l \times 2l$ matrix Z_S over $\mathbb{R}$ with $Z_S C(p_0(x)^l)Z_S^{-1} = J_S(C(p_0(x))^T, l)$. On completing the square we obtain $p_0(x) = x^2 + a_1 x + a_0 = (x - a)^2 + b^2$ where $a = -a_1/2$ and $b = \sqrt{a_0 - a_1^2/4}$. The reader can check that $Z_0 = \left(\begin{smallmatrix} 1 & a+b \\ 1 & a-b \end{smallmatrix}\right)$ is invertible over $\mathbb{R}$ and satisfies $Z_0 C(p_0(x))^T Z_0^{-1} = B$ where $B = \left(\begin{smallmatrix} a & b \\ -b & a \end{smallmatrix}\right)$ as in Definition 6.24. Let $Z_1 = Z_0 \oplus Z_0 \oplus \cdots \oplus Z_0$ with l terms. Then the $2l \times 2l$ matrix $Z = Z_1 Z_S$ is invertible over $\mathbb{R}$ and satisfies $ZC((p_0(x))^l)Z^{-1} = J_S(B, l)$.

Finally the direct sum of the m matrices Z, as described above, one for each direct summand $C((x - \lambda)^l)$ or $C((p_0(x))^l)$ in the pcf of A, is an invertible $t \times t$ matrix over $\mathbb{R}$ which transforms the pcf of A into a matrix in real Jordan form. As A is similar to its pcf Definition 6.14 we conclude that A is similar to a matrix in real Jordan form. $\square$

EXERCISES 6.2

1. (a) For each of the matrices A below over the rational field $\mathbb{Q}$ find invertible matrices $P(x)$ and $Q(x)$ over $\mathbb{Q}[x]$ such that $P(x)(xI - A)Q(x)^{-1}$ is in Smith normal form. Hence find a 3×3 invertible matrix X over $\mathbb{Q}$ with XAX^{-1} in rational canonical form. Determine $\mathbb{Q}$-bases of the primary components $M(A)_x$ and $M(A)_{x+2}$ of the $\mathbb{Q}[x]$-module $M(A)$. Specify a 3×3 invertible matrix Y over $\mathbb{Q}$ with YAY^{-1} in primary canonical form:

$$
\text{(i)} \quad \begin{pmatrix} 1 & -1 & 2 \\ 1 & -1 & 2 \\ -1 & 1 & -2 \end{pmatrix}; \qquad
\text{(ii)} \quad \begin{pmatrix} -1 & 1 & 1 \\ 1 & -2 & -1 \\ -1 & 1 & 1 \end{pmatrix};
$$

$$
\text{(iii)} \quad \begin{pmatrix} -3 & 1 & 2 \\ -1 & -1 & 2 \\ -1 & 1 & 0 \end{pmatrix}; \qquad
\text{(iv)} \quad \begin{pmatrix} -1 & 1 & 1 \\ 1 & 0 & 1 \\ -1 & -1 & -3 \end{pmatrix}.
$$

   Are any two of these matrices similar?

   (b) Let A be a $t \times t$ matrix over a field F. Let $p(x)$ be an irreducible polynomial over F. Show $p(x)|\chi_A(x) \Leftrightarrow \det p(A) = 0$.
   Hint: Argue by contradiction using Corollaries 4.6 and 6.11.

(c) Let

$$A = \begin{pmatrix} 2 & 1 & -1 \\ 1 & 0 & -1 \\ 2 & 2 & -1 \end{pmatrix}$$

over $\mathbb{Q}$. Calculate $A^2 + I$. Hence find the factorisation of $\chi_A(x)$ using trace A and (b) above. Determine an invertible 3×3 matrix Y over $\mathbb{Q}$ with YAY^{-1} in primary canonical form.

2. (a) Let A be a $t \times t$ matrix over a field F having s invariant factors. Suppose $\chi_A(x) = p_1(x)^{n_1} p_2(x)^{n_2} \cdots p_k(x)^{n_k}$ where $p_1(x), p_2(x), \ldots, p_k(x)$ are distinct and irreducible over F. Using the notation introduced after Corollary 6.13 write $\mu_A(x) = p_1(x)^{t_{s1}} p_2(x)^{t_{s2}} \cdots p_k(x)^{t_{sk}}$. Show $M(A)_{p_j(x)} = \{v \in M(A) : p_j(x)^{t_{sj}} v = 0\}$ for $1 \le j \le k$.

(b) Let A be a $t \times t$ matrix over a field F. Suppose A has t distinct eigenvalues $\lambda_1, \lambda_2, \ldots, \lambda_t$ in F. Show that A is similar to $\operatorname{diag}(\lambda_1, \lambda_2, \ldots, \lambda_t)$. Is $M(A)$ cyclic?

Hint: Use Theorem 6.12 and bases (row eigenvectors of A) of the primary components $M(A)_{x-\lambda_j}$ for $1 \le j \le t$.

More generally show that A is similar to a diagonal matrix over F if and only if $\mu_A(x)$ is a product of distinct factors of degree 1 over F.

True or false? The $t \times t$ matrix A over the field F is similar to a diagonal matrix over F if and only if

(i) all its elementary divisors are of degree 1 over F,

(ii) each invariant factor of A splits into distinct factors of degree 1 over F.

(c) Let s and t be positive integers. Let $L(s, t)$ denote the set of sequences $(l_1, l_2, \ldots, l_s)$ of s non-negative integers l_j with $l_1 + l_2 + \cdots + l_s = t$. Let $M(s, t)$ denote the set of sequences $(m_1, m_2, \ldots, m_{t+s-1})$ where either $m_j = 0$ or $m_j = 1$ for $1 \le j < s + t$ and $m_1 + m_2 + \cdots + m_{s+t-1} = t$. Write $(l_1, l_2, \ldots, l_s)\alpha = (m_1, m_2, \ldots, m_{s+t-1})$ where $m_j = 0$ for $j = l_1 + l_2 + \cdots + l_i + i$ ($1 \le i < s$) and $m_j = 1$ otherwise ($1 \le j < s + t$). Show that $\alpha : L(s, t) \to M(s, t)$ is a bijection by constructing α^{-1}. Deduce $|L(s, t)| = \binom{s+t-1}{t}$.

Hint: To get the idea, list the images under α of the 10 elements of $L(3, 3)$.

(d) Let $c_1, c_2, \ldots, c_s$ be distinct elements of a field F and write $f(x) = (x - c_1)(x - c_2) \cdots (x - c_s)$. Use Theorem 6.12 to show that $\binom{s+t-1}{t}$ is the number of similarity classes of $t \times t$ matrices A over F satisfying $f(A) = 0$. Are all such matrices diagonalisable (similar to a diagonal matrix)?

Hint: Let $l_j = \dim U_j$ where $U_j = \{v \in F^t : vA = c_j v\}$ and use (c) above.

(e) Let F be a field with $\chi(F) \neq 2$, that is, $1 \neq -1$ in F. Determine the number of similarity classes of $t \times t$ matrices A over F with (i) $A^2 = I$; (ii) $A^3 = A$.

Hint: Use (d) above.

(f) Use Corollary 3.17 to show $x^2 + x + 1$ is irreducible over the field $\mathbb{Z}_p$ (p prime) if and only if $p \equiv -1 \pmod 3$. Determine the number of similarity classes of $t \times t$ matrices A over $\mathbb{Z}_p$ with $p \neq 3$ satisfying $A^3 = I$.

Hint: Treat the cases $p \equiv 1 \pmod 3$ and $p \equiv -1 \pmod 3$ separately.

(g) Let A be a $t \times t$ matrix over the finite field $\mathbb{F}_q$. Show that A is similar to a diagonal matrix over $\mathbb{F}_q$ if and only if $A^q = A$. Find the number of similarity classes of $t \times t$ matrices A over $\mathbb{F}_q$ satisfying $A^q = A$. List diagonal representatives of these similarity classes in the cases (i) $t = 4$, $q = 2$; (ii) $t = 2$, $q = 4$.

Hint: The polynomial $x^q - x$ splits into q distinct factors over $\mathbb{F}_q$.

3. (a) Let A be a $t \times t$ matrix over a field F. Suppose $\chi_A(x) = p(x)^n q(x)$ where $p(x)$ is a monic irreducible polynomial over F, n is a positive integer and $\gcd\{p(x), q(x)\} = 1$. Show that the primary component $M(A)_{p(x)} = \{v \in M(A) : p(x)^n v = 0\}$ is a submodule of the $F[x]$-module $M(A)$. Using Corollary 6.11 show $M(A)_{p(x)}$ to be non-trivial.

(b) Let $p(x)$ be a monic irreducible polynomial over a field F and let n be a positive integer. Use Theorem 5.28 to show that the $F[x]$-module $M = M(C(p(x)^n))$ is *indecomposable*, that is, M is non-zero and does *not* have non-zero submodules N_1 and N_2 with $M = N_1 \oplus N_2$. Conversely let N be an indecomposable submodule of the $F[x]$-module $M(A)$ where A is a $t \times t$ matrix over a field F. Show $N \cong M(C(p(x)^n))$ where $p(x)^n | \chi_A(x)$, $p(x)$ is irreducible over F and n is a positive integer.

Hint: Use Exercises 5.1, Question 5, Theorems 6.5, 6.12 and (a) above.

(c) Let A be a $t \times t$ matrix over a field F. Suppose

$$M(A) = N_1 \oplus N_2 \oplus \cdots \oplus N_r$$

where N_j is a non-zero submodule of $M(A)$ for $1 \leq j \leq r$. Show $r \leq t$.

Suppose r is as large as possible. Show that N_j is indecomposable for $1 \leq j \leq r$. Deduce from (b) above that r is the number of elementary divisors of A.

(d) Construct a proof of Theorem 6.12 using Theorem 3.10 as a guide.

4. (a) Let p be prime and let A be a $t \times t$ matrix over $\mathbb{Z}_p$. The image of the evaluation homomorphism $\varepsilon_A : \mathbb{Z}_p[x] \to \mathfrak{M}_t(\mathbb{Z}_p)$ is the commutative subring $\operatorname{im} \varepsilon_A = \{f(A) \in \mathfrak{M}_t(\mathbb{Z}_p) : f(x) \in \mathbb{Z}_p[x]\}$ of $\mathfrak{M}_t(\mathbb{Z}_p)$. Show that $\varphi : \operatorname{im} \varepsilon_A \to \operatorname{im} \varepsilon_A$, given by $(X)\varphi = X^p$ for all $X \in \operatorname{im} \varepsilon_A$, is a ring homomorphism. Show also that φ is $\mathbb{Z}_p$-linear.

   Hint: Use the method of Exercises 2.3, Question 5(a).

   Show $\mu_{A^p}(x) | \mu_A(x)$. Suppose $\mu_A(x)$ is irreducible over $\mathbb{Z}_p$. Deduce $A^p \sim A$ and find the connection between φ, $\tilde{\varepsilon}_A$ and the Frobenius automorphism θ of the field $F[x]/\langle \mu_A(x) \rangle$.

   Let $A = C(x^2 + x + 1) \oplus C(x^2 + x + 1)$ and $B = C((x^2 + x + 1)^2)$ over $\mathbb{Z}_2$. Decide whether or not $A^2 \sim A$ and $B^2 \sim B$.

   (b) Let p be prime and let A be a $t \times t$ matrix over $\mathbb{Z}_p$. Suppose $\mu_A(x) = p_0(x)^m$ where $p_0(x)$ is monic and irreducible of degree n over $\mathbb{Z}_p$ and $m \geq 2$. Use φ above to find a non-zero polynomial $g(x)$ over $\mathbb{Z}_p$ with $g(A^p) = 0$ and $\deg g(x) < mn$. Deduce $A^p \not\sim A$.

   Hint: Consider first the case $1 < m \leq p$.

   (c) Let A be a $t \times t$ matrix over $\mathbb{Z}_p$. Use Corollary 6.13 to show $\chi_A(x) = \chi_{A^p}(x)$. Show also $A \sim A^p$ if and only if the minimum polynomial $\mu_A(x)$ is a product of distinct irreducible factors over $\mathbb{Z}_p$.

   Hint: Use (a) and (b) above.

5. (a) Determine the number of similarity classes of 12×12 matrices A over the rational field $\mathbb{Q}$ having characteristic polynomial $\chi_A(x) = (x-1)^3 x^4 (x+1)^5$. How many different minimum polynomials $\mu_A(x)$ are there among these classes? Specify the partitions of 3, 4 and 5 such that (i) A is cyclic, (ii) A is diagonalisable (similar to a diagonal matrix over $\mathbb{Q}$).

   Construct the table of $\cong M(A)$ specified by the partitions $(1, 2)$, $(1, 3)$, $(1, 1, 1, 2)$ of 3, 4, 5. List the invariant factors of A. Find $\mu_A(x)$ and express the rcf and pcf of A as direct sums of companion matrices.

   (b) Determine the number of similarity classes of 18×18 matrices A over the rational field $\mathbb{Q}$ having $\chi_A(x) = (x-1)^5 (x^2 + 1)^4 (x+1)^5$. Construct the table of $\cong M(A_0)$ specified by the partitions $(1, 2, 2)$, $(1, 1, 2)$, $(1, 1, 3)$ of the exponents of $x - 1$, $x^2 + 1$, $x + 1$ respectively. List the invariant factors of A_0. Using Exercises 6.1, Question 2(a), list the invariant factors of $-A_0$ and construct the table of $\cong M(-A_0)$. Is $-A_0 \sim A_0$? How many of these classes are such that $-A \sim A$? How many of these classes are such that $A \sim A^{-1}$?

   (c) Determine the number of similarity classes of 22×22 matrices A over the rational field $\mathbb{Q}$ having

$$\chi_A(x) = (x-1)^5 (x+1)^5 (x-2)^6 (x-1/2)^6.$$

How many of these classes are such that $-A \sim A$? How many of these classes are such that $A \sim A^{-1}$?

6. (a) Let $A = (a_{ij})$ be a $t \times t$ matrix over a field F such that $a_{ij} = 0$ for $i + 1 < j$ where $t \geq 2$. Show that the $F[x]$-module $M(A)$ is generated by e_1 if and only if $a_{ii+1} \neq 0$ for $1 \leq i < t$.

   (b) Working over the rational field $\mathbb{Q}$, use Theorem 6.16 to find invertible matrices Z_1 and Z_2 satisfying

$$Z_1 C((x+1)^3) Z_1^{-1} = J(x+1, 3) \quad \text{and}$$

$$Z_2 C((x^2+1)^2) Z_2^{-1} = J(x^2+1, 2).$$

   Hence find Z with $Z(C((x+1)^3) \oplus C((x^2+1)^2)) Z^{-1}$ in Jnf (Definition 6.15).

   (c) Which (if any) pairs of the following 6×6 matrices over a field F are similar? $C((x+1)^6)$, $J((x+1)^2, 3)$, $J((x+1)^3, 2)$, $J(x+1, 6)$, $J((x+1)^6, 1)$. Which of these matrices is in Jnf?

   (d) Write $A = C(x^3(x+1)^2)$ over an arbitrary field F. Specify invertible matrices Y and Z over F such that YAY^{-1} is in pcf and ZAZ^{-1} is in Jnf. Find the Jnf of A^2 and the Jnf of $A + A^2$.

   (e) Let λ be an element of a field F. Show that there are $p(t)$ similarity classes of $t \times t$ matrices A over F with $\chi_A(x) = (x - \lambda)^t$, where $p(t)$ is the number of partitions Definition 3.13 of the positive integer t. Taking $t = 5$, list representatives of these classes in (i) pcf, (ii) Jnf. Express the rank of the $t \times t$ matrix $J - \lambda I$ in terms of s and t where

$$J = \sum_{j=1}^{s} \oplus J(x - \lambda, t_j).$$

   Write $B = J(x - 1, t)$ over a field F where $t \geq 2$. Show $B \sim B^2$ if and only if $\chi(F) \neq 2$. Find the Jnf of B^2 in the case $\chi(F) = 2$. Hint: $B = I + C(x^t)$.

7. (a) The polynomials

$$f(x) = \sum_{i=0}^{m} a_i x^i \quad \text{and} \quad g(x) = \sum_{j=0}^{n} b_j x^j$$

   are over an arbitrary field F. Establish the rules of formal differentiation Definition 6.18:

   (i) $(f(x) + g(x))' = f'(x) + g'(x)$, $(cf(x))' = cf'(x)$ for $c \in F$,

   (ii) $(f(x)g(x))' = f'(x)g(x) + f(x)g'(x)$.

   Using induction on the positive integer (i) and (ii) above, show $(g(x)^i)' = ig(x)^{i-1}g'(x)$. Deduce the 'function of a function' rule: $(f(g(x)))' = f'(g(x))g'(x)$.

(b) Let p be a prime and write $F = \mathbb{Z}_p(y)$ for the field of fractions of $\mathbb{Z}_p[x]$ (see the discussion after Definition 6.17). Let E be an extension field of F containing a zero c of the polynomial $x^p - y$ over F. Show $x^p - y = (x - c)^p$ over E and deduce that the monic irreducible factors of $x^p - y$ over F are equal. Show $c \notin F$ and hence show that $x^p - y$ is irreducible over F. Is $x^p - y$ inseparable over F? Are your conclusions unchanged on replacing $\mathbb{Z}_p$ by an arbitrary field F_0 of characteristic p?

(c) Let i be a non-negative integer and let

$$f(x) = a_n x^n + a_{n-1} x^{n-1} + \cdots + a_1 x + a_0$$

be a polynomial over the field F. The polynomial

$$H_i(f(x)) = \sum_{j \geq i}^{n} a_j \binom{j}{i} x^{j-i}$$

over F is called the ith *Hasse derivative* of $f(x)$. Show $f(x) \to H_i(f(x))$ is a linear mapping of the vector space $F[x]$ over F.

Which $f(x) \in \mathbb{Z}_2[x]$ satisfy $f''(x) = 0$ and which satisfy $H_2(f(x)) = 0$?

Hint: Consider $f(x) = x^j$. Describe $H_2(H_2(f(x)))$ for $f(x) \in \mathbb{Z}_2[x]$. Show $i! H_i(f(x)) = f^{(i)}(x)$ and $H_i'(f(x)) = (i+1)H_{i+1}(f(x))$ for all $f(x) \in F[x]$. Suppose the characteristic $\chi(F)$ of F is not a divisor of $i!$. Show $H_i(f(x)) = f^{(i)}(x)/i!$. Suppose $\chi(F)$ is not a divisor of $i+1$. Show $H_{i+1}(f(x)) = H_i(f(x))/(i+1)$.

(d) Let B be an $n \times n$ matrix over a field F. Calculate the 4×4 matrices $J_S(B, 4)^2$, $J_S(B, 4)^3$, $J_S(B, 4)^4$ having entries in $\mathfrak{M}_n(F)$ where $J_S(B, 4)$ is the separable Jordan block matrix (Definition 6.20). Let l be a positive integer and let $\tilde{B}$ and $\tilde{N}$ be the $ln \times ln$ matrices defined in the proof of Theorem 6.21. Show

$$f(J_S(B, l)) = \sum_{i=0}^{l-1} H_i(f(\tilde{B}))\tilde{N}^i$$

for all $f(x) \in F[x]$ where $f(x)$ and

$$H_i(f(\tilde{B})) = \sum_{j \geq i}^{n} a_j \binom{j}{i} \tilde{B}^{j-i}$$

are as in (c) above. Hence find a formula for the $n \times n$ matrix entries ? in the partitioned $ln \times ln$ matrix $\chi_B(J(B, l))$ constructed in the proof of Theorem 6.21.

8. (a) Find the minimum polynomial $\mu_A(x)$ where $A = J_S(C(x^2), l)$ is the separable Jordan block matrix (Definition 6.20) over a field F and $l \geq 2$. (The answer depends on whether or not $\chi(F)$ is a divisor of l.) *Hint*: Write $A = \tilde{C} + \tilde{N}$ as in Theorem 6.21 where $\tilde{C}$ is the direct sum of l matrices $C(x^2)$.
   What is the value of nullity A? Find the rcf of A.

   (b) Let A and A' be respectively an $n \times n$ matrix and an $n' \times n'$ matrix over a field F and let l be a positive integer. Find an invertible $l(n + n') \times l(n + n')$ matrix X over F with

   $$X(J_S(A, l) \oplus J_S(A', l)) = J_S(A \oplus A', l)X.$$

   Hint: To get started take $n = n' = 1$ and $l = 2$. Then try n, n' arbitrary, $l = 3$, before tackling the general case.
   Let A, B, X be $n \times n$ matrices over F with $AX = XB$, X invertible over F. Use $\tilde{X} = X \oplus X \oplus \cdots \oplus X$ (the direct sum of l matrices X) to show $J_S(A, l) \sim J_S(B, l)$.
   Let $f(x)$ and $g(x)$ be monic polynomials of positive degree over F with $\gcd\{f(x), g(x)\} = 1$. Using Theorem 5.31 deduce $J_S(C(f(x)g(x)), l) \sim J_S(C(f(x)), l) \oplus J_S(C(g(x)), l)$.

9. (a) Let M be an R-module where R is a non-trivial commutative ring. An additive mapping $\alpha : M \to M$ is called *semi-linear* if there is an automorphism θ of R such that $(av)\alpha = (a)\theta(v)\alpha$ for all $a \in R$, $v \in M$. Let α and β be semi-linear mappings of M. Show that $\alpha\beta$ is semi-linear. Show that α bijective implies α^{-1} semi-linear.
   Let $\theta \in \operatorname{Aut} R$ and let t be a positive integer. Show that $\hat{\theta} : R^t \to R^t$, defined by $(a_1, a_2, \ldots, a_t)\hat{\theta} = ((a_1)\theta, (a_2)\theta, \ldots, (a_t)\theta)$ for all $(a_1, a_2, \ldots, a_t) \in R^t$, is semi-linear.

   (b) Let E be a field containing c, c' where $c \neq c'$ and let l be positive integer. Let v_{j0} and v_{j1} belong to E^{2l} for $1 \leq j \leq l$. Let Y denote the $2l \times 2l$ matrix over E with $e_i Y = v_{i0} + cv_{i1}$ for $1 \leq i \leq l$ and $e_i Y = v_{i-l0} + c'v_{i-l1}$ for $l < i \leq 2l$. Let Z_S be the $2l \times 2l$ matrix over E with $e_{2j}Z_S = v_{j0}, e_{2j-1}Z_S = v_{j1}$ for $1 \leq j \leq l$ (see the proof of Theorem 6.23). Show $\det Y = (-1)^{l(l+1)/2}(c - c')^l \det Z_S$.

   (c) Write $p_0(x) = x^2 - 2x + 2$ over the real field $\mathbb{R}$. Is $p_0(x)$ irreducible and separable over $\mathbb{R}$? Find the integer entries in the 6×6 matrix Z_S of Theorem 6.23 satisfying $Z_S C(p_0(x)^3) = J_S(C(p_0(x))^T, 3)Z_S$. Find the value of $\det Z_S$. Is Z_S invertible over $\mathbb{Q}$? Find an invertible 6×6 matrix Z over $\mathbb{R}$ with $ZC(p_0(x)^3)Z^{-1}$ in real Jordan form.

6.3 Endomorphisms and Automorphisms of $M(A)$

Let $M(A)$ denote, as usual, the $F[x]$-module determined (Definition 5.8) by the $t \times t$ matrix A over the field F. We study here the endomorphism ring $\operatorname{End} M(A)$, that is, the ring of $F[x]$-linear mappings $\beta : M(A) \to M(A)$, the ring multiplication being composition of mappings. Each β in $\operatorname{End} M(A)$ is an additive mapping (Definition 2.3) of the abelian group $(F^t, +)$ and so $\operatorname{End} M(A)$ is a subring of the ring $\operatorname{End}(F^t, +)$ discussed in Lemma 3.14. Now $\operatorname{End}(F^t, +)$ is a t^2-dimensional vector space over F having $\operatorname{End} M(A)$ as a subspace. The dimension of $\operatorname{End} M(A)$ is determined by *Frobenius' theorem* (Corollary 6.34) in terms of the invariant factors (Definition 6.8) of A. We'll see that $\operatorname{End} M(A)$ is isomorphic (both as a ring and a vector space) to the *centraliser* $Z(A)$ where

$$Z(A) = \{B \in \mathfrak{M}_t(F) : AB = BA\}$$

consists of all matrices B which commute with the given matrix A. The index of the group $GL_t(F) \cap Z(A) = U(Z(A))$ in $GL_t(F)$ is the size of the similarity class of A. Finally, knowing the irreducible factorisation of $\chi_A(x)$ over F, we analyse the group $\operatorname{Aut} M(A) \cong U(Z(A))$ of invertible endomorphisms of $M(A)$.

We start the details by reminding the reader that $\mu_A(x)$ denotes the minimum polynomial (Corollary 6.11) of A. Notice $\mu_A(x) = x - c$ if and only if $A = cI$ in which case $F(c) = F$ as $c \in F$ and $\operatorname{End} M(A) \cong \mathfrak{M}_t(F)$ as discussed in Example 5.9c. Our first theorem, which is the polynomial analogue of Theorem 3.16, shows that $M(A)$ and its endomorphism ring are easily described should $\mu_A(x)$ be irreducible over F.

Theorem 6.26

Let A denote a $t \times t$ matrix over a field F and suppose $\mu_A(x)$ is irreducible over F. Then $\chi_A(x) = \mu_A(x)^s$ where $s = t/m$ and $m = \deg \mu_A(x)$. Let $F(c)$ be the extension field of F obtained by adjoining a zero c of $\mu_A(x)$ to F. Writing

$$f(c)v = f(x)v \quad \text{for all } v \in F^t \text{ and all } f(x) \in F[x]$$

gives the $F[x]$-module $M(A)$ the structure of a vector space of dimension s over $F(c)$. Also N is a submodule of the $F[x]$-module $M(A)$ if and only if N is a subspace of the $F(c)$-module $M(A)$. Further $\operatorname{End} M(A) \cong \mathfrak{M}_s(F(c))$ and $\operatorname{Aut} M(A) \cong GL_s(F(c))$.

Let the $F(c)$-module $M(A)$ have basis $v_1, v_2, \ldots, v_s$. Denote by X the $t \times t$ matrix over F with $e_i X = x^r v_j$ where $i - 1 = (j - 1)m + r, 0 \leq r < m, 1 \leq i \leq t$. Then X is invertible over F and

$$XAX^{-1} = C(\mu_A(x)) \oplus C(\mu_A(x)) \oplus \cdots \oplus C(\mu_A(x)) \quad (s \text{ terms}).$$

Proof

The proof of Theorem 6.26, except the last paragraph, is omitted as it is closely analogous to that of Theorem 3.16 (Exercises 6.3, Question 1(e)). As for the last paragraph, let $j - 1$ and r be the quotient and remainder on dividing $i - 1$ by m and so $1 \le j \le s$ as $0 \le i - 1 < ms$. Write $p(x) = \mu_A(x)$ to remind ourselves that $\mu_A(x)$ is irreducible over F. All non-zero vectors in F^t have order $p(x)$ in the $F[x]$-module $M(A)$ as $p(x)v = vp(A) = v\mu_A(A) = v \times 0 = 0$. So the cyclic submodule $N_j = \langle v_j \rangle$ has F-basis $\mathcal{B}_{v_j}$ consisting of $v_j, xv_j, \ldots, x^{m-1}v_j$ for $1 \le j \le s$ by Corollary 5.29. The restriction of the linear mapping determined by A to N_j has matrix $C(p(x))$ relative to $\mathcal{B}_{v_j}$ by Corollary 5.27 for $1 \le j \le s$. As $v_1, v_2, \ldots, v_s$ is a basis of the $F(c)$-module $M(A)$ we see that the $F[x]$-module $M(A)$ decomposes $M(A) = N_1 \oplus N_2 \oplus \cdots \oplus N_s$. Therefore $\mathcal{B} = \mathcal{B}_{v_1} \cup \mathcal{B}_{v_2} \cup \cdots \cup \mathcal{B}_{v_s}$ is a basis of F^t by Lemma 5.18. The matrix X, as specified above, has the vectors of $\mathcal{B}$ as its rows and so is invertible over F by Corollary 2.23. Finally X satisfies

$$XAX^{-1} = C(p(x)) \oplus C(p(x)) \oplus \cdots \oplus C(p(x))$$

the direct sum of s companion matrices $C(p(x))$, by Corollary 5.20. □

To illustrate Theorem 6.26 consider

$$A = \begin{pmatrix} 1 & 1 & 0 & 1 \\ 1 & 1 & -1 & 2 \\ -4 & -1 & 1 & -3 \\ -3 & -2 & 1 & -3 \end{pmatrix}$$

over $\mathbb{Q}$. By direct calculation the reader can check $A^2 = -I$. So $\mu_A(x)$ is a monic and non-constant divisor over $\mathbb{Q}$ of $x^2 + 1$ by Corollary 6.10. As $x^2 + 1$ is irreducible over $\mathbb{Q}$ we deduce $\mu_A(x) = x^2 + 1$. Without further calculation the invariant factors of A are $x^2 + 1, x^2 + 1$ (what else could they be?) and hence $\chi_A(x) = (x^2 + 1)^2$. In this case reducing $xI - A$ to Smith normal form is *not* the easiest way to find X with XAX^{-1} in rcf. Rather first pick any non-zero vector as $v_1 \in \mathbb{Q}^4$ in Theorem 6.26, say $v_1 = e_1$, and secondly pick any vector v_2 not in $\langle v_1, xv_1 \rangle = \langle (1,0,0,0), (1,1,0,1) \rangle$, for instance $v_2 = e_2$. Then v_1, xv_1, v_2, xv_2 is a basis $\mathcal{B}$ of $\mathbb{Q}^4$ and

$$X = \begin{pmatrix} v_1 \\ \hline xv_1 \\ \hline v_2 \\ \hline xv_2 \end{pmatrix} = \begin{pmatrix} 1 & 0 & 0 & 0 \\ 1 & 1 & 0 & 1 \\ 0 & 1 & 0 & 0 \\ 1 & 1 & -1 & 2 \end{pmatrix}$$

satisfies

$$XAX^{-1} = C(x^2 + 1) \oplus C(x^2 + 1).$$

Incidentally this matrix XAX^{-1} is in pcf and Jnf as well as rcf. Writing $iv = xv$ for all $v \in M(A)$ turns $M(A)$ into a 2-dimensional vector space over $\mathbb{Q}(i) = \{a + ib : a, b \in \mathbb{Q}\}$ where, as usual, $i^2 = -1$. The submodule $\langle v_1, xv_1 \rangle$ of the $\mathbb{Q}[x]$-module $M(A)$ is a 2-dimensional subspace over $\mathbb{Q}$ and also a 1-dimensional subspace over $\mathbb{Q}(i)$. Finally, there is one matrix X with XAX^{-1} in rcf for each basis v_1, v_2 of $M(A)$ over $\mathbb{Q}(i)$.

Now suppose A to be an arbitrary $t \times t$ matrix over a field F. There is a close connection between endomorphisms of $M(A)$ and matrices B which commute with A as we now demonstrate.

Theorem 6.27

Let A be a $t \times t$ matrix over a field F and let the F-linear mapping $\beta : F^t \to F^t$ have matrix B relative to the standard basis $\mathcal{B}_0$ of F^t. Then β is an endomorphism of $M(A)$ if and only if $AB = BA$.

Suppose $M(A)$ to be a cyclic $F[x]$-module. Then each matrix which commutes with A is of the type $B = f(A)$ where $f(x) \in F[x]$, $\deg f(x) < t$. Also β is an automorphism of $M(A)$ if and only if $\gcd\{f(x), \chi_A(x)\} = 1$.

Proof

Let β be an endomorphism of $M(A)$. Then $(xv)\beta = x((v)\beta)$ for all $v \in F^t$ by Definition 2.24. As $xv = vA$ and $(v)\beta = vB$ for all $v \in F^t$, on taking $v = e_i$ we obtain $e_i AB = (xe_i)\beta = x((e_i)\beta) = e_i BA$ for $1 \leq i \leq t$. So AB and BA have equal rows showing $AB = BA$.

Conversely suppose $AB = BA$. Then $(xv)\beta = vAB = vBA = x((v)\beta)$ for all $v \in F^t$. From Lemma 5.15 with β in place of θ, we deduce that β is an endomorphism of $M(A)$.

Suppose $M(A)$ is cyclic with generator v_0 and let B be a $t \times t$ matrix over F satisfying $AB = BA$. Let β be the endomorphism of $M(A)$ defined by $(v)\beta = vB$ for all $v \in F^t$. We use the basis $\mathcal{B}_{v_0}$ of F^t consisting of $v_0, xv_0, x^2 v_0, \ldots, x^{t-1} v_0$ introduced in (5.24). There are unique scalars a_{i-1} where $1 \leq i \leq t$ such that $(v_0)\beta = a_0 v_0 + a_1 xv_0 + a_2 x^2 v_0 + \cdots + a_{t-1} x^{t-1} v_0$. Write $f(x) = a_0 + a_1 x + a_2 x^2 + \cdots + a_{t-1} x^{t-1}$ and let $\gamma : F^t \to F^t$ be the F-linear mapping defined by $(v)\gamma = vf(A)$ for all $v \in F^t$. As A and $f(A)$ commute, γ is an endomorphism of $M(A)$ by the first part of the proof. As $(v_0)\beta = (v_0)\gamma$ and v_0 generates $M(A)$ we deduce $\beta = \gamma$. As the matrices of β and γ relative to $\mathcal{B}_0$ are respectively B and $f(A)$ we conclude $B = f(A)$ where $\deg f(x) < t$.

Suppose that β is an automorphism of $M(A)$. Then β^{-1} is also an automorphism of $M(A)$ and B^{-1} is its matrix relative to $\mathcal{B}_0$. So there is $a(x) \in F[x]$ with $B^{-1} = a(A)$ by the above paragraph. On dividing $a(x)f(x)$ by $\chi_A(x)$ we obtain $a(x)f(x) =$

$q(x)\chi_A(x) + r(x)$ where $q(x), r(x) \in F[x]$ and $\deg r(x) < t = \deg \chi_A(x)$. Therefore $I = a(A)f(A) = q(A)\chi_A(A) + r(A) = r(A)$ by Corollary 6.11. So $r'(x) = r(x) - 1$ satisfies $r'(A) = 0$ with $\deg r'(x) < t = \deg \mu_A(x)$ as $\mu_A(x) = \chi_A(x)$ by Corollary 6.10. The conclusion is: $r'(x)$ is the zero polynomial, that is, $r(x) = 1$. From $f(x)g(x) = q(x)\chi_A(x) + 1$ we deduce $\gcd\{f(x), \chi_A(x)\} = 1$.

Conversely suppose $\gcd\{f(x), \chi_A(x)\} = 1$. By Corollary 4.6 there are $a(x), b(x) \in F[x]$ with $a(x)f(x) + b(x)\chi_A(x) = 1$. Evaluating this polynomial equality at A gives $a(A)f(A) = I$ as $\chi_A(A) = 0$ by Corollary 6.11. So $B = f(A)$ is invertible over F with inverse $B^{-1} = a(A)$. Therefore β is an automorphism of $M(A)$. □

There are three important facts in Theorem 6.27. First, endomorphisms β of $M(A)$ correspond to matrices B which commute with A.

Definition 6.28

Let A be a $t \times t$ matrix over a field F. Then

$$Z(A) = \{B \in \mathfrak{M}_t(F) : AB = BA\}$$

is called *the centraliser of* A.

The centraliser $Z(A)$ is a subring of the ring $\mathfrak{M}_t(F)$ of $t \times t$ matrices over F and $Z(A)$ is also a subspace of the t^2-dimensional vector space $\mathfrak{M}_t(F)$ over F. For this reason $Z(A)$ is called an *algebra over* F. Denote by $(\beta)\theta$ the matrix of β relative to the standard basis $\mathcal{B}_0$ of F^t for all $\beta \in \operatorname{End} M(A)$. Then $\theta : \operatorname{End} M(A) \cong Z(A)$ is *an algebra isomorphism* (an isomorphism of rings and vector spaces) (Exercises 6.3, Question 2(b)).

In the case of a cyclic $M(A)$ the matrices $I, A, A^2, \ldots, A^{t-1}$ are F-independent and from Theorem 6.27 these matrices span $Z(A)$. So

$$t = \dim Z(A) = \dim \operatorname{End} M(A) \quad \text{for cyclic } F[x]\text{-modules } M(A)$$

which is the second important fact contained in Theorem 6.27.

The order $|\operatorname{Aut} M(A)|$ of the automorphism group of a cyclic module $M(A)$ can be found, in the case of a finite field F of order q, using a polynomial version of the Euler ϕ-function as we now explain. So suppose $|F| = q$ and for each non-zero polynomial $g(x)$ over F let

$$\Phi_q(g(x)) \text{ denote the number of polynomials } f(x) \text{ over } F$$

$$\text{with } \gcd\{f(x), g(x)\} = 1, \deg f(x) < \deg g(x).$$

Then thirdly from Theorem 6.27 we obtain:

Let A be a $t \times t$ matrix over a finite field F of order q with $M(A)$ cyclic.

Then $|\operatorname{End} M(A)| = q^t$ and $|\operatorname{Aut} M(A)| = \Phi_q(\chi_A(x))$.

The value of $\Phi_q(g(x))$ can be calculated in the same way as $\phi(n)$. Specifically

$$\Phi_q(g(x)h(x)) = \Phi_q(g(x))\Phi_q(h(x)) \quad \text{for } \gcd\{g(x), h(x)\} = 1,$$

that is, Φ_q is *multiplicative*. Also

$$\Phi_q(p(x)^n) = q^{mn} - q^{m(n-1)}$$

where $p(x)$ is irreducible of degree m over the finite field F of order q (Exercises 6.3, Question 2(c)).

For example let $A = C(x^3(x^2 + x + 1))$ over $\mathbb{Z}_2$. Then $M(A)$ is cyclic with generator e_1 and $\chi_A(x) = x^3(x^2 + x + 1)$ by Theorem 5.26. As $\Phi_2(x^3) = 2^3 - 2^2$ (the four relevant polynomials are $1, x + 1, x^2 + 1, x^2 + x + 1$) and $\Phi_2(x^2 + x + 1) = 2^2 - 1$ (the three relevant polynomials are $1, x, x + 1$) we see $\Phi_2(\chi_A(x)) = 4 \times 3 = 12$. So $|\operatorname{End} M(A)| = 2^5 = 32$ and $|\operatorname{Aut} M(A)| = 12$ by Theorem 6.27. Knowing the order of $\operatorname{Aut} M(A)$ the number of 5×5 matrices similar to A can be found using our next theorem. In this case the size of the similarity class of A is

$$|GL_5(\mathbb{Z}_2)|/|\operatorname{Aut} M(A)| = (2^5 - 1)(2^5 - 2)(2^5 - 2^2)(2^5 - 2^3)(2^5 - 2^4)/12$$

$$= 833280.$$

Theorem 6.29

Let A be a $t \times t$ matrix over a field F. Write $G = GL_t(F)$ and $H_A = G \cap Z(A)$. Then $H_A = U(Z(A))$ is a subgroup of the multiplicative group G and $\operatorname{Aut} M(A) \cong H_A$. The correspondence

$$H_A X \to X^{-1} A X \quad \text{for } X \in G$$

between the set of left cosets $H_A X$ of H_A in G and the similarity class of A is unambiguous and bijective. For a finite field F there are $|GL_t(F)|/|H_A|$ matrices similar to A.

Proof

We leave the routine verification that $H_A = U(Z(A))$ is a subgroup of G to the reader and also that $\theta : \operatorname{Aut} M(A) \cong H_A$ is a group isomorphism, where $(\beta)\theta$ is the matrix of β relative to the standard basis $\mathcal{B}_0$ of F^t for all $\beta \in \operatorname{Aut} M(A)$ (Exercises 6.3, Question 2(b)).

For $X, Y \in G$ write $X \equiv Y$ if there is $B \in H_A$ with $BX = Y$. Then $\equiv$ is an equivalence relation on G (it is the analogue of Lemma 2.9 for multiplicative groups). The equivalence class of X is $H_A X = \{BX : B \in H_A\}$ which we call the (left) coset of H_A in G with representative X. The set of all such cosets of H_A in G constitutes a partition of G as each is non-empty and each matrix in G belongs to exactly one of them.

Suppose $H_A X = H_A Y$. Then $Y \in H_A X$ and there is $B \in H_A$ with $BX = Y$. As $B \in Z(A)$ we know $AB = BA$ and so $B^{-1}AB = A$. Therefore

$$Y^{-1}AY = (BX)^{-1}ABX = X^{-1}B^{-1}ABX = X^{-1}AX$$

showing that the correspondence $H_A X \to X^{-1}AX$ is unambiguously defined, as it does not depend on the choice of coset representative X or Y.

Conversely suppose $X^{-1}AX = Y^{-1}AY$ where $X, Y \in G$. Reversing the above steps we obtain $B \in Z(A)$ where $B = YX^{-1}$ and so $B \in G \cap Z(A) = H_A$. So $X \equiv Y$, that is, $H_A X = H_A Y$. We have now shown that $H_A X \to X^{-1}AX$ is injective. From Definition 5.4 we conclude that this correspondence is surjective.

Let F be a finite field. Then G and H_A are finite groups. Each coset of H_A in G consists of $|H_A|$ elements. As these cosets partition G we see that their number is $|G|/|H_A|$, the index of H_A in G. Using the above correspondence $|GL_t(F)|/|\operatorname{Aut} M(A)|$ is the size of the similarity class of A. $\square$

The correspondence of Theorem 6.29 is a special case of a general fact, *the orbit-stabiliser theorem*, in the theory of permutation representations. Here each $X \in GL_t(F) = G$ gives rise to a permutation (bijection) $\pi(X)$ of the set $\mathfrak{M}_t(F)$ defined by: $(A)^{\pi(X)} = X^{-1}AX$ for all $A \in \mathfrak{M}_t(F)$, that is, the image of each A by $\pi(X)$ is $X^{-1}AX$. Notice

$$\pi(XX') = \pi(X)\pi(X') \quad \text{for all } X, X' \in G$$

showing that π is a *permutation representation* of G, that is, π is a homomorphism from G to a group of permutations. Notice $(A')^{\pi(X)} = A$ where A and A' are similar Definition 5.3.

The *orbit of* A is

$$\{(A)^{\pi(X)} : X \in G\} = \{X^{-1}AX : X \in G\},$$

which is the similarity class of A.

 The stabiliser of A is

$$\{X \in G : (A)^{\pi(X)} = A\} = \{X \in G : X^{-1}AX = A\}$$

$$= \{X \in G : AX = XA\} = G \cap Z(A) = H_A.$$

The orbit-stabiliser theorem asserts that, in a general context, the length of the orbit of A equals the index of the stabiliser of A in G (Exercises 6.3, Question 3(a)).

Consider the particular case $\mathfrak{M}_2(\mathbb{Z}_3)$. By Exercises 6.1, Question 6(a) the similarity classes of 2×2 matrices A over $\mathbb{Z}_3$ are $12 = 3 + 9$ in number since they correspond to the 12 possible minimum polynomials $\mu_A(x)$. The 3 similarity classes of scalar matrices 0, I, $-I$ over $\mathbb{Z}_3$ are *singletons* (they each consist of one matrix only) and have minimum polynomials x, $x - 1$, $x + 1$ respectively.

The 9 similarity classes with $\deg \mu_A(x) = 2$ are such that $M(A)$ is cyclic and so the second part of Theorem 6.27 can be used. Consider first $\mu_A(x) = x^2 + 1$ which is irreducible over $\mathbb{Z}_3$. There are $9 - 1 = 8$ polynomials $f(x)$ of degree at most 1 over $\mathbb{Z}_3$ with $\gcd\{f(x), x^2 + 1\} = 1$, namely all $f(x) = a_1 x + a_0$ except the zero polynomial, and so $\Phi_3(x^2 + 1) = 8$. As $|GL_2(\mathbb{Z}_3)| = (3^2 - 1)(3^2 - 3) = 48$ we see that the similarity class of $C(x^2 + 1)$ over $\mathbb{Z}_3$ consists of exactly $48/8 = 6$ matrices by Theorem 6.29. The reader can check

$$\pm\begin{pmatrix} 0 & 1 \\ -1 & 0 \end{pmatrix}, \quad \pm\begin{pmatrix} 1 & -1 \\ -1 & -1 \end{pmatrix}, \quad \pm\begin{pmatrix} -1 & -1 \\ -1 & 1 \end{pmatrix}$$

all satisfy $\det = 1$, $\text{trace} = 0$ and so are the 6 matrices in the similarity class of $C(x^2 + 1)$ over $\mathbb{Z}_3$. As $x^2 + x - 1$ and $x^2 - x - 1$ are irreducible over $\mathbb{Z}_3$ (they have no zeros in $\mathbb{Z}_3$, – now use Lemma 4.8(ii)) we see that the similarity classes of $C(x^2 + x - 1)$ and $C(x^2 - x - 1)$ over $\mathbb{Z}_3$ also each contain exactly 6 matrices.

Now consider $\mu_A(x) = x^2$. There are $3^2 - 3 = 6$ polynomials $f(x)$ of degree at most 1 over $\mathbb{Z}_3$ with $\gcd\{f(x), x^2\} = 1$ namely all $f(x) = a_1 x + a_0$ except those with $a_0 = 0$. So $\Phi_3(x^2) = 6$ and by Theorem 6.29 there are precisely $48/6 = 8$ matrices similar to $C(x^2)$ over $\mathbb{Z}_3$ (they all have rank 1, trace 0). In the same way the similarity class of $C((x + 1)^2)$ over $\mathbb{Z}_3$ contains exactly 8 matrices and the similarity class of $C((x - 1)^2)$ over $\mathbb{Z}_3$ also contains exactly 8 matrices.

Finally consider $\mu_A(x) = x(x + 1)$. The polynomials ± 1, $\pm(x - 1)$ and only these contribute to $\Phi_3(x(x + 1)) = 4$. By Theorem 6.29 there are precisely $48/4 = 12$ matrices similar to $C(x(x + 1))$ over $\mathbb{Z}_3$ (they all have rank $= 0$, trace $= -1$). In the same way the similarity class of $C(x(x - 1))$ over $\mathbb{Z}_3$ contains exactly 12 matrices and the similarity class of $C((x + 1)(x - 1))$ over $\mathbb{Z}_3$ also contains exactly 12 matrices.

We have now accounted for the similarity classes of all $3^4 = 81$ matrices in $\mathfrak{M}_2(\mathbb{Z}_3)$. It is reassuring that $81 = 1 + 1 + 1 + 6 + 6 + 6 + 8 + 8 + 8 + 12 + 12 + 12$ as the similarity classes partition $\mathfrak{M}_2(\mathbb{Z}_3)$. Now $A \in GL_t(F) \Leftrightarrow \mu_A(0) \neq 0$. So the 48 matrices of $GL_2(\mathbb{Z}_3)$ partition into 8 similarity (conjugacy) classes and $48 = 1 + 1 + 6 + 6 + 6 + 8 + 8 + 12$. In the same way the number and sizes of the similarity classes of matrices in $\mathfrak{M}_2(\mathbb{F}_q)$ can be found (Exercises 6.3, Question 2(e)).

Our next task is to study the algebra $\text{End } M(A)$ in the case of an arbitrary $t \times t$ matrix A over a field. For cyclic $M(A)$ this has already been done in Theorem 6.27. The general case can be 'cracked' by decomposing $M(A)$ as in Theorem 6.5 and using certain $s \times s$ matrices over $F[x]$ which represent endomorphisms of $M(A)$. We are used to specifying linear mappings of F^t via their matrices relative to a convenient basis. Square matrices over $\mathbb{Z}$ are used to describe endomorphisms of finite abelian groups (Exercises 3.3, Question 5(a)). The same method, with a few minor adjustments, does the trick yet again!

Definition 6.30

Let A be a $t \times t$ matrix over a field F having s invariant factors $d_i(x)$ for $1 \le i \le s$. By Theorem 6.5 there are vectors v_i of order $d_i(x)$ in $M(A)$ such that

$$M(A) = \langle v_1 \rangle \oplus \langle v_2 \rangle \oplus \cdots \oplus \langle v_s \rangle.$$

Let $\beta \in \text{End}\, M(A)$. Then

$$(v_i)\beta = b_{i1}(x)v_1 + b_{i2}(x)v_2 + \cdots + b_{is}(x)v_s$$

$$= \sum_{j=1}^{s} b_{ij}(x)v_j \quad \text{for } 1 \le i, j \le s,$$

where $b_{ij}(x) \in F[x]$. The $s \times s$ matrix $B(x) = (b_{ij}(x))$ is said to *represent* β relative to the ordered set of generators $v_1, v_2, \ldots, v_s$ of $M(A)$. If $\deg b_{ij}(x) < \deg d_j(x)$ for $1 \le i, j \le s$ then $B(x)$ is said to be *reduced*.

It is not correct to refer to $B(x)$ as *the* matrix of β relative to $v_1, v_2, \ldots, v_s$ as there are many matrices $B(x)$ as in Definition 6.30. For example the zero endomorphism of $M(A)$ is represented by the zero $s \times s$ matrix over $F[x]$ as well as (for instance) $B(x) = (b_{ij}(x))$ where $b_{i1}(x) = d_1(x)$, $b_{ij}(x) = 0$ for $1 \le i \le s$, $1 < j \le s$ since $d_1(x)v_1 = 0$. However we will see in Theorem 6.32 that each endomorphism β of $M(A)$ is represented by a *unique reduced* matrix $B(x)$. Notice that $B(x)$ in $\mathfrak{M}_s(F[x])$ is reduced if and only if each entry in col j is of smaller degree than $d_j(x)$ for $1 \le j \le s$.

Except in the special case $d_1(x) = d_s(x)$, that is, all the invariant factors of A are equal, there are some $s \times s$ matrices over $F[x]$ which do not represent any endomorphism of $M(A)$ as we now explain.

Definition 6.31

Let $d_1(x), d_2(x), \ldots, d_s(x)$ denote the invariant factors of the $t \times t$ matrix A over the field F. The $s \times s$ matrix $B(x) = (b_{ij}(x))$ over $F[x]$ is said to satisfy *the endomorphism condition* relative to $M(A)$ (e.c.rel. $M(A)$) if

$$(dj(x)/d_i(x))|b_{ij}(x) \quad \text{for } 1 \le i < j \le s.$$

For example suppose $A = C(x) \oplus C(x^2)$ and $F = \mathbb{Z}_p$. So $s = 2$, $t = 3$ and $d_1(x) = x$, $d_2(x) = x^2$. In this case the 2×2 matrix $B(x) = (b_{ij}(x))$ satisfies e.c.rel. $M(A)$ if and only if $x|b_{12}(x)$. There are p^5 reduced matrices satisfying e.c.rel. $M(A)$ namely those of the type

$$B(x) = \begin{pmatrix} c_{11} & c_{12}x \\ c_{21} & c_{22}x + c'_{22} \end{pmatrix} \quad \text{for } c_{11}, c_{12}, c_{21}, c_{22}, c'_{22} \in \mathbb{Z}_p$$

as the entries in col 1 are of degree less than $1 = \deg x$ and the entries in col 2 are of degree less than $2 = \deg x^2$. By Corollary 6.34 we will see that $M(A)$ has p^5 endomorphisms corresponding to the above p^5 matrices. Further $M(A)$ has $p^3(p-1)^2$ automorphisms corresponding to reduced matrices $B(x)$ with $c_{11} \neq 0$, $c'_{22} \neq 0$, that is, the matrix $\begin{pmatrix} c_{11} & 0 \\ c_{21} & c'_{22} \end{pmatrix}$ of remainders on division by x is invertible over $\mathbb{Z}_p$ (see Lemma 6.35).

Theorem 6.32

Let A be a $t \times t$ matrix over a field F and let $\beta \in \operatorname{End} M(A)$. Using the notation of Definition 6.30 let $B(x) = (b_{ij}(x))$ represent β relative to $v_1, v_2, \ldots, v_s$. Then $B(x)$ satisfies e.c.rel. $M(A)$. There is a unique reduced matrix $B(x)$ which represents β relative to $v_1, v_2, \ldots, v_s$. The set R_A of all $s \times s$ matrices $B(x)$ over $F[x]$ satisfying e.c.rel. $M(A)$ is a subring of $\mathfrak{M}_s(F[x])$.

Proof

We begin by reformulating the endomorphism condition in a more usable way. Suppose $B(x) = (b_{ij}(x))$ satisfies Definition 6.31, that is, $(d_j(x)/d_i(x))|b_{ij}(x)$ for $1 \leq i < j \leq s$. On multiplying through by $d_i(x)$ we obtain $d_j(x)|d_i(x)b_{ij}(x)$ which can be written as $d_i(x)b_{ij}(x) \equiv 0 \pmod{d_j(x)}$. These latter conditions hold for all i and j in the range $1 \leq i, j \leq s$ (they hold for $i \geq j$ as then $d_j(x)|d_i(x)$). Therefore

$$B(x) = (b_{ij}(x)) \text{ satisfies e.c.rel. } M(A) \quad \text{if and only if}$$
$$d_i(x)b_{ij}(x) \equiv 0 \pmod{d_j(x)} \quad \text{for } 1 \leq i, j \leq s.$$

Suppose $B(x) = (b_{ij}(x))$ represents β relative to $v_1, v_2, \ldots, v_s$. Then $(v_i)\beta = b_{i1}(x)v_1 + b_{i2}(x)v_2 + \cdots + b_{is}(x)v_s$ for $1 \leq i \leq s$. Multiplying this equation by the order $d_i(x)$ of v_i in $M(A)$ gives

$$0 = (d_i(x)v_i)\beta = d_i(x)((v_i)\beta)$$

$$= d_i(x)b_{i1}(x)v_1 + d_i(x)b_{i2}(x)v_2 + \cdots + d_i(x)b_{is}(x)v_s$$

for $1 \leq i \leq s$. Using the independence Definition 2.14 of the submodules $\langle v_i \rangle$ in the direct sum $M(A) = \langle v_1 \rangle \oplus \langle v_2 \rangle \oplus \cdots \oplus \langle v_s \rangle$, we deduce $d_i(x)b_{ij}(x)v_j = 0$ for

$1 \leq i, j \leq s$ from the above equation. As $d_j(x)$ is the order of v_j in $M(A)$ we conclude $d_i(x)b_{ij}(x) \equiv 0 \pmod{d_j(x)}$ for $1 \leq i, j \leq s$, that is, the $s \times s$ matrix $B(x) = (b_{ij}(x))$ over $F[x]$ satisfies e.c.rel. $M(A)$.

For $1 \leq i, j \leq s$ let $r_{ij}(x)$ be the remainder on dividing $b_{ij}(x)$ by $d_j(x)$. Then $b_{ij}(x)v_j = r_{ij}(x)v_j$ as $d_j(x)v_j = 0$ for $1 \leq i, j \leq s$. From Definition 6.30 we see that the $s \times s$ matrix $(r_{ij}(x))$ is reduced and represents β relative to $v_1, v_2, \ldots, v_s$. Conversely suppose the $s \times s$ matrix $(r'_{ij}(x))$ to be reduced and represent β relative to $v_1, v_2, \ldots, v_s$. Subtracting $(v_i)\beta = \sum_{j=1}^{s} r'_{ij}(x)v_j$ from $(v_i)\beta = \sum_{j=1}^{s} r_{ij}(x)v_j$ gives $0 = \sum_{j=1}^{s}(r_{ij}(x) - r'_{ij}(x))v_j$ for $1 \leq i \leq s$. The independence property of the internal direct sum $M(A) = \langle v_1 \rangle \oplus \langle v_2 \rangle \oplus \cdots \oplus \langle v_s \rangle$ now gives $(r_{ij}(x) - r'_{ij}(x))v_j = 0$ for $1 \leq i, j \leq s$. From Definition 5.11 we deduce $d_j(x) | (r_{ij}(x) - r'_{ij}(x))$ and so $r_{ij}(x) = r'_{ij}(x)$ using Theorem 4.1 for $1 \leq i, j \leq s$ as both $r_{ij}(x)$ and $r'_{ij}(x)$ have lower degree than $d_j(x)$. The conclusion is: each endomorphism of $M(A)$ is represented as in Definition 6.30 by a unique reduced matrix.

We denote by R_A the subset of $\mathfrak{M}_s(F[x])$ consisting of all matrices $B(x)$ satisfying e.c.rel. $M(A)$. Consider $B(x) = (b_{ij}(x))$ and $B'(x) = (b'_{ij}(x))$ in R_A. There are $q_{ij}(x), q'_{ij}(x) \in F[x]$ with $d_i(x)b_{ij}(x) = q_{ij}(x)d_j(x)$ and $d_i(x)b'_{ij}(x) = q'_{ij}(x)d_j(x)$ for $1 \leq i, j \leq s$. Adding these equations gives

$$d_i(x)(b_{ij}(x) + b'_{ij}(x)) = (q_{ij}(x) + q'_{ij}(x))d_j(x) \equiv 0 \pmod{d_j(x)}$$

for $1 \leq i, j \leq s$, which shows $B(x) + B'(x) \in R_A$. For $1 \leq i, j, k \leq s$ we have

$$d_i(x)b_{ij}(x)b'_{jk}(x) = q_{ij}(x)d_j(x)b'_{jk}(x)$$

$$= q_{ij}(x)q'_{jk}(x)d_k(x) \equiv 0 \pmod{d_k(x)}.$$

Summing over j produces

$$d_i(x)\left(\sum_{j=1}^{s} b_{ij}(x)b'_{jk}(x) \right) \equiv 0 \pmod{d_k(x)} \quad \text{for } 1 \leq i, k \leq s,$$

showing $B(x)B'(x) \in R_A$. So R_A is closed under addition and multiplication. We leave the reader to check $-B(x) \in R_A$ and also $0, I \in R_A$, that is, the $s \times s$ zero and identity matrices over $F[x]$ satisfy e.c.rel. $M(A)$ as in Definition 6.31 and so belong to R_A (Exercises 6.3, Question 6(b)). Therefore R_A is a subring of $\mathfrak{M}_s(F[x])$. $\square$

By Theorem 6.32 each endomorphism β of $M(A)$ gives rise to a unique reduced matrix $B(x)$ satisfying e.c.rel. $M(A)$. We now address the question: does each matrix in R_A arise from some endomorphism of $M(A)$ as in Definition 6.30? The next theorem tells us, amongst other things, that the answer is: Yes!

Theorem 6.33

Let A be a $t \times t$ matrix over a field F with invariant factors $d_1(x), d_2(x), \ldots, d_s(x)$. Let $M(A) = \langle v_1 \rangle \oplus \langle v_2 \rangle \oplus \cdots \oplus \langle v_s \rangle$ where v_i has order $d_i(x)$ in $M(A)$ for $1 \le i \le s$. Let $B(x)$ belong to the ring R_A of Theorem 6.32. There is a unique endomorphism β of $M(A)$ such that $B(x)$ represents β relative to $v_1, v_2, \ldots, v_s$. Write $\beta = (B(x))\varphi$ for all $B(x) \in R_A$. Then $\varphi : R_A \to \operatorname{End} M(A)$ is a surjective ring homomorphism.

Let $K = \{(b_{ij}(x)) \in \mathfrak{M}_s(F[x]) : d_j(x)|b_{ij}(x) \text{ for } 1 \le i, j \le s\}$. Then $K = \ker \varphi$ and $\tilde{\varphi} : R_A/K \cong \operatorname{End} M(A)$ where $(K + B(x))\tilde{\varphi} = (B(x))\varphi$ for all $B(x) \in R_A$.

Proof

Consider the $s \times s$ matrix $B(x) = (b_{ij}(x))$ in the ring R_A. Then $d_i(x)b_{ij}(x) \equiv 0 \pmod{d_j(x)}$ for $1 \le i, j \le s$. As $v_1, v_2, \ldots, v_s$ generate $M(A)$ for each $v \in M(A)$ there are polynomials $f_i(x)$ over F for $1 \le i \le s$ with $v = f_1(x)v_1 + f_2(x)v_2 + \cdots + f_s(x)v_s$. We want to construct $\beta \in \operatorname{End} M(A)$ so that Definition 6.30 holds, that is, $(v_i)\beta = \sum_{j=1}^{s} b_{ij}(x)v_j$ for $1 \le i \le s$. There is no choice for the image of v by β as

$$(v)\beta = f_1(x)(v_1)\beta + f_2(x)(v_2)\beta + \cdots + f_s(x)(v_s)\beta = \sum_{i,j=1}^{s} f_i(x)b_{ij}(x)v_j \qquad (\blacklozenge)$$

and so there cannot be two different β as in Definition 6.30.

To show that we are not chasing a will-o'-the-wisp we start again and *define* $\beta : M(A) \to M(A)$ by $(\blacklozenge)$. We must first check that this definition of $(v)\beta$ is unambiguous, in other words, does $(v)\beta$ remain unchanged on changing the above polynomials $f_i(x)$ for $1 \le i \le s$? So suppose also $v = g_1(x)v_1 + g_2(x)v_2 + \cdots + g_s(x)v_s$ where $g_i(x) \in F[x]$ for $1 \le i \le s$. Then $0 = v - v = \sum_{i=1}^{s}(f_i(x) - g_i(x))v_i$ and so $(f_i(x) - g_i(x))v_i = 0$ using the independence Definition 2.14 of the submodules $\langle v_i \rangle$ of $M(A)$ for $1 \le i \le s$. As v_i has order $d_i(x)$ in $M(A)$ we see $f_i(x) - g_i(x) = q_i(x)d_i(x)$ where $q_i(x) \in F[x]$ for $1 \le i \le s$. On multiplying this equation by $b_{ij}(x)$ we obtain

$$(f_i(x) - g_i(x))b_{ij}(x) = q_i(x)d_i(x)b_{ij}(x) \equiv 0 \pmod{d_j(x)} \quad \text{for } 1 \le i, j \le s$$

as $B(x) = (b_{ij}(x))$ belongs to R_A and so satisfies e.c.rel. $M(A)$. Therefore $(f_i(x) - g_i(x))b_{ij}(x)v_j = 0$ as $d_j(x)$ is the order of v_j in $M(A)$, that is, $f_i(x)b_{ij}(x)v_j = g_i(x)b_{ij}(x)v_j$ for $1 \le i, j \le s$. So the r.h.s. of $(\blacklozenge)$ remains unchanged on replacing $f_i(x)$ by $g_i(x)$ and $(v)\beta$ is unambiguously defined by $(\blacklozenge)$. It is now 'plain sailing' to verify that β is an $F[x]$-linear mapping of $M(A)$, that is, $\beta \in \operatorname{End} M(A)$, and $B(x)$ represents β relative to $v_1, v_2, \ldots, v_s$ (Exercises 6.3, Question 6(c)).

As β is specified uniquely by $B(x)$ it is legitimate to introduce

$$\varphi : R_A \to \operatorname{End} M(A) \quad \text{by } \beta = (B(x))\varphi \text{ for all } B(x) \in R_A.$$

(The reader may find it helpful to compare this proof with that of Theorem 3.15: the ring isomorphism θ^{-1} of Theorem 3.15 is analogous to φ although φ is a long way from being injective.) Each β in $\operatorname{End} M(A)$ arises from some $B(x)$ in R_A by Definition 6.30 and Theorem 6.32, showing φ to be surjective.

To show that φ is a ring homomorphism consider $B(x) = (b_{ij}(x))$ and $B'(x) = (b'_{ij}(x))$ in R_A. Write $\beta = (B(x))\varphi$ and $\beta' = (B'(x))\varphi$. Using ($\blacklozenge$)

$$(v)((B(x))\varphi + (B'(x))\varphi) = (v)(\beta + \beta') = (v)\beta + (v)\beta'$$

$$= \sum_{i,j=1}^{s} f_i(x)b_{ij}(x)v_j + \sum_{i,j=1}^{s} f_i(x)b'_{ij}(x)v_j$$

$$= \sum_{i,j=1}^{s} f_i(x)(b_{ij}(x) + b'_{ij}(x))v_j$$

$$= (v)((B(x) + B'(x))\varphi) \quad \text{for all } v \in M(A)$$

and so φ respects addition: $(B(x))\varphi + (B'(x))\varphi = (B(x) + B'(x))\varphi$. Using ($\blacklozenge$) again

$$(v)(((B(x))\varphi)((B'(x))\varphi)) = (v)(\beta\beta') = ((v)\beta)\beta'$$

$$= \left(\sum_{i,j=1}^{s} f_i(x)b_{ij}(x)v_j \right)\beta'$$

$$= \sum_{k=1}^{s} \left(\sum_{i,j=1}^{s} f_i(x)b_{ij}(x)b'_{jk}(x) \right) v_k$$

$$= \sum_{i,k=1}^{s} f_i(x) \left(\sum_{j=1}^{s} b_{ij}(x)b'_{jk}(x) \right) v_k$$

$$= (v)((B(x)B'(x))\varphi) \quad \text{for all } v \in M(A)$$

showing that φ respects multiplication, that is, $(B(x))\varphi(B'(x))\varphi = (B(x)B'(x))\varphi$. The 1-element of R_A is the $s \times s$ identity matrix $I(x) = (\delta_{ij}(x))$ over $F[x]$ and so $\delta_{ij}(x) = 1(x)$ or $0(x)$ according as $i = j$ or $i \neq j$ ($1 \leq i, j \leq s$). Using ($\blacklozenge$) gives

$$(v)((I(x))\varphi) = \sum_{i,j=1}^{s} f_i(x)\delta_{ij}(x)v_j = \sum_{i=1}^{s} f_i(x)\delta_{ii}(x)v_i = \sum_{i=1}^{s} f_i(x)v_i = v$$

for all $v \in M(A)$ showing that $(I(x))\varphi = \iota$ the identity mapping of $M(A)$. As ι is the 1-element of $\operatorname{End} M(A)$ we conclude that $\varphi : R_A \to \operatorname{End} M(A)$ is a surjective ring homomorphism.

Suppose $B(x) = (b_{ij}(x))$ in $\mathfrak{M}_s(F[x])$ satisfies $d_j(x)|b_{ij}(x)$ for $1 \leq i, j \leq s$. Then $b_{ij}(x) \equiv 0 \pmod{d_j(x)}$ and so $d_i(x)b_{ij}(x) \equiv 0 \pmod{d_j(x)}$ for $1 \leq i, j \leq s$

showing $B(x) \in R_A$ and $K \subseteq R_A$. As $d_j(x)$ is the order of v_j in $M(A)$ we see $b_{ij}(x)v_j = 0$ for $1 \le i, j \le s$. Hence

$$(v)((B(x))\varphi) = \sum_{i,j=1}^{s} f_i(x)b_{ij}(x)v_j = \sum_{i,j=1}^{s} f_i(x) \times 0 = 0 \quad \text{for all } v \in M(A)$$

by ($\blacklozenge$) showing $(B(x))\varphi = 0$, that is, $B(x) \in \ker \varphi$ and $K \subseteq \ker \varphi$.

Conversely suppose $B(x) = (b_{ij}(x))$ belongs to $\ker \varphi$. As the endomorphism $(B(x))\varphi$ of $M(A)$ is represented Definition 6.30 by $B(x)$ we have $0 = (v_i)((B(x))\varphi) = \sum_{j=1}^{s} b_{ij}(x)v_j$ for $1 \le i \le s$. By the independence Definition 2.14 of the submodules $\langle v_j \rangle$ for $1 \le j \le s$ we obtain $b_{ij}(x)v_j = 0$ for $1 \le i, j \le s$. As $d_j(x)$ is the order of v_j in $M(A)$ we conclude $d_j(x)|b_{ij}(x)$ for $1 \le i, j \le s$ from Definition 5.11. Therefore $\ker \varphi \subseteq K$ and so $K = \ker \varphi$.

From the first isomorphism theorem for rings (Exercises 2.3, Question 3(b)) we deduce $\tilde{\varphi} : R_A/K \cong \operatorname{End} M(A)$ where $(K + B(x))\tilde{\varphi} = (B(x))\varphi$ for all $B(x) \in R_A$. $\square$

The integer $\dim Z(A)$ is a measure of the number of matrices which commute Definition 6.28 with a given $t \times t$ matrix A over a field F. From Theorem 6.27 we know $\dim Z(A) = t$ in the case of $M(A)$ cyclic, that is, of A having only one invariant factor $d_1(x) = \mu_A(x) = \chi_A(x)$ of degree t. We now combine Theorems 6.32 and 6.33 to deal with the general case.

Corollary 6.34 (Frobenius' theorem)

Let A be a $t \times t$ matrix over a field F with invariant factors $d_1(x), d_2(x), \ldots, d_s(x)$. Let $Z(A) = \{B \in \mathfrak{M}_t(F) : AB = BA\}$ denote the centraliser of A. Then

$$\dim \operatorname{End} M(A) = \dim Z(A) = \sum_{i=1}^{s}(2s - 2i + 1)\deg d_i(x).$$

Proof

We use the notation of Theorem 6.33. The algebra isomorphism θ of Definition 6.28 shows that $\operatorname{End} M(A)$ and $Z(A)$ are vector spaces of equal dimension over F. Write

$$V_A = \{B(x) \in R_A : B(x) \text{ is reduced}\}.$$

Then V_A is a subspace of the vector space R_A over F and we leave the reader to verify that the mapping $\varphi : R_A \to \operatorname{End} M(A)$ of Theorem 6.33 is F-linear. The restriction $\varphi' = \varphi|_{V_A}$ of φ to V_A is an F-linear mapping $\varphi' : V_A \to \operatorname{End} M(A)$. In fact φ' is a vector space isomorphism, $(\beta)(\varphi')^{-1}$ being the unique reduced matrix representing β relative to $v_1, v_2, \ldots, v_s$ by Theorem 6.32. So $\varphi' : V_A \cong \operatorname{End} M(A)$ and $\dim V_A = \dim \operatorname{End} M(A)$.

We now determine $\dim V_A$. For $1 \le i, j \le s$ let $V_A(i, j)$ denote the subspace of all matrices in V_A having zero entries except possibly for their (i, j)-entry. Then $V_A = \sum_{i,j=1}^{s} \oplus V_A(i, j)$, that is, V_A is the direct sum of its s^2 subspaces $V_A(i, j)$. Consider $B(x) = (b_{ij}(x))$ in $V_A(i, j)$. Then $\deg b_{ij}(x) < \deg d_j(x)$ for $1 \le i, j \le s$. For $1 \le i < j \le s$ we have $b_{ij}(x) = q_{ij}(x)(d_j(x)/d_i(x))$ by Definition 6.31 where $q_{ij}(x)$ is an arbitrary polynomial over F with $\deg q_{ij}(x) < \deg d_i(x)$. So $\dim V_A(i, j) = \deg d_i(x)$ for $i < j$. For $1 \le j \le i \le s$ we see that $b_{ij}(x)$ is an arbitrary polynomial over F with $\deg b_{ij}(x) < \deg d_j(x)$ and so $\dim V_A(i, j) = \deg d_j(x)$ for $j \le i$. The general rule is therefore

$$\dim V_A(i, j) = \deg d_{\min\{i, j\}}(x) \quad \text{for } 1 \le i, j \le s$$

and the $s \times s$ matrix with (i, j)-entry $\dim V_A(i, j)$ is

$$\begin{pmatrix} \deg d_1(x) & \deg d_1(x) & \deg d_1(x) & \cdots & \deg d_1(x) \\ \deg d_1(x) & \deg d_2(x) & \deg d_2(x) & \cdots & \deg d_2(x) \\ \deg d_1(x) & \deg d_2(x) & \deg d_3(x) & \cdots & \deg d_3(x) \\ \vdots & \vdots & \vdots & & \vdots \\ \deg d_1(x) & \deg d_2(x) & \deg d_3(x) & \cdots & \deg d_s(x) \end{pmatrix}.$$

There are $2s - 1$ entries $\deg d_1(x)$ in the above matrix, $2s - 3$ entries $\deg d_2(x)$ and more generally there are $2s - 2i + 1$ entries $\deg d_i(x)$ for $1 \le i \le s$. As $\dim V_A$ is the sum of the entries in the above matrix we obtain

$$\dim V_A = \sum_{i=1}^{s} (2s - 2i + 1) \deg d_i(x)$$

which is also the dimension of $Z(A)$.   □

For example suppose a 16×16 matrix A over a field F has invariant factors $x^2, x^2(x + 1), x^3(x + 1)^2, x^4(x + 1)^2$. So $s = 4$ and

$$\dim Z(A) = 7 \times 2 + 5 \times 3 + 3 \times 5 + 1 \times 6 = 50$$

by Corollary 6.34 showing that the matrices which commute with A are the elements of the 50-dimensional algebra $Z(A)$. The structure of a typical matrix in $Z(A)$ for A in rcf is the subject of Exercises 6.3, Question 5(d). However, as we'll see, it's necessary to use the primary decomposition Theorem 6.12 of $M(A)$ to fully describe the similarity class of A. In this case $A \sim A_1 \oplus A_2$ where A_1 is 11×11 with invariant factors x^2, x^2, x^3, x^4 and A_2 is 5×5 with invariant factors $x + 1$, $(x + 1)^2$, $(x + 1)^2$ by Definition 6.14. Then $Z(A) \cong Z(A_1) \oplus Z(A_2)$ and using Corollary 6.34 we see $\dim Z(A_1) = 37$, $\dim Z(A_2) = 13$. Using Theorem 6.37 we'll be able to specify the invertible matrices in $Z(A_1)$ and $Z(A_2)$ and hence construct the subgroup

$$H_A = G \cap Z(A) = U(Z(A)) \cong U(Z(A_1)) \times U(Z(A_2)) \quad \text{of Theorem 6.29.}$$

We now discuss automorphisms of $M(A)$ in detail. Our next lemma involves the adjugate matrix adj $B(x)$ introduced in Section 1.3.

Lemma 6.35

Let A be a $t \times t$ matrix over a field F with invariant factors $d_1(x), d_2(x), \ldots, d_s(x)$. Let $B(x) = (b_{ij}(x))$ represent the endomorphism β of $M(A)$ as in Definition 6.30. Then adj $B(x)$ belongs to the ring R_A.

Suppose $\gcd\{\det B(x), \chi_A(x)\} = 1$. Then β is an automorphism of $M(A)$.

Conversely suppose β is an automorphism of $M(A)$ and $\chi_A(x) = p(x)^n$ where $p(x)$ is irreducible over F. Then $\gcd\{\det B(x), \chi_A(x)\} = 1$.

Proof

Write $S = \{1, 2, \ldots, s\}$ and let $i_0, j_0 \in S$. The (i_0, j_0)-entry in adj $B(x)$ is the cofactor $B(x)_{j_0 i_0}$ which is (apart from sign) the determinant of the $(s - 1) \times (s - 1)$ matrix which remains on deleting row j_0 and column i_0 from $B(x)$. We show $d_{i_0}(x) B(x)_{j_0 i_0} \equiv 0 \pmod{d_{j_0}(x)}$ which is the endomorphism condition, as in Theorem 6.32, for adj $B(x)$. Now $B(x)_{j_0 i_0}$ is the sum of $(s - 1)!$ terms $\pm t_\pi(x)$, one for each permutation $\pi : S \to S$ with $(j_0)\pi = i_0$, where $t_\pi(x) = \prod_{j \neq j_0} b_{j(j)\pi}(x)$. As $d_{i_0}(x) B(x)_{j_0 i_0} \equiv 0 \pmod{d_{j_0}(x)}$ holds for $i_0 \geq j_0$, since $d_{i_0}(x) \equiv 0 \pmod{d_{j_0}(x)}$, we assume $i_0 < j_0$. As $(j_0)\pi = i_0$ the integers i_0 and j_0 belong to the same cycle in the cycle decomposition of the permutation π (we trust that the reader is familiar with the resolution of a permutation of S into disjoint cycles). Let l be the smallest positive integer with $(i_0)\pi^l = j_0$. Then $i_0, i_1, \ldots, i_l$ are $l + 1$ distinct integers in S where $i_k = (i_0)\pi^k$ for $1 \leq k \leq l$. So $(i_{k-1})\pi = i_k$ for $1 \leq k \leq l$ and $i_l = j_0$, that is, $(i_l)\pi = i_0$. Then $t_\pi(x)$ has factor $f_\pi(x) = b_{i_0 i_1}(x) b_{i_1 i_2}(x) \cdots b_{i_{l-1} i_l}(x)$. The matrix $B(x)$ belongs to the ring R_A and so $d_{i_{k-1}}(x) b_{i_{k-1} i_k}(x) \equiv 0 \pmod{d_{i_k}(x)}$ for $1 \leq k \leq l$ by Theorem 6.32. Using induction on k we obtain $d_{i_0}(x) b_{i_0 i_1}(x) b_{i_1 i_2}(x) \cdots b_{i_{k-1} i_k}(x) \equiv 0 \pmod{d_{i_k}(x)}$ which is a local climax! It is now downhill: taking $k = l$ gives $d_{i_0}(x) f_\pi(x) \equiv 0 \pmod{d_{j_0}(x)}$. Therefore $d_{i_0}(x) t_\pi(x) \equiv 0 \pmod{d_{j_0}(x)}$ for all $(s - 1)!$ permutations π of S with $(j_0)\pi = i_0$ since $f_\pi(x) | t_\pi(x)$. As $B(x)_{k_0 j_0}$ is a sum of terms $\pm t_\pi(x)$ we conclude $d_{i_0}(x) B(x)_{j_0 i_0} \equiv 0 \pmod{d_{j_0}(x)}$ as we set out to show. So adj $B(x) \in R_A$ by Theorem 6.32.

Suppose $\gcd\{\det B(x), \chi_A(x)\} = 1$. By Corollary 4.6 there are $a(x), b(x) \in F[x]$ with $a(x)\det B(x) + b(x)\chi_A(x) = 1$. By the first part of the proof $a(x)\operatorname{adj}B(x)$ belongs to R_A. Write $(a(x)\operatorname{adj}B(x))\varphi = \gamma$ where $\varphi : R_A \to \operatorname{End}M(A)$ is the ring homomorphism of Theorem 6.33. So the endomorphism γ of $M(A)$ satisfies $\beta\gamma = (B(x))\varphi(a(x)\operatorname{adj}B(x))\varphi = (a(x)B(x)\operatorname{adj}B(x))\varphi = (a(x)\det B(x)I)\varphi$ using the familiar property of the adjugate matrix (see before Theorem 1.18). Now $\chi_A(x)I = d_1(x)d_2(x)\cdots d_s(x)I \in \ker\varphi$ by Corollary 6.7 and Theorem 6.33. Therefore $(a(x)\det B(x)I)\varphi = ((1 - b(x)\chi_A(x)I))\varphi = (I)\varphi - (b(x)\chi_A(x)I)\varphi = (I)\varphi = \iota$ showing $\beta\gamma = \iota$, the identity automorphism of $M(A)$. In the same way $\gamma\beta = \iota$ and so β has inverse γ, that is, $\beta \in \operatorname{Aut}M(A)$.

Let the automorphism $\beta = (B(x))\varphi$ have inverse $\gamma = (C(x))\varphi$. Then $B(x)C(x)$ belongs to the coset $I + \ker\varphi$. So $B(x)C(x) = I + K(x)$ where $K(x) \in \ker\varphi$ and so each entry in $K(x)$ is divisible by $d_1(x)$ using Theorem 6.33. By hypothesis $\chi_A(x) = p(x)^n$. We see $\det B(x)\det C(x) = \det(I + K(x)) \equiv 1 \pmod{p(x)}$ as $p(x)|d_1(x)$ and so $p(x)$ cannot be a divisor of $\det B(x)$. As $p(x)$ is irreducible over F all monic divisors of $p(x)^n$ except 1 are divisible by $p(x)$. So $\gcd\{\det B(x), \chi_A(x)\} = \gcd\{\det B(x), p(x)^n\} = 1$. $\qquad\square$

We now suppose that the s invariant factors of A are all equal to $p(x)^l$ where $p(x)$ is irreducible over F. A consequence of Theorem 6.33 is that $\operatorname{End}M(A)$ is isomorphic to the ring $\mathfrak{M}_s(F[x]/\langle p(x)^l\rangle)$. Here we ask: how can the automorphisms β of $M(A)$ be found? The reader should take heart from the fact that the case $l = 1$ has already been dealt with completely in Theorem 6.26 and involved the extension of F to $E = F[x]/\langle p(x)\rangle$. As explained after Theorem 4.9 it is convenient to write $c = x + \langle p(x)\rangle$ and so $E = F(c)$ where $p(c) = 0$. A typical element of E becomes $f(c) = f(x) + \langle p(x)\rangle$ where $f(x) \in F[x]$ and so at the wave of a magic wand the coset notation miraculously disappears! Also the natural ring homomorphism $\eta : F[x] \to E$ and the 'evaluation at c' homomorphism $\varepsilon_c : F[x] \to E$ coincide as

$$(f(x))\eta = f(x) + \langle p(x)\rangle = f(c) = (f(x))\varepsilon_c \quad \text{for all } f(x) \in F[x],$$

that is, $\eta = \varepsilon_c$. We extend ε_c to $\varepsilon_c : \mathfrak{M}_s(F[x]) \to \mathfrak{M}_s(E)$, a surjective homomorphism of matrix rings, by writing $(B(x))\varepsilon_c = (b_{jk}(c))$ for all $s \times s$ matrices $B(x) = (b_{jk}(x))$ over $F[x]$.

To get used to these ideas and as preparation for the next theorem we work through two examples. First suppose $s = 2$, $d_1(x) = d_2(x) = x^3$ and so $p(x) = x$ the field F being arbitrary (any field whatsoever) and $A = C(x^3) \oplus C(x^3)$. Notice $f(x) \equiv f(0) \pmod{x}$ as $x|(f(x) - f(0))$. In this case $E = F[x]/\langle x\rangle = F$ on identifying the coset $f(x) + \langle x\rangle$ with the scalar $f(0)$. Then $\eta : F[x] \to F$ is given by $(f(x))\eta = f(0) = (f(x))\varepsilon_0$ for all $f(x) \in F[x]$ and so $\eta = \varepsilon_0$. Every 2×2 matrix over $F[x]$

satisfies the endomorphism condition Definition 6.31 as the invariant factors of the 6×6 matrix A are equal. A typical reduced matrix in $R_A = \mathfrak{M}_2(F[x])$ is

$$B(x) = \begin{pmatrix} a_{11} + b_{11}x + c_{11}x^2 & a_{12} + b_{12}x + c_{12}x^2 \\ a_{21} + b_{21}x + c_{21}x^2 & a_{22} + b_{22}x + c_{22}x^2 \end{pmatrix}$$

and

$$(B(x))\varepsilon_0 = \begin{pmatrix} a_{11} & a_{12} \\ a_{21} & a_{22} \end{pmatrix} \in \mathfrak{M}_2(F).$$

As $\chi_A(x) = x^6$ it follows from Lemma 6.35 that $(B(x))\varphi$ is an automorphism of $M(A)$ if and only if $\gcd\{\det B(x), x\} = 1$. But $\det B(x) \equiv \det((B(x))\varepsilon_0) \pmod{x}$ and so

$$(B(x))\varphi \in \operatorname{Aut} M(A) \quad \Leftrightarrow \quad \begin{vmatrix} a_{11} & a_{12} \\ a_{21} & a_{22} \end{vmatrix} \neq 0.$$

Notice that the scalars b_{ij} and c_{ij} do not feature in the above condition and so are arbitrary. In the case of a finite field F of order q we see

$$|\operatorname{Aut} M(A)| = q^8 |GL_2(F)| = q^8(q^2 - 1)(q^2 - q)$$

as there are q choices for each of the 8 scalars b_{ij}, c_{ij} and $|GL_2(F)|$ choices for the invertible matrix $(B(x))\varepsilon_0$ over F. By Theorem 6.29 the size of the similarity class of A is

$$|GL_6(F)|/|\operatorname{Aut} M(A)|$$

$$= (q^6 - 1)(q^6 - q)(q^6 - q^2)(q^6 - q^3)(q^6 - q^4)(q^6 - q^5)$$

$$/(q^8(q^2 - 1)(q^2 - q))$$

$$= (q^6 - 1)(q^6 - q)(q^6 - q^2)(q^6 - q^3).$$

As a second example let $A = C(p(x)^2) \oplus C(p(x)^2)$ where $p(x) = x^2 + x + 1$ over $F = \mathbb{Z}_2$. So $t = 8$, $s = 2$, $d_1(x) = d_2(x) = x^4 + x^2 + 1$ and $\chi_A(x) = p(x)^4 = x^8 + x^4 + 1$. In this case $E = \mathbb{Z}_2[x]/\langle x^2 + x + 1 \rangle$ is a field of order 4, the elements of E corresponding to the 4 remainders on division of $f(x) \in \mathbb{Z}_2[x]$ by $p(x)$. Here $E = F(c) = \{0, 1, c, 1 + c\}$ where $p(c) = c^2 + c + 1 = 0$. As above, every matrix $B(x) \in \mathfrak{M}_2(\mathbb{Z}_2[x])$ satisfies the endomorphism condition. A typical reduced matrix in $\mathfrak{M}_2(\mathbb{Z}_2[x]) = R_A$ is

$$B(x) = \begin{pmatrix} q_{11}(x)p(x) + r_{11}(x) & q_{12}(x)p(x) + r_{12}(x) \\ q_{21}(x)p(x) + r_{21}(x) & q_{22}(x)p(x) + r_{22}(x) \end{pmatrix}$$

where $\deg q_{ij}(x) < 2$ and $\deg r_{ij}(x) < 2$ for $1 \le i, j \le 2$ and

$$(B(x))\varepsilon_c = \begin{pmatrix} r_{11}(c) & r_{12}(c) \\ r_{21}(c) & r_{22}(c) \end{pmatrix} \in \mathfrak{M}_2(E).$$

As $\chi_A(x) = p(x)^4$ we see from Lemma 6.35 that $(B(x))\varphi$ is an automorphism of $M(A)$ if and only if $\gcd\{\det B(x), p(x)\} = 1$, that is $\det B(x) \not\equiv 0 \pmod{p(x)}$. But

$$\det B(x) \equiv \begin{vmatrix} r_{11}(x) & r_{12}(x) \\ r_{21}(x) & r_{22}(x) \end{vmatrix} \pmod{p(x)}$$

and

$$\begin{vmatrix} r_{11}(x) & r_{12}(x) \\ r_{21}(x) & r_{22}(x) \end{vmatrix} \not\equiv 0 \pmod{p(x)} \quad \Leftrightarrow \quad \det((B(x))\varepsilon_c) \ne 0.$$

Therefore

$$(B(x))\varphi \in \operatorname{Aut} M(A) \quad \Leftrightarrow \quad \det((B(x))\varepsilon_c) \ne 0.$$

For example $\left(\begin{smallmatrix} x & x \\ x & x^2 \end{smallmatrix}\right)\varphi$ is an automorphism of $M(A)$ as

$$\begin{vmatrix} c & c \\ c & c^2 \end{vmatrix} = c \ne 0,$$

but $\left(\begin{smallmatrix} x & 1+x \\ 1 & x \end{smallmatrix}\right)\varphi$ is not an automorphism of $M(A)$ as

$$\begin{vmatrix} c & 1+c \\ 1 & c \end{vmatrix} = 1 + c + c^2 = 0.$$

The polynomials $q_{ij}(x)$ are not mentioned in the above condition and so are arbitrary. In the case of an automorphism $(B(x))\varphi$ of $M(A)$ there are 4 possibilities for each of the 4 polynomials $q_{ij}(x)$ and $|GL_2(E)| = (4^2 - 1)(4^2 - 4)$ possibilities for $(B(x))\varepsilon_c \in GL_2(E)$. Therefore

$$|\operatorname{Aut} M(A)| = 4^4(4^2 - 1)(4^2 - 4) = 2^{10} \times 3^2 \times 5.$$

By Theorem 6.29 the number of matrices in the similarity class of A is

$$|GL_8(\mathbb{Z}_2)|/|\operatorname{Aut} M(A)|$$

$$= (2^8 - 1)(2^8 - 2)(2^8 - 2^2)(2^8 - 2^3)(2^8 - 2^4)$$

$$\times (2^8 - 2^5)(2^8 - 2^6)(2^8 - 2^7)/|\operatorname{Aut} M(A)|$$

$$= 2^{18} \times 3^3 \times 5 \times 7^2 \times 17 \times 31 \times 127.$$

Theorem 6.36

Let A be a $t \times t$ matrix over a field F having s equal invariant factors $p(x)^l$ where $p(x)$ is irreducible of degree m over F. Then $m = t/(ls)$ and $R_A = \mathfrak{M}_s(F[x])$. Write $E = F(c)$ where $p(c) = 0$. Let $B(x) = (b_{ij}(x))$ represent the endomorphism β of $M(A)$ as in Definition 6.30 and write $B(c) = (b_{ij}(c))$. Then

$$\beta \in \operatorname{Aut} M(A) \quad \Leftrightarrow \quad \det B(c) \neq 0 \quad \Leftrightarrow \quad B(c) \in GL_s(E).$$

Let F be a finite field of order q. Then $|\operatorname{End} M(A)| = q^{ms^2 l}$ and

$$|\operatorname{Aut} M(A)| = q^{ms^2(l-1)}|GL_s(E)|$$

$$= q^{ms^2(l-1)}(q^{ms} - 1)(q^{ms} - q^m) \cdots (q^{ms} - q^{m(s-1)}).$$

There are exactly $|GL_t(F)|/|\operatorname{Aut} M(A)|$ matrices similar to A.

Proof

As the invariant factors of A are equal the endomorphism condition Definition 6.31 is satisfied by all $s \times s$ matrices $B(x)$ over $F[x]$, that is, $R_A = \mathfrak{M}_s(F[x])$. From Theorem 6.5 the characteristic polynomial is the product of the invariant factors. Here $\chi_A(x) = (p(x)^l)^s = p(x)^{ls}$. Equating degrees gives $t = mls$ and so $m = t/(ls)$. We know $\beta \in \operatorname{Aut} M(A) \Leftrightarrow \gcd\{\det B(x), p(x)\} = 1$ by Lemma 6.35 and $\gcd\{\det B(x), p(x)\} = 1 \Leftrightarrow \det B(x) \not\equiv 0 \pmod{p(x)}$ as $p(x)$ is irreducible over F. The field $E = F(c)$ where $p(c) = 0$ is tailor-made for the job of determining whether or not a polynomial $f(x)$ over F is divisible by $p(x)$: in fact

$$f(x) \not\equiv 0 \pmod{p(x)} \quad \Leftrightarrow \quad f(c) \neq 0.$$

The 'evaluation at c' ring homomorphism $\varepsilon_c : F[x] \to F(c) = E$ has $\operatorname{im}\varepsilon_c = E$ and $\ker\varepsilon_c = \langle p(x) \rangle$. Taking $f(x) = \det B(x)$ and using ε_c gives $(\det B(x))\varepsilon_c = (\det(b_{ij}(x)))\varepsilon_c = \det((b_{ij}(x))\varepsilon_c) = \det B(c)$. Putting the pieces together shows $\beta \in \operatorname{Aut} M(A) \Leftrightarrow \det B(x) \not\equiv 0 \pmod{p(x)} \Leftrightarrow \det B(c) \neq 0$. As $B(c)$ is an $s \times s$ over the field E we conclude $\det B(c) \neq 0 \Leftrightarrow B(c) \in GL_s(E)$.

Assume now that $B(x)$ is reduced, that is, $\deg b_{ij}(x) < \deg p(x)^l = ml$ for $1 \leq i, j \leq s$. On dividing $b_{ij}(x)$ by $p(x)$ using Theorem 4.1, there are polynomials $q_{ij}(x)$ and $r_{ij}(x)$ over F with $b_{ij}(x) = q_{ij}(x)p(x) + r_{ij}(x)$ where $\deg q_{ij}(x) < m(l-1)$ and $\deg r_{ij}(x) < m$ for $1 \leq i, j \leq s$. As $p(c) = 0$ we see $b_{ij}(c) = r_{ij}(c) \in E$ for $1 \leq i, j \leq s$. So whether or not β is an automorphism of $M(A)$ depends on the remainders $r_{ij}(x)$ but not on the quotients $q_{ij}(x)$.

$$\text{Let } S(m) = \{(r_{ij}(x)) \in \mathfrak{M}_s(F[x]) : \deg r_{ij}(x) < m\}.$$

So the elements of $S(m)$ are all $s \times s$ matrices over $F[x]$ having entries of degree less than $m = \deg p(x)$. It is straightforward to show that $S(m)$ is a subgroup of the additive group of $\mathfrak{M}_s(F[x])$. The ring homomorphism $\varepsilon_c : \mathfrak{M}_s(F[x]) \to \mathfrak{M}_s(E)$ has kernel consisting of $s \times s$ matrices over $F[x]$ having all entries divisible by $p(x)$. Therefore $S(m) \cap \ker \varepsilon_c = \{0\}$, that is, the only matrix belonging to $S(m)$ and $\ker \varepsilon_c$ is the zero matrix. As a consequence the restriction of ε_c to $S(m)$ is an isomorphism $\varepsilon_c|_{S(m)} : S(m) \cong \mathfrak{M}_s(E)$ of additive abelian groups.

Suppose $|F| = q$. By Corollary 6.34 we know

$$\dim \operatorname{End} M(A) = \sum_{j=1}^{s} (2s - 2j + 1)ml = s^2 ml$$

as the terms are in arithmetic progression. So $|\operatorname{End} M(A)| = q^{s^2 ml}$.

Let $\beta \in \operatorname{Aut} M(A)$. There are $q^{m(l-1)}$ possibilities for each of the s^2 polynomials $q_{ij}(x)$, namely any polynomial of degree less than $m(l-1)$ over F. By the above paragraph and the first part of the proof there are $|GL_s(E)|$ possibilities for the $s \times s$ matrix $(r_{ij}(x))$ of remainders, namely those with $(r_{ij}(x))\varepsilon_c = (r_{ij}(c)) \in GL_s(E)$. These choices are independent of each other and so $|\operatorname{Aut} M(A)| = q^{ms^2(l-1)}|GL_s(E)|$ since β is represented by a unique reduced matrix using Theorem 6.32. As $|E| = q^m$ the formula for $|GL_s(E)|$ after Lemma 2.18 and Theorem 6.29 combine to complete the proof. $\qquad\square$

The proof of our next theorem is due to the Japanese mathematician K. Shoda who in 1928 analysed the automorphisms of an arbitrary finite abelian p-group (Exercises 3.3, Question 5(c)). We now study the polynomial analogue, namely $\operatorname{Aut} M(A)$ where $\chi_A(x)$ is a power of an irreducible polynomial $p(x)$ over F. The case $A = C(x) \oplus C(x^2)$ over an arbitrary field F, discussed after Definition 6.31, is the smallest example covered by Theorem 6.37 but not by Theorem 6.36.

Theorem 6.37

For $1 \le i \le r$ write $A_i = C(p(x)^{l_i}) \oplus C(p(x)^{l_i}) \oplus \cdots \oplus C(p(x)^{l_i})$ (s_i terms) where $p(x)$ is irreducible of degree m over a field F and $l_1 < l_2 < \cdots < l_r$. Write $A = A_1 \oplus A_2 \oplus \cdots \oplus A_r$ and $s = s_1 + s_2 + \cdots + s_r$. Each $s \times s$ matrix $B(x)$ in the ring R_A Theorem 6.32 partitions into $s_i \times s_j$ submatrices $B_{ij}(x)$ ($1 \le i, j \le r$) as shown:

$$B(x) = \begin{pmatrix} B_{11}(x) & B_{12}(x) & \cdots & B_{1r}(x) \\ B_{21}(x) & B_{22}(x) & \cdots & B_{2r}(x) \\ \vdots & \vdots & \ddots & \vdots \\ B_{r1}(x) & B_{r2}(x) & \cdots & B_{rr}(x) \end{pmatrix}$$

where each entry in $B_{ij}(x)$ is divisible by $p(x)$ for $i < j$. Also

$$(B(x))\varphi \in \operatorname{Aut} M(A) \quad \Leftrightarrow \quad (B_{ii}(x))\varphi \in \operatorname{Aut} M(A_i) \quad \text{for } 1 \le i \le r.$$

The dimension of $\operatorname{End} M(A)$ over F is me where

$$e = \sum_{j=1}^{r} l_j s_j (s_j + 2(s_{j+1} + s_{j+2} + \cdots + s_r)).$$

Let F be a finite field of order q and let $E = F(c)$ where $p(c) = 0$. Then

$$|\operatorname{End} M(A)| = |E|^e.$$

Write $k_i = |GL_{s_i}(E)|/q^{s_i^2 m}$ for $1 \le i \le r$. Then

$$|\operatorname{Aut} M(A)| = k_1 k_2 \cdots k_r |\operatorname{End} M(A)|.$$

Proof

Write $n = l_1 s_1 + l_2 s_2 + \cdots + l_r s_r$. Then A is the $mn \times mn$ matrix over F in rcf with s_1 invariant factors $p(x)^{l_1}$, s_2 invariant factors $p(x)^{l_2}, \ldots, s_r$ invariant factors $p(x)^{l_r}$ and $\chi_A(x) = p(x)^n$. Consider the (i', j')-entry $b_{i'j'}(x)$ in $B(x)$. Where is this entry located in the above partition of $B(x)$? As $1 \le i', j' \le s$ there are i, j with $1 \le i, j \le r$ such that $i' = s_1 + s_2 + \cdots + s_{i-1} + i''$, $1 \le i'' \le s_i$ and $j' = s_1 + s_2 + \cdots + s_{j-1} + j''$, $1 \le j'' \le s_j$. Therefore the (i', j')-entry in $B(x)$ is the (i'', j'')-entry of $B_{ij}(x)$. Then $d_{i'}(x) = p(x)^{l_i}$, that is, the i'th invariant factor of A is $p(x)^{l_i}$. Also $d_{j'}(x) = p(x)^{l_j}$. The endomorphism condition Corollary 6.34

$$d_{i'}(x) b_{i'j'}(x) \equiv 0 \pmod{d_{j'}(x)}$$

is satisfied for $i \ge j$ as then $l_i \ge l_j$ and so $d_{i'}(x) | d_{j'}(x)$. For $i < j$ the above endomorphism condition gives $p(x)^{l_j - l_i} | b_{i'j'}(x)$, and so from $l_i < l_j$ we deduce that all entries $b_{i'j'}(x)$ in $B_{ij}(x)$ are divisible by $p(x)$. Therefore $B(x)$ partitions as described above.

The next step in the proof is crucial as it produces an important factorisation of $\det B(x) \pmod{p(x)}$. In fact

$$\det B(x) \equiv \begin{vmatrix} B_{11}(x) & 0 & \cdots & 0 \\ B_{21}(x) & B_{22}(x) & \ddots & \vdots \\ \vdots & \vdots & \ddots & 0 \\ B_{r1}(x) & B_{r2}(x) & \cdots & B_{rr}(x) \end{vmatrix} \pmod{p(x)}$$

where all submatrices $B_{ij}(x)$ for $i < j$ have been replaced by zero matrices. By Exercises 5.1, Question 2(b)

$$\det B(x) \equiv \det B_{11}(x) \det B_{22}(x) \cdots \det B_{rr}(x) \pmod{p(x)}. \qquad (\heartsuit)$$

Each $F[x]$-module $M(A_i)$ is of the type covered by Theorem 6.36 and so

$$(B_{ii}(x))\varphi \in \operatorname{Aut} M(A_i) \quad \Leftrightarrow \quad \det B_{ii}(x) \not\equiv 0 \pmod{p(x)} \quad \text{for } 1 \le i \le r.$$

Suppose $(B(x))\varphi \in \operatorname{Aut} M(A)$. As $\chi_A(x) = p(x)^n$ we see

$$\det B(x) \not\equiv 0 \pmod{p(x)}$$

by Lemma 6.35. Therefore $\det B_{ii}(x) \not\equiv 0 \pmod{p(x)}$ for all i with $1 \le i \le r$ by ($\heartsuit$). So $(B_{ii}(x))\varphi \in \operatorname{Aut} M(A_i)$ for all $1 \le i \le r$. The converse is proved by reversing the steps.

The algebra $\operatorname{End} M(A)$ is a vector space over F and has dimension

$$\dim \operatorname{End} M(A) = \sum_{j'=1}^{s} (2s - 2j' + 1) \deg d_{j'}(x)$$

by Corollary 6.34. The s_j terms in this sum, for $j' = s_1 + s_2 + \cdots + s_{j-1} + j''$, $1 \le j'' \le s_j$, are in arithmetic progression as $\deg d_{j'}(s) = \deg p(x)^{l_j} = ml_j$ and the average term is $s_j + 2(s_{j+1} + s_{j+2} + \cdots + s_r)$. So the sum of these s_j terms is $ml_j s_j(s_j + 2(s_{j+1} + s_{j+2} + \cdots + s_r))$ and hence $\dim \operatorname{End} M(A) = me$ where $e = \sum_{j=1}^{r} l_j s_j(s_j + 2(s_{j+1} + s_{j+2} + \cdots + s_r))$.

Suppose now that F is a finite field with $|F| = q$. Then $|E| = |F(c)| = q^m$ by the discussion following Theorem 4.9. From the preceding paragraph

$$|\operatorname{End} M(A)| = q^{me} = |E|^e.$$

Suppose $B(x)$ to be reduced and $(B(x))\varphi \in \operatorname{Aut} M(A)$. By the earlier part of the proof $(B_{ii}(x))\varphi \in \operatorname{Aut} M(A_i)$ for $1 \le i \le r$. From Theorem 6.36

$$|\operatorname{Aut} M(A_i)|/|\operatorname{End} M(A_i)| = q^{ms_i^2(l_i-1)}|GL_{s_i}(E)|/q^{ms_i^2 l_i}$$

$$= |GL_{s_i}(E)|/q^{s_i^2 m} = k_i$$

and so there are $k_i q^{ms_i^2 l_i}$ choices for each matrix $B_{ii}(x)$. The number of choices for each $B_{ij}(x)$ with $i \ne j$ is the same, whether or not $(B(x))\varphi \in \operatorname{Aut} M(A)$, namely $q^{ms_i s_j l_j}$ for $i > j$ and $q^{ms_i s_j l_i}$ for $i < j$. As these r^2 choices are independent the number of reduced $B(x)$ with $(B(x))\varphi \in \operatorname{Aut} M(A)$ is their product, that is,

$$|\operatorname{Aut} M(A)| = k_1 k_2 \cdots k_r |\operatorname{End} M(A)|. \qquad \square$$

As an illustration we calculate the number of 6×6 matrices over $\mathbb{Z}_3$ which are similar to $A = C(x^2 + 1) \oplus C((x^2 + 1)^2)$. So $F = \mathbb{Z}_3$, $q = 3$, $p(x) = x^2 + 1$, $m = 2$, $c^2 = -1$ and $E = \mathbb{Z}_3(c)$ is a field of order 9. Also $r = 2$, $s = 2$, $s_1 = s_2 = 1$, $l_1 = 1$, $l_2 = 2$ and so $e = 3 + 2 = 5$. On constructing a 2×2 reduced matrix

$$B(x) = \left(\begin{array}{c|c} B_{11}(x) & B_{12}(x) \\ \hline B_{21}(x) & B_{22}(x) \end{array} \right)$$

the number of choices for the 1×1 matrices $B_{ij}(x)$ are the entries in

$$\begin{pmatrix} 3^2 & 3^2 \\ 3^2 & 3^4 \end{pmatrix} \quad \text{and} \quad |\operatorname{End} M(A)| = 3^{10}.$$

As $|GL_1(E)| = 9 - 1 = 8$ we see $k_1 = k_2 = 8/9$ giving $|\operatorname{Aut} M(A)| = (8/9)(8/9) \times 3^{10} = 2^6 \times 3^6$. By Theorem 6.29 the size of the similarity class of A is the index of $\operatorname{Aut} M(A)$ in $GL_6(\mathbb{Z}_3)$ which is

$$|GL_6(\mathbb{Z}_3)|/|\operatorname{Aut} M(A)|$$

$$= (3^6 - 1)(3^6 - 3)(3^6 - 3^2)(3^6 - 3^3)(3^6 - 3^4)(3^6 - 3^5)/(2^6 \times 3^6)$$

$$= 13^2 \times 11^2 \times 7 \times 5 \times 3^9 \times 2^8.$$

Notice that the numbers are unchanged on replacing $p(x)$ by any other monic irreducible quadratic over $\mathbb{Z}_3$, that is, by $x^2 + x - 1$ or $x^2 - x - 1$. More generally any two monic irreducible polynomials of the same degree over F are interchangeable in this sense.

As a second example we calculate the sizes of the six conjugacy (similarity) classes which make up the group $GL_3(\mathbb{Z}_2)$. There are four cyclic classes having as $\chi_A(x)$ the four polynomials of degree 3 over $\mathbb{Z}_2$ with non-zero constant terms, namely

$$x^3 + 1 = (x^2 + x + 1)(x + 1), \qquad x^3 + x + 1, \qquad x^3 + x^2 + 1,$$

$$x^3 + x^2 + x + 1 = (x + 1)^3$$

on factoring into irreducible polynomials over $\mathbb{Z}_2$. In these cases $|\operatorname{Aut} M(A)| = \Phi_2(\chi_A(x))$ using the theory following Definition 6.28. You can check:

$$\Phi_2((x^2 + x + 1)(x + 1)) = 3, \qquad \Phi_2(x^3 + x + 1) = 7,$$

$$\Phi_2(x^3 + x^2 + 1) = 7, \qquad \Phi_2((x + 1)^3) = 4.$$

Using the formula $|GL_2(\mathbb{Z}_2)|/|\operatorname{Aut} M(A)|$ of Theorem 6.29 these classes have sizes:

$$168/3 = 56, \qquad 168/7 = 24, \qquad 168/7 = 24, \qquad 168/4 = 42$$

as $|GL_2(\mathbb{Z}_2)| = (8 - 1)(8 - 2)(8 - 4) = 168$. There remain just two non-cyclic classes in $GL_3(\mathbb{Z}_2)$ namely the class with invariant factors $x + 1$, $(x + 1)^2$ and the class with invariant factors $x + 1$, $x + 1$, $x + 1$. By Theorem 6.37 the corresponding groups Aut $M(A)$ have orders 8 and 168 and so the class sizes are $168/8 = 21$ and $168/168 = 1$ respectively. The conjugacy classes of every group partition the group: in this case it is comforting to check $56 + 24 + 24 + 42 + 21 + 1 = 168$ showing that all the similarity classes in $GL_3(\mathbb{Z}_2)$ have been accounted for. This type of analysis can be carried out on all groups $GL_3(\mathbb{F}_q)$ (Exercises 6.3, Question 7(c)).

Our next (and last) theorem completes the theory. It tells us that the primary decomposition Theorem 6.12 of $M(A)$ leads to decompositions of both End $M(A)$ and Aut $M(A)$. It should present no problem to the diligent reader who has completed the analogous exercises, namely Exercises 3.2, Question 5(b) and Exercises 3.3, Question 1(f), for finite abelian groups. Taken together with Theorem 6.37 the structure of every group Aut $M(A)$ can be analysed provided the resolution of $\chi_A(x)$ into irreducible polynomials is known.

We remind the reader of the terminology introduced in Section 6.2: let A be a $t \times t$ matrix over a field F with $\chi_A(x) = p_1(x)^{n_1} p_2(x)^{n_2} \cdots p_k(x)^{n_k}$ where $p_1(x), p_2(x), \ldots, p_k(x)$ are k distinct monic irreducible polynomials over F and $n_1, n_2, \ldots, n_k$ are positive integers. For $1 \leq j \leq k$ the submodule $M(A)_{p_j(x)} = \{v \in M(A) : p_j(x)^{n_j}v = 0\}$ is the $p_j(x)$-component of $M(A)$. Then $M(A) = M(A)_{p_1(x)} \oplus M(A)_{p_2(x)} \oplus \cdots \oplus M(A)_{p_k(x)}$ by Theorem 6.12, that is, $M(A)$ is the internal direct sum of its primary components.

Theorem 6.38

Using the above notation let $\beta_j \in$ End $M(A)_{p_j(x)}$ for $1 \leq j \leq k$. Then $\beta = \beta_1 \oplus \beta_2 \oplus \cdots \oplus \beta_k$ is an endomorphism of $M(A)$ where $(u)\beta = \sum_{j=1}^{k}(u_j)\beta_j$ and $u = \sum_{j=1}^{k} u_j$, $u_j \in M(A)_{p_j(x)}$ for $1 \leq j \leq k$. Each $\beta \in$ End $M(A)$ satisfies $(u_j)\beta \in M(A)_{p_j(x)}$ for all $u_j \in M(A)_{p_j(x)}$ and is uniquely expressible as $\beta = \beta_1 \oplus \beta_2 \oplus \cdots \oplus \beta_k$ where $\beta_j \in$ End $M(A)_{p_j(x)}$ for $1 \leq j \leq k$.

Write $(\beta)\sigma = (\beta_1, \beta_2, \ldots, \beta_k)$. Then

$$\sigma : \text{End } M(A) \cong \text{End } M(A)_{p_1(x)} \oplus \text{End } M(A)_{p_2(x)} \oplus \cdots \oplus \text{End } M(A)_{p_k(x)}$$

is an algebra isomorphism. Also

$$\sigma|_{\text{Aut } M(A)} : \text{Aut } M(A) \cong \text{Aut } M(A)_{p_1(x)} \times \text{Aut } M(A)_{p_2(x)} \times \cdots \times \text{Aut } M(A)_{p_k(x)}$$

is a group isomorphism.

Proof

Consider $u, u' \in M(A)$. By Theorem 6.12 there are $u_j, u'_j \in M(A)_{p_j(x)}$ for $1 \leq j \leq k$ with

$$u = \sum_{j=1}^{k} u_j \quad \text{and} \quad u' = \sum_{j=1}^{k} u'_j.$$

So $u_j + u'_j \in M(A)_{p_j(x)}$ for $1 \leq j \leq k$ and $u + u' = \sum_{j=1}^{k}(u_j + u'_j)$. Therefore

$$(u + u')\beta = \sum_{j=1}^{k}(u_j + u'_j)\beta_j = \sum_{j=1}^{k}((u_j)\beta_j + (u'_j)\beta_j)$$

$$= \sum_{j=1}^{k}(u_j)\beta_j + \sum_{j=1}^{k}(u'_j)\beta_j = (u)\beta + (u')\beta$$

showing $\beta : M(A) \rightarrow M(A)$ to be an additive mapping of the additive abelian group $M(A)$. In the same way for $f(x) \in F[x]$ we have $f(x)u = \sum_{j=1}^{k} f(x)u_j$ and $f(x)u_j \in M(A)_{p_j(x)}$ for $1 \leq j \leq k$. Therefore

$$(f(x)u)\beta = \sum_{j=1}^{k}(f(x)u_j)\beta_j$$

$$= \sum_{j=1}^{k} f(x)((u_j)\beta_j)$$

$$= f(x)\left(\sum_{j=1}^{k}(u_j)\beta_j\right) = f(x)((u)\beta)$$

showing β to be $F[x]$-linear. So $\beta \in \text{End}\,M(A)$. Notice $(u_j)\beta = (u_j)\beta_j$ for all $u_j \in M(A)_{p_j(x)}$, that is, β_j is the restriction of β to $M(A)_{p_j(x)}$ for $1 \leq j \leq k$.

Conversely let $\beta \in \text{End}\,M(A)$. For $u_j \in M(A)_{p_j(x)}$ we have $p_j(x)^{n_j}u_j = 0$ and so $p_j(x)^{n_j}((u_j)\beta) = (p_j(x)^{n_j}u_j)\beta = (0)\beta = 0$ showing $(u_j)\beta \in M(A)_{p_j(x)}$ for $1 \leq j \leq k$. We have established, in one line, an important fact:

The endomorphisms of $M(A)$ respect the primary decomposition of $M(A)$.

Therefore the restriction of β to $M(A)_{p_j(x)}$ is an endomorphism β_j of $M(A)_{p_j(x)}$ for $1 \leq j \leq k$. We've now come full circle as $\beta = \beta_1 \oplus \beta_2 \oplus \cdots \oplus \beta_k$ by the first part of the proof. As each β_j is uniquely determined by β it's legitimate to define

$$\sigma : \text{End}\,M(A) \rightarrow \text{End}\,M(A)_{p_1(x)} \oplus \text{End}\,M(A)_{p_2(x)} \oplus \cdots \oplus \text{End}\,M(A)_{p_k(x)}$$

by $(\beta)\sigma = (\beta_1, \beta_2, \ldots, \beta_k)$ for all $\beta \in \text{End}\, M(A)$. By the first part of the proof σ is a bijection from $\text{End}\, M(A)$ to the external direct sum of the endomorphism rings of the primary components of $M(A)$. It is straightforward to verify that σ is F-linear and also that σ respects addition and multiplication (composition) of endomorphisms as in Exercises 3.3, Question 1(f). Therefore σ is an algebra isomorphism. Further $\beta = \beta_1 \oplus \beta_2 \oplus \cdots \oplus \beta_k$ is an invertible element of $\text{End}\, M(A)$ if and only if β_j is an invertible element of $\text{End}\, M(A)_{p_j(x)}$ for $1 \leq j \leq k$ as in Exercises 3.2, Question 5(b). In other words

$$\beta \in \text{Aut}\, M(A) \quad \Leftrightarrow \quad \beta_j \in \text{Aut}\, M(A)_{p_j(x)} \quad \text{for } 1 \leq j \leq k,$$

and so

$$\sigma|_{\text{Aut}\, M(A)} : \text{Aut}\, M(A) \cong \text{Aut}\, M(A)_{p_1(x)} \times \text{Aut}\, M(A)_{p_2(x)} \times \cdots \times \text{Aut}\, M(A)_{p_k(x)}$$

is a group isomorphism between $\text{Aut}\, M(A)$ and the direct product of the automorphism groups of the primary components of $M(A)$. □

As an example let A be an 8×8 matrix over the finite field $\mathbb{F}_q$ having invariant factors x, $x(x+1)$, $x^2(x+1)^3$. Then $A \sim P_1 \oplus P_2$ as in Definition 6.14 where P_1 has invariant factors x, x, x^2 and P_2 has invariant factors $x+1$, $(x+1)^3$. So $M(P_1) \cong M(A)_x$ and $M(P_2) \cong M(A)_{x+1}$. From the algebra isomorphism σ of Theorem 6.38 we deduce $\dim \text{End}\, M(A) = \dim \text{End}\, M(P_1) + \dim \text{End}\, M(P_2)$. But by Corollary 6.34 this equation is $16 = 10 + 6$ and so we are none the wiser. However the group isomorphism $\sigma|_{\text{Aut}\, M(A)}$ of Theorem 6.38 gives

$$|\text{Aut}\, M(A)| = |\text{Aut}\, M(P_1)| \times |\text{Aut}\, M(P_2)|$$

and the factors on the right of this equation can be found using Theorem 6.37. So

$$|\text{Aut}\, M(P_1)| = k_1 k_2 |\text{End}\, M(P_1)| = (|GL_2(\mathbb{F}_q)|/q^4)(|GL_1(\mathbb{F}_q)|/q)q^{10}$$

$$= (q+1)(q-1)^3 q^6$$

and

$$|\text{Aut}\, M(P_2)| = ((q-1)/q)^2 q^6 = (q-1)^2 q^4$$

giving $|\text{Aut}\, M(A)| = (q+1)(q-1)^5 q^{10}$. By Theorem 6.29 the number of matrices similar to A is

$$|GL_8(\mathbb{F}_q)|/|\text{Aut}\, M(A)|$$

$$= (q^8 - 1)(q^8 - q)(q^8 - q^2)(q^8 - q^3)(q^8 - q^4)(q^8 - q^5)$$

$$\times (q^8 - q^6)(q^8 - q^7)/((q+1)(q-1)^5 q^{10})$$

$$= (q^8 - 1)(q^7 - 1)(q^6 - 1)(q^4 + q^3 + q^2 + q + 1)$$

$$\times (q^3 + q^2 + q + 1)(q^2 + q + 1)q^{18}.$$

As a finale we partition the group $GL_4(\mathbb{Z}_2)$ into conjugacy classes. The list of irreducible polynomials $p(x)$ of degree at most 4 over $\mathbb{Z}_2$ with $p(x) \neq x$ is: $x + 1$, $x^2 + x + 1$, $x^3 + x + 1$, $x^3 + x^2 + 1$, $x^4 + x + 1$, $x^4 + x^3 + 1$, $x^4 + x^3 + x^2 + x + 1$. The invariant factors of matrices A in $GL_4(\mathbb{Z}_2)$ have these polynomials as their factors. There are 8 cyclic classes (classes with a single invariant factor $\chi_A(x)$) listed next together with the number $|\operatorname{Aut} M(A)| = \Phi_2(\chi_A(x))$ as in Theorem 6.27:

$$(x + 1)^4, 8; \qquad (x + 1)^2(x^2 + x + 1), 6; \qquad (x + 1)(x^3 + x + 1), 7;$$

$$(x + 1)(x^3 + x^2 + 1), 7; \qquad (x^2 + x + 1)^2, 12; \qquad x^4 + x + 1, 15;$$

$$x^4 + x^3 + 1, 15; \qquad x^4 + x^3 + x^2 + x + 1, 15.$$

There are 6 non-cyclic classes listed next by their invariant factor sequence, together with $|\operatorname{Aut} M(A)|$ calculated using Theorem 6.37 and Definition 6.28:

$$(x + 1, x + 1, x + 1, x + 1), |GL_4(\mathbb{Z}_2)| = 20160; \qquad (x + 1, x + 1, (x + 1)^2), 192;$$

$$(x + 1, (x + 1)^3), 16; \qquad ((x + 1)^2, (x + 1)^2), 96;$$

$$(x + 1, (x + 1)(x^2 + x + 1)), 18; \qquad (x^2 + x + 1, x^2 + x + 1), |GL_2(\mathbb{F}_4)| = 180.$$

By Theorem 6.29 the 14 conjugacy classes of elements of $GL_4(\mathbb{Z}_2)$ have sizes: 2520, 3360, 2880, 2880, 1680, 1344, 1344, 1344, 1, 105, 1260, 210, 1120, 112 and their sum is 20160 as the reader can check as a final act.

EXERCISES 6.3

1. (a) Let

$$A = \begin{pmatrix} 1 & 1 & -2 & 2 \\ -5 & -2 & 6 & -4 \\ 2 & 1 & -4 & 3 \\ 3 & 1 & -5 & 3 \end{pmatrix}$$

over $\mathbb{Q}$. Verify $A^2 = -A - I$ and hence find, without further calculation, $\mu_A(x)$ and $\chi_A(x)$. State the invariant factors of A and specify an invertible matrix X over $\mathbb{Q}$ with XAX^{-1} in rcf.

(b) Show that $\varphi : \mathfrak{M}_2(\mathbb{Q}(i)) \rightarrow \mathfrak{M}_4(\mathbb{Q})$, given by

$$\begin{pmatrix} a_0 + ia_1 & b_0 + ib_1 \\ c_0 + ic_1 & d_0 + id_1 \end{pmatrix} \varphi = \begin{pmatrix} a_0 & a_1 & b_0 & b_1 \\ -a_1 & a_0 & -b_1 & b_0 \\ c_0 & c_1 & d_0 & d_1 \\ -c_1 & c_0 & -d_1 & d_0 \end{pmatrix}$$

for all $a_0, a_1, b_0, b_1, c_0, c_1, d_0, d_1 \in \mathbb{Q}$,

is a ring homomorphism where $i^2 = -1$. Show $\ker \varphi = 0$ and $\operatorname{im} \varphi = Z(A)$, the centraliser (Definition 6.28) of $A = C(x^2 + 1) \oplus C(x^2 + 1)$ over $\mathbb{Q}$. Show $\det(B')\varphi = |\det B'|^2$ for $B' \in \mathfrak{M}_2(\mathbb{Q}(i))$.

(c) Let A be a $t \times t$ matrix over the real field $\mathbb{R}$ with minimum polynomial $\mu_A(x) = x^2 + 1$ and let $r(x) = ax + b$ where $a, b \in \mathbb{R}$. Show that $r(A)$ has $t/2$ invariant factors $x^2 - 2bx + a^2 + b^2$ in the case $a \neq 0$. What are the invariant factors of $r(A)$ in the case $a = 0$? Describe the invariant factors of $f(A)$ for an arbitrary polynomial $f(x)$ over $\mathbb{R}$. Are $A^{10} - A^9$ and $A^{11} + A^{10}$ similar?

(d) Let A be a $t \times t$ matrix over a finite field F with $|F| = q$. Suppose the minimum polynomial $\mu_A(x)$ is irreducible of degree n. Use Theorem 6.26 to find a formula for the number of invertible matrices X over F with XAX^{-1} in rcf. Hence find a formula for the size of the similarity class of A using Theorem 6.29.

Taking $F = \mathbb{Z}_3$ calculate the number of matrices similar to each of the following matrices A over F: $C(x^2 + 1)$, $C(x^3 - x + 1)$, $C(x^2 + 1) \oplus C(x^2 + 1)$.

(e) Complete the proof of Theorem 6.26 using Theorem 3.16 as a guide.

2. (a) Let $A = C(x^3)$ over an arbitrary field F. Describe the matrices B belonging to the centraliser $Z(A)$. Is $Z(A)$ a ring? (Yes/No) Is $Z(A)$ a vector space over F? If so what is $\dim Z(A)$? Which matrices B belong to the group $U(Z(A))$ of invertible elements of $Z(A)$? Are $\operatorname{End} M(A)$ and $Z(A)$ isomorphic rings? (Yes/No) Are $\operatorname{Aut} M(A)$ and $U(Z(A))$ isomorphic groups? (Yes/No)

Taking $F = \mathbb{Z}_2$ find the integers $|\operatorname{End} M(A)|$ and $|\operatorname{Aut} M(A)|$. Is $\operatorname{Aut} M(A)$ a cyclic group? In the case $F = \mathbb{F}_q$ state formulae for $|\operatorname{End} M(A)|$ and $|\operatorname{Aut} M(A)|$.

(b) Let A be a $t \times t$ matrix over a field F. For each $\beta \in \operatorname{End} M(A)$ write $(\beta)\theta$ for the matrix of β relative to the standard basis $\mathcal{B}_0$ of F^t. Using Theorem 6.27 show that $\theta : \operatorname{End} M(A) \cong Z(A)$ is an algebra isomorphism, that is, θ is both a ring isomorphism and a vector space isomorphism. Deduce that the restriction $\theta|$ of θ to $\operatorname{Aut} M(A)$ is a group isomorphism $\theta| : \operatorname{Aut} M(A) \cong U(Z(A))$.

(c) Let $g(x)$ be a non-zero polynomial over a finite field F of order q. Show $\Phi_q(g(x)) = |U(F[x]/\langle g(x)\rangle)|$ where $\Phi_q : F[x] \to \mathbb{Z}$ is the function introduced after Definition 6.28. Using (c) above deduce the multiplicative property of Φ_q, that is, $\Phi_q(g(x)h(x)) = \Phi_q(g(x))\Phi_q(h(x))$ where $\gcd\{g(x), h(x)\} = 1$. Show also $\Phi_q(p(x)^n) = q^{mn} - q^{m(n-1)}$ where $p(x)$ is irreducible of degree m over F.

(d) Determine $|\operatorname{End} M(A)|$ and $|\operatorname{Aut} M(A)|$ in the following cases where $A = C(f(x))$, $F = \mathbb{Z}_2$ and $f(x)$ is

    (i) x^6;   (ii) $(x^2 + x + 1)^3$;

    (iii) $x^2(x^2 + x + 1)^2$;   (iv) $x^6 + x^2$.

In each case use Theorem 6.29 to find the number of matrices similar to A.

(e) Let $\mathbb{F}_q$ denote a finite field having prime power q elements. Determine the sizes of the $q + q^2$ similarity classes of 2×2 matrices A over $\mathbb{F}_q$ and verify that their sum is $|\mathfrak{M}_2(\mathbb{F}_q)|$. Specify those classes which belong to $GL_2(\mathbb{F}_q)$ and verify that the sum of their sizes is $|GL_2(\mathbb{F}_q)|$.

3. (a) Let Ω be a non-empty set and let G be a multiplicative group. A *permutation representation* θ *of* G *on* Ω is a group homomorphism $\theta : G \to S(\Omega)$ where $S(\Omega)$ is *the symmetric group on* Ω, that is, $S(\Omega)$ is the group of all bijections $\beta : \Omega \to \Omega$ the group operation being composition of bijections. Let x^β denote the image of $x \in \Omega$ by $\beta \in S(\Omega)$. Write $x \sim y$ if there is $g \in G$ with $x^{(g)\theta} = y$. Show that $\sim$ is an equivalence relation on Ω. The equivalence class of x is called the *orbit* of x and denoted by O_x where $x \in \Omega$. Show $O_x = \{x^{(g)\theta} : g \in G\}$. For $x \in \Omega$ show that $G_x = \{g \in G : x^{(g)\theta} = x\}$ is a subgroup of G. G_x is called *the stabiliser of* x. Prove the *orbit-stabiliser theorem*, namely that for each $x \in \Omega$ the correspondence $G_x g \to x^{(g)\theta}$ is a bijection from the set of all (left) cosets $G_x g$ of G_x in G to the orbit O_x. In the case G finite deduce $|G|/|G_x| = |O_x|$ for all $x \in \Omega$.

Describe O_x and G_x in the two cases:
(i) θ trivial, that is, $(g)\theta$ is the identity mapping of Ω for all $g \in G$,
(ii) $G = S(\Omega)$ and $\theta : S(\Omega) \to S(\Omega)$ the identity automorphism, that is, $(\beta)\theta = \beta$ for all $\beta \in S(\Omega)$.

(b) Let F be a field and t a positive integer. Write $\Omega = \mathfrak{M}_t(F)$ and $G = GL_t(F)$. For each $X \in G$ show that the mapping $(X)\theta : \Omega \to \Omega$, given by $A^{(X)\theta} = X^{-1}AX$ for all $A \in \Omega$, is a bijection, that is, $(X)\theta \in S(\Omega)$. Show that $\theta : G \to S(\Omega)$ is a permutation representation of G on Ω. What is the connection between the orbit O_A of A and the similarity class of the $t \times t$ matrix A over F? Is $G_A = U(Z(A))$?

4. (a) Let M and M' be R-modules where R is a commutative ring. Let $\alpha : M \to M'$ and $\beta : M \to M'$ be R-linear mappings (module homomorphisms). Show that $\alpha + \beta : M \to M'$ and $r\alpha : M \to M'$ are R-linear mappings for $r \in R$ where $(v)(\alpha + \beta) = (v)\alpha + (v)\beta$ and $(v)(r\alpha) = r((v)\alpha)$ for all $v \in M$. Hence show that the set $\mathrm{Hom}(M, M')$ of all R-linear mappings $\alpha : M \to M'$ can be given the structure of an R-module.

   Hint: Adapt the first part of Exercises 5.1, Question 2(e) and use Exercises 3.3, Question 6(a).

   (b) Let $M, M_1, M_2, M', M_1', M_2'$ be R-modules where R is a commutative ring. Establish the module isomorphisms

   $$\mathrm{Hom}(M_1 \oplus M_2, M') \cong \mathrm{Hom}(M_1, M') \oplus \mathrm{Hom}(M_2, M')$$

   and

   $$\mathrm{Hom}(M, M_1' \oplus M_2') \cong \mathrm{Hom}(M, M_1') \oplus \mathrm{Hom}(M, M_2').$$

   Hint: Adapt the answer to Exercises 3.3, Question 6(b).

   (c) Let M and M' be cyclic $F[x]$-modules where F is a field. Suppose $M = \langle v_0 \rangle$ where v_0 has order $d_0(x)$ in M. Suppose also $M' = \langle v_0' \rangle$ where v_0' has order $d_0'(x)$ in M'. Show that, except for the case $d_0(x) = 0(x)$, $d_0'(x) \neq 0(x)$, the $F[x]$-module $\mathrm{Hom}(M, M')$ is cyclic with generator β_0 of order $\gcd\{d_0(x), d_0'(x)\}$. Describe $\mathrm{Hom}(M, M')$ in the exceptional case.

   (d) Let $(d_1(x), d_2(x), \ldots, d_s(x))$ be the invariant factor sequence of the $F[x]$-module M and let $(d_1'(x), d_2'(x), \ldots, d_t'(x))$ be the invariant factor sequence of the $F[x]$-module M' where $d_s(x) \neq 0(x)$ and $d_t'(x) \neq 0(x)$. Generalise Frobenius' theorem (Corollary 6.34) by showing that $\mathrm{Hom}(M, M')$ is a vector space over F of dimension

   $$\sum_{i=1}^{s} \sum_{j=1}^{t} \deg(\gcd\{d_i(x), d_j'(x)\}).$$

   Find a necessary and sufficient condition on the invariant factors of M and M' so that $\mathrm{Hom}(M, M')$ is cyclic.

5. (a) Let A be an $m \times m$ matrix over a field F and let A' be a $n \times n$ matrix over F. Let B be an $m \times n$ matrix over F and let $\beta : F^m \to F^n$ be the F-linear mapping determined by B. Generalise the first part of Theorem 6.27 by showing $\beta \in \mathrm{Hom}(M(A), M(A'))$ if and only if B intertwines A and A', that is, $AB = BA'$.

(b) Working over an arbitrary field F determine the matrices B intertwining $C(x^2)$ and $C(x^3)$. Find a matrix B_0 such that the linear mapping β_0 determined by B_0 generates the $F[x]$-module $\mathrm{Hom}(M(C(x^2)),$ $M(C(x^3)))$ and find the order of β_0 in this module. Are the $F[x]$-modules $\mathrm{Hom}(M(C(x^2)), M(C(x^3)))$ and $M(C(x^2))$ isomorphic? Answer the same question with the roles of $C(x^2)$ and $C(x^3)$ interchanged.

(c) Let $d(x)$ and $d'(x)$ be monic polynomials of positive degrees m and n respectively over a field F and write $A = C(d(x))$, $A' = C(d'(x))$. Let

$$\gcd\{d(x), d'(x)\} = a_0 + a_1 x + \cdots + a_{r-1}x^{r-1} + x^r$$

and

$$d'(x)/\gcd\{d(x), d'(x)\} = b_0 + b_1 x + \cdots + b_{n-r-1}x^{n-r-1} + x^{n-r}.$$

Write $u_0 = (d'(x)/\gcd\{d(x), d'(x)\})e_1'$ working in the $F[x]$-module $M(A')$ and construct the $m \times n$ matrix B_0 over F with $e_i B_0 = x^{i-1}u_0$ for $1 \le i \le m$ where $e_1, e_2, \ldots, e_m$ and $e_1', e_2', \ldots, e_n'$ denote the standard bases of F^m and F^n respectively. Show that u_0 has order $\gcd\{d(x), d'(x)\}$ in $M(A')$ and verify $AB_0 = B_0 A'$. Show

$$e_i B_0 = b_0 e_i' + b_1 e_{i+1}' + \cdots + b_{n-r-1}e_{i+n-r-1}' + e_{i+n-r}'$$

for $1 \le i \le r$,

$$e_i B_0 = -(a_0 e_{i-r} B_0 + a_1 e_{i-r+1} B_0 + \cdots + a_{r-1}e_{i-1}B_0)$$

for $r < i \le m$.

What is rank B_0?
Write $(v)\beta_0 = vB_0$ for all $v \in F^m$. Use Question 4(c) above to show that β_0 generates $\mathrm{Hom}\{M(A), M(A')\}$.
Discuss the simplifications to B_0 in the cases

    (i) $d'(x)|d(x);$ (ii) $d(x)|d'(x);$

    (iii) $\gcd\{d(x), d(x)'\} = 1.$

(d) Let $C = C(d_1(x)) \oplus C(d_2(x)) \oplus \cdots \oplus C(d_s(x))$ be a $t \times t$ matrix over a field F in rcf Definition 6.4 where $t = \sum_{i=1}^{s} \deg d_i(x)$. The $t \times t$ matrix

$$B = \begin{pmatrix} B_{11} & B_{12} & \cdots & B_{1s} \\ B_{21} & B_{22} & \cdots & B_{2s} \\ \vdots & \vdots & \ddots & \vdots \\ B_{s1} & B_{s2} & \cdots & B_{ss} \end{pmatrix}$$

over F is 'sympathetically' partitioned, that is, the B_{ij} are $\deg d_i(x) \times \deg d_j(x)$ submatrices for $1 \le i, j \le s$. Show $B \in Z(C)$ if and only if $C(d_i(x))B_{ij} = B_{ij}C(d_j(x))$ for $1 \le i, j \le s$. For $B \in Z(C)$ use (c) above to describe the matrices B_{ij} for $1 \le i, j \le s$.

(e) Let C and C' be square matrices over F which are both in rcf. Explain how the theory in (d) above can be modified to determine all matrices B which intertwine C and C'. Suppose $A = X^{-1}CX$ and $A' = (X')^{-1}C'X'$ where X and X' are invertible over F. How are the matrices which intertwine A and A' related to the matrices B which intertwine C and C'? Under what condition does there exist an invertible matrix B over F which intertwines C and C'?

(f) Working over the rational field $\mathbb{Q}$ find a $\mathbb{Q}$-basis for the matrices B which intertwine A and A' in the following cases:

(i) $A = C((x+1)(x^2+1))$, $\qquad A' = C((x+1)^3)$;

(ii) $A = C((x+1)(x^2+1))$, $\qquad A' = C((x^2+1)^2)$;

(iii) $A = C(x+1) \oplus C((x+1)^2)$,
$\qquad A' = C(x+1) \oplus C((x^3+x^2+x+1))$.

How would your answers change on replacing $\mathbb{Q}$ by $\mathbb{Z}_2$?

6. (a) Let $A = C(x) \oplus C(x^2) \oplus C(x^2)$ over a field F. Verify that

$$B(x) = \begin{pmatrix} b_{11}(x) & b_{12}(x) & b_{13}(x) \\ b_{21}(x) & b_{22}(x) & b_{23}(x) \\ b_{31}(x) & b_{32}(x) & b_{33}(x) \end{pmatrix} \quad \text{in } \mathfrak{M}_3(F[x])$$

satisfies the endomorphism condition Definition 6.31 relative to $M(A)$ if and only if $x|b_{12}(x)$ and $x|b_{13}(x)$. Verify directly that the sum and product of two matrices $B(x)$ and $B'(x)$ satisfying e.c.rel. $M(A)$ also satisfies e.c.rel. $M(A)$. Show further that the set R_A of all $B(x)$ satisfying e.c.rel. $M(A)$ is a subring (Exercises 2.3, Question 3(b)) of $\mathfrak{M}_3(F[x])$. Let $K = \{B(x) = (b_{ij}(x)) \in \mathfrak{M}_3(F[x]) : x|b_{i1}(x), x^2|b_{i2}(x), x^2|b_{i3}(x), i = 1, 2, 3\}$. Is $K \subseteq R_A$? (Yes/No). Show that K is an ideal (Exercises 2.3, Question 3(a)) of R_A. Is K an ideal of $\mathfrak{M}_3(F[x])$?

Taking $F = \mathbb{Z}_2$ use reduced matrices to find $|R_A/K|$ and use Lemma 6.35 to find $|U(R_A/K)|$. Find the size of the similarity class of A.

(b) Let A be a $t \times t$ matrix over a field F with invariant factor sequence $(d_1(x), d_2(x), \ldots, d_s(x))$. Let R_A denote the set of matrices $B(x) = (b_{ij}(x)) \in \mathfrak{M}_s(F[x])$ satisfying e.c.rel. $M(A)$, that is, $d_i(x)b_{ij}(x) \equiv 0 \pmod{d_j(x)}$ for $1 \le i, j \le s$. Complete the proof Theorem 6.32 that R_A is a subring of $\mathfrak{M}_s(F[x])$ by showing

(i) R_A is closed under negation,

(ii) R_A contains the zero and identity matrices of $\mathfrak{M}_s(F[x])$.

(c) Let $B(x) = (b_{ij}(x))$ belong to the ring R_A. Complete the proof of Theorem 6.33 by showing that there is an endomorphism β of $M(A)$ represented by $B(x)$ as in Definition 6.30.

7. (a) Show that the set of 344 non-invertible 3×3 matrices over $\mathbb{Z}_2$ partitions into 8 similarity classes. Find the number of matrices in each similarity class.

   Hint: Start by listing the invariant factor sequences.

(b) List the 8 monic irreducible cubic polynomials over $\mathbb{Z}_3$. Determine the number of 3×3 matrices over $\mathbb{Z}_3$ in each similarity class and verify that the sum of these numbers is 3^9. Find the number of matrices in each of the 24 conjugacy classes in $GL_3(\mathbb{Z}_3)$. Show that $GL_3(\mathbb{Z}_3)$ contains an element of multiplicative order 26. Find the multiplicative order of each element of $SL_3(\mathbb{Z}_3) = \{A \in GL_3(\mathbb{Z}_3) : \det A = 1\}$ and deduce that $SL_3(\mathbb{Z}_3)$ does not contain an element of multiplicative order 26.

   Hint: Use Exercises 4.1, Question 3(c).

(c) Determine the number of 3×3 matrices over the finite field $\mathbb{F}_q$ in each of the $q^3 + q^2 + q$ similarity classes partitioning $\mathfrak{M}_3(\mathbb{F}_q)$ (Exercises 6.1, Question 7(a)). Verify that their sum is q^9. Specify the $q^2 + 2q$ similarity classes *not* in $GL_3(\mathbb{F}_q)$ and verify that the size of their union is $q^9 - |GL_3(\mathbb{F}_q)| = q^8 + q^7 - q^5 - q^4 + q^3$.

   Hint: There are $q(q-1)/2$ monic irreducible quadratic polynomials over $\mathbb{F}_q$ and $(q+1)q(q-1)/3$ monic irreducible cubic polynomials over $\mathbb{F}_q$.

(d) Find the number of monic irreducible quartic (degree 4) polynomials over $\mathbb{Z}_3$ by factorising $x^{81} - x$ over $\mathbb{Z}_3$. List the numbers of invariant factor sequences of 4×4 matrices A over $\mathbb{Z}_3$ according to their irreducible factorisations over $\mathbb{Z}_3$ (there are 10 cyclic types and 11 non-cyclic types). Find the number of matrices in each of the 129 similarity class in $\mathfrak{M}_4(\mathbb{Z}_3)$ (there are 20 different numbers). Which similarity classes belong to $GL_4(\mathbb{Z}_3)$? Check that the sums of these numbers are 3^{16} and $|GL_4(\mathbb{Z}_3)|$.

Solutions to Selected Exercises

EXERCISES (page 7)

Solution 1:

$$\binom{\rho_1}{\rho_2}^{-1} = \begin{pmatrix} -5 & 4 \\ 4 & -3 \end{pmatrix} \quad \text{as} \quad \begin{vmatrix} 3 & 4 \\ 4 & 5 \end{vmatrix} = -1.$$

So ρ_1, ρ_2 is a $\mathbb{Z}$-basis of $\mathbb{Z}^2$ and

$$(m_1, m_2) = (10, 7) \begin{pmatrix} -5 & 4 \\ 4 & -3 \end{pmatrix} = (-22, 19).$$

No, as

$$\begin{vmatrix} 3 & 5 \\ 5 & 6 \end{vmatrix} = -7 \neq \pm 1.$$

Solution 2: The 6 elements of $\mathbb{Z}^2/K$ are: $1g_0 = K + e_1 + e_2$, $2g_0 = K + 2e_2$, $3g_0 = K + e_1$, $4g_0 = K + e_2$, $5g_0 = K + e_1 + 2e_2$, $6g_0 = K$. So $\mathbb{Z}^2/K = \langle g_0 \rangle$ is cyclic with generator g_0 and invariant factor 6.

Solution 3: Each of $(\bar{1}, \bar{0})$, $(\bar{1}, \bar{2})$, $(\bar{0}, \bar{2})$ has order 2 and generates a C_2 type subgroup. Each of $(\bar{0}, \bar{1})$, $(\bar{0}, \bar{3})$, $(\bar{1}, \bar{1})$, $(\bar{1}, \bar{3})$ has order 4. $\langle (\bar{0}, \bar{1}) \rangle = \langle (\bar{0}, \bar{3}) \rangle$ and $\langle (\bar{1}, \bar{1}) \rangle = \langle (\bar{1}, \bar{3}) \rangle$ are C_4 type subgroups. $H = \langle (\bar{1}, \bar{0}) \rangle \oplus \langle (\bar{0}, \bar{2}) \rangle$ has isomorphism type $C_2 \oplus C_2$. $\langle (\bar{0}, \bar{0}) \rangle$ and G' are subgroups of type C_1 and $C_2 \oplus C_4$ respectively. H_1 is either $\langle (\bar{1}, \bar{2}) \rangle$ or $\langle (\bar{1}, \bar{0}) \rangle$, H_2 is either $\langle (\bar{0}, \bar{1}) \rangle$ or $\langle (\bar{1}, \bar{1}) \rangle$.

C. Norman, *Finitely Generated Abelian Groups and Similarity of Matrices over a Field*,   339
Springer Undergraduate Mathematics Series,
DOI 10.1007/978-1-4471-2730-7, © Springer-Verlag London Limited 2012

EXERCISES 1.1

Solution 5(a): Apply $c_1 + c_2$, $c_2 - c_1$.

Solution 5(b): Applying the sequence to $A = (a, b)$ gives $(b, -a)$, to $(b, -a)$ gives $(-a, -b)$ and to $(-a, -b)$ gives $(-b, a)$. One of these has non-negative entries.

EXERCISES 1.2

Solution 1(b): $e_j P_1 = e_j + le_k$, $e_i P_1 = e_i$ ($i \neq j$). Postmultiply by $A : e_j P_1 A = e_j A + le_k A$, i.e. row j of $P_1 A$ is row j of $A + l$ (row k of A). Also $e_i P_1 A = e_i A$, that is row i of $P_1 A$ is row i of A for $i \neq j$. So $P_1 A$ is the result of applying $r_j + lr_k$ to A.

Solution 1(c): $Q_1 e_j^T = e_k^T$, $Q_1 e_k^T = e_j^T$, $Q_1 e_i^T = e_i^T$ ($i \neq j, k$). Premultiply by $A : AQ_1 e_j^T = Ae_k^T$, $AQ_1 e_k^T = Ae_j^T$, $AQ_1 e_i^T = Ae_i^T$, i.e. cols j and k of AQ_1 are cols k and j respectively of A, col i of AQ_1 is col i of A ($i \neq j, k$). So AQ_1 is the result of applying $c_j \leftrightarrow c_k$ to A.

Solution 1(d): Let I_s and I_t denote the $s \times s$ and $t \times t$ identity matrices over $\mathbb{Z}$. Then $I_s A I_t^{-1} = A$ showing (i) $A \equiv A$ for all $s \times t$ matrices A over $\mathbb{Z}$. Suppose $A \equiv B$. There are invertible P and Q over $\mathbb{Z}$ with $PAQ^{-1} = B$. Then $P^{-1} B (Q^{-1})^{-1} = A$ showing (ii) $A \equiv B \Rightarrow B \equiv A$ as P^{-1} and Q^{-1} are invertible over $\mathbb{Z}$. Suppose $A \equiv B$ and $B \equiv C$. There are invertible P_1, P_2, Q_1, Q_2 over $\mathbb{Z}$ with $P_1 A Q_1^{-1} = B$ and $P_2 B Q_2^{-1} = C$. Then $(P_2 P_1) A (Q_2 Q_1)^{-1} = C$ showing (iii) $A \equiv B$ and $B \equiv C \Rightarrow A \equiv C$ as $P_2 P_1$ and $Q_2 Q_1$ are invertible over $\mathbb{Z}$. So $\equiv$ is an equivalence relation.

EXERCISES 1.3

Solution 4(d): $BC = B'C'$ and so $\det BC = \det B'C' = (\det B')(\det C') = 0 \times 0 = 0$ by Theorem 1.18 as col l of B' is zero and row l of C' is zero.

Solution 5(c): By Corollary 1.20 $d_1 d_2 \cdots d_s = g_s(A) = 1$. So each $d_i = 1$ and $S(A) = (I_s \mid 0)$ where I_s is the $s \times s$ identity matrix. There are invertible matrices P_1 and Q_1 over $\mathbb{Z}$ with $P_1 A Q_1^{-1} = S(A) = (I_s \mid 0)$. So

$$AQ_1^{-1} = P_1^{-1}(I_s \mid 0) = (P_1^{-1} \mid 0) = (I_s \mid 0) \left(\begin{array}{c|c} P_1^{-1} & 0 \\ \hline 0 & I_{t-s} \end{array} \right)$$

where I_{t-s} is the $(t - s) \times (t - s)$ identity matrix. So

$$Q = \left(\begin{array}{c|c} P_1^{-1} & 0 \\ \hline 0 & I_{t-s} \end{array} \right) Q_1$$

is invertible over $\mathbb{Z}$ and satisfies $A = (I_s \mid 0)Q$. So A can be reduced to $S(A) = (I_s \mid 0)$ using *ecos* only and A is the submatrix of Q consisting of the first s rows.

EXERCISES 2.1

Solution 4(a): Let $g_1, g_2 \in G$. Then $(g_1 + g_2)\theta = c(g_1 + g_2) = cg_1 + cg_2 = (g_1)\theta + (g_2)\theta$. For $g \in G$, $m \in \mathbb{Z}$, $(mg)\theta = c(mg) = (cm)g = (mc)g = m(cg) = m((g)\theta)$. So θ is $\mathbb{Z}$-linear.

As $(g_0)\theta \in G = \langle g_0 \rangle$ there is an integer c with $(g_0)\theta = cg_0$. For $g \in G$ there is $m \in \mathbb{Z}$ with $g = mg_0$. Hence $(g)\theta = (mg_0)\theta = m((g_0)\theta) = m(cg_0) = c(mg_0) = cg$. So there is an integer c as stated. Let $c' \in \mathbb{Z}$ satisfy $(g)\theta = c'g$ for all $g \in G$. Then $(c - c')g_0 = cg_0 - c'g_0 = (g_0)\theta - (g_0)\theta = 0$ showing that $c - c' \in \langle n \rangle$. Hence $n|(c - c')$, i.e. $c \equiv c' \pmod{n}$, i.e. c is unique modulo n. In particular c is unique for $n = 0$ and c is arbitrary for $n = 1$. Suppose that θ is an automorphism of G. As θ is surjective there is $a \in \mathbb{Z}$ with $(ag_0)\theta = g_0$, i.e. $cag_0 = g_0$, i.e. $(ca - 1)g_0 = 0$, i.e. $ca - 1 \in \langle n \rangle$, i.e. $ca - 1 = bn$ for some $b \in \mathbb{Z}$. Hence $ca - bn = 1$ showing that $\gcd\{c, n\} = 1$. Conversely suppose that $\gcd\{c, n\} = 1$. There are integers a, b with $ca - bn = 1$. Reversing the above steps gives $(ag_0)\theta = g_0$ and hence $(mag_0)\theta = mg_0$ for all $m \in \mathbb{Z}$, showing θ to be surjective. Suppose $(mg_0)\theta = (m'g_0)\theta$ for some $m, m' \in \mathbb{Z}$. Then $cmg_0 = cm'g_0$ and so $cm - cm' \in \langle n \rangle$, i.e. $n|c(m - m')$. Hence $n|m - m'$ as $\gcd\{c, n\} = 1$. So $m - m' \in \langle n \rangle$. As $\langle n \rangle$ is the order ideal of g_0 we conclude $(m - m')g_0 = 0$, i.e. $mg_0 = m'g_0$ showing that θ is injective. So θ is an automorphism being bijective. The additive group $\mathbb{Z}$ is generated by the integer 1 with order ideal $\langle 0 \rangle$; so $n = 0$ and $\gcd\{c, 0\} = 1 \Leftrightarrow c = \pm 1$. So $\mathbb{Z}$ has exactly two automorphisms namely $m \to m$ and $m \to -m$ for all $m \in \mathbb{Z}$.

For $n > 0$ the $\mathbb{Z}$-module $\mathbb{Z}_n$ is cyclic being generated by $\overline{1}$ with order ideal $\langle n \rangle$. By the first part every $\mathbb{Z}$-linear mapping $\theta : \mathbb{Z}_n \to \mathbb{Z}_n$ is of the form $(\overline{m})\theta = c\overline{m}$ for some integer c and all $\overline{m} \in \mathbb{Z}_n$. As c is unique modulo n we may write $c\overline{m} = \overline{cm}$ unambiguously. It follows directly from the first part with $G = \mathbb{Z}_n$, $g_0 = \overline{1}$, that θ is an automorphism of $\mathbb{Z}_n \Leftrightarrow \gcd\{c, n\} = 1$. So the additive group $\mathbb{Z}_9$ has 6 automorphisms corresponding to the 6 invertible elements $\overline{c}$ of $\mathbb{Z}_9$ namely $\overline{1}, \overline{2}, \overline{4}, \overline{5}, \overline{7}, \overline{8}$, i.e. the elements $\overline{c}$ with $\gcd\{c, 9\} = 1$. Yes, all these automorphisms are powers of θ_2 since $(\overline{2})^1 = \overline{2}$, $(\overline{2})^2 = \overline{4}$, $(\overline{2})^3 = \overline{8}$, $(\overline{2})^4 = \overline{16} = \overline{7}$, $(\overline{2})^5 = \overline{32} = \overline{5}$, $(\overline{2})^6 = \overline{64} = \overline{1}$, i.e. $\overline{2}$ generates the multiplicative group of invertible elements of $\mathbb{Z}_9$. So $(\overline{m})\theta_2^3 = \overline{8m}$, $(\overline{m})\theta_2^4 = \overline{7m}$ etc.

Solution 4(b): As $ng_0 = 0$, applying φ gives $n((g_0)\varphi) = (ng_0)\varphi = (0)\varphi = 0'$ showing $n \in \langle n' \rangle$, i.e. $n'|n$. Suppose first that $\theta : G \to G'$ is $\mathbb{Z}$-linear and $(g_0)\theta = g_0'$. Then $(mg_0)\theta = m((g_0)\theta) = mg_0'$ for all $m \in \mathbb{Z}$ and so there is at most one such θ. Consider $\theta : G \to G'$ given by $(mg_0)\theta = mg_0'$ for all $m \in \mathbb{Z}$. Let $m_1g_0 = m_2g_0$. Then $n|(m_1 - m_2)$ as $m_1 - m_2 \in \langle n \rangle$ since $(m_1 - m_2)g_0 = 0$. As $d|n$ we deduce $d|(m_1 - m_2)$. So $(m_1 - m_2)g_0' = 0$ as $\langle d \rangle$ is the order ideal of g_0'. So $m_1g_0' = m_2g_0'$ showing that θ is unambiguously defined. Also θ is additive as $(mg_0 + m'g_0)\theta = ((m + m')g_0)\theta = (m + m')g_0' = mg_0' + m'g_0' = (mg_0)\theta + (m'g_0)\theta$ for $m, m' \in \mathbb{Z}$.

As $(m(m'g_0))\theta = ((mm')g_0)\theta = (mm')g_0' = m(m'g_0') = m((m'g_0)\theta)$ we see θ is $\mathbb{Z}$-linear.

Solution 4(c): With $g_1 = g_2 = 0$ in Definition 2.3 we obtain $(0+0)\theta = (0)\theta + (0)\theta$, i.e. $(0)\theta = (0)\theta + (0)\theta$ as $0 + 0 = 0$. Add $-(0)\theta$, the negative in G' of $(0)\theta$, to both sides obtaining $0' = -(0)\theta + (0)\theta = -(0)\theta + (0)\theta + (0)\theta = 0' + (0)\theta = (0)\theta$. Apply θ to $-g + g = 0$ and use Definition 2.3 to obtain $(-g)\theta + (g)\theta = (-g + g)\theta = (0)\theta = 0'$ which means $-(g)\theta = (-g)\theta$ for all $g \in G$. The integer m is in the order ideal of

$$\bar{r} \quad \Leftrightarrow \quad m\bar{r} = \bar{0} \text{ in } \mathbb{Z}_n \quad \Leftrightarrow \quad mr = qn$$

for some

$$q \in \mathbb{Z} \quad \Leftrightarrow \quad m(r/\gcd\{r, n\}) = q(n/\gcd\{r, n\}) \quad \Leftrightarrow \quad (n/\gcd\{r, n\})|m.$$

Therefore $\langle n/\gcd\{r, n\}\rangle$ is the order ideal of $\bar{r}$ in $\mathbb{Z}_n$. For $\bar{r} \in \mathbb{Z}_n$ there is a unique $\mathbb{Z}$-linear mapping $\theta : \mathbb{Z}_m \to \mathbb{Z}_n$ with $(\bar{1})\theta = \bar{r} \Leftrightarrow (n/\gcd\{r, n\})|m$. Hence $(n/\gcd\{m, n\})|\gcd\{r, n\}$ and so $(n/\gcd\{m, n\})|r$ as $\gcd\{r, n\}|r$. Conversely $(n/\gcd\{m, n\})|r \Rightarrow (n/\gcd\{r, n\})|m$ in the same way. So there are $\gcd\{m, n\}$ choices for $\bar{r} \in \mathbb{Z}_n$ namely $r = l(n/\gcd\{m, n\})$ for $1 \leq l \leq \gcd\{m, n\}$.

Solution 4(d): $(g_1 + g_2)\theta\theta' = ((g_1)\theta + (g_2)\theta)\theta' = (g_1)\theta\theta' + (g_2)\theta\theta' \; \forall g_1, g_2 \in G$ and $(mg)\theta\theta' = ((mg)\theta)\theta' = (m((g)\theta))\theta' = m(((g)\theta)\theta') = m((g)\theta\theta') \; \forall m \in \mathbb{Z}$, $g \in G$. So $\theta\theta'$ is $\mathbb{Z}$-linear. Suppose θ bijective. Then $(g_1' + g_2')\theta^{-1}\theta = g_1' + g_2' = (g_1')\theta^{-1}\theta + (g_2')\theta^{-1}\theta = ((g_1')\theta^{-1} + (g_2')\theta^{-1})\theta$ as θ is $\mathbb{Z}$-linear. As θ is injective $(g_1' + g_2')\theta^{-1} = (g_1')\theta^{-1} + (g_2')\theta^{-1} \; \forall g_1', g_2' \in G'$. Also $((mg')\theta^{-1})\theta = (mg')\theta^{-1}\theta = mg' = m((g')\theta^{-1}\theta) = (m((g')\theta^{-1}))\theta$ and as θ is injective $(mg')\theta^{-1} = m((g')\theta^{-1}) \; \forall m \in \mathbb{Z}$, $g' \in G'$. So θ^{-1} is $\mathbb{Z}$-linear. Let θ, φ, ψ be automorphisms of G. Then $\theta\varphi \in \text{Aut } G$ by the above theory with $G' = G'' = G$ and $\theta' = \varphi$. Also $(\theta\varphi)\psi = \theta(\varphi\psi)$ as composition of mappings is associative. The identity $\iota : G \to G$ is in $\text{Aut } G$ and $\iota\theta = \theta = \theta\iota$ for all θ in $\text{Aut } G$. For each $\theta \in \text{Aut } G$ we see $\theta^{-1} \in \text{Aut } G$ and $\theta^{-1}\theta = \iota = \theta\theta^{-1}$. Hence $\text{Aut } G$ is a group.

Solution 7(a): (i) Let $h, h' \in H_1 \cap H_2$. Then $h, h' \in H_i$ $(i = 1, 2)$ and so $h + h' \in H_i$ as H_i is closed under addition. So $h + h' \in H_1 \cap H_2$ showing that $H_1 \cap H_2$ is closed under addition. $0 \in H_i$ $(i = 1, 2)$ and so $0 \in H_1 \cap H_2$. $-h \in H_i$ $(i = 1, 2)$ as H_i is closed under negation and so $-h \in H_1 \cap H_2$. Therefore $H_1 \cap H_2$ is a subgroup of G.

(ii) $(h_1 + h_2) + (h_1' + h_2') = (h_1 + h_1') + (h_2 + h_2') \in H_1 + H_2$ for all $h_i, h_i' \in H_i$ $(i = 1, 2)$. $0 = 0 + 0 \in H_1 + H_2$. $-(h_1 + h_2) = (-h_1) + (-h_2) \in H_1 + H_2$. So $H_1 + H_2$ is a subgroup of G.

Solution 8(a): For $n = 3$ we have $s_3 = (g_1 + g_2) + g_3 = g_1 + (g_2 + g_3)$ by the associative law. Take $n > 3$ and suppose inductively the result to be true for all ordered sets of less than n elements of G. Each summation of $g_1, g_2, \ldots, g_n$ in order decomposes $h_i + h'_{n-i}$ for some i with $1 \le i < n$ where h_i is a summation of $g_1, g_2, \ldots, g_i$ in order and h'_{n-i} is a summation of $g_{i+1}, g_{i+2}, \ldots, g_n$ in order. By induction

$$h_i = s_i \quad \text{and} \quad h'_{n-i} = s'_{n-i}$$

where $s'_{n-i} = (\cdots((g_{i+1} + g_{i+2}) + g_{i+3}) \cdots) + g_n = s'_{n-i-1} + g_n$ say. Hence $h_i + h'_{n-i} = s_i + (s'_{n-i-1} + g_n) = (s_i + s'_{n-i-1}) + g_n$. As $s_i + s'_{n-i-1}$ is a summation of $g_1, g_2, \ldots, g_{n-1}$ we deduce $s_i + s'_{n-i-1} = s_{n-1}$ by induction. Therefore $h_i + h'_{n-i} = s_{n-1} + g_n = s_n$ which completes the induction. Each summation of $g_1, g_2, \ldots, g_n$ in order is equal to s_n. So the generalised associative law of addition holds.

Solution 8(b): By the commutative law $g_1 + g_2 = g_2 + g_1$. Take $n > 2$ and suppose the result is true for all sets of less than n elements of G. Each summation of $g_1, g_2, \ldots, g_n$ decomposes $h_i + h'_{n-i}$ for some i with $1 \le i < n$ where h_i is a summation of g_j, $j \in X$, $|X| = i$ and h'_{n-i} is a summation of g_j, $j \in Y$, $|Y| = n - i$, $X \cap Y = \emptyset$. Interchanging h_i and h'_{n-i} if necessary, we may assume $n \in Y$. By induction $h'_{n-i} = h'_{n-i-1} + g_n$ where h'_{n-i-1} is a summation of g_j for $j \in Y/\{n\}$ and so $h_i + h'_{n-i-1} = s_{n-1}$ by induction. The induction is completed by $h_i + h'_{n-i} = h_i + (h'_{n-i-1} + g_n) = (h_i + h'_{n-i-1}) + g_n = s_{n-1} + g_n = s_n$.

Solution 8(c): For $m \ge 0$ by (b) above $m(g_1 + g_2) = mg_1 + mg_2$ on adding up the $2m$ elements $g_i, g_i, \ldots, g_i$ ($i = 1, 2$) in two ways. For $m < 0$ write $m = -n$. Then $m(g_1 + g_2) = -n(g_1 + g_2) = -ng_1 + (-ng_2) = mg_1 + mg_2$. If $m_1 m_2 = 0$ then $(m_1 + m_2)g = m_1 g + m_2 g$. By symmetry we may assume $m_1 \ge m_2$. For $m_1 > 0$, $m_2 > 0$ using (a) above with $g_i = g$, $(m_1 + m_2)g = s_{m_1 + m_2} = s_{m_1} + s_{m_2} = m_1 g + m_2 g$. For $m_1 = -n_1 < 0$, $m_2 = -n_2 < 0$ we have $(m_1 + m_2)g = -(n_1 + n_2)g = -n_1 g + (-n_2 g) = m_1 g + m_2 g$. For $m_1 > 0$, $m_2 = -n_2 < 0$, $m_1 + m_2 > 0$, $(m_1 + m_2)g = s_{m_1 + m_2} = s_{m_1} - s_{n_2} = m_1 g - n_2 g = m_1 g + m_2 g$. For $m_1 > 0$, $m_2 = -n_2 < 0$, $m_1 + m_2 = -n < 0$, $(m_1 + m_2)g = -ng = -s_n = -s_{n_2 - m_1} = -(s_{n_2} - s_{m_1}) = s_{m_1} - s_{n_2} = m_1 g - n_2 g = m_1 g + m_2 g$. Now $(m_1 m_2)g = 0 = m_1(m_2 g)$ for $m_1 m_2 = 0$. For $m_1 > 0$, $m_2 > 0$, by (a) above, $(m_1 m_2)g = s_{m_1 m_2} = m_1(m_2 g)$. Hence for $m_1 = -n_1 < 0$, $m_2 = -n_2 < 0$, $(m_1 m_2)g = ((-n_1)(-n_2))g = (n_1 n_2)g = n_1(n_2 g) = (-n_1)(-n_2 g) = m_1(m_2 g)$. For $m_1 > 0$, $m_2 = -n_2 < 0$, $(m_1 m_2)g = (-m_1 n_2)g = -((m_1 n_2)g) = -(m_1(n_2 g)) = m_1(-n_2 g) = m_1(m_2 g)$. For $m_1 = -n_1 < 0$, $m_2 > 0$, $(m_1 m_2)g = (-n_1 m_2)g = -((n_1 m_2)g) = -(n_1(m_2 g)) = (-n_1)(m_2 g) = m_1(m_2 g)$.

EXERCISES 2.2

Solution 4(e): As $mn(g + h) = nmg + mnh = n0 + m0 = 0$ we see that $g + h$ has finite order l where $l | mn$. Now $l(g + h) = 0$ and so $lg = -lh$. Hence $nlg = n(-lh) = -lnh = -l0 = 0$ showing that the order m of g is a divisor of nl, i.e. $m|nl$. As $\gcd\{m, n\} = 1$ we deduce $m|l$. In the same way we obtain $n|l$ and so $mn|l$ using $\gcd\{m, n\} = 1$ again. Therefore $mn = l$. Note that $|G| = |K| \times |G/K| = mn$. Replacing φ in Exercises 2.1, Question 4(b) by the natural homomorphism $\eta : G \to G/K$, we see that the order s of h_0 is a divisor of the order n of $(h_0)\eta = K + h_0$. So $h = (s/n)h_0$ has order n. By the above $g + h$ has order mn, as g has order m where $K = \langle g \rangle$. Therefore $g + h$ generates G, i.e. $G = \langle g + h \rangle$ is cyclic.

Solution 4(f): Let $g_1, g_1', g_1'' \in G_1$ and $g_2, g_2', g_2'' \in G_2$. Addition in $G_1 \oplus G_2$ is associative as

$$((g_1, g_2) + (g_1', g_2')) + (g_1'', g_2'') = (g_1 + g_1', g_2 + g_2') + (g_1'', g_2'')$$
$$= ((g_1 + g_1') + g_1'', (g_2 + g_2') + g_2'')$$
$$= (g_1 + (g_1' + g_1''), g_2 + (g_2' + g_2''))$$
$$= (g_1, g_2) + (g_1' + g_1'', g_2' + g_2'')$$
$$= (g_1, g_2) + ((g_1', g_2') + (g_1'', g_2'')).$$

The zero element of $G_1 \oplus G_2$ is $(0_1, 0_2)$ since $(0_1, 0_2) + (g_1, g_2) = (0_1 + g_1, 0_2 + g_2) = (g_1, g_2)$. The negative of (g_1, g_2) is $(-g_1, -g_2)$ as $(-g_1, -g_2) + (g_1, g_2) = (-g_1 + g_1, -g_2 + g_2) = (0_1, 0_2)$. Addition in $G_1 \oplus G_2$ is commutative as

$$(g_1, g_2) + (g_1', g_2') = (g_1 + g_1', g_2 + g_2') = (g_1' + g_1, g_2' + g_2)$$
$$= (g_1', g_2') + (g_1, g_2).$$

So $G_1 \oplus G_2$ is an additive abelian group. Consider $\alpha : G_1 \oplus G_2 \to G_2 \oplus G_1$ defined by $(g_1, g_2)\alpha = (g_2, g_1)$ for all $g_1 \in G_1, g_2 \in G_2$. Then $\alpha : G_1 \oplus G_2 \cong G_2 \oplus G_1$.

Solution 6(b): Suppose to the contrary that the additive group $\mathbb{Z}$ has non-trivial subgroups H_1 and H_2 such that $\mathbb{Z} = H_1 \oplus H_2$. As H_1 and H_2 are ideals of the ring $\mathbb{Z}$, by Theorem 1.15 there are positive integers n_1 and n_2 with $H_1 = \langle n_1 \rangle$ and $H_2 = \langle n_2 \rangle$. Then $0 = 0 \times n_1 + 0 \times n_2 = n_2 \times n_1 + (-n_1) \times n_2$, i.e. the integer zero is expressible in two different ways as a sum of integers from H_1 and H_2. So $\mathbb{Z}$ is indecomposable.

EXERCISES 2.3

Solution 1(a): (i) Write $K = \ker\theta$ and let $k, k' \in K$. Then $(k + k')\theta = (k)\theta + (k')\theta = 0' + 0' = 0'$ showing $k + k' \in K$. Also $(-k)\theta = -(k)\theta = -0' = 0'$ and $(0)\theta = 0'$ showing $-k \in K$ and $0 \in K$. As $(mk)\theta = m((k)\theta) = m0' = 0'$ for $m \in \mathbb{Z}$, we conclude $mk \in K$ and so K is a submodule of the $\mathbb{Z}$-module G. Suppose $K = \{0\}$ and let $g_1, g_2 \in G$ satisfy $(g_1)\theta = (g_2)\theta$. Then $(g_1 - g_2)\theta = (g_1)\theta - (g_2)\theta =$

$(g_1)\theta - (g_1)\theta = 0'$ showing $g_1 - g_2 \in K$. So $g_1 - g_2 = 0$, i.e. $g_1 = g_2$ and θ is injective. Conversely suppose that θ is injective and let $k \in K$. Then $(k)\theta = 0' = (0)\theta$. So $k = 0$ as θ is injective, giving $K = \{0\}$.

(ii) Let $g_1', g_2' \in \text{im}\,\theta$. Then $g_1' = (g_1)\theta$ and $g_2' = (g_2)\theta$ for some $g_1, g_2 \in G$. Then $g_1' + g_2' = (g_1)\theta + (g_2)\theta = (g_1 + g_2)\theta \in \text{im}\,\theta$ as $g_1 + g_2 \in G$. Also $-g_1' = -(g_1)\theta = (-g_1)\theta \in \text{im}\,\theta$ since $-g_1 \in G$. As $0' = (0)\theta \in \text{im}\,\theta$ and $mg_1' = m((g_1)\theta) = (mg_1)\theta \in \text{im}\,\theta$ for all $m \in \mathbb{Z}$, we see $\text{im}\,\theta$ is a submodule of the $\mathbb{Z}$-module G'. Yes, $\text{im}\,\theta = G'$ is the same as θ being surjective.

Solution 3(a): There are $k_1, k_2 \in K$ such that $r_1 = r_1' + k_1$, $r_2 = r_1' + k_2$. Therefore $r_1 r_2 = (r_1' + k_1)(r_2' + k_2) = r_1' r_2' + k_3$ where $k_3 = r_1' k_2 + k_1 r_2' + k_1 k_2$. As K is an ideal of R we see that $r_1' k_2, k_1 r_2', k_1 k_2 \in K$, and so $k_3 \in K$ as K is closed under addition. Hence $r_1 r_2 \equiv r_1' r_2' \pmod{K}$ and so $K + r_1 r_2 = K + r_1' r_2'$ by Lemma 2.9, showing that coset multiplication is unambiguously defined. Write $\bar{r} = K + r$ and then R/K has binary operations $\bar{r_1} + \bar{r_2} = \overline{r_1 + r_2}$ and $(\bar{r_1})(\bar{r_2}) = \overline{r_1 r_2}$ where $r_1, r_2 \in R$. Now $(R/K, +)$ is an abelian group by Lemma 2.10. Let $r_1, r_2, r_3 \in R$. Then $((\bar{r_1})(\bar{r_2}))(\bar{r_3}) = (\overline{r_1 r_2})(\bar{r_3}) = \overline{(r_1 r_2) r_3} = \overline{r_1 (r_2 r_3)} = (\bar{r_1})(\overline{r_2 r_3}) = (\bar{r_1})((\bar{r_2})(\bar{r_3}))$ showing that coset multiplication is associative. Coset multiplication is distributive because $((\bar{r_1}) + (\bar{r_2}))(\bar{r_3}) = (\overline{r_1 + r_2})(\bar{r_3}) = \overline{(r_1 + r_2) r_3} = \overline{r_1 r_3 + r_2 r_3} = \overline{r_1 r_3} + \overline{r_2 r_3} = (\bar{r_1})(\bar{r_3}) + (\bar{r_2})(\bar{r_3})$ and similarly $(\bar{r_1})((\bar{r_2}) + (\bar{r_3})) = (\bar{r_1})(\bar{r_2}) + (\bar{r_1})(\bar{r_3})$. Also $(\bar{e})(\bar{r}) = \overline{er} = \bar{r} = \overline{re} = (\bar{r})(\bar{e})$ for all $r \in R$, and so R/K is a ring with 1-element $\bar{e} = K + e$. By Lemma 2.10 η is additive. As $(r_1 r_2)\eta = \overline{r_1 r_2} = (\bar{r_1})(\bar{r_2}) = (r_1)\eta (r_2)\eta$ for all $r_1, r_2 \in R$ and $(e)\eta = \bar{e}$ we see η is a ring homomorphism. Also $\text{im}\,\eta = R/K$ and $\ker \eta = K$.

Solution 3(b): By Exercises 2.3, Question 1(a)(ii), $\text{im}\,\theta$ is a subgroup of $(R', +)$. As $(r_1)\theta (r_2)\theta = (r_1 r_2)\theta$ for all $r_1, r_2 \in R$ we see $\text{im}\,\theta$ is closed under multiplication. The 1-element e' of R' belongs to $\text{im}\,\theta$ as $e' = (e)\theta$. So $\text{im}\,\theta$ is a subring of R' and hence $\text{im}\,\theta$ is itself a ring. By Exercises 2.3, Question 1(a)(i), $\ker \theta$ is a subgroup of $(R, +)$. Consider $r \in R$, $k \in K$; then $(rk)\theta = (r)\theta (k)\theta = (r)\theta \times 0 = 0$ and so $rk \in K = \ker \theta$. Similarly $kr \in K$ and so K is an ideal of R. *Kernels of ring homomorphisms are ideals.* By Theorem 2.16 $\tilde{\theta} : R/K \cong \text{im}\,\theta$ is an isomorphism of additive abelian groups. Also $((\bar{r_1})(\bar{r_2}))\tilde{\theta} = (\overline{r_1 r_2})\tilde{\theta} = (r_1 r_2)\theta = (r_1)\theta (r_2)\theta = (\bar{r_1})\tilde{\theta}(\bar{r_2})\tilde{\theta}$ for all $r_1, r_2 \in R$. So $\tilde{\theta}$ is a ring isomorphism as $(\bar{e})\tilde{\theta} = e'$. Therefore $\tilde{\theta} : R/K \cong \text{im}\,\theta$.

Solution 3(c): By (b) above $\ker \theta$ is an ideal of the ring $\mathbb{Z}$. By Theorem 1.15 there is a non-negative integer d with $\ker \theta = \langle d \rangle$. By (b) above $\tilde{\theta} : \mathbb{Z}/\langle d \rangle \cong \text{im}\,\theta$, showing that the rings $\mathbb{Z}_d = \mathbb{Z}/\langle d \rangle$ are, up to isomorphism, the (ring) homomorphic images of $\mathbb{Z}$.

Solution 3(d): For $r_1, r_2 \in R$ we see $(r_1 + r_2)\theta\theta' = ((r_1)\theta + (r_2)\theta)\theta' = (r_1)\theta\theta' + (r_2)\theta\theta'$ and $(r_1 r_2)\theta\theta' = ((r_1)\theta (r_2)\theta)\theta' = (r_1)\theta\theta' (r_2)\theta\theta'$ showing that $\theta\theta'$ is additive and multiplicative. As $(e)\theta = e'$ and $(e')\theta' = e''$ we see $(e)\theta\theta' = e''$

where e, e', e'' are the 1-elements of R, R', R''. So $\theta\theta' : R \rightarrow R''$ is a ring homomorphism. Suppose that θ is a ring isomorphism. Consider $r_1', r_2' \in R'$ and write $r_1 = (r_1')\theta^{-1}$, $r_2 = (r_2')\theta^{-1}$. Then $((r_1')\theta^{-1} + (r_2')\theta^{-1})\theta = (r_1 + r_2)\theta = (r_1)\theta + (r_2)\theta = r_1' + r_2' = (r_1' + r_2')\theta^{-1}\theta$ and so $(r_1')\theta^{-1} + (r_2')\theta^{-1} = (r_1' + r_2')\theta^{-1}$ as θ is injective, and θ^{-1} is additive. Similarly $((r_1')\theta^{-1}(r_2')\theta^{-1})\theta = (r_1 r_2)\theta = (r_1)\theta(r_2)\theta = r_1' r_2' = (r_1' r_2')\theta^{-1}\theta$ and so $(r_1')\theta^{-1}(r_2')\theta^{-1} = (r_1' r_2')\theta^{-1}$ as θ is additive, and θ^{-1} is multiplicative. As $(e')\theta^{-1} = e$ we conclude $\theta^{-1} : R \rightarrow R'$ is a ring isomorphism. Take $R = R' = R''$ and suppose θ, θ' are bijective, i.e. suppose $\theta, \theta' \in \text{Aut } R$. By the above theory $\theta\theta', \theta^{-1} \in \text{Aut } R$. As the identity mapping ι_R of R belongs to Aut R we see that Aut R is a group (it's a subgroup of the group of all bijections of $R \rightarrow R$).

Solution 3(f): By Exercises 2.1, Question 7(a)(i) and (ii) both $K \cap L$ and $K + L$ are additive abelian groups. Consider $r \in R$ and $m \in K \cap L$. Then $m \in K$ and $m \in L$. As K is an ideal of R we see $rm, mr \in K$. As L is an ideal of R we see $rm, mr \in L$. So $rm, mr \in K \cap L$ and $K \cap L$ is an ideal of R. Consider $r \in R$, $m \in K + L$. Then $m = k + l$ where $k \in K$ and $l \in L$. So $rm = r(k+l) = rk + rl \in K + L$ since $rk \in K$ and $rl \in L$ as before. Also $mr = (k+l)r = kr + lr \in K + L$ since $kr \in K$ and $lr \in L$. So $K + L$ is an ideal of R. For $r_1, r_2 \in R$ using addition and multiplication in the rings R/K, R/L and $R/K \oplus R/L$

$$(r_1 + r_2)\alpha = (r_1 + r_2 + K, r_1 + r_2 + L)$$

$$= ((r_1 + K) + (r_2 + K), (r_1 + L) + (r_2 + L))$$

$$= (r_1 + K, r_1 + L) + (r_2 + K, r_2 + L) = (r_1)\alpha + (r_2)\alpha$$

and

$$(r_1 r_2)\alpha = (r_1 r_2 + K, r_1 r_2 + L)$$

$$= ((r_1 + K)(r_2 + K), (r_1 + L)(r_2 + L))$$

$$= (r_1 + K, r_1 + L)(r_2 + K, r_2 + L) = (r_1)\alpha(r_2)\alpha.$$

Let e be the 1-element of R. As $(e)\alpha = (e + K, e + L)$ is the 1-element of $R/K \oplus R/L$ we see that α is a ring homomorphism. The 0-element of $R/K \oplus R/L$ is (K, L). As

$$(r)\alpha = (K, L) \quad \Leftrightarrow \quad (r + K, r + L) \quad \Leftrightarrow \quad r \in K, r \in L \quad \Leftrightarrow \quad r \in K \cap L$$

we see ker $\alpha = K \cap L$. Now use $K + L = R$ to find im α: there are elements $k_0 \in K$ and $l_0 \in L$ with $k_0 + l_0 = e$. Consider an arbitrary element $(s + K, t + L)$ of $R/K \oplus R/L$ and so $s, t \in R$. Write $r = sl_0 + tk_0$. Then $r - s = r - se = s(l_0 - e) + tk_0 = s(-k_0) + tk_0 = (t - s)k_0 \in K$ and so $r + K = s + K$. Also $r - t = r - te = sl_0 + t(k_0 - e) = sl_0 + t(-l_0) = (s - t)l_0 \in L$ and so $r + L = t + L$.

Therefore $(r)\alpha = (r+K, r+L) = (s+K, t+L)$ and $\operatorname{im}\alpha = R/K \oplus R/L$. By (b) above $\tilde{\alpha} : R/(K \cap L) \cong R/K \oplus R/L$ is a ring isomorphism where $(r+K \cap L)\tilde{\alpha} = (r+K, r+L)$ for all $r \in R$.

Solution 4(a): Suppose that K is normal in G. Let $g \in G$ and consider $kg \in Kg$ where $k \in K$. Then $kg = g(g^{-1}kg) \in gK$ as $g^{-1}kg \in K$. So $Kg \subseteq gK$. Replacing g by g^{-1} in the normality condition gives $gkg^{-1} \in K$ for all $k \in K$. Consider $gk \in gK$. Then $gk = (gkg^{-1})g \in Kg$. So $gK \subseteq Kg$ and hence $Kg = gK$. Conversely suppose $Kg = gK$ for all $g \in G$. For $k \in K$, $g \in G$ we have $kg \in Kg$ and so $kg \in gK$. There is $k' \in K$ with $kg = gk'$. Hence $g^{-1}kg = k' \in K$, i.e. $g^{-1}kg \in K$ for all $k \in K$, $g \in G$. Suppose $Kg_1 = Kg_1'$ and $Kg_2 = Kg_2'$. Using the above theory we obtain $Kg_1 g_2 = Kg_1' g_2 = g_1' g_2 K = g_1' g_2' K = Kg_1' g_2'$, showing that coset multiplication is unambiguously defined. So G/K is closed under coset multiplication. Let e denote the identity element of G and let $g, g_1, g_2, g_3 \in G$. Then $(Kg_1 Kg_2)Kg_3 = K(g_1 g_2)g_3 = Kg_1(g_2 g_3) = Kg_1(Kg_2 Kg_3)$ showing that coset multiplication is associative. As $KeKg = Keg = Kg = Kge = KgKe$ we see that $K = Ke$ is the identity element of G/K. As $Kg^{-1}Kg = Kg^{-1}g = Ke = Kgg^{-1} = KgKg^{-1}$ we see that Kg^{-1} is the inverse of Kg. So G/K is a group.

Solution 4(c): Write $x = (e)\theta$. Then $x^2 = (e)\theta(e)\theta = (e^2)\theta = (e)\theta = x$. As $x \in G'$ there is $x^{-1} \in G'$ with $xx^{-1} = e'$. Hence $e' = xx^{-1} = x^2 x^{-1} = x$, i.e. $(e)\theta = e'$. Applying θ to $g^{-1}g = e = gg^{-1}$ gives $(g^{-1})\theta(g)\theta = e' = (g)\theta(g^{-1})\theta$ showing that $(g^{-1})\theta$ is the inverse of $(g)\theta$, i.e. $(g^{-1})\theta = ((g)\theta)^{-1}$ for all $g \in G$. As $(e)\theta = e'$ we see $e \in K$. Let $k_1, k_2 \in K$. Then $(k_1 k_2)\theta = (k_1)\theta(k_2)\theta = e'e' = e'$ showing $k_1 k_2 \in K$. As $(k_1^{-1})\theta = ((k_1)\theta)^{-1} = e'^{-1} = e'$ we see $k_1^{-1} \in K$. Therefore $K = \ker\theta$ is a subgroup of G. As $(e)\theta = e'$ we see $e' \in \operatorname{im}\theta$. As $(g_1)\theta(g_2)\theta = (g_1 g_2)\theta \in \operatorname{im}\theta$ for $g_1, g_2 \in G$ and so $\operatorname{im}\theta$ is closed under multiplication. As $((g_1)\theta)^{-1} = (g_1^{-1})\theta \in \operatorname{im}\theta$ for all $g_1 \in G$ we see $\operatorname{im}\theta$ is a subgroup of G'. Let $k \in K$, $g \in G$. Then $(g^{-1}kg)\theta = (g^{-1})\theta(k)\theta(g)\theta = ((g)\theta)^{-1}e'(g)\theta = ((g)\theta)^{-1}(g)\theta = e'$ showing that $g^{-1}kg \in K$. So $K = \ker\theta$ is normal in G. *Kernels of group homomorphisms are normal subgroups.* As $(kg)\theta = (k)\theta(g)\theta = e'(g)\theta = (g)\theta$ all elements of the coset Kg are mapped by θ to the same element $(g)\theta$. So $\tilde{\theta} : G/K \to \operatorname{im}\theta$ defined by $(Kg)\tilde{\theta} = (g)\theta$ is unambiguous and surjective. Suppose $(Kg_1)\tilde{\theta} = (Kg_2)\tilde{\theta}$. Then $(g_1)\theta = (g_2)\theta$ and so $(g_1 g_2^{-1})\theta = (g_1)\theta((g_2)\theta)^{-1} = (g_1)\theta((g_1)\theta)^{-1} = e'$ showing $g_1 g_2^{-1} = k \in K$. So $g_1 = kg_2$ and hence $Kg_1 = Kg_2$, that is, $\tilde{\theta}$ is injective. As $((Kg_1)(Kg_2))\tilde{\theta} = (Kg_1 g_2)\tilde{\theta} = (g_1 g_2)\theta = (g_1)\theta(g_2)\theta = (Kg_1)\tilde{\theta}(Kg_2)\tilde{\theta}$ we see that $\tilde{\theta}$ is a group isomorphism and so $\tilde{\theta} : G/K \cong \operatorname{im}\theta$, the first isomorphism theorem for groups.

Solution 4(d): Let $g_1, g_1', g_1'' \in G_1$ and $g_2, g_2', g_2'' \in G_2$. Then

$$((g_1, g_2)(g_1', g_2'))(g_1'', g_2'') = (g_1 g_1', g_2 g_2')(g_1'', g_2'')$$

$$= ((g_1 g_1')g_1'', (g_2 g_2')g_2'')$$

$$= (g_1(g_1' g_1''), g_2(g_2' g_2''))$$

$$= (g_1, g_2)(g_1'g_1'', g_2'g_2'')$$

$$= (g_1, g_2)((g_1', g_2')(g_1'', g_2''))$$

showing that componentwise multiplication on $G_1 \times G_2$ is associative. The pair (e_1, e_2) consisting of the identity elements e_1 of G_1 and e_2 of G_2 is the identity element of $G_1 \times G_2$ because $(g_1, g_2)(e_1, e_2) = (g_1e_1, g_2e_2) = (g_1, g_2) = (e_1g_1, e_2g_2) = (e_1, e_2)(g_1, g_2)$ for all $(g_1, g_2) \in G_1 \times G_2$. The inverse of (g_1, g_2) is (g_1^{-1}, g_2^{-1}) as $(g_1, g_2)(g_1^{-1}, g_2^{-1}) = (g_1g_1^{-1}, g_2g_2^{-1}) = (e_1, e_2) = (g_1^{-1}g_1, g_2^{-1}g_2) = (g_1^{-1}, g_2^{-1})(g_1, g_2)$. So $G_1 \times G_2$ is a group, the external direct product of G_1 and G_2.

Solution 5(a): Let $m_1, m_2 \in \mathbb{Z}$. Then $(m_1 + m_2)\chi = (m_1 + m_2)e = m_1e + m_2e = (m_1)\chi + (m_2)\chi$ applying the result of Exercises 2.1, Questions 8(c) to the additive group of F. Also $(m_1m_2)\chi = (m_1m_2)e = (m_1m_2)e^2 = (m_1e)(m_2e) = (m_1)\chi(m_2)\chi$. So χ is a ring homomorphism as $(1)\chi = e$. By Theorem 1.15 there is a unique non-negative integer d with $\ker \chi = \langle d \rangle$. By the first isomorphism theorem for rings $\tilde{\chi} : \mathbb{Z}/\ker \chi \cong \mathrm{im}\,\chi$, i.e. $\tilde{\chi} : \mathbb{Z}_d \cong \mathrm{im}\,\chi$ defined by $(\bar{i})\tilde{\chi} = ie$ for all $\bar{i} \in \mathbb{Z}_d = \mathbb{Z}/\langle d \rangle$ is a ring isomorphism. Suppose $d > 0$. As $e \neq 0$ we see $d \neq 1$. Suppose that d is not prime. Then $d = d_1d_2$ for positive integers d_1, d_2. Hence $(d_1)\chi(d_2)\chi = (d_1d_2)\chi = (d)\chi = 0$. As F has no divisors of zero, either $(d_1)\chi = 0$ or $(d_2)\chi = 0$, i.e. either $d_1 \in \langle d \rangle$ or $d_2 \in \langle d \rangle$ both of which are impossible as d is not a divisor of either d_1 or d_2. So either $d = 0$ or $d = p$ a prime. For $d = 0$ we have $\tilde{\chi} : \mathbb{Z}_0 \cong \mathrm{im}\,\chi$ and so $\mathrm{im}\,\chi$ has an infinite number of elements. So for each finite field F there is a prime p (*the characteristic of F*) such that $\tilde{\chi} : \mathbb{Z}_p \cong \mathrm{im}\,\chi$. As $\mathbb{Z}_p$ is a field we see that $\mathrm{im}\,\chi = F_0$ is a subfield of F. Regard the elements v, v' of F as vectors and the elements a of F_0 as scalars; then $v + v' \in F$ and $av \in F$ satisfy the vector space laws as these laws follow directly from the laws of a field. In short F is a vector space over F_0. As there are only a finite number of vectors, this vector space is finitely generated and so has a basis $v_1, v_2, \ldots, v_s$. Each element of F can be uniquely expressed in the form $a_1v_1 + a_2v_2 + \cdots + a_sv_s$ where $a_1, a_2, \ldots, a_s \in F_0$. As $\mathbb{Z}_p \cong F_0$ there are p independent choices for each of the s scalars $a_1, a_2, \ldots, a_s$. Hence $|F| = p^s$. For $0 < r < p$ the binomial coefficient $\binom{p}{r}$ is divisible by the prime p. As $pe = 0$ by the first paragraph, we see $\binom{p}{r}a^{p-r}b^r = ((\binom{p}{r})/p)(pe)a^{p-r}b^r = 0$. By the binomial theorem

$$(a+b)^p = \sum_{r=0}^{p} \binom{p}{r} a^{p-r}b^r = a^p + b^p$$

as only the first and last terms contribute to the sum. Therefore $(a+b)\theta = (a+b)^p = a^p + b^p = (a)\theta + (b)\theta$ showing that θ is additive. As $\ker \theta = \{a \in F : a^p = 0\} = \{0\}$ we see that θ is injective by Exercises 2.3, Question 1(a)(i). As $\theta : F \to F$ and F has only a finite number of elements we deduce that θ is also surjective. Finally $(ab)\theta = (ab)^p = a^p b^p = (a)\theta(b)\theta$ showing that θ is multiplicative. As $(e)\theta = e^p = e$ we conclude that θ is an automorphism of the finite field F.

Solution 5(b): Let $A = (a_{ij})$ and $B = (b_{ij})$ be $t \times t$ matrices over $\mathbb{Z}_n$. Then

$$(A + B)\delta_t = ((a_{ij} + b_{ij})\delta_1) = ((a_{ij})\delta_1 + (b_{ij})\delta_1)$$

$$= ((a_{ij})\delta_1) + ((b_{ij})\delta_1) = (A)\delta_t + (B)\delta_t$$

and

$$(AB)\delta_t = \left(\left(\sum_{j=1}^{t} a_{ij} b_{jk} \right) \delta_1 \right) = \left(\left(\sum_{j=1}^{t} (a_{ij})\delta_1 (b_{jk})\delta_1 \right) \right)$$

$$= ((a_{ij})\delta_1)((b_{jk})\delta_1) = (A)\delta_t (B)\delta_t.$$

As $(\bar{1}_n)\delta_1 = \bar{1}_d$ and $(\bar{0}_n)\delta_1 = \bar{0}_d$ we see that δ_t maps the 1-element of the ring $\mathfrak{M}_t(\mathbb{Z}_n)$ to the 1-element of $\mathfrak{M}_t(\mathbb{Z}_d)$ and so δ_t is a ring homomorphism. As δ_1 is surjective so also is δ_t, i.e. $\text{im}\,\delta_t = \mathfrak{M}_t(\mathbb{Z}_d)$. Let $A = (a_{ij}) \in \ker \delta_t$. Then $(a_{ij})\delta_1 = \bar{0}_d$ and so $a_{ij} = \bar{m}_n$ where $d|m$. There are therefore n/d independent choices for each of the t^2 entries in A. Hence $|\ker \delta_t| = (n/d)^{t^2}$. Using the multiplicative property of determinants

$$A \in GL_t(\mathbb{Z}_n) \quad \Leftrightarrow \quad \det A = \bar{m}_n \in U(\mathbb{Z}_n) \quad \Leftrightarrow \quad \gcd\{m, p^s\} = 1$$

$$\Leftrightarrow \quad \gcd\{m, p\} = 1$$

$$\Leftrightarrow \quad \det(A)\delta_t = \bar{m}_p \in U(\mathbb{Z}_p) = \mathbb{Z}_p^*$$

$$\Leftrightarrow \quad (A)\delta_t \in GL_t(\mathbb{Z}_p).$$

Hence $\delta_t| : GL_t(\mathbb{Z}_{p^s}) \to GL_t(\mathbb{Z}_p)$ makes sense and is surjective. As δ_t respects products so also does the restriction $\delta_t|$, i.e. it is a homomorphism of multiplicative groups. Now $A \in \ker \delta_t| \Leftrightarrow (A)\delta_t = (I)\delta_t \Leftrightarrow A \in \ker \delta_t + I$ and so $\ker \delta_t| = \ker \delta_t + I$ is a normal subgroup of $GL_t(\mathbb{Z}_{p^s})$ having $|\ker \delta_t| = p^{(s-1)t^2}$ elements. By the first isomorphism theorem for groups $GL_t(\mathbb{Z}_{p^s})/\ker \delta_t| \cong GL_t(\mathbb{Z}_p)$. So $GL_t(\mathbb{Z}_{p^s})$ partitions into $|GL_t(\mathbb{Z}_p)|$ cosets of $\ker \delta_t$. Hence $|GL_t(\mathbb{Z}_{p^s})| = p^{(s-1)t^2}(p^t - 1)(p^t - p) \cdots (p^t - p^{t-1})$ using the formula following Lemma 2.18.

Solution 7(a): Suppose $v_1', v_2', \ldots, v_t'$ generate M'. Let $v' \in M'$. By Definition 2.19(i) there are $r_1, r_2, \ldots, r_t \in R$ with $r_1 v_1' + r_2 v_2' + \cdots + r_t v_t' = v'$. Write $v = r_1 v_1 + r_2 v_2 + \cdots + r_t v_t \in M$. Then

$$(v)\theta = (r_1 v_1 + r_2 v_2 + \cdots + r_t v_t)\theta$$

$$= r_1(v_1)\theta + r_2(v_2)\theta + \cdots + r_t(v_t)\theta$$

$$= r_1 v_1' + r_2 v_2' + \cdots + r_t v_t' = v'$$

showing θ to be surjective. Conversely suppose that θ is surjective. Let $v' \in M'$. There is $v \in M$ with $(v)\theta = v'$. As $v_1, v_2, \ldots, v_t$ generate M there are $r_1, r_2, \ldots, r_t \in R$ with $v = r_1 v_1 + r_2 v_2 + \cdots + r_t v_t$. As θ is R-linear we have

$$v' = (v)\theta = (r_1 v_1 + r_2 v_2 + \cdots + r_t v_t)\theta$$

$$= r_1(v_1)\theta + r_2(v_2)\theta + \cdots + r_t(v_t)\theta$$

$$= r_1 v'_1 + r_2 v'_2 + \cdots + r_t v'_t$$

showing that $v'_1, v'_2, \ldots, v'_t$ generate M'. Suppose $v'_1, v'_2, \ldots, v'_t$ are R-independent elements of M'. Consider $u \in \ker\theta$. As $v_1, v_2, \ldots, v_t$ generate M there are $r_1, r_2, \ldots, r_t \in R$ with $u = r_1 v_1 + r_2 v_2 + \cdots + r_t v_t$. As θ is R-linear

$$0 = (u)\theta = (r_1 v_1 + r_2 v_2 + \cdots + r_t v_t)\theta$$

$$= r_1(v_1)\theta + r_2(v_2)\theta + \cdots + r_t(v_t)\theta$$

$$= r_1 v'_1 + r_2 v'_2 + \cdots + r_t v'_t$$

and so $r_1 = r_2 = \cdots = r_t = 0$. Hence $u = 0v_1 + 0v_2 + \cdots + 0v_t = 0$ showing $\ker\theta = \{0\}$ and so θ is injective by Exercises 2.3, Question 1(a)(i). Conversely suppose θ is injective. Then $\ker\theta = \{0\}$ by Exercises 2.3, Question 1(a)(i). Consider $r_1 v'_1 + r_2 v'_2 + \cdots + r_t v'_t = 0$ where $r_1, r_2, \ldots, r_t \in R$. Then

$$(r_1 v_1 + r_2 v_2 + \cdots + r_t v_t)\theta = 0,$$

i.e. $r_1 v_1 + r_2 v_2 + \cdots + r_t v_t \in \ker\theta$. So $r_1 v_1 + r_2 v_2 + \cdots + r_t v_t = 0$. As $v_1, v_2, \ldots, v_t$ are R-independent we conclude $r_1 = r_2 = \cdots = r_t = 0$ and so $v'_1, v'_2, \ldots, v'_t$ are R-independent elements of M'. Now suppose $\theta : M \to M'$ is an isomorphism and M is free of rank t. Then M has R-basis $v_1, v_2, \ldots, v_t$. Let $v'_i = (v_i)\theta$ for $1 \le i \le t$. As θ is surjective and injective $v'_1, v'_2, \ldots, v'_t$ generate M' and are R-independent using the above theory. So M' has R-basis $v'_1, v'_2, \ldots, v'_t$ and so is free of rank $t = t'$. Conversely suppose M and M' to be free R-modules of the same rank t. Let M have R-basis $v_1, v_2, \ldots, v_t$ and let M' have R-basis $v'_1, v'_2, \ldots, v'_t$. Consider $\theta : M \to M'$ defined by $(r_1 v_1 + r_2 v_2 + \cdots + r_t v_t)\theta = r_1 v'_1 + r_2 v'_2 + \cdots + r_t v'_t$ for all $r_1, r_2, \ldots, r_t \in R$. Let $u = r_1 v_1 + r_2 v_2 + \cdots + r_t v_t$, $v = s_1 v_1 + s_2 v_2 + \cdots + s_t v_t$ where $s_i \in R$ for $1 \le i \le t$. Then

$$(u + v)\theta = ((r_1 + s_1)v_1 + (r_2 + s_2)v_2 + \cdots + (r_t + s_t)v_t)\theta$$

$$= (r_1 + s_1)v'_1 + (r_2 + s_2)v'_2 + \cdots + (r_t + s_t)v'_t$$

$$= (r_1 v'_1 + r_2 v'_2 + \cdots + r_t v'_t) + (s_1 v'_1 + s_2 v'_2 + \cdots + s_t v'_t)$$

$$= (r_1 v_1 + r_2 v_2 + \cdots + r_t v_t)\theta + (s_1 v_1 + s_2 v_2 + \cdots + s_t v_t)\theta$$

$$= (u)\theta + (v)\theta$$

and so θ is additive. Also for $r \in R$ we see

$$(ru)\theta = (r(r_1v_1 + r_2v_2 + \cdots + r_tv_t))\theta$$

$$= (rr_1v_1 + rr_2v_2 + \cdots + rr_tv_t)\theta$$

$$= rr_1v_1' + rr_2v_2' + \cdots + rr_tv_t'$$

$$= r(r_1v_1' + r_2v_2' + \cdots + r_tv_t') = r((u)\theta).$$

So θ is R-linear and as $(v_i)\theta = v_i'$ we can apply the above theory: since $v_1', v_2', \ldots, v_t'$ generate M' and are R-independent, θ is surjective and injective, i.e. $\theta : M \cong M'$.

Solution 7(b): By hypothesis M has R-basis $v_1, v_2, \ldots, v_t$. There are $t \times t$ matrices $P = (p_{ij})$ and $Q = (q_{jk})$ over R such that $v_i = \sum_{j=1}^{t} p_{ij}u_j (1 \le i \le t)$ and $u_j = \sum_{k=1}^{t} q_{jk}v_k$ $(1 \le j \le t)$ as in the proof of Theorem 2.20. Then $PQ = I$ on comparing coefficients in the equation

$$v_i = \sum_{j=1}^{t} p_{ij}\left(\sum_{k=1}^{t} q_{jk}v_k\right) = \sum_{k=1}^{t}\left(\sum_{j=1}^{t} p_{ij}q_{jk}\right)v_k.$$

So Q is invertible over R by Lemma 2.18. Hence $u_1, u_2, \ldots, u_t$ is an R-basis of M by Corollary 2.21.

Solution 7(c): Regarding M and M' as $\mathbb{Z}$-modules, by Exercises 2.3, Question 1(a)(i) and (ii) above, $\ker\theta$ and $\operatorname{im}\theta$ are additive subgroups of M and M' respectively. For $u \in \ker\theta$, $r \in R$, we have $(ru)\theta = r((u)\theta) = r \times 0 = 0$ showing that $ru \in \ker\theta$. So $\ker\theta$ is a submodule of the R-module M by Definition 2.26. For $u' \in \operatorname{im}\theta$, $r \in R$, there is $u \in M$ with $(u)\theta = u'$. So $(ru)\theta = r((u)\theta) = ru'$ showing $ru' \in \operatorname{im}\theta$ as $ru \in M$. So $\operatorname{im}\theta$ is a submodule of the R-module M' by Definition 2.26. Let θ be bijective. Then $\theta^{-1} : M' \to M$ is additive by Exercises 2.1, Question 4(d). Let $r \in R$ and $v' \in M'$ and write $v = (v')\theta^{-1}$. Then $(rv)\theta = r((v)\theta) = rv'$. Applying θ^{-1} gives $(rv')\theta^{-1} = rv = r((v')\theta^{-1})$ showing that θ^{-1} is R-linear. Yes, the inverse of an isomorphism of R-modules is bijective and R-linear and so is itself an isomorphism of R-modules by Definition 2.24.

Solution 7(d): Consider $r_1, r_2 \in R$ and $v \in M$. Using coset addition, before Lemma 2.10, and law 5 (part 2), before Definition 2.19, we see $(r_1 + r_2)(N + v) = N + (r_1 + r_2)v = N + r_1v + r_2v = (N + r_1) + (N + r_2)$ which shows that law 5 (part 2) holds in M/N. As law 6 holds in M we see $(r_1r_2)(N + v) = N + (r_1r_2)v = N + r_1(r_2v) = r_1(N + r_2v) = r_1(r_2(N + v))$ showing that law 6 holds in M/N. The 1-element 1 of R satisfies $1(N + v) = N + 1v = N + v$ and so law 7 holds in M/N. Therefore M/N is an R-module.

Solution 7(e): There is an element $u_0 \in N$. Then $z_0 = 0u_0 \in N$. As $0 + 1 = 1$ in R we see $z_0 + u_0 = 0u_0 + 1u_0 = (0 + 1)u_0 = 1u_0 = u_0$ using the distributive law in M. Hence $-u_0 + (z_0 + n_0) = -u_0 + u_0 = 0$ and so $z_0 = 0$ (the 0-element of M) on using the associative and commutative laws of addition in M. So N contains the 0-element of M. For $u \in N$ we have $(-1)u \in N$. Then $(-1)u + u = (-1)u + 1u = (-1+1)u = 0u = 0$ on replacing u_0 by u in the above paragraph. So $-u = (-1)u$ and so N is closed under negation. N is a subgroup of the additive group of M. So N is a submodule of M.

Consider $v, v' \in N_1 + N_2$. There are $u_1, u'_1 \in N_1$ and $u_2, u'_2 \in N_2$ with

$$v = u_1 + u_2, \qquad v' = u'_1 + u'_2.$$

So $v + v' = u_1 + u_2 + u'_1 + u'_2 = (u_1 + u'_1) + (u_2 + u'_2) \in N_1 + N_2$ since $u_1 + u'_1 \in N_1$ and $u_2 + u'_2 \in N_2$. So $N_1 + N_2$ is closed under addition. Also $rv = r(u_1 + u_2) = ru_1 + ru_2 \in N_1 + N_2$ as $ru_1 \in N_1$ and $ru_2 \in N_2$ for all $r \in R$. By the above theory $N_1 + N_2$ is a submodule of M. Consider $u, u' \in N_1 \cap N_2$. Then $u, u' \in N_1$ and so $u + u' \in N_1$. Also $u, u' \in N_2$ and so $u + u' \in N_2$. Therefore $u + u' \in N_1 \cap N_2$ and so $N_1 \cap N_2$ is closed under addition. For $r \in R$ we have $ru \in N_1$ as $u \in N_1$. Also $ru \in N_2$ as $u \in N_2$. So $ru \in N_1 \cap N_2$. So $N_1 \cap N_2$ is a submodule of M.

Solution 7(f): By Exercises 2.2, Question 4(f) we know $M_1 \oplus M_2$ is an additive abelian group. We next check that the R-module laws 5, 6 and 7 (before Definition 2.19) hold in $M_1 \oplus M_2$ given that these laws hold in M_1 and M_2. Consider $v = (v_1, v_2) \in M_1 \oplus M_2$, $v' = (v'_1, v'_2) \in M_1 \oplus M_2$ and $r, r' \in R$. Then

$$r(v + v') = r((v_1, v_2) + (v'_1, v'_2)) = r(v_1 + v'_1, v_2 + v'_2)$$

$$= (r(v_1 + v'_1), r(v_2 + v'_2))$$

$$= (rv_1 + rv'_1, rv_2 + rv'_2) = (rv_1, rv_2) + (rv'_1, rv'_2)$$

$$= r(v_1, v_2) + r(v'_1, v'_2) = rv + rv'$$

and

$$(r + r')v = (r + r')(v_1, v_2) = ((r + r')v_1, (r + r')v_2)$$

$$= (rv_1 + r'v_1, rv_2 + r'v_2)$$

$$= r(v_1, v_2) + r'(v_1, v_2) = rv + r'v$$

which shows that law 5 holds. Also

$$(rr')v = (rr')(v_1, v_2) = ((rr')v_1, (rr')v_2) = (r(r'v_1), r(r'v_2))$$

$$= r(r'v_1, r'v_2) = r(r'(v_1, v_2)) = r(r'v)$$

showing that law 6 holds. As $1v = 1(v_1, v_2) = (1v_1, 1v_2) = (v_1, v_2) = v$ we see that law 7 holds and so $M_1 \oplus M_2$ is an R-module.

EXERCISES 3.1

Solution 5(b): As G is generated by t say of its elements there is a surjective $\mathbb{Z}$-linear mapping $\theta : \mathbb{Z}^t \to G$. Consider $k_1, k_2 \in K'$ and so $(k_1)\theta = h_1 \in H$ and $(k_2)\theta = h_2 \in H$. Then $(k_1 + k_2)\theta = (k_1)\theta + (k_2)\theta = h_1 + h_2 \in H$ showing $k_1 + k_2 \in K'$. Also $(-k_1)\theta = -(k_1)\theta = -h_1 \in H$ showing $-k_1 \in K'$. As $(0)\theta = 0 \in H$ we see $0 \in K'$ and so K' is an additive subgroup of $\mathbb{Z}^t$, i.e. K' is a submodule of $\mathbb{Z}^t$. So K' is free with $\mathbb{Z}$-basis $z_1, z_2, \ldots, z_s$, $s \leq t$ by Theorem 3.1. Let $h \in H$. There is $k \in K'$ with $(k)\theta = h$. There are integers $m_1, m_2, \ldots, m_s$ with $k = m_1 z_1 + m_2 z_2 + \cdots + m_s z_s$. As θ is $\mathbb{Z}$-linear we obtain

$$h = (k)\theta = (m_1 z_1 + m_2 z_2 + \cdots + m_s z_s)\theta$$

$$= m_1 (z_1)\theta + m_2 (z_2)\theta + \cdots + m_s (z_s)\theta$$

which shows that the s elements $(z_1)\theta, (z_2)\theta, \ldots, (z_s)\theta$ generate H. So H is finitely generated.

Solution 5(c): Consider $g = (g_1, g_2, \ldots, g_s) \in G_1 \oplus G_2 \oplus \cdots \oplus G_s$. Then

$$ng \in n(G_1 \oplus G_2 \oplus \cdots \oplus G_s)$$

and also

$$ng = (ng_1, ng_2, \ldots, ng_s) \in nG_1 \oplus nG_2 \oplus \cdots \oplus nG_s.$$

So the $\mathbb{Z}$-modules $n(G_1 \oplus G_2 \oplus \cdots \oplus G_s)$ and $nG_1 \oplus nG_2 \oplus \cdots \oplus nG_s$ are identical.

Consider $g = (g_1, g_2, \ldots, g_s) \in G_1 \oplus G_2 \oplus \cdots \oplus G_s$. Then

$$g \in (G_1 \oplus G_2 \oplus \cdots \oplus G_s)_{(n)} \quad \Leftrightarrow \quad ng = 0 \Leftrightarrow ng_i = 0 \text{ for } 1 \leq i \leq s$$

$$\Leftrightarrow \quad g_i \in (G_i)_{(n)} \text{ for } 1 \leq i \leq s.$$

So the $\mathbb{Z}$-modules $(G_1 \oplus G_2 \oplus \cdots \oplus G_s)_{(n)}$ and $(G_1)_{(n)} \oplus (G_2)_{(n)} \oplus \cdots \oplus (G_s)_{(n)}$ are equal.

Solution 5(d): Consider $m\overline{1}$ in $(\mathbb{Z}_d)_{(n)}$. Then $mn\overline{1} = \overline{0}$, the 0-element of $\mathbb{Z}_d$. As $\overline{1}$ has order d in the additive group $(\mathbb{Z}_d, +)$ we deduce $d | mn$ from the discussion preceding Theorem 2.5. Therefore $d/\gcd\{n, d\} | m(n/\gcd\{n, d\})$ and so $d/\gcd\{n, d\} | m$ as $d/\gcd\{n, d\}$ and $n/\gcd\{n, d\}$ are *coprime* integers. So $m\overline{1} = q(d/\gcd\{n, d\})\overline{1}$ where $q \in \mathbb{Z}$ showing that $(d/\gcd\{n, d\})\overline{1}$ generates the $\mathbb{Z}$-module $(\mathbb{Z}_d)_{(n)}$.

Solution 7(a): (i) Let e denote the 1-element of R. Then $a \equiv a$ for all $a \in R$ as $a = ae$ and $e \in U(R)$. Suppose $a \equiv b$; then $a = bu$ for some $u \in U(R)$; hence $b \equiv a$ as $u^{-1} \in U(R)$ and $b = au^{-1}$. Suppose $a \equiv b$ and $b \equiv c$ where $a, b, c \in R$; there are $u, v \in U(R)$ with $a = bu$, $b = cv$; hence $a \equiv c$ as $a = (cv)u = c(vu)$ and $vu \in U(R)$. So $\equiv$ is an equivalence relation on R.

EXERCISES 3.2

Solution 5(b): Let $g, g' \in G$. By Theorem 3.10 there are unique elements $g_j, g'_j \in G_{p_j}$ $(1 \leq j \leq k)$ with $g = g_1 + g_2 + \cdots + g_k$ and $g' = g'_1 + g'_2 + \cdots + g'_k$. Then

$$(g + g')\alpha = \left(\sum_{j=1}^{k} (g_j + g'_j) \right) \alpha = \sum_{j=1}^{k} (g_j + g'_j)\alpha_j = \sum_{j=1}^{k} ((g_j)\alpha_j + (g'_j)\alpha_j)$$

$$= \sum_{j=1}^{k} (g_j)\alpha_j + \sum_{j=1}^{k} (g'_j)\alpha_j = (g)\alpha + (g')\alpha$$

showing that $\alpha : G \to G$ is a homomorphism. As $\alpha^{-1} = \alpha_1^{-1} \oplus \alpha_2^{-1} \oplus \cdots \oplus \alpha_k^{-1}$ we see that α is bijective and so $\alpha \in \text{Aut } G$. Consider $\beta \in \text{Aut } G$. As $(G_{p_j})\beta = G_{p_j}$ the mapping $\beta_j : G_{p_j} \to G_{p_j}$ defined by $(g_j)\beta_j = (g_j)\beta$ for all $g_j \in G_{p_j}$ is an automorphism of G_{p_j} for $1 \leq j \leq k$. Then

$$(g)\beta = \left(\sum_{j=1}^{k} g_j \right) \beta = \sum_{j=1}^{k} (g_j)\beta = \sum_{j=1}^{k} (g_j)\beta_j$$

$$= (g)(\beta_1 \oplus \beta_2 \oplus \cdots \oplus \beta_k)$$

for all $g \in G$ showing $\beta = \beta_1 \oplus \beta_2 \oplus \cdots \oplus \beta_k$. As β_j is the restriction of β to G_{p_j} for $1 \leq j \leq k$ we see that the β_j are uniquely determined by β. Hence the mapping $\text{Aut } G \to \text{Aut } G_{p_1} \times \text{Aut } G_{p_2} \times \cdots \times \text{Aut } G_{p_k}$, defined by $\beta \to (\beta_1, \beta_2, \ldots, \beta_k)$ for all $\beta \in \text{Aut } G$, is bijective. Let $\beta, \beta' \in \text{Aut } G$. There are $\beta'_j \in \text{Aut } G_{p_j}$ $(1 \leq j \leq k)$ with $\beta' = \beta'_1 \oplus \beta'_2 \oplus \cdots \oplus \beta'_k$. Then

$$(g)\beta\beta' = ((g)\beta)\beta' = \left(\left(\sum_{j=1}^{k} g_j \right) \beta \right) \beta' = \left(\sum_{j=1}^{k} (g_j)\beta_j \right) \beta'$$

$$= \sum_{j=1}^{k} ((g_j)\beta_j)\beta'_j = \sum_{j=1}^{k} (g_j)\beta_j\beta'_j$$

showing that the correspondence $\beta \leftrightarrow (\beta_1, \beta_2, \ldots, \beta_k)$ is a group homomorphism. So $\text{Aut } G \cong \text{Aut } G_{p_1} \times \text{Aut } G_{p_2} \times \cdots \times \text{Aut } G_{p_k}$ is a group isomorphism.

Solution 6(a): Suppose H is indecomposable. Let H have t' invariant factors. Then $t' = 1$ by Definition 3.8 as otherwise $H = H_1 \oplus (H_2 \oplus \cdots \oplus H_{t'})$ with both H_1 and $H_2 \oplus \cdots \oplus H_{t'}$ non-trivial. So H is cyclic of isomorphism type C_d where $d \neq 1$. Either $d = 0$ or $d \geq 2$. In the latter case $|H| = d$ is divisible by just one prime, as otherwise the primary decomposition Theorem 3.10 of H would be a non-trivial decomposition contradicting the fact that H is indecomposable. So $d = p^n$ where p is prime. Conversely suppose H has isomorphism type C_d where either $d = 0$ or $d = p^n$ where p is prime. In the case $d = 0$ we see $H \cong \mathbb{Z}$, that is, H is isomorphic to the additive group $\mathbb{Z}$ of integers; but $\mathbb{Z}$ (and hence H) is indecomposable by Exercises 2.2, Question 6(b). By the discussion following Corollary 3.12, cyclic groups of prime power order $d = p^n$ are indecomposable.

Solution 6(b): The submodule H_i of the f.g. $\mathbb{Z}$-module G is itself f.g. by Exercises 3.1, Question 5(b) for $1 \le i \le m$. Let r_i be the torsion-free rank of H_i and let the torsion subgroup T_i of H_i have l_i elementary divisors. Then $l_i + r_i \ge 1$ and, using Theorem 3.4, Corollary 3.5 and Theorem 3.10 we see H_i is the direct sum of $l_i + r_i$ non-trivial indecomposable submodules for $1 \le i \le m$. Substituting for each H_i we obtain a decomposition $G = H_1' \oplus H_2' \oplus \cdots \oplus H_{m'}'$ where $m \le m'$ and each $H_{i'}'$ is indecomposable for $1 \le i' \le m'$. By (a) above each $H_{i'}'$ has isomorphism type C_0 or C_{p^n}. The number of $H_{i'}'$ having isomorphism type C_0 is r the torsion-free rank of G. Now $H_{i'}'$ has isomorphism type C_{p^n} if and only if p^n is an elementary divisor of the torsion subgroup T of G; so the number of such $H_{i'}'$ is l. Therefore $m' = l + r$ and so $m \le l + r$. From above we see $l_1 + l_2 + \cdots + l_m = l$ as $T_1 \oplus T_2 \oplus \cdots \oplus T_m = T$. Also $r_1 + r_2 + \cdots + r_m = r$ on comparing torsion-free ranks in $H_1 \oplus H_2 \oplus \cdots \oplus H_m = G$.

Suppose $m = l + r$. Then $m = m'$ which means $l_i + r_i = 1$ for $1 \le i \le m$. So either $r_i = 1$ or $l_i = 1$. In the first case H_i is indecomposable being of isomorphism type C_0. In the second case H_i is indecomposable being of isomorphism type C_{p^n}. So $m = l + r$ implies that H_i is indecomposable for $1 \le i \le m$. Conversely suppose H_i is indecomposable for $1 \le i \le m$. Then $H_i' = H_i$ for $1 \le i \le m$ and so $m = m' = l + r$.

EXERCISES 3.3

Solution 1(a): For g_1, g_2 in G by Definition 2.3 we have

$$(g_1 + g_2)(\alpha + \alpha') = (g_1 + g_2)\alpha + (g_1 + g_2)\alpha'$$

$$= (g_1)\alpha + (g_2)\alpha + (g_1)\alpha' + (g_2)\alpha'$$

$$= (g_1)\alpha + (g_1)\alpha' + (g_2)\alpha + (g_2)\alpha'$$

$$= (g_1)(\alpha + \alpha') + (g_2)(\alpha + \alpha')$$

showing that $\alpha + \alpha'$ is an endomorphism of G. We verify the axioms of an additive abelian group (Section 2.1). Consider $\alpha, \alpha', \alpha''$ in End G. Then $(\alpha + \alpha') + \alpha'' = \alpha + (\alpha' + \alpha'')$ as

$$(g)((\alpha + \alpha') + \alpha'') = ((g)\alpha + (g)\alpha') + (g)\alpha''$$

$$= (g)\alpha + ((g)\alpha' + (g)\alpha'') = (g)(\alpha + (\alpha' + \alpha''))$$

for all $g \in G$. The zero endomorphism 0 satisfies $0 + \alpha = \alpha$ as $(g)(0 + \alpha) = (g)0 + (g)\alpha = 0 + (g)\alpha = (g)\alpha$ for all $g \in G$. Write $-\alpha : G \to G$ where $(g)(-\alpha) = -(g)\alpha$ for all $g \in G$. Then $-\alpha \in$ End G as

$$(g_1 + g_2)(-\alpha) = -(g_1 + g_2)\alpha = -((g_1)\alpha + (g_2)\alpha)$$

$$= -(g_1)\alpha - (g_2)\alpha = (g_1)(-\alpha) + (g_2)(-\alpha)$$

for all $g_1, g_2 \in G$. Also $-\alpha + \alpha = 0$ since $(g)(-\alpha + \alpha) = (g)(-\alpha) + (g)\alpha = -(g)\alpha + (g)\alpha = 0 = (g)0$ for all $g \in G$. Finally $\alpha + \alpha' = \alpha' + \alpha$ as $(g)(\alpha + \alpha') = (g)\alpha + (g)\alpha' = (g)\alpha' + (g)\alpha = (g)(\alpha' + \alpha)$ for all $g \in G$. So $(\text{End } G, +)$ is an abelian group. As

$$(g)(\alpha(\alpha' + \alpha'')) = ((g)\alpha)(\alpha' + \alpha'') = ((g)\alpha)\alpha' + ((g)\alpha)\alpha''$$

$$= (g)\alpha\alpha' + (g)\alpha\alpha'' = (g)(\alpha\alpha' + \alpha\alpha'')$$

for all $g \in G$, the distributive law $\alpha(\alpha' + \alpha'') = \alpha\alpha' + \alpha\alpha''$ holds.

EXERCISES 4.1

Solution 2(a): Consider $f(x) = \sum a_i x^i$ and $g(x) = \sum b_i x^i$ in $F[x]$. Then

$$(f(x) + g(x))\varepsilon_a = \left(\sum (a_i + b_i)x^i\right)\varepsilon_a = \sum (a_i + b_i)a^i.$$

As a_i, b_i, a^i belong to the field F we obtain

$$\sum (a_i + b_i)a^i = \sum a_i a^i + \sum b_i a^i$$

from the commutative and distributive laws. As

$$\sum_{i \geq 0} a_i a^i + \sum_{i \geq 0} b_i a^i = f(a) + g(a) = (f(x))\varepsilon_a + (g(x))\varepsilon_a$$

we obtain

$$(f(x) + g(x))\varepsilon_a = (f(x))\varepsilon_a + (g(x))\varepsilon_a,$$

showing that ε_a is additive. Similarly

$$(f(x)g(x))\varepsilon_a = \left(\sum (a_0 b_i + a_1 b_{i-1} + \cdots + a_i b_0)x^i\right)\varepsilon_a$$

$$= \sum (a_0 b_i + a_1 b_{i-1} + \cdots + a_i b_0)a^i$$

which is the result of collecting together terms involving $a^j a^k$ where $j + k = i$ in the product

$$\left(\sum a_j a^j\right)\left(\sum b_k a^k\right) = f(a)g(a) = (f(x))\varepsilon_a(g(x))\varepsilon_a.$$

So $(f(x)g(x))\varepsilon_a = (f(x))\varepsilon_a(g(x))\varepsilon_a$ showing that ε_a is multiplicative. For all $c \in F$ we have $(c)\varepsilon_a = c$ showing that $\varepsilon_a : F[x] \to F$ is surjective and ε_a maps the 1-element of $F[x]$ to the 1-element of F. Therefore ε_a is a surjective ring homomorphism.

Solution 2(b): Suppose $h(x) \in \langle f(x), g(x) \rangle$. There are $a(x), b(x) \in F[x]$ with

$$h(x) = a(x)f(x) + b(x)g(x).$$

Hence

$$h(x) = (a(x)q(x) + b(x))g(x) + a(x)r(x) \in \langle g(x), r(x) \rangle$$

showing $\langle f(x), g(x) \rangle \subseteq \langle g(x), r(x) \rangle$. As $r(x) = f(x) - q(x)g(x)$ we obtain $\langle g(x), r(x) \rangle \subseteq \langle f(x), g(x) \rangle$ and so $\langle f(x), g(x) \rangle = \langle g(x), r(x) \rangle$. Comparing monic generators of these ideals of $F[x]$ gives $\gcd\{f(x), g(x)\} = \gcd\{g(x), r(x)\}$ by Theorem 4.4. Write $d(x) = \gcd\{d_1(x), d_2(x), \ldots, d_t(x)\}$, $d'(x) = \gcd\{d_1(x), \gcd\{d_2(x), \ldots, d_t(x)\}\}$. Then $d(x)|d_1(x)$ and $d(x)|d_i(x)$ for $2 \le i \le t$. So $d(x)|d_1(x)$ and $d(x)|\gcd\{d_2(x), \ldots, d_t(x)\}$ which combine to give $d(x)|d'(x)$. Conversely $d'(x)|d_1(x)$ and $d'(x)|\gcd\{d_2(x), \ldots, d_t(x)\}$ which combine to give $d'(x)|d_1(x)$ and $d'(x)|d_i(x)$ for $2 \le i \le t$. So $d'(x)|d(x)$. Therefore $d(x) = d'(x)$.

Solution 2(c): Consider (a_i) and (b_i) in $P(R)$. There are non-negative integers m and n with $a_i = 0$ for all $i > m$ and $b_i = 0$ for all $i > n$; the least such integers are denoted $\deg(a_i)$ and $\deg(b_i)$, the degrees of the non-zero sequences (a_i) and (b_i) respectively. Then $a_i + b_i = 0$ for $i > \max\{m, n\}$ and

$$\sum_{j+k=i} a_j b_k = a_0 b_i + a_1 b_{i-1} + \cdots + a_i b_0 = 0$$

for $i > m + n$ as each term in the sum is zero. So $(a_i) + (b_i), (a_i)(b_i) \in P(R)$. As the elements of R form an additive abelian group and addition of sequences is carried out entry-wise, the elements of $P(R)$ also form an additive abelian group: e.g. $(a_i) + (b_i) = (a_i + b_i) = (b_i + a_i) = (b_i) + (a_i)$ shows addition of sequences to be commutative. The zero sequence (0) having all entries zero is the 0-element of $P(R)$. Consider $(a_i), (b_i), (c_i) \in P(R)$. As the elements of R obey the ring laws we obtain

$$\sum_{l+k=t} \left(\sum_{i+j=l} a_i b_j \right) c_k = \sum_{i+j+k=t} (a_i b_j) c_k = \sum_{i+j+k=t} a_i (b_j c_k)$$

$$= \sum_{i+s=t} a_i \left(\sum_{j+k=s} b_j c_k \right)$$

showing $((a_i)(b_i))(c_i) = (a_i)((b_i)(c_i))$ as these sequences have the same entry t, i.e. multiplication in $P(R)$ is associative. Similarly $\sum_{j+k=i}(a_j + b_j)c_k = \sum_{j+k=i} a_j c_k + \sum_{j+k=i} b_j c_k$ shows that entry i in the sequence $((a_i) + (b_i))(c_i)$ is equal to entry i in the sequence $(a_i)(c_i) + (b_i)(c_i)$ for all $i \ge 0$. So $((a_i) + (b_i))(c_i) = (a_i)(c_i) + (b_i)(c_i)$ showing that the right distributive law holds in $P(R)$. In the same way the left distributive law $(a_i)((b_i) + (c_i)) =$

$(a_i)(b_i) + (a_i)(c_i)$ holds in $P(R)$. The sequence $e_0 = (1, 0, 0, \ldots, 0, \ldots)$ is the
1-element of $P(R)$ as $e_0(a_i) = (a_i) = (a_i)e_0$. So $P(R)$ is a ring. Let $a_0, b_0 \in R$.
Then $(a_0 + b_0)\iota' = (a_0 + b_0, 0, 0, \ldots) = (a_0, 0, 0, \ldots) + (b_0, 0, 0, \ldots) =$
$(a_0)\iota' + (b_0)\iota'$, $(a_0 b_0)\iota' = (a_0 b_0, 0, 0, \ldots) = (a_0, 0, 0, \ldots)(b_0, 0, 0, \ldots) =$
$(a_0)\iota'(b_0)\iota'$ and $(1)\iota' = e_0$ showing that $\iota' : R \to P(R)$ is a ring homomorphism.
As $(a_0)\iota' = (b_0)\iota'$, i.e. $(a_0, 0, 0, \ldots) = (b_0, 0, 0, \ldots)$ implies $a_0 = b_0$ we see that ι'
is injective. Hence $a_0 \to (a_0)\iota'$ is a ring isomorphism between R and $\operatorname{im} \iota' = R'$,
showing that R' is a subring of $P(R)$ (Exercises 2.3, Question 3(b)) and $R' \cong R$.
Let R be an integral domain, i.e. R is commutative, non-trivial and has no zero-
divisors. Consider $(a_i), (b_i) \in P(R)$. Then

$$\sum_{j+k=i} a_j b_k = \sum_{k+j=i} b_k a_j$$

showing $(a_i)(b_i) = (b_i)(a_i)$, i.e. the ring $P(R)$ is commutative. As R' is non-
trivial, being isomorphic to R, we see that $P(R)$ is also non-trivial, as it contains the subring R'. Suppose $(a_i) \neq (0)$, $(b_i) \neq (0)$. Then $\sum_{j+k=i} a_j b_k = 0$ for
$i > m + n$ where $m = \deg(a_i)$, $n = \deg(b_i)$ as each term in the sum is zero.
But $\sum_{j+k=m+n} a_j b_k = a_m b_n \neq 0$, only one term in the sum being non-zero
as R has no zero-divisors. So $(a_i)(b_i) \neq 0$, showing that $P(R)$ has no zero-
divisors and in fact $\deg(a_i)(b_i) = m + n = \deg(a_i) + \deg(b_i)$. So $P(R)$ is an
integral domain. Conversely $P(R)$ an integral domain implies that its subring R'
is an integral domain and hence R is an integral domain as $R \cong R'$. Using the
multiplication rule in $P(R)$ we obtain $(a_0)\iota' x = (a_0, 0, 0, \ldots)(0, 1, 0, 0, \ldots) =$
$(0, a_0, 0, 0, \ldots) = (0, 1, 0, 0, \ldots)(a_0, 0, 0, \ldots) = x(a_0)\iota'$ for all $a_0 \in R$. By convention $x^0 = e_0$ the 1-element of $P(R)$, and $x^1 = x = e_1$. Suppose $x^{i-1} = e_{i-1}$
for some $i > 1$. Then $x^i = x x^{i-1} = e_1 e_{i-1}$ and using the multiplication rule in
$P(R)$ we see $e_1 e_{i-1} = e_i$, showing $x^i = e_i$ and completing the induction. Hence
$(0, 0, \ldots, 0, a_i, 0, 0, \ldots) = (a_i, 0, 0, \ldots)e_i = (a_i)\iota' x^i$ and so $(a_0, a_1, \ldots, a_i, \ldots) =$
$\sum_{i \geq 0}(a_i)\iota' x^i$ which is a polynomial in the indeterminate x over R'.

Solution 3(c): First note that $b = 0$ is a zero of $x^{q^n} - x$. For $b \neq 0$ we see b belongs to the multiplicative group E^* of non-zero elements of the field E. As E
is an n-dimensional vector space over F we see that $|E| = |F|^n = q^n$ and so
$|E^*| = q^n - 1$. By the $|G|$-lemma, in multiplicative notation, $b^{q^n-1} = 1$ and so
$b^{q^n} - b = 0$ showing that b is a zero of $x^{q^n} - x$. The q^n elements of E are therefore the q^n zeros of $x^{q^n} - x$. So $x - b$ is a monic, irreducible over E, factor of
$x^{q^n} - x$ for each $b \in E$. Hence $x^{q^n} - x = \prod_{b \in E}(x - b)$ is the factorisation of
$x^{q^n} - x$ into monic irreducible polynomials over E. So $x^{q^n} - x$ splits over E into
distinct monic factors.
Write $d(x) = \gcd\{p(x), x^{q^n} - x\}$. As $p(x)$ and $x^{q^n} - x$ are polynomials over F,
$d(x)$ is also a polynomial over F. Either $d(x) = p(x)$ or $d(x) = 1$ as $p(x)$
is monic and irreducible over F and $d(x) | p(x)$. Let $c = \langle p(x) \rangle + x \in E$.

The discussion following Theorem 4.9 shows $p(c) = 0$ and $c^{q^n} - c = 0$ from above. Suppose $d(x) = 1$. By Corollary 4.6 there are $a_1(x), a_2(x) \in F[x]$ with $a_1(x)p(x) + a_2(x)(x^{q^n} - x) = 1$. Applying ε_c gives $0 = a_1(c)0 + a_2(c)0 = a_1(c)p(c) + a_2(c)(c^{q^n} - c) = 1$ which is a contradiction, showing $d(x) = p(x)$ and so $p(x)|(x^{q^n} - x)$. As $x^{q^n} - x$ splits over E into a product of distinct factors so also does its divisor $p(x)$. Substituting $p'(x)$ for $p(x)$ in the foregoing part of the question we see that $p'(x)|(x^{q^n} - x)$ and so there is $c' \in E$ with $p'(c') = 0$. By Theorem 4.4 $\varepsilon_{c'} : F[x] \to E$ has kernel $\langle p'(x) \rangle$. As $1, c', (c')^2, \ldots, (c')^{n-1}$ is a basis of the n-dimensional vector space E over F we see $\varepsilon_{c'}$ is surjective. Hence $\tilde{\varepsilon}_{c'} : F[x]/\langle p'(x) \rangle \cong E$ where $(\langle p'(x) \rangle + f(x))\tilde{\varepsilon}_{c'} = f(c')$ for all $f(x) \in F[x]$. Let $p(x)$ be irreducible over F and have zero c in an extension field E of F. As above $p(x)|(x^{q^n} - x)$ and so $p(x)$ has no squared factor of positive degree, i.e. it is impossible for an irreducible polynomial over a finite field to have a repeated zero in an extension field.

Solution 4(c): Consider $e = a + bc$, $e' = a' + b'c \in F(c)$. Then

$$(e + e')\theta = ((a + a') + (b + b')c)\theta$$

$$= a + a' - (b + b')a_1 - (b + b')c$$

$$= (a - ba_1 - bc) + (a' - b'a_1 - b'c)$$

$$= (a + bc)\theta + (a' + b'c)\theta = (e)\theta + (e')\theta$$

showing that θ respects addition. Now $p(c) = 0$ gives $c^2 = -a_0 - a_1c$. Therefore

$$(ee')\theta = ((a + bc)(a' + b'c))\theta$$

$$= (aa' - bb'a_0 + (ab' + ba' - bb'a_1)c)\theta$$

$$= aa' - bb'a_0 - (ab' + ba' - bb'a_1)a_1$$

$$- (ab' + ba' - bb'a_1)c.$$

But

$$(e)\theta(e')\theta = (a - ba_1 - bc)(a' - b'a_1 - b'c)$$

$$= (a - ba_1)(a' - b'a_1) - bb'a_0$$

$$- ((a - ba_1)b' + b(a' - b'a_1) + bb'a_1)c.$$

Comparison of the above expressions gives $(ee')\theta = (e)\theta(e')\theta$ showing that θ respects multiplication. As $(a)\theta = a$ for all $a \in F$ we see $(1)\theta = 1$ showing that θ respects the 1-element of F which is also the 1-element of $F(c)$. Finally

$$(e)\theta^2 = ((e)\theta)\theta = (a + bc)\theta = (a - ba_1 - bc)\theta$$

$$= a - ba_1 - (-b)a_1 - (-b)c = a + bc = e$$

for all $e \in F(c)$ and so θ is self-inverse. So θ is an automorphism of $F(c)$.

Solution 8(d): To show α is unambiguously defined (u.d.), consider $f_1(x), f_2(x) \in F[x]$ such that $\langle g(x)h(x) \rangle + f_1(x) = \langle g(x)h(x) \rangle + f_2(x)$. Then $f_1(x) - f_2(x) \in \langle g(x)h(x) \rangle$ and so $g(x)h(x)|(f_1(x) - f_2(x))$. Hence $g(x)|(f_1(x) - f_2(x))$ and $h(x)|(f_1(x) - f_2(x))$, i.e. $f_1(x) - f_2(x) \in \langle g(x) \rangle$ and $f_1(x) - f_2(x) \in \langle h(x) \rangle$. Therefore $\langle g(x) \rangle + f_1(x) = \langle g(x) \rangle + f_2(x)$ and $\langle h(x) \rangle + f_1(x) = \langle h(x) \rangle + f_2(x)$ showing that α is u.d. as we set out to prove. Consider elements $a, a' \in F[x]/\langle g(x)h(x) \rangle$. There are $f(x), f'(x) \in F[x]$ with $a = \langle g(x)h(x) \rangle + f(x)$ and $a' = \langle g(x)h(x) \rangle + f'(x)$. Then

$$(a + a')\alpha = (\langle g(x)h(x) \rangle + f(x) + f'(x))\alpha$$

$$= (\langle g(x) \rangle + f(x) + f'(x), \langle h(x) \rangle + f(x) + f'(x))$$

$$= (\langle g(x) \rangle + f(x), \langle h(x) \rangle + f(x))$$

$$+ (\langle g(x) \rangle + f'(x), \langle h(x) \rangle + f'(x))$$

$$= (a)\alpha + (a')\alpha$$

and

$$(aa')\alpha = (\langle g(x)h(x) \rangle + f(x)f'(x))\alpha$$

$$= (\langle g(x) \rangle + f(x)f'(x), \langle h(x) \rangle + f(x)f'(x))$$

$$= (\langle g(x) \rangle + f(x), \langle h(x) \rangle + f(x))$$

$$\cdot (\langle g(x) \rangle + f'(x), \langle h(x) \rangle + f'(x))$$

$$= (a)\alpha(a')\alpha$$

showing that α respects addition and multiplication. As $(\langle g(x)h(x) \rangle + 1)\alpha = (\langle g(x) \rangle + 1, \langle h(x) \rangle + 1)$ we see that α respects 1-elements. So α is a ring homomorphism. Suppose $\gcd\{g(x), h(x)\} = 1$. To show that α is injective suppose $(a)\alpha = (a')\alpha$. Using the above notation we obtain

$$(\langle g(x) \rangle + f(x), \langle h(x) \rangle + f(x)) = (\langle g(x) \rangle + f'(x), \langle h(x) \rangle + f'(x))$$

which gives $\langle g(x) \rangle + f(x) = \langle g(x) \rangle + f'(x)$ and $\langle h(x) \rangle + f(x) = \langle h(x) \rangle + f'(x)$. Therefore $f(x) - f'(x)$ is divisible by $g(x)$ and by $h(x)$. So $f(x) - f'(x)$ is divisible by $g(x)h(x)$ giving $a = \langle g(x)h(x) \rangle + f(x) = \langle g(x)h(x) \rangle + f'(x) = a'$. So α is injective. Consider a typical element $(\langle g(x) \rangle + s(x), \langle h(x) \rangle + t(x))$ of $(F[x]/\langle g(x) \rangle) \oplus (F[x]/\langle h(x) \rangle)$. By Corollary 4.6 there are $a(x), b(x) \in F[x]$ with $a(x)g(x) + b(x)h(x) = 1$. Let $r(x) = t(x)a(x)g(x) + s(x)b(x)h(x)$. Then

$$r(x) \equiv s(x) \pmod{g(x)} \quad \text{and} \quad r(x) \equiv t(x) \pmod{h(x)}$$

and so $(\langle g(x)\rangle + s(x), \langle h(x)\rangle + t(x)) = (\langle g(x)h(x)\rangle + r(x))\alpha$ showing α surjective. The conclusion is:

$$\alpha : F[x]/\langle g(x)h(x)\rangle \cong (F[x]/\langle g(x)\rangle) \oplus (F[x]/\langle h(x)\rangle).$$

EXERCISES 5.1

Solution 2(a): Write $xI - A = (b_{ij})$. The entries in $xI - A$ are $x - a_{ii} = b_{ii}$ of degree 1 over R and $-a_{ij} = b_{ij}$ for $i \neq j$ is a constant polynomial over R. So $(\text{sign } \pi)b_{1(1)\pi}b_{2(2)\pi} \cdots b_{t(t)\pi}$ has degree $< t - 1$ over R unless $(i)\pi = i$ for at least $t - 1$ integers $i \in \{1, 2, \ldots, t\}$. In this case $(i)\pi = i$ for all $i \in \{1, 2, \ldots, t\}$ as π is a permutation of $\{1, 2, \ldots, t\}$, i.e. $\pi = \iota$, the identity. Therefore the coefficient of x^{t-1} in $\chi_A(x)$ is the coefficient of x^{t-1} in $(\text{sign } \iota)b_{11}b_{22} \cdots b_{tt} = (x - a_{11})(x - a_{22}) \cdots (x - a_{tt})$ which is $-a_{11} - a_{22} - \cdots - a_{tt} = -\text{trace } A$. The constant term in $\chi_A(x)$ is $\chi_A(0) = (\chi_A(x))\varepsilon_0$ where $\varepsilon_0 : R[x] \to R$ is 'evaluation at 0'. As $\det(xI - A) \in R[x]$ we see $(\chi_A(x))\varepsilon_0 = (\det(xI - A))\varepsilon_0 = \det(0I - A) = \det(-A) = (-1)^t \det A$ since $-A$ is the result of changing the sign of each of the t rows of A.

Solution 2(b): Consider a permutation π of $\{1, \ldots, t_1, t_1 + 1, ., ., t_1 + t_2\}$ and suppose first there is $i \in \{t_1 + 1, \ldots, t_1 + t_2\}$ with $(i)\pi \notin \{t_1 + 1, \ldots, t_1 + t_2\}$. The $(i, (i)\pi)$-entry in A is zero and so $(\text{sign } \pi) \times a_{1(1)\pi}a_{2(2)\pi} \cdots a_{t(t)\pi} = 0$. Now suppose π satisfies $i \in \{t_1 + 1, \ldots, t_1 + t_2\} \Rightarrow (i)\pi \in \{t_1 + 1, \ldots, t_1 + t_2\}$ and so, as π is a permutation, $i \in \{1, \ldots, t_1\} \Rightarrow (i)\pi \in \{1, \ldots, t_1\}$. Let π_1 be the restriction of π to $\{1, \ldots t_1\}$; then π_1 is a permutation of $\{1, \ldots, t_1\}$. Let π_2 be the restriction of π to $\{t_1 + 1, \ldots, t_1 + t_2\}$; then π_2 is a permutation of $\{t_1 + 1, \ldots, t_1 + t_2\}$. Also $\text{sign } \pi = (\text{sign } \pi_1)(\text{sign } \pi_2)$. Conversely each pair of permutations π_1, π_2 as above arises from a unique permutation π. So $\det A = \sum_{\pi} (\text{sign } \pi)a_{1(1)\pi}a_{2(2)\pi} \cdots a_{t(t)\pi}$ where the summation is restricted to π with $i \in \{t_1 + 1, \ldots, t_1 + t_2\} \Rightarrow (i)\pi \in \{t_1 + 1, \ldots, t_1 + t_2\}$ all other terms being zero. Hence

$$\det A = \sum_{\pi_1, \pi_2} (\text{sign } \pi_1)$$

$$\cdot (\text{sign } \pi_2)a_{1(1)\pi_2} \cdots a_{t_1(t_1)\pi_1}a_{t_1+1(t_1+1)\pi_2} \cdots a_{t_2(t_2)\pi_2}$$

$$= \left(\sum_{\pi_1} (\text{sign } \pi_1)a_{1(1)\pi_1} \cdots a_{t_1(t_1)\pi_1}\right)$$

$$\cdot \left(\sum_{\pi_2} (\text{sign } \pi_2)a_{t_1+1(t_1+1)\pi_2} \cdots a_{t_2(t_2)\pi_2}\right)$$

$$= (\det A_1)(\det A_2).$$

Taking $B = 0$ we obtain $A = A_1 \oplus A_2$. So $\det(A_1 \oplus A_2) = (\det A_1)(\det A_2)$.

Solution 2(c): Let

$$X = \left(\begin{array}{c|c} 0 & I_{t_2} \\ \hline I_{t_1} & 0 \end{array}\right),$$

that is, X is a partitioned $(t_2+t_1) \times (t_1+t_2)$ matrix with entries the $t_1 \times t_1$ identity matrix I_{t_1} and the $t_2 \times t_2$ matrix I_{t_2} as indicated together with rectangular zero matrices. Then $\det X = (-1)^{t_1 t_2}$ as the operation of $t_1 t_2$ interchanges of consecutive rows changes X into I_t. So X is invertible over F. Also

$$X(A_1 \oplus A_2) = \left(\begin{array}{c|c} 0 & I_{t_2} \\ \hline I_{t_1} & 0 \end{array}\right)\left(\begin{array}{c|c} A_1 & 0 \\ \hline 0 & A_2 \end{array}\right) = \left(\begin{array}{c|c} 0 & A_2 \\ \hline A_1 & 0 \end{array}\right)$$

$$= \left(\begin{array}{c|c} A_2 & 0 \\ \hline 0 & A_1 \end{array}\right)\left(\begin{array}{c|c} 0 & I_{t_2} \\ \hline I_{t_1} & 0 \end{array}\right) = (A_2 \oplus A_1)X.$$

Therefore $X(A_1 \oplus A_2)X^{-1} = A_2 \oplus A_1$ showing $A_1 \oplus A_2 \sim A_2 \oplus A_1$.

Solution 2(d): Suppose $f(x) = a$, i.e. $f(x)$ is a constant polynomial over R. Then $f(A_1 \oplus A_2) = a(A_1 \oplus A_2) = (aA_1) \oplus (aA_2) = f(A_1) \oplus f(A_2)$. Let $\deg f(x) = t > 0$ and suppose the equation $g(A_1 \oplus A_2) = g(A_1) \oplus g(A_2)$ holds for all $g(x) \in R[x]$ with $\deg g(x) < t$. Then $f(x) = ax^t + g(x)$ and so

$$f(A_1 \oplus A_2) = a(A_1 \oplus A_2)^t + g(A_1 \oplus A_2)$$

$$= a(A_1^t \oplus A_2^t) + g(A_1) \oplus g(A_2)$$

as $\oplus$ respects matrix multiplication. As $\oplus$ respects matrix addition we obtain

$$f(A_1 \oplus A_2) = (aA_1^t) \oplus (aA_2^t) + g(A_1) \oplus g(A_2)$$

$$= (aA_1^t + g(A_1)) \oplus (aA_2^t + g(A_2)) = f(A_1) \oplus f(A_2)$$

which completes the induction and the proof in the case $s = 2$. Suppose $s > 2$ and inductively $f(A_1 \oplus A_2 \oplus \cdots \oplus A_{s-1}) = f(A_1) \oplus f(A_2) \oplus \cdots \oplus f(A_{s-1})$. Using the case $s = 2$ with A_1 and A_2 replaced by $A_1 \oplus A_2 \oplus \cdots \oplus A_{s-1}$ and A_s respectively the proof is finished by

$$f(A_1 \oplus A_2 \oplus \cdots \oplus A_{s-1} \oplus A_s)$$

$$= f(A_1 \oplus A_2 \oplus \cdots \oplus A_{s-1}) \oplus f(A_s)$$

$$= f(A_1) \oplus f(A_2) \oplus \cdots \oplus f(A_{s-1}) \oplus f(A_s).$$

Solution 2(e): Modify the solution of Exercises 4.1, Question 2(a).

Solution 3(d): Let $f_1(x), f_2(x) \in K$. Then $(v)f_1(\alpha) = 0$ and $(v)f_2(\alpha) = 0$. Adding gives $(v)(f_1(\alpha) + f_2(\alpha)) = (v)f_1(\alpha) + (v)f_2(\alpha) = 0 + 0 = 0$ which shows $f_1(x) + f_2(x) \in K$. Also $-f_1(x) \in K$ as $(v)(-f_1(\alpha)) = -(v)f_1(\alpha) = -0 = 0$.

As $0(x)v = 0$ the zero polynomial $0(x)$ belongs to K. Therefore K is a subgroup of the additive group of $F[x]$. For $f(x) \in F[x]$ we have $f(x)f_1(x) \in K$ as $(v)f(\alpha)f_1(\alpha) = (v)f_1(\alpha)f(\alpha) = (0)f(\alpha) = 0$. So K is an ideal of $F[x]$ by Definition 4.3. Suppose V is t-dimensional. Then $v, (v)\alpha, (v)\alpha^2, \ldots, (v)\alpha^t$ are $t+1$ vectors of V. These vectors are linearly dependent and so there are $a_0, a_1, \ldots, a_t \in F$, not all zero, with $a_0 v + a_1(v)\alpha + \cdots + a_t(v)\alpha^t = 0$, i.e. $(v)f_0(\alpha) = 0$ where $f_0(x) = a_0 + a_1 x + \cdots + a_t x^t \neq 0(x)$ and $f_0(x) \in K$. So K is a non-zero ideal of $F[x]$.

Solution 3(e): Suppose $\overline{f(x)} = \overline{f'(x)}$. Then $f(x) - f'(x) = q(x)m(x)$ for some $q(x) \in F[x]$. So $f(A) - f'(A) = q(A)m(A) = q(A) \times 0 = 0$, i.e. $f(A) = f'(A)$. Therefore $\overline{f(x)}v = vf(A) = vf'(A) = \overline{f'(x)}v$ showing that the product $\overline{f(x)}v$ is unambiguously defined. As $M' = F^t$ as sets and addition in M' is the usual vector addition, we see module laws 1, 2, 3 and 4 (stated before Definition 2.19) are obeyed by M'. For $f(x), f_1(x), f_2(x) \in F[x]$ and $v, v_1, v_2 \in M'$ we have $\overline{f(x)}(v_1 + v_2) = (v_1 + v_2)f(A) = v_1 f(A) + v_2 f(A) = \overline{f(x)}v_1 + \overline{f(x)}v_2$ and

$$(\overline{f_1(x)} + \overline{f_2(x)})v = (\overline{f_1(x) + f_2(x)})v = v(f_1(A) + f_2(A))$$

$$= vf_1(A) + vf_2(A) = \overline{f_1(x)}v + \overline{f_2(x)}v$$

showing that M' obeys module law 5. Also

$$(\overline{f_1(x)f_2(x)})v = \overline{f_1(x)f_2(x)}v = vf_1(A)f_2(A)$$

$$= vf_2(A)f_1(A) = (\overline{f_2(x)}v)f_1(A) = \overline{f_1(x)}(\overline{f_2(x)}v)$$

showing that M' obeys module law 6. The 1-element of $F[x]/\langle m(x)\rangle$ is $\overline{1(x)}$ and $\overline{1(x)}v = v1(A) = vI = v$ for all $v \in M'$ showing that M' obeys module law 7. So M' is an $F[x]/\langle m(x)\rangle$-module. Suppose $m(x)$ irreducible over F and let $r = \deg m(x)$. Then $F' = F[x]/\langle m(x)\rangle$ is a field and so M' is a vector space over F'. Consider $v \in M'$. There are $f_j(x) \in F[x]$ for $1 \le j \le t'$ with $v = \sum_{j=1}^{t'} \overline{f_j(x)}v_j$ as $v_1, v_2, \ldots, v_{t'}$ span M' over F'. Further we may assume $\deg f_j(x) < r$. Write $f_j(x) = \sum_{i=0}^{r-1} a_{ji}x^i$. Substituting for each $f_j(x)$ gives $v = \sum_{i,j} a_{ji}\overline{x^i}v_j$ where the summation is for $0 \le i < r$, $1 \le j \le t'$. The rt' vectors $\overline{x^i}v_j$ span M' over F as $a_{ji} \in F$. Suppose $\sum_{i,j} a_{ji}\overline{x^i}v_j = 0$. As $v_1, v_2, \ldots, v_{t'}$ are linearly independent over F' we deduce $\overline{f_j(x)} = \overline{0(x)}$, i.e. $m(x)|f_j(x)$. So $f_j(x) = 0(x)$ for $1 \le j \le t'$. The rt' vectors $\overline{x^i}v_j$ are linearly independent and so form an F-basis of F^t. Hence $t/t' = r = \deg m(x)$.

Solution 5: Consider $(u)\beta = (a_1, a_2, \ldots, a_s) \in F^s$, $(u')\beta = (a_1', a_2', \ldots, a_s') \in F^s$ for $u, u' \in N$ which gives

$$u = a_1 u_1 + a_2 u_2 + \cdots + a_s u_s \quad \text{and} \quad u' = a_1' u_1 + a_2' u_2 + \cdots + a_s' u_s.$$

So $u + u' = (a_1 + a'_1)u_1 + (a_2 + a'_2)u_2 + \cdots + (a_s + a'_s)u_s$ which shows $(u + u')\beta = (a_1 + a'_1, a_2 + a'_2, \ldots, a_s + a'_s) = (u)\beta + (u')\beta$, i.e. β is additive. For $a \in F$ we have $au = aa_1u_1 + aa_2u_2 + \cdots + aa_su_s$ showing $(au)\beta = a((u)\beta)$ and so β is F-linear. Also β is bijective as $u_1, u_2, \ldots, u_s$ is an F-basis of N. So $\beta : N \cong F^s$ is a vector space isomorphism with $(u_i)\beta = e_i$ for $1 \leq i \leq s$. Write $B = (b_{ij})$ for $1 \leq i, j \leq s$. Then

$$(xu_i)\beta = ((u_i)\alpha)\beta = (b_{i1}u_1 + b_{i2}u_2 + \cdots + b_{is}u_s)\beta$$

$$= (b_{i1}, b_{i2}, \ldots, b_{is}) = e_i B = ((u_i)\beta)B = x((u_i)\beta)$$

for $1 \leq i \leq s$. Using the F-linearity of β we see $(xu)\beta = x((u)\beta)$ for all $u \in N$. By Lemma 5.15 with $\theta = \beta$ we conclude $\beta : N \cong M(B)$ is an isomorphism of $F[x]$-modules.

EXERCISES 5.2

Solution 1(d): Suppose $C(f(x)) \sim D$ where $D = \mathrm{diag}(\lambda_1, \lambda_2, \ldots, \lambda_t)$. Comparing characteristic polynomials using Lemma 5.5 and Theorem 5.26 gives $f(x) = (x - \lambda_1)(x - \lambda_2) \cdots (x - \lambda_t)$. Suppose $\lambda_i = \lambda_j$ for i, j with $1 \leq i < j \leq t$. Then $N = \langle e_i, e_j \rangle$ is a non-cyclic submodule of $M(D)$: for all $v \in N$ we have $xv = x(a_ie_i + a_je_j) = (a_ie_i + a_je_j)D = a_i\lambda_ie_i + a_j\lambda_je_j = \lambda_i(a_ie_i + a_je_j) = \lambda_iv$ showing that v does not generate the 2-dimensional subspace N. But $M(D)$ is cyclic being isomorphic to the cyclic $F[x]$-module $C(f(x))$ by Theorem 5.13. So N is cyclic by Theorem 5.28. This contradiction shows $\lambda_i \neq \lambda_j$. Therefore $f(x)$ has t distinct zeros in F. Conversely suppose $f(x)$ has t distinct zeros $\lambda_1, \lambda_2, \ldots, \lambda_t$ in F. So $\lambda_1, \lambda_2, \ldots, \lambda_t$ are t distinct eigenvalues of $C(f(x))$ by Theorem 5.26. Denote by v_i a row eigenvector of $C(f(x))$ corresponding to λ_i for $1 \leq i \leq t$. Let X be the $t \times t$ matrix with $e_i X = v_i$ for $1 \leq i \leq t$. Then X is invertible over F ($\lambda_1, \lambda_2, \ldots, \lambda_t$ distinct implies $v_1, v_2, \ldots, v_t$ linearly independent) and $XC(f(x))X^{-1} = \mathrm{diag}(\lambda_1, \lambda_2, \ldots, \lambda_t)$.

Solution 3(a): Write $d_0(x) = \gcd\{f(x), d(x)\}$. Then $(d(x)/d_0(x))f(x)v = (f(x)/d_0(x))d(x)v = (f(x)/d_0(x))0 = 0$ which shows that the monic polynomial $d(x)/d_0(x)$ belongs to the order ideal K of $f(x)v$ in M. So K is non-zero and so has a unique monic generator $g(x)$ by Theorem 4.4. By Definition 5.11 $g(x)$ is the order of $f(x)v$ in M and $g(x)|d(x)/d_0(x)$. But $g(x)f(x)v = 0$ shows $g(x)f(x)$ belongs to the order ideal $\langle d(x) \rangle$ of v in M. So $g(x)f(x) = h(x)d(x)$ where $h(x) \in F[x]$. Hence $g(x)(f(x)/d_0(x)) = h(x)(d(x)/d_0(x))$. As $\gcd\{f(x)/d_0(x), d(x)/d_0(x)\} = 1$ we conclude $d(x)/d_0(x)|g(x)$. Therefore

$$g(x) = d(x)/d_0(x),$$

i.e. $f(x)v$ has order $d(x)/\gcd\{f(x), d(x)\}$ in M.

Solution 4(a): Consider $f_1(x)$, $f_2(x) \in K_N$. Then $f_1(x)v_0 \in N$ and $f_2(x)v_0 \in N$. As N is closed under addition we see $(f_1(x) + f_2(x))v_0 = f_1(x)v_0 + f_2(x)v_0 \in N$, i.e. $f_1(x) + f_2(x) \in K_N$ showing K_N to be closed under addition. As $0(x)v_0 = 0 \in N$ we see $0(x) \in K_N$, i.e. K_N contains the zero polynomial. As N is closed under negation we have $(-f_1(x))v_0 = -f_1(x)v_0 \in N$ showing that $-f_1(x) \in K_N$, i.e. K_N is closed under negation. As N is closed under polynomial multiplication we obtain $(g(x)f_1(x))v_0 = g(x)(f_1(x)v_0) \in N$, i.e. $g(x)f_1(x) \in K_N$ showing K_N to be closed under polynomial multiplication. So K_N is an ideal Definition 4.3 of $F[x]$.

As $d_0(x)v_0 = 0 \in N$ we see $d_0(x) \in K_N$ and so $\langle d_0(x) \rangle \subseteq K_N$. Suppose $K_N = \langle d_0(x) \rangle$ and $v \in \langle v_0 \rangle \cap N$. Then $v = f(x)v_0 \in N$ for some $f(x) \in F[x]$. Therefore $f(x) \in K_N$ and so $d_0(x)|f(x)$. Hence $f(x)v_0 = 0$, i.e. $\langle v_0 \rangle \cap N = 0$. Conversely suppose $\langle v_0 \rangle \cap N = 0$ and consider $f(x) \in K_N$. So $f(x)v_0 \in N$. As $f(x)v_0 \in \langle v_0 \rangle$ we see $f(x)v_0 = 0$, i.e. $f(x) \in \langle d_0(x) \rangle$ and so $K_N = \langle d_0(x) \rangle$. Suppose $N_1 \subseteq N_2$ and consider $f(x) \in K_{N_1}$. Then $f(x)v_0 \in N_1$ and so $f(x)v_0 \in N_2$. Therefore $f(x) \in K_{N_2}$ showing $K_{N_1} \subseteq K_{N_2}$. Suppose $M = \langle v_0 \rangle$ and $K_{N_1} \subseteq K_{N_2}$. Let $v \in N_1$. There is $f(x) \in F[x]$ with $v = f(x)v_0$. Therefore $f(x) \in K_{N_1}$ and so $f(x) \in K_{N_2}$ also. This means $f(x)v_0 \in N_2$, i.e. $v \in N_2$. So $N_1 \subseteq N_2$.

Solution 6(d): The formula $\det T = (-1)^{t(t-1)/2}$ holds for $t = 1$. Take $t \geq 2$ and suppose inductively that the $t \times t$ matrix

$$T = \left(\begin{array}{c|c} 0 & 1 \\ \hline T' & 0 \end{array} \right)$$

is such that $\det T' = (-1)^{(t-1)(t-2)/2}$. The inductive step is completed by $\det T = (-1)^{t-1} \det T' = (-1)^{t-1}(-1)^{(t-1)(t-2)/2} = (-1)^{t-1+(t-1)(t-2)/2} = (-1)^{t(t-1)/2}$.

EXERCISES 6.1

Solution 4(b):

$$R_f C(f(x)) = \begin{pmatrix} a_1 & a_2 & a_3 & 1 \\ a_2 & a_3 & 1 & 0 \\ a_3 & 1 & 0 & 0 \\ 1 & 0 & 0 & 0 \end{pmatrix} \begin{pmatrix} 0 & 1 & 0 & 0 \\ 0 & 0 & 1 & 0 \\ 0 & 0 & 0 & 1 \\ -a_0 & -a_1 & -a_2 & -a_3 \end{pmatrix}$$

$$= \begin{pmatrix} -a_0 & 0 & 0 & 0 \\ 0 & a_2 & a_3 & 1 \\ 0 & a_3 & 1 & 0 \\ 0 & 1 & 0 & 0 \end{pmatrix}$$

which is symmetric. As R_f is invertible and symmetric we obtain

$$R_f C(f(x)) = (R_f C(f(x)))^T = C(f(x))^T R_f^T = C(f(x))^T R_f$$

and so $R_f C(f(x)) R_f^{-1} = C(f(x))^T$.

Let $f(x) = a_0 + a_1 x + a_2 x^2 + \cdots + a_{t-1} x^{t-1} + a_t x^t$ where $a_t = 1$. Let R_f denote the $t \times t$ matrix over F with (i, j)-entry a_{i+j-1} for $1 \le i + j - 1 \le t$ and (i, j)-entry 0 for $i + j - 1 > t$. Write $g(x) = (f(x) - a_0)/x = a_1 + a_2 x + \cdots + a_{t-1} x^{t-2} + a_t x^{t-1}$. Then

$$R_f C(f(x)) = \left(\begin{array}{c|c} a & 1 \\ \hline R_g & 0^T \end{array} \right) \left(\begin{array}{c|c} 0 & I \\ \hline -a_0 & -a \end{array} \right)$$

where $a = (a_1, a_2, \ldots, a_{t-1})$, 0 is the $1 \times (t - 1)$ zero matrix and I is the $(t - 1) \times (t - 1)$ identity matrix. Therefore

$$R_f C(f(x)) = \left(\begin{array}{c|c} -a_0 & 0 \\ \hline 0^T & R_g \end{array} \right)$$

on multiplying out the indicated partitioned matrices. So $R_f C(f(x))$ is symmetric as R_g is symmetric. As $\det R_f = (-1)^{t(t-1)/2}$ (use induction on t here) we see that R_f is invertible and so $R_f C(f(x)) R_f^{-1} = C(f(x))^T$ as before.

Solution 6(a): Suppose $\mu_A(x) = \mu_B(x) = x - c$ for some $c \in F$. Then $\mu_A(A) = 0$ gives $A - cI = 0$, i.e. $A = cI$. In the same way $\mu_B(B) = 0$ gives $B = cI$ and so $A \sim B$ as $A = B$ in this case. Suppose $\mu_A(x) = \mu_B(x) \ne x - c$ for any $c \in F$. As $\mu_A(x) | \chi_A(x)$ and $\deg \chi_A(x) = 2$ we see $\mu_A(x) = \chi_A(x)$. By Corollary 6.10 there is v_0 having order $\mu_A(x)$ in $M(A)$. So $M(A) = \langle v_0 \rangle$ and $A \sim C(\mu_A(x))$ by Corollary 5.27. For the same reason $B \sim C(\mu_B(x))$ and so $A \sim B$ as $C(\mu_A(x)) = C(\mu_B(x))$.

Solution 8(b): Consider the t-tuples $v(x) = (f_1(x), f_2(x), \ldots, f_t(x)) \in F[x]^t$ and $v'(x) = (f_1'(x), f_2'(x), \ldots, f_t'(x)) \in F[x]^t$ and let $f(x) \in F[x]$. Then θ_A is additive as

$$(v(x) + v'(x))\theta_A = \sum_{i=1}^{t} (f_i(x) + f_i'(x)) e_i$$

$$= \sum_{i=1}^{t} f_i(x) e_i + \sum_{i=1}^{t} f_i'(x) e_i = (v(x))\theta_A + (v'(x))\theta_A$$

and

$$(f(x) v(x))\theta_A = \sum_{i=1}^{t} (f(x) f_i(x)) e_i = f(x) \left(\sum_{i=1}^{t} f_i(x) e_i \right)$$

$$= f(x)((v(x))\theta_A)$$

using the module laws which hold in $M(A)$ by Lemma 5.7 and Definition 5.8. So θ_A is $F[x]$-linear.

Solution 8(c): Take $d(x) = d_1(x)$ and consider the isomorphism $\alpha| : M_{(d_1)} \cong M'_{(d_1)}$. Then $(F[x]/\langle d_i(x)\rangle)_{(d_1(x))} \cong F[x]/\langle d_1(x)\rangle$ since $\gcd\{d_1(x), d_i(x)\} = d_1(x)$ for $1 \leq i \leq s$. From $M \cong F[x]/\langle d_1(x)\rangle \oplus F[x]/\langle d_2(x)\rangle \oplus \cdots \oplus F[x]/\langle d_s(x)\rangle$ we deduce

$$M_{(d_1(x))} \cong (F[x]/\langle d_1(x)\rangle)_{(d_1(x))} \oplus (F[x]/\langle d_2(x)\rangle)_{(d_1(x))} \oplus \cdots$$

$$\oplus (F[x]/\langle d_s(x)\rangle)_{(d_1(x))}$$

$$\cong F[x]/\langle d_1(x)\rangle \oplus F[x]/\langle d_1(x)\rangle \oplus \cdots \oplus F[x]/\langle d_1(x)\rangle$$

$$= (F[x]/\langle d_1(x)\rangle)^s.$$

The $F[x]$-module $M_{(d_1(x))}$ is therefore isomorphic to the free $Fx]/\langle d_1(x)\rangle$-module $(F[x]/\langle d_1(x)\rangle)^s$ of rank s. By Lemma 2.25 both $M_{(d_1(x))}$ and $M'_{(d_1(x))}$ are free $F[x]$-modules of rank s. Combining

$$(F[x]/\langle d'_i(x)\rangle)_{(d_1(x))} \cong F[x]/\langle \gcd\{d_1(x), d'_i(x)\}\rangle \text{ for } 1 \leq i \leq s'$$

and

$$M' \cong F[x]/\langle d'_1(x)\rangle \oplus F[x]/\langle d'_2(x)\rangle \oplus \cdots \oplus F[x]/\langle d'_{s'}(x)\rangle$$

gives $M'_{(d_1(x))} \cong \oplus \sum_{i=1}^{s'} F[x]/\langle \gcd\{d_1(x), d'_i(x)\}\rangle$ showing that $M'_{(d_1(x))}$ is the direct sum of s' cyclic submodules and so is generated by s' of its elements. From Theorem 2.20 we deduce $s' \geq s$. As $\alpha^{-1} : M' \cong M$ the preceding theory 'works' with M and M' interchanged. Using $\alpha^{-1}| : M'_{(d'_1(x))} \cong M_{(d'_1(x))}$ the $F[x]/\langle d'_1(x)\rangle$-module $M'_{(d'_1(x))}$ is isomorphic to the free $F[x]/\langle d'_1(x)\rangle$-module $(F[x]/\langle d'_1(x)\rangle)^{s'}$ of rank s'. Hence $M_{(d'_1(x))} \cong \oplus \sum_{i=1}^{s} F[x]/\langle \gcd\{d'_1(x), d_i(x)\}\rangle$ is a free $F[x]/\langle d'_1(x)\rangle$-module of rank s' and is generated by s of its elements. Therefore $s \geq s'$ by Theorem 2.20 and so $s = s'$. From Lemma 2.18 and Corollary 2.21 the s generators of $M'_{(d_1(x))}$ form an $F[x]/\langle d_1(x)\rangle$-basis of the $F[x]/\langle d_1(x)\rangle$-module $M'_{(d_1(x))}$ (Exercises 2.3, Question 7(b)). Therefore each of these s generators has order 0 in the $F[x]/\langle d_1(x)\rangle$-module $M'_{(d_1(x))}$ and order $d_1(x)$ in the $F[x]$-module $M'_{(d_1(x))}$. From the first of these generators we deduce $\gcd\{d_1(x), d'_1(x)\} = d_1(x)$ showing $d_1(x)|d'_1(x)$. Interchanging the roles of M and M' gives $d'_1(x)|d_1(x)$ and so $d_1(x) = d'_1(x)$. Let m_1 denote the number of i with $d_i(x) = d_1(x)$ and let m'_1 denote the number of i with $d'_i(x) = d_1(x)$. As $d_1(x)(F[x]/\langle d_i(x)\rangle) \cong F[x]/\langle d_i(x)/\gcd\{d_1(x), d_i(x)\}\rangle = F[x]/\langle d_i(x)/d_1(x)\rangle$ we obtain $d_1(x)M \cong \sum_{m_1 < i \leq s} \oplus F[x]/\langle d_i(x)/d_1(x)\rangle$. So $d_1(x)M$ is the direct sum of $s - m_1$ non-trivial cyclic submodules. As $d_i(x)/d_1(x)|d_j(x)/d_1(x)$ for $m_1 < i \leq j \leq s$ this decomposition of $d_1(x)M$ is again as in Theorem 6.6. In the same way $d_1(x)M' \cong \sum_{m'_1 < i \leq s} \oplus F[x]/\langle d'_i(x)/d_1(x)\rangle$ which is a decomposition of $d_1(x)M'$ into $s' - m'_1$ non-trivial cyclic submodules as in Theorem 6.6. As

$\alpha| : d_1(x)M \cong d_1(x)M'$ the proof can be completed by induction on the number, r say, of different polynomials among $d_1(x), d_2(x), \ldots, d_s(x)$. Take $r = 1$. Then $m_1 = s$ and $d_1(x)M$ is trivial. So $d_1(x)M'$ is also trivial and $m_1' = s$. Therefore $d_i(x) = d_1(x) = d_i'(x)$ for $1 \le i \le s$. Now take $r > 1$. There are $r - 1$ different polynomials among $d_{m_1+1}(x)/d_1(x), d_{m_1+2}(x)/d_1(x), \ldots, d_s(x)/d_1(x)$ and so the conclusion of Theorem 6.6 holds on replacing $\alpha : M \cong M'$ by $\alpha| : d_1(x)M \cong d_1(x)M'$, i.e. $s - m_1 = s - m_1'$ (showing $m_1 = m_1'$) and also $d_i(x)/d_1(x) = d_i'(x)/d_1(x)$ for $m_1 < i \le s$. Therefore $s = s'$, $d_i(x) = d_1(x) = d_i'(x)$ for $1 \le i \le m_1 = m_1'$ and $d_i(x) = d_i'(x)$ for $m_1 < i \le s'$ on multiplying by $d_1(x)$. The induction is now complete.

EXERCISES 6.2

Solution 3(a): Write $N = M(A)_{p(x)}$. Consider $u, v \in N$. Then $p(x)^n u = 0$ and $p(x)^n v = 0$. Using the distributive law in $M(A)$ we have

$$p(x)^n (u + v) = p(x)^n u + p(x)^n v = 0 + 0 = 0$$

showing $u + v \in N$. Also $-v \in N$ as

$$p(x)^n (-v) = -p(x)^n v = -0 = 0.$$

The zero row vector 0 in F^t belongs to N as $p(x)^n 0 = 0$ and so N is a subgroup of the additive group $(F^t, +)$. For

$$f(x) \in F[x] p(x)^n (f(x)v) = (p(x)^n f(x))v = (f(x)p(x)^n)v = f(x)(p(x)^n v)$$

$$= f(x)0 = 0$$

showing that $f(x)v \in N$. By Definition 2.26 we conclude that N is a submodule of $M(A)$.

Suppose $N = \{0\}$ and look for a contradiction. By the Cayley–Hamilton theorem $p(x)^n (q(x)v) = (p(x)^n q(x))v = (\chi_A(x)v) = v\chi_A(A) = v0 = 0$ for all $v \in M(A)$. So $q(x)v \in N$, i.e. $q(x)v = 0$, for all $v \in M(A)$. Therefore $e_i q(A) = q(x)e_i = 0$ for $1 \le i \le t$ showing that all the rows of the $t \times t$ matrix $q(A)$ are zero. So $q(A) = 0$ which means $\mu_A(x)|q(x)$ by Corollary 6.10. But $p(x)|\mu_A(x)$ as $\chi_A(x)$ and $\mu_A(x)$ have equal irreducible factors by Corollary 6.11. So $p(x)|q(x)$ contrary to $\gcd\{p(x), q(x)\} = 1$. So $N \ne \{0\}$, i.e. the primary components of $M(A)$ are non-trivial.

Solution 3(b): Suppose $M = M(C(p(x)^n))$ has submodules N_1 and N_2 such that $M = N_1 \oplus N_2$. Then M is cyclic with generator e_1, the first element of the standard basis of F^t where $t = \deg p(x)^n$, by Theorem 5.26. The monic divisors of $p(x)^n$ are the $n + 1$ polynomials $p(x)^m$ for $0 \le m \le n$. By Theorem 5.28 the submodules N_1 and N_2 are also cyclic with generators $p(x)^{m_1} e_1$ and $p(x)^{m_2} e_1$

respectively and we may assume $0 \leq m_1 \leq m_2 \leq n$. So $N_2 \subseteq N_1$ as the generator $p(x)^{m_2} e_1$ of N_2 is a polynomial multiple $p(x)^{m_2-m_1}$ of the generator $p(x)^{m_1} e_1$ of N_1. As $M = N_1 \oplus N_2$ we know N_1 and N_2 are independent, i.e. $(N_1, +)$ and $(N_2, +)$ are independent subgroups (Definition 2.14) of $(M, +)$. So $N_1 \cap N_2 = \{0\}$. But $N_1 \cap N_2 = N_2$ as $N_2 \subseteq N_1$ and so $N_2 = \{0\}$. As M is non-zero we see that M is indecomposable. Let N be an indecomposable submodule of $M(A)$. As $N \neq \{0\}$ by Exercises 5.1, Question 5 there is an $s \times s$ matrix B over F with $N \cong M(B)$. So $M(B)$ is indecomposable. Should $\chi_B(x)$ be divisible by two or more irreducible polynomials over F the primary decomposition Theorem 6.12 of $M(B)$ would contradict its indecomposability by (a) above. So $\chi_B(x) = p(x)^n$ where $p(x)$ is a monic irreducible polynomial over F and n is a positive integer. Also $M(B)$ has only one invariant factor, since otherwise the invariant factor decomposition Theorem 6.5 of $M(B)$ would contradict its indecomposability. So $M(B)$ is cyclic. By Corollary 5.27 we conclude $N \cong M(C(p(x)^n))$.

Solution 3(c): As $\dim N_j \geq 1$ for $1 \leq j \leq r$, on comparing dimensions of subspaces of

$$F^t t = \dim F^t = \dim M(A) = \dim(N_1 \oplus N_2 \oplus \cdots \oplus N_r)$$

$$= \dim N_1 + \dim N_2 + \cdots + N_r \geq r$$

showing $r \leq t$. Consider a decomposition $M(A) = N_1 \oplus N_2 \oplus \cdots \oplus N_r$ into a direct sum of r non-zero submodules N_j with r as large as possible (as r is bounded above by t there is such an r). Suppose N_1 is decomposable and so $N_1 = N_1' \oplus N_1''$ where N_1' and N_1'' are non-zero. Then $M(A) = N_1' \oplus N_1'' \oplus N_2 \oplus \cdots \oplus N_r$ is a decomposition with $r + 1$ summands (terms) contrary to the choice of r. So N_1 is indecomposable and in the same way each N_j is indecomposable for $1 \leq j \leq r$. Therefore $N_j \cong M(C(p_j(x)^{n_j}))$ where $p_j(x)$ is monic and irreducible over F and n_j is a positive integer by (b) above. From Corollary 5.20 we conclude $A \sim C(p_1(x)^{n_1}) \oplus C(p_2(x)^{n_2}) \oplus \cdots \oplus C(p_r(x)^{n_r})$ showing that the polynomials $p_j(x)^{n_j}$ for $1 \leq j \leq r$ are the elementary divisors of A. So r is the number of elementary divisors of A.

Solution 3(d): Write $m_j(x) = \chi_A(x)/p_j(x)^{n_j} = \prod_{i=1, i \neq j}^{k} p_i(x)^{n_i}$ which is a monic polynomial of degree $t - n_j \deg p_j(x)$ over F. Then

$$\mathrm{lcm}\{m_1(x), m_2(x), \ldots, m_k(x)\} = \chi_A(x)$$

and (more to the point) $\gcd\{m_1(x), m_2(x), \ldots, m_k(x)\} = 1$ as $p_j(x)$ is not a divisor of $m_j(x)$ for $1 \leq j \leq k$, i.e. the polynomials $m_1(x), m_2(x), \ldots, m_k(x)$ have no common irreducible divisor. By Corollary 4.6 there are $a_j(x) \in F[x]$ for $1 \leq j \leq k$ such that $a_1(x)m_1(x) + a_2(x)m_2(x) + \cdots + a_k(x)m_k(x) = 1$. We are now ready to prove Theorem 6.12.

Consider $v \in M(A)$. Then

$$v = 1v = \left(\sum_{j=1}^{k} a_j(x)m_j(x)\right)v = \sum_{j=1}^{k} a_j(x)m_j(x)v = \sum_{j=1}^{k} v_j$$

where $v_j = a_j(x)m_j(x)v$ for $1 \le j \le k$. Now $v_j \in M(A)_{p_j(x)}$ as

$$p_j(x)^{n_j}v_j = p_j(x)^{n_j}a_j(x)m_j(x)v_j = a_j(x)\chi_A(x)v$$

$$= a_j(x)(v\chi_A(A)) = a_j(x)0 = 0$$

for $1 \le j \le k$. Therefore

$$M(A) = M(A)_{p_1(x)} + M(A)_{p_2(x)} + \cdots + M(A)_{p_k(x)}.$$

Suppose $v_1 + v_2 + \cdots + v_k = 0$ where $v_j \in M(A)_{p_j(x)}$ for $1 \le j \le k$. We concentrate on one particular term v_j. For $i \ne j$, $1 \le i \le k$ we have $m_j(x)v_i = 0$ as $p_i(x)^{n_i}|m_j(x)$ and $p_i(x)^{n_i}v_i = 0$. Inserting $k - 1$ zero terms $m_j(x)v_i = 0$ produces

$$m_j(x)v_j = m_j(x)v_j + \sum_{i=1,i\ne j}^{k} m_j(x)v_i = \sum_{i=1}^{k} m_j(x)v_i$$

$$= m_j(x)\left(\sum_{i=1}^{k} v_i\right) = m_j(x)0 = 0.$$

The polynomial $1 - a_j(x)m_j(x) = \sum_{i=1,i\ne j}^{k} a_i(x)m_i(x)$ is divisible by $p_j(x)^{n_j}$ since $p_j(x)^{n_j}|m_i(x)$ for $i \ne j$, $1 \le i \le k$. So $a'_j(x) = (1 - a_j(x)m_j(x))/p_j(x)^{n_j}$ is a polynomial over F. Then

$$v_j = 1v_j = (a_j(x)m_j(x) + (1 - a_j(x)m_j(x)))v_j$$

$$= a_j(x)m_j(x)v_j + (1 - a_j(x)m_j(x))v_j$$

$$= a_j(x)m_j(x)v_j + a'_j(x)p_j(x)^{n_j}v_j$$

$$= a_j(x)0 + a'_j(x)0 = 0 \quad \text{for } 1 \le j \le k.$$

So $v_1 + v_2 + \cdots + v_k = 0$ implies $v_1 = v_2 = \cdots = v_k = 0$. Therefore the primary components of $M(A)$ are independent Definition 2.14. By Lemma 2.15

$$M(A) = M(A)_{p_1(x)} \oplus M(A)_{p_2(x)} \oplus \cdots \oplus M(A)_{p_k(x)}.$$

Solution 7(c): Let $g(x) = b_n x^n + b_{n-1}x^{n-1} + b_1 x + b_0$ be a polynomial over F and let $c \in F$. For $i \le j \le n$ the coefficients of x^{j-i} in $H_i(f(x) + g(x))$ and in

$H_i(f(x)) + H_i(g(x))$ are equal as

$$(a_j + b_j)\binom{j}{i} = a_j\binom{j}{i} + b_j\binom{j}{i}.$$

Therefore $H_i(f(x) + g(x)) = H_i(f(x)) + H_i(g(x))$. Also $cH_i(f(x)) = H_i(cf(x))$ as the coefficient of x^{j-i} in both these polynomials is $ca_j\binom{j}{i}$ for $i \le j \le n$. So $f(x) \to H_i(f(x))$ is a linear mapping of the (infinite dimensional) vector space $F[x]$.

With $f(x) = x^j$ we have $f''(x) = 0$ for $0 \le j < 2$ and $f''(x) = j(j-1)x^{j-2}$ for $j \ge 2$. As $j(j-1)$ is even we see $f''(x) = 0$ for all $j \ge 0$ in the case $F = \mathbb{Z}_2$. Applying twice in succession part (a)(i) above, we conclude: $f''(x) = 0$ for all $f(x) \in \mathbb{Z}_2[x]$. Now $H_2(x^j) = 0$ for $0 \le j < 2$ and $H_2(x^j) = \binom{j}{2}x^{j-2}$ for $j \ge 2$. Also $\binom{j}{2} = j(j-1)/2$ is even if and only if either $j \equiv 0 \pmod 4$ or $j \equiv 1 \pmod 4$. So $f(x) = a_n x^n + a_{n-1}x^{n-1} + a_1 x + a_0 \in \mathbb{Z}_2[x]$ satisfies $H_2(f(x)) = 0$ if and only if $a_j = 0$ for $j \equiv 2 \pmod 4$ and $a_j = 0$ for $j \equiv 3 \pmod 4$. So

$$\ker H_2 = \langle 1, x, x^4, x^5, x^8, x^9, \ldots \rangle,$$

$$\operatorname{im} H_2 = \langle 1, x, x^4, x^5, x^8, x^9, \ldots \rangle \quad \text{and} \quad \ker H_2 = \operatorname{im} H_2.$$

Hence $H_2^2 = 0$, i.e. $H_2(H_2(f(x)))$ is the zero polynomial for all $f(x) \in \mathbb{Z}_2[x]$. As $i!\binom{j}{i} = j!/(j-i)!$ we see $i!H_i(x^j) = j(j-1)(j-2)\cdots(j-i+1)x^{j-i}$ which is the result of formally differentiating i times the polynomial x^j for $j \ge i$. As both $i!H_i$ and formally differentiating i times are linear mappings of $F[x]$ with x^j in their kernels for $0 \le j < i$, we conclude $i!H_i(f(x)) = f^{(i)}(x)$ for all $f(x) \in F[x]$. As

$$(j-i)\binom{j}{i} = (i+1)\binom{j}{i+1}$$

we deduce

$$H_i'(x^j) = (j-i)\binom{j}{i}x^{j-i-1} = (i+1)\binom{j}{i+1}x^{j-i-1}$$

$$= (i+1)H_{i+1}(x^j)$$

for all j. As above we conclude $H_i'(f(x)) = (i+1)H_{i+1}(f(x))$ for all $f(x) \in F[x]$.

Suppose $\chi(F)$ is not a divisor of $i!$, i.e. $i! \ne 0$ in F. Dividing the equation $i!H_i(f(x)) = f^{(i)}(x)$ through by $i!$ gives $H_i(f(x)) = f^{(i)}(x)/i!$.

Suppose $\chi(F)$ is not a divisor of $i+1$. Then $i+1 \ne 0$ in F. Dividing the equation $H_i'(f(x)) = (i+1)H_{i+1}(f(x))$ through by $i+1$ gives $H_{i+1}(f(x)) = H_i(f(x))/(i+1)$.

Solution 9(a): The composition $\alpha\beta$ of the additive mappings α and β is itself additive by Exercises 2.1, Question 4(d). As α and β are semi-linear there are $\theta, \varphi \in \text{Aut } R$ with $(av)\alpha = (a)\theta(v)\alpha$ and $(bw)\beta = (b)\varphi(w)\beta$ for all $a, b \in R$ and $v, w \in M$. Setting $b = (a)\theta$, $w = (v)\alpha$ gives

$$(av)\alpha\beta = ((av)\alpha)\beta = ((a)\theta(v)\alpha)\beta = (bw)\beta = (b)\varphi(w)\beta$$

$$= (a)\theta\varphi(v)\alpha\beta$$

for all $a \in R$, $v \in M$ and so $\alpha\beta$ is semilinear as $\theta\varphi \in \text{Aut } R$ by Exercises 2.3, Question 3(d).

Let α be bijective. Then α^{-1} is additive by Exercises 2.1, Question 4(d). For $a \in R$, $v \in M$ we see $av \in M$ and so there is $w \in M$ with $(w)\alpha = av$. By Exercises 2.3, Question 3(d) we know $\theta^{-1} \in \text{Aut } R$. Also $((a)\theta^{-1}(v)\alpha^{-1})\alpha = ((a)\theta^{-1}\theta)((v)\alpha^{-1}\alpha) = av = (w)\alpha$. Therefore $(av)\alpha^{-1} = w = (a)\theta^{-1}(v)\alpha^{-1}$ showing that α^{-1} is semi-linear.

Denote two elements of R^t by

$$v = (a_1, a_2, \ldots, a_t) \quad \text{and} \quad w = (b_1, b_2, \ldots, b_t).$$

As θ is additive we have $(a_i + b_i)\theta = (a_i)\theta + (b_i)\theta$ for $1 \le i \le t$ showing that the ith entries in $(v + w)\hat{\theta}$ and $(v)\hat{\theta} + (w)\hat{\theta}$ are equal. So $(v + w)\hat{\theta} = (v)\hat{\theta} + (w)\hat{\theta}$ for all $v, w \in R^t$ showing $\hat{\theta}$ to be additive. For $a \in R$ we have $(aa_i)\theta = (a)\theta(a_i)\theta$ showing that the ith entries in $(av)\hat{\theta}$ and $(a)\theta(v)\hat{\theta}$ are equal for $1 \le i \le t$. Therefore $(av)\hat{\theta} = (a)\theta(v)\hat{\theta}$ for all $a \in R$, $v \in R^t$ showing that $\hat{\theta}$ is semi-linear.

Solution 9(b): Applying the eros $r_i - r_{l+i}$ for $1 \le i \le l$ over E to Y followed by $(c - c')^{-1}r_i$ for $1 \le i \le l$ and finally $r_{l+i} - c'r_i$ for $1 \le i \le l$ produces the matrix Z'_S with rows $e_i Z'_S = v_{1i}$, $e_{l+i} Z'_S = v_{0i}$ for $1 \le i \le l$. Comparing determinants gives $\det Y = (c - c')^l \det Z'_S$. For $i = 1, 2, \ldots, l$ in turn we apply the $l - i + 1$ eros $r_j \leftrightarrow r_{j-1}$ to Z'_S for $j = l+i, l+i-1, \ldots, 2i$ which produces Z_S. Therefore $\det Z'_S = (-1)^{l(l+1)/2} \det Z_S$ as each of these

$$l + l - 1 + l - 2 + \cdots + 2 + 1 = l(l + 1)/2$$

eros gives a sign change in the determinant. So $\det Y = (-1)^{l(l+1)/2}(c - c')^l \det Z_S$.

EXERCISES 6.3

Solution 1(e): As $\mu_A(x)$ is irreducible over F we see $\chi_A(x) = \mu_A(x)^s$ by Corollary 6.13. Comparing degrees gives $s = t/m$ where $m = \deg \mu_A(x)$. A typical element of the field $F(c)$, where $\mu_A(c) = 0$, is $f(c)$ where $f(x) \in F[x]$. Also $f(c) = g(c) \Leftrightarrow f(x) \equiv g(x) \pmod{\mu_A(x)}$ where $f(x), g(x) \in F[x]$ (see the discussion after Theorem 4.9). So the product of $f(c)$ and the element v of the

$F[x]$-module $M(A)$ is unambiguously defined by $f(c)v = f(x)v = vf(A)$. The seven module laws in $M(A)$ immediately give rise to the seven laws of a vector space over $F(c)$, i.e. F^t has the structure of a vector space over $F(c)$ which we denote by $M(A)'$. Then $\dim M(A)' = s'$ where $s' \le t$ as $F(c)$ is an extension field of F. Let $v_1, v_2, \ldots, v_{s'}$ be a basis of $M(A)'$. The ms' vectors $x^{i-1}v_j$ for $1 \le i \le m$, $1 \le j \le s'$ form a basis of F^t and so $\dim M(A)' = s' = s = t/m$. Let N be a submodule of $M(A)$. Then $f(x)v \in N$ where $f(x) \in F[x]$, $v \in N$. So $f(c)v \in N$ where $f(c) \in F(c)$, $v \in N$. So N is a subspace of $M(A)'$. Finally N a subspace of $M(A)' \Rightarrow N$ a submodule of $M(A)$. Let $\beta \in \operatorname{End} M(A)$. Then $(f(c)v)\beta = (f(x)v)\beta = f(x)((v)\beta) = f(c)((v)\beta)$ showing that β is a linear mapping of $M(A)'$. Conversely each linear mapping of $M(A)'$ belongs to $\operatorname{End} M(A)$. Let $\mathcal{B}$ denote a basis of $M(A)'$. For each $\beta \in \operatorname{End} M(A)$ write $(\beta)\theta$ for the matrix of the linear mapping β of $M(A)'$ relative to $\mathcal{B}$. Then $(\beta)\theta \in \mathfrak{M}_s(F(c))$ and, mimicking the proof of Theorem 3.15 with $\mathbb{Z}$, t, replaced by $F(c)$, s, we see that $\theta : \operatorname{End} M(A) \cong \mathfrak{M}_s(F(c))$ is a ring isomorphism. Restricting θ to $\operatorname{Aut} M(A)$, the group of invertible elements of $\operatorname{End} M(A)$, produces the group isomorphism $\theta| : \operatorname{Aut} M(A) \cong GL_s(F(c))$.

Solution 2(b): Consider $\beta, \beta' \in \operatorname{End} M(A)$ with matrices $B = (b_{ij})$, $B' = (b'_{ij})$ respectively relative to the standard basis $\mathcal{B}_0$ of F^t. So $(e_i)\beta = \sum_{j=1}^t b_{ij}e_j$ and $(e_i)\beta' = \sum_{j=1}^t b'_{ij}e_j$ for $1 \le i \le t$. Adding these equations gives

$$(e_i)(\beta + \beta') = (e_i)\beta + (e_i)\beta' = \sum_{j=1}^t b_{ij}e_j + \sum_{j=1}^t b'_{ij}e_j$$

$$= \sum_{j=1}^t (b_{ij} + b'_{ij})e_j \quad \text{for } 1 \le i \le t$$

which shows that $\beta + \beta'$ has matrix $B + B' = (b_{ij} + b'_{ij})$ relative to $\mathcal{B}_0$, i.e. $(\beta + \beta')\theta = B + B' = (\beta)\theta + (\beta')\theta$. So θ respects addition. Now $(e_j)\beta' = \sum_{k=1}^t b'_{jk}e_k$ for $1 \le j \le t$. So applying β' to $(e_i)\beta = \sum_{j=1}^t b_{ij}e_j$ gives

$$(e_i)(\beta\beta') = ((e_i)\beta)\beta' = \left(\sum_{j=1}^t b_{ij}e_j \right)\beta' = \sum_{j=1}^t b_{ij}(e_j)\beta'$$

$$= \sum_{j=1}^t b_{ij}\left(\sum_{k=1}^t b'_{jk}e_k \right) = \sum_{k=1}^t \left(\sum_{j=1}^t b_{ij}b'_{jk} \right)e_k$$

for $1 \le j \le t$ which shows that $\beta\beta'$ has matrix $BB' = (\sum_{j=1}^t b_{ij}b'_{jk})$ relative to $\mathcal{B}_0$, i.e. $(\beta\beta')\theta = BB' = (\beta)\theta(\beta')\theta$. So θ respects multiplication. The identity mapping ι of F^t is the 1-element of the ring $\operatorname{End} M(A)$ and $(\iota)\theta = I$, i.e. the matrix of ι relative to $\mathcal{B}_0$ is the $t \times t$ identity matrix I over F. As I is the 1-element of $\mathfrak{M}_t(F)$ we conclude that $\theta' : \operatorname{End} M(A) \to \mathfrak{M}_t(F)$, defined by $(\beta)\theta' = (\beta)\theta$ for all $\beta \in \operatorname{End} M(A)$, is a ring homomorphism. Let $\beta \in \ker \theta'$. Then $(\beta)\theta = B = 0$, the zero $t \times t$ matrix over F, i.e. $B = (b_{ij})$ where $b_{ij} = 0$ for $1 \leq i, j \leq t$. This gives $(e_i)\beta = \sum_{j=1}^{t} b_{ij} e_j = \sum_{j=1}^{t} 0 e_j = 0$ for $1 \leq i \leq t$. For $v \in F^t$ we have $v = \sum_{i=1}^{t} a_i e_i$ and so $(v)\beta = \sum_{i=1}^{t} a_i (e_i)\beta = \sum_{i=1}^{t} a_i 0 = 0$ showing that $\beta = 0$. Therefore $\ker \theta' = 0$ showing that θ' is injective by Exercises 2.3, Question 1(a)(i). By Theorem 6.27 and Definition 6.28 we see $\operatorname{im} \theta' = Z(A)$. So $Z(A)$ is a subring of $\mathfrak{M}_t(F)$ by Exercises 2.3, Question 3(b). Write $\theta = \theta' \iota'$ where $\iota' : Z(A) \to \mathfrak{M}_t(F)$ is the inclusion. As ι' is an injective ring homomorphism we see that $\theta : \operatorname{End} M(A) \to Z(A)$ is a ring isomorphism, i.e. $\theta : \operatorname{End} M(A) \cong Z(A)$. Now $\operatorname{Aut} M(A) = U(\operatorname{End} M(A))$, i.e. the automorphisms of the $F[x]$-module $M(A)$ are exactly the invertible elements of the ring $\operatorname{End} M(A)$. Therefore $\theta| : \operatorname{Aut} M(A) \cong U(Z(A))$, i.e. $\theta|$ is a group isomorphism (Exercises 2.3, Question 4(c)) between the corresponding groups of invertible elements of these rings.

Solution 2(c): Each element of the ring $F[x]/\langle g(x) \rangle$ is uniquely expressible $\langle g(x) \rangle + f(x)$ where $\deg f(x) < \deg g(x)$. Also $\langle g(x) \rangle + f(x)$ is an invertible element of $F[x]/\langle g(x) \rangle$ if and only if $\gcd\{f(x), g(x)\} = 1$. Therefore in the case of a finite field F of order q we have $\Phi_q(g(x)) = |U(F[x]/\langle g(x) \rangle)|$, i.e. the number of polynomials $f(x)$ over F with $\gcd\{f(x), g(x)\} = 1$ and $\deg f(x) < \deg g(x)$ is the order of the multiplicative group $U(F[x]/\langle g(x) \rangle)$. For non-zero $g(x), h(x) \in F[x]$ with $\gcd\{g(x), h(x)\} = 1$

$$\Phi_q(g(x)h(x)) = |U(F[x]/\langle g(x)h(x) \rangle)|$$

$$= |U(F[x]/\langle g(x) \rangle) \times U(F[x]/\langle h(x) \rangle)|$$

$$= |U(F[x]/\langle g(x) \rangle)| \times |U(F[x]/\langle h(x) \rangle)|$$

$$= \Phi_q(g(x))\Phi_q(h(x))$$

showing that Φ_q has the multiplicative property. As $\deg p(x)^n = mn$ there are q^{mn} polynomials $f(x)$ over F with $\deg f(x) < \deg p(x)^n$ as there are q choices for each of the mn coefficients of x^i in $f(x)$ for $0 \leq i < mn$. Suppose $\gcd\{f(x), p(x)^n\} \neq 1$. Then $p(x)| \gcd\{f(x), p(x)^n\}$ as $p(x)$ is irreducible over F. So $p(x)|f(x)$ and so $f(x)/p(x)$ is a polynomial of degree less than $mn - m = m(n - 1)$ over F. There are $q^{m(n-1)}$ such polynomials over F and hence there are exactly $q^{m(n-1)}$ polynomials $f(x)$ as above with $\gcd\{f(x), p(x)^n\} \neq 1$. Therefore

$$\Phi_q(p(x)^n) = q^{mn} - q^{m(n-1)} = q^{m(n-1)}(q^m - 1).$$

Solution 2(e): Each of the q scalar matrices aI for $a \in \mathbb{F}_q$ belongs to a singleton similarity class $\{aI\}$. Let A be a non-scalar 2×2 matrix over $\mathbb{F}_q$. Then $M(A)$ is cyclic with quadratic minimum polynomial $\mu_A(x)$ by Exercises 6.1, Question 6(a). There are q polynomials $\mu_A(x) = (x - a)^2$ giving $\Phi_q((x - a)^2) = q^2 - q$, and so the similarity class of A has size $(q^2 - 1)(q^2 - q)/(q^2 - q) = q^2 - 1$ by Theorem 6.29. There are $q(q - 1)/2$ polynomials $\mu_A(x) = (x - a)(x - b)$, $a \neq b$, giving $\Phi_q((x - a)(x - b)) = (q - 1)^2$ and so the size of the similarity class of A is $(q^2 - 1)(q^2 - q)/(q - 1)^2 = (q + 1)q$. There remain $q^2 - q - q(q - 1)/2 = q(q - 1)/2$ monic quadratics which are the irreducible $\mu_A(x)$ and so $\Phi_q(\mu_A(x)) = q^2 - 1$ and $(q^2 - 1)(q^2 - q)/(q^2 - 1) = q(q - 1)$ is the size of the similarity class of A. Adding the sizes of similarity classes in $\mathfrak{M}_2(\mathbb{F}_q)$ gives

$$q + q(q^2 - 1) + (q(q - 1)/2)(q + 1)q + (q(q - 1)/2)q(q - 1)$$

$$= q + q^3 - q + q^4 - q^3 = q^4$$

as expected since $|\mathfrak{M}_2(\mathbb{F}_q)| = q^4$. There are $q - 1$ scalar matrices aI in $GL_2(\mathbb{F}_q)$ namely those with $a \neq 0$. The $q - 1$ similarity classes of matrices A with $\mu_A(x) = (x - a)^2, a \neq 0$ are contained in $GL_2(\mathbb{F}_q)$. The $(q - 1)(q - 2)/2$ similarity classes of matrices A with $\mu_A(x) = (x - a)(x - b), a \neq 0, b \neq 0, a \neq b$ are contained in $GL_2(\mathbb{F}_q)$. All the $q(q - 1)/2$ similarity classes of matrices A with $\mu_A(x)$ irreducible are contained in $GL_2(\mathbb{F}_q)$ as $\mu_A(0) \neq 0$. Adding up the number of matrices in $GL_2(\mathbb{F}_q)$ according to their $q^2 - 1$ conjugacy classes gives

$$q - 1 + (q - 1)(q^2 - 1) + ((q - 1)(q - 2)/2)(q + 1)q$$

$$+ (q(q - 1)/2)q(q - 1)$$

$$= (q^2 - 1)(q^2 - q) = |GL_2(\mathbb{F}_q)|.$$

Solution 3(a): Let e denote the identity element of G. Then $(e)\theta$ is the identity mapping ι of Ω. So $x^{(e)\theta} = x^{\iota} = x$ showing $x \sim x$ for all $x \in \Omega$. Suppose $x \sim y$ for $x, y \in \Omega$. There is $g \in G$ with $x^{(g)\theta} = y$. As $\theta : G \to S(\Omega)$ is a group homomorphism we see $((g)\theta)^{-1} = (g^{-1})\theta$ and so $y^{(g^{-1})\theta} = x$ showing $y \sim x$ as $g^{-1} \in G$. Suppose $x \sim y$ and $y \sim z$ where $x, y, z \in \Omega$. There are $g, h \in G$ with $x^{(g)\theta} = y$ and $y^{(h)\theta} = z$. As $(g)\theta(h)\theta = (gh)\theta$ we see $x^{(gh)\theta} = x^{(g)\theta(h)\theta} = (x^{(g)\theta})^{(h)\theta} = y^{(h)\theta} = z$ showing $x \sim z$ as $gh \in G$. We conclude that $\sim$ is an equivalence relation on Ω. The equivalence class of x is

$$O_x = \{y \in \Omega : y \sim x\} = \{y \in \Omega : y = x^{(g)\theta} \text{ for } g \in G\} = \{x^{(g)\theta} : g \in G\}.$$

We verify that G_x is a subgroup of G by showing that G_x contains the identity e of G and G_x is closed under multiplication and inversion. As $(e)\theta = \iota$ we see $x^{(e)\theta} = x^{\iota} = x$ showing $e \in G_x$. Consider $g, h \in G_x$. Then $x^{(g)\theta} = x$ and $x^{(h)\theta} = x$. As above we see $x^{(gh)\theta} = x^{(g)\theta(h)\theta} = (x^{(g)\theta})^{(h)\theta} = x^{(h)\theta} = x$ showing $gh \in G_x$. Also $x^{(g)\theta} = x$ gives $x^{((g)\theta)^{-1}} = x$ and so $x^{(g^{-1})\theta} = x$ showing

$g^{-1} \in G_x$. Therefore G_x is a subgroup of G. To show that the correspondence $G_x g \to x^{(g)\theta}$ is unambiguously defined suppose $G_x g = G_x h$ for $g, h \in G$. Then $gh^{-1} \in G_x$ which means $x^{(gh^{-1})\theta} = x$. As $x^{(g)\theta(h^{-1})\theta} = x^{(g)\theta((h)\theta)^{-1}}$ on applying $(h)\theta$ we obtain $x^{(g)\theta} = x^{(h)\theta}$ showing that the above correspondence from the set of left cosets of G_x to O_x is indeed ambiguously defined. This correspondence is surjective directly from the definition of O_x. This correspondence is injective because $x^{(g)\theta} = x^{(h)\theta}$ implies $x^{(gh^{-1})\theta} = x$ which implies $gh^{-1} \in G_x$ and hence $G_x g = G_x h$, on reversing the above steps. So this correspondence is bijective, completing the proof of the orbit-stabiliser theorem. In the case of G_x having finite index n in G we see $|O_x| = n$ as there are exactly n distinct left cosets of G_x in G. If G is finite then $n = |G|/|G_x|$ as the n left cosets each consist of $|G_x|$ elements and partition G. So $|G|/|G_x| = |O_x|$.

Solution 5(c): The $1 \times n$ vector e'_1 has order $d'(x)$ in $M(A')$ by Theorem 5.26 as $A' = C(d'(x))$. Write $f(x) = d'(x)/\gcd\{d(x), d'(x)\}$. Then $\gcd\{f(x), d'(x)\} = f(x)$ as $f(x)|d'(x)$ and so $u_0 = f(x)e'_1$ has order $d'(x)/\gcd\{f(x), d'(x)\} = d'(x)/f(x) = \gcd\{d(x), d'(x)\}$ by Lemma 5.23. For $1 \le i < m$ we have

$$e_i A B_0 = e_{i+1} B_0 = x^i u_0 = x^{i-1} u_0 A' = e_i B_0 A'.$$

Let $d(x) = c_0 + c_1 x + \cdots + c_{m-1}x^{m-1} + x^m$. By Theorem 5.26 we know e_1 has order $d(x)$ in $M(A)$ as $A = C(d(x))$. So

$$e_m A B_0 = -(c_0 e_1 + c_1 e_2 + \cdots + c_{m-1}e_m)B_0$$

$$= -(c_0 + c_1 x + \cdots + c_{m-1}x^{m-1})u_0$$

$$= x^m u_0 = x^{m-1} u_0 A' = e_m B_0 A'$$

since $d(x)u_0 = 0$ as $\gcd\{d(x), d'(x)\}|d(x)$. Therefore $A B_0 = B_0 A'$ as these $m \times n$ matrices have equal rows, i.e. B_0 intertwines A and A'. For $1 \le i \le r$ we have

$$e_i B_0 = x^{i-1}(b_0 + b_1 x + \cdots + b_{n-r-1}x^{n-r-1} + x^{n-r})e'_1$$

$$= b_0 e'_i + b_1 e'_{i+1} + \cdots + b_{n-r-1}e'_{i+n-r-1} + b_{n-r}e'_{i+n-r}$$

as $x^{j-1}e'_1 = e'_j$ for $1 \le j \le n$ since A' is a companion matrix. For $r < i \le m$ we show that row i of B_0 is a certain linear combination of the preceding r rows of B_0. In fact $(a_0 + a_1 x + \cdots + a_{r-1}x^{r-1} + x^r)u_0 = \gcd\{d(x), d'(x)\}u_0 = 0$. Multiplying this equation by x^{i-r-1} and rearranging gives

$$e_i B_0 = x^{i-1} u_0 = -(a_0 x^{i-r-1} + a_1 x^{i-r} + \cdots + a_{r-1}x^{i-2})u_0$$

$$= -(a_0 e_{i-r}B_0 + a_1 e_{i-r+1}B_0 + \cdots + a_{r-1}e_{i-1}B_0).$$

The first r rows of B_0 are the vectors of the basis $\mathcal{B}_{u_0}$ as in Theorem 5.24 and so are linearly independent over F. By the above equation the remaining rows of B_0 are linear combinations of the first r rows of B_0. Therefore rank $B_0 = r$.

As B_0 intertwines A and A', by (a) above $\beta_0 \in \mathrm{Hom}(M(A), M(A'))$. As $(e_1)\beta_0 = e_1 B_0 = u_0 = (d_0'(x)/\gcd\{d_0(x), d_0'(x)\})e_1'$, taking $v_0 = e_1$, $v_0' = e_1'$ in Question 4(c) above we see that β_0 generates $\mathrm{Hom}(M(A), M(A'))$.

(i) In the case $d'(x)|d(x)$ we see $\gcd\{d(x), d'(x)\} = d'(x)$ and so $d'(x)/\gcd\{d(x), d'(x)\} = 1$. So $u_0 = 1 \times e_1' = e_1'$ and $r = n = \deg d'(x) \le m$. Therefore $e_i B_0 = e_i'$ for $1 \le i \le n$ and $e_i B_0 = e_n'(A')^{i-n}$ for $n < i \le m$.

(ii) In the case $d(x)|d'(x)$ we see $\gcd\{d(x), d'(x)\} = d(x)$ and so $r = m = \deg d(x) \le n$. Therefore all the rows of B_0 are given by the above formula $e_i B_0 = b_0 e_i' + b_1 e_{i+1}' + \cdots + b_{n-r-1} e_{i+n-r-1}' + b_{n-r} e_{i+n-r}'$ for $1 \le i \le m$.

(iii) In the case $\gcd\{d(x), d'(x)\} = 1$ we have $r = 0$ and $u_0 = d'(x)e_1' = 0$. So $B_0 = 0$ showing that the only matrix which intertwines $C(d(x))$ and $C(d'(x))$ is the zero $m \times n$ matrix.

Solution 5(d): The $t \times t$ matrix CB is partitioned, in the same way as B, into $\deg d_i(x) \times \deg d_j(x)$ submatrices $C(d_i(x))B_{ij}$ for $1 \le i, j \le s$. Also the $t \times t$ matrix BC is partitioned, in the same way as B, into $\deg d_i(x) \times \deg d_j(x)$ submatrices $B_{ij}C(d_j(x))$ for $1 \le i, j \le s$. Therefore

$$B \in Z(C) \quad \Leftrightarrow \quad CB = BC \quad \Leftrightarrow \quad C(d_i(x))B_{ij} = B_{ij}C(d_j(x))$$

for all $1 \le i, j \le s$.

So $B \in Z(C)$ if and only if B_{ij} intertwines $C(d_i(x))$ and $C(d_j(x))$ for all $1 \le i, j \le s$. By (c) above there is a $\deg d_i(x) \times \deg d_j(x)$ matrix $(B_{ij})_0$ such that $B_{ij} = f_{ij}(x)(B_{ij})_0$ where $f_{ij}(x)$ is a polynomial of degree less than $\deg \gcd\{d_i(x), d_j(x)\}$ over F for all $1 \le i, j \le s$. Therefore the $F[x]$-module $\mathrm{End}\, M(C)$ is the direct sum of s^2 cyclic submodules M_{ij} generated by $(\beta_{ij})_0$ determined by $(B_{ij})_0$ for $1 \le i, j \le s$.

Solution 6(b): (i) Let $B(x) = (b_{ij}(x))$ belong to the ring R_A. Multiplying

$$d_i(x)b_{ij}(x) \equiv 0 \pmod{d_j(x)}$$

by -1 gives

$$d_i(x)(-b_{ij}(x)) \equiv 0 \pmod{d_j(x)} \quad \text{for } 1 \le i, j \le s.$$

Therefore $-B(x) \in R_A$, i.e. R_A is closed under negation.

(ii) As

$$d_i(x)0(x) \equiv 0 \pmod{d_j(x)} \quad \text{for } 1 \le i, j \le s$$

the zero matrix of $\mathfrak{M}_s(F[x])$ is in R_A. As

$$d_i(x)0(x) \equiv 0 \pmod{d_j(x)} \quad \text{for } 1 \le i, j \le s, i \ne j$$

and

$$d_i(x)1(x) \equiv 0 \pmod{d_i(x)} \quad \text{for } 1 \le i \le s$$

we see that the identity matrix of $\mathfrak{M}_s(F[x])$ is in R_A. Therefore R_A is a subring of $\mathfrak{M}_s(F[x])$ completing the proof of Theorem 6.32.

Solution 6(c): Let $B(x) = (b_{ij}(x))$ belong to the ring R_A and suppose $v, w \in M(A)$. For $1 \le i \le s$ there are $f_i(x), g_i(x) \in F[x]$ with $v = \sum_{i=1}^s f_i(x)v_i$, $w = \sum_{i=1}^s g_i(x)v_i$ and so $v + w = \sum_{i=1}^s (f_i(x) + g_i(x))v_i$ as module law 5 holds in $M(A)$. As

$$(v)\beta = \sum_{i,j=1}^s f_i(x)b_{ij}(x)v_j \quad \text{and} \quad (w)\beta = \sum_{i,j=1}^s g_i(x)b_{ij}(x)v_j,$$

on adding and using the module laws

$$(v)\beta + (w)\beta = \sum_{i,j=1}^s (f_i(x) + g_i(x))b_{ij}(x)v_j = (v+w)\beta$$

which shows β to be additive. For $f(x) \in F[x]$ we have

$$f(x)v = \sum_{i=1}^s f(x)f_i(x)v_i$$

and so

$$(f(x)v)\beta = \sum_{i,j=1}^s f(x)f_i(x)b_{ij}(x)v_j = f(x)((v)\beta)$$

as the module laws are obeyed by $M(A)$. So β is $F[x]$-linear, i.e. $\beta \in \text{End}\, M(A)$. For $1 \le i \le s$ on taking $f_i(x) = 1(x)$, $f_j(x) = 0(x)$ for $j \ne i$ we obtain

$$v_i = \sum_{i=1}^s f_i(x)v_i \quad \text{and} \quad (v_i)\beta = \sum_{i,j=1}^s f_i(x)b_{ij}(x)v_j = \sum_{j=1}^s b_{ij}(x)v_j$$

showing that $B(x)$ represents β (Definition 6.30).

Index